Truth, Lies, and O-Rings

UNIVERSITY PRESS OF FLORIDA

Florida A&M University, Tallahassee
Florida Atlantic University, Boca Raton
Florida Gulf Coast University, Ft. Myers
Florida International University, Miami
Florida State University, Tallahassee
New College of Florida, Sarasota
University of Central Florida, Orlando
University of Florida, Gainesville
University of North Florida, Jacksonville
University of South Florida, Tampa
University of West Florida, Pensacola

TRUTH, LIES,
AND
O-RINGS

Inside the Space Shuttle *Challenger* Disaster

Allan J. McDonald with James R. Hansen

University Press of Florida
GAINESVILLE · TALLAHASSEE · TAMPA · BOCA RATON
PENSACOLA · ORLANDO · MIAMI · JACKSONVILLE · FT. MYERS · SARASOTA

14 13 12 11 10 09 c 6 5 4 3 2

17 16 15 14 13 12 p 6 5 4 3 2 1

LIBRARY OF CONGRESS CATALOGING-IN-PUBLICATION DATA
McDonald, Allan J.
Truth, lies, and o-rings : inside the Space Shuttle Challenger disaster / Allan J.
McDonald with James R. Hansen.
p. cm.
Includes bibliographical references and index.
ISBN 978-0-8130-3326-6 (alk. paper)
ISBN 978-0-8130-4193-3 (pbk.)
    1. Challenger (Spacecraft)—Accidents. 2. Whistle blowing—United States.
3. Space shuttles—Accidents—Investigation. 4. United States. National Aeronautics
and Space Administration. 5. United States—Politics and government. I. Hansen,
James R. II. Title.
TL867.M3625   2009
363.12'416—dc22   2008047644

The substance of all quoted material in this book is accurate but may not, in all cases,
reflect the exact words used by the speakers involved. Whenever possible, for literary
fluency, quotations were taken verbatim from primary sources, especially when the
material involved sworn testimony before the Presidential Commission or congressional hearings on the *Challenger* accident investigation.

The University Press of Florida is the scholarly publishing agency for the State Uni-
versity System of Florida, comprising Florida A&M University, Florida Atlantic
University, Florida Gulf Coast University, Florida International University, Florida
State University, New College of Florida, University of Central Florida, University of
Florida, University of North Florida, University of South Florida, and University of
West Florida.

University Press of Florida
15 Northwest 15th Street
Gainesville, FL 32611-2079
http://www.upf.com

To my entire family: Linda, Greg, Lisa and Ted, Lora and John, Meghan and Dave, and all of my grandchildren—Robbie, Remy, Molly, Isabelle, Christopher, Kate, and Meg

The future is not free: the story of all human progress is one of a struggle against all odds. We learned again that this America, which Abraham Lincoln called the last, best hope of man on Earth, was built on heroism and noble sacrifice. It was built by men and women like our seven star voyagers, who answered a call beyond duty, who gave more than was expected or required and who gave it little thought of worldly reward.

—President Ronald Reagan, January 31, 1986

Contents

Foreword

The Place of the McDonald Memoir
in *Challenger* History

JAMES R. HANSEN

The fiery destruction of the Space Shuttle *Challenger*, with seven American astronauts aboard, including the first ordinary citizen, a beloved school-teacher, happened more than two decades ago. One might think that histo-rians have duly recorded—and that the technical aerospace community has fully comprehended for some time—exactly why the U.S. space program's first fatal in-flight accident occurred on the cold, heartless morning of Janu-ary 28, 1986. Surely, the facts underlying the horrible tragedy of *Challenger* (STS-51L) must have all surfaced by now, all the evidence fully examined and reexamined, all the expert testimony scrutinized, all the critical failure points analyzed and digested, all the penetrating engineering studies performed, all the revelatory books and articles written, all the important lessons learned.

It is not the case.

Surprisingly, until Allan J. McDonald, former director of the Space Shut-tle Solid Rocket Motor Project for Morton Thiokol, Inc., came forward with his remarkable first-person story, no one *directly* involved in the decision to launch *Challenger* had published a memoir about the experience. No ac-count—firsthand or otherwise—had penetrated all of the factors leading to the accident. No book had critically evaluated all the testimony given to the Presidential Commission on the Space Shuttle *Challenger* Accident (popu-larly known as the Rogers Commission, after its chair, former secretary of state William P. Rogers), and none had done a comprehensively effective job of pointing out the conflicts of testimony and other evidence that dem-onstrated that some form of cover-up had taken place.

Allan McDonald's memoir now contributes all of that and more. It ad-dresses what happened to the people at the National Aeronautics and Space Administration (NASA) and Morton Thiokol who were involved in the ill-fated decision to launch *Challenger*, and it contains McDonald's own deeply personal recollections of his traumatic travails as he fought to draw attention to the real reasons behind the disaster and to the fact that some of the responsible individuals in NASA and his own company, Morton Thiokol, were doing their very best to cover up those reasons. McDonald's remarkable story—one revealing the character of the very best type of "whistleblower"—uncovers the acts of retribution directed against him by his company and by NASA, as well as the protective measures that had to be taken by the Rogers Commission and by Congress to preserve his job at Thiokol. The memoir lays bare Thiokol's reluctance in giving him the job of leading the redesign of the Space Shuttle's defective solid rocket motor following his controversial role in the *Challenger* investigation, and it addresses the many serious difficulties encountered in that critically important redesign program, whose goal was to restore the Space Shuttle as soon as possible to safe flight. This two-and-a-half-year solid rocket motor (SRM) redesign program—done under the bright spotlights of a hyperactive media, with technical oversight provided by the National Research Council (NRC), along with input solicited by NASA from the entire U.S. propulsion community—was arguably the most scrutinized engineering project in American history. The result, initially called the redesigned solid rocket motor (and which later came to be known as the *reusable* solid rocket motor to retain its original acronym of RSRM), became the safest and most reliable propulsion system currently flying on the Shuttle—so successful that NASA in 2005 selected an upgraded five-segment version of the reusable solid rocket motor as the booster propulsion system of choice for the Ares I crew launch vehicle, which will replace the Shuttle in 2010, and in conjunction with the larger Ares V heavy-lift cargo vehicle that will possibly take American astronauts back to the Moon and eventually on to Mars.

McDonald's memoir is also critically important because it explores early warnings from the first two flights *after Challenger* (in 1988) of the very same serious debris problems that would later result in the tragic loss of another Shuttle, *Columbia* (STS-107), on February 1, 2003.

As soon as Allan McDonald contacted me in early 2006 and I read a draft of his manuscript, I knew that I wanted to get involved as coauthor. His memoir so vividly captured what it was like for him to have found himself in the middle of the *Challenger* tragedy—first as a skilled engineer and executive voicing his doubts about a launch that should never have taken place, and then as an honest servant of his company and the American public when politicians and others sought to hijack the *Challenger* investigations. Al's book was the most coherent description I'd ever read of what happened with the infamous Space Shuttle O-rings and, in particular, of how events unfolded in the hours before, and the weeks and months following, the launch of *Challenger*. His narrative lent great urgency—and sometimes an appropriate measure of humor when circumstances allowed—to what was a milestone event in my own life and in the life of all Americans and supporters of the American space program. I came into the project not to change or improve McDonald's voice, because it didn't need that, but simply to make his narrative as cogent, clearly presented, and forceful as possible.

An insider account can be a two-edged sword. On the negative side, such a memoir can be self-serving, petulant, and petty. It can be nasty, condescending, and mean-spirited to everyone but the author and his friends. Fortunately, Allan McDonald is not the sort of man ever to tell such a one-sided or egocentric story. It was not the way he was brought up or lived his life. In the case of his memoir, the sword cuts much deeper to the positive side. Allan has very carefully used the transcripts of the Rogers Commission and Congress, and his narrative of the events conforms very precisely to the history of the *Challenger* disaster that is commonly accepted. His account of private conversations is based not just on memories now twenty years old, but also on 1,400 pages of detailed handwritten notes he had made at the time of these events; all of the conversations recounted in his narrative ring loud and true given the context of the official testimony on these subjects. The account that he has put together is, yes, detailed to a fault and tremendously meticulous in its telling, because that is who he is, that is what his engineering has been all about, and that is the only way the real truth behind the *Challenger* accident can ever finally come out. The result is a version of the events surrounding *Challenger* that is clearly as insightful and trustworthy as Allan himself is thoughtful and honest. NASA's side of the story is only told through the testimony that he chose to include, it is true, but his narrative in no way comes across as being manipulative or unfairly stacked against his antagonists, which, after all, were never NASA as a whole but individual NASA officials—and Morton Thiokol officials— against whom he had legitimate complaints. In my professional judgment, Allan makes a very strong case for his opinion that a few people in NASA engaged in a cover-up of the pressure that was placed on Morton Thiokol and others to agree to the launch of *Challenger*. He does not believe that NASA as a whole was involved in a cover-up.

The story that Allan McDonald tells is a story that most adults in America are relatively familiar with, but from the unique perspective of a person who recognized the potential disaster and tried to prevent it. It is a story with all the elements of a Greek tragedy put into a modern-day technological perspective. Even though his book deals with the failure of one part of a complex machine, his narrative never loses sight of the human element in technological failure and its aftermath. Clearly, Allan still feels a gnawing anguish about the chain of events that led to the *Challenger* launch that cold January morning, and his book makes clear that the memory of that event still sears his soul. He identifies heroes and excoriates villains, but always has a compelling explanation of the judgments that he renders. Readers may not always agree with his conclusions, but they will be impressed by how painstakingly he has re-created the causes of the event that prematurely took the lives of the seven astronauts.

Clearly, the *Challenger* disaster could have been avoided if only the voice of reasonable caution had been heeded. In a story that should never be forgotten, Allan McDonald tells the heartbreaking tale of how he saw his words of warning ignored, and the horrible, fateful consequences of that terribly bad decision.

Preface

I initially wrote most of the material for this book some twenty years ago as if it were an engineering report augmented with sworn testimony from the hearings of the Presidential Commission on the Space Shuttle *Challenger* Accident and the congressional hearings on the results of the *Challenger* accident investigation. After the first closed executive hearing of the Presidential Commission, I decided I needed to document everything I knew, everything I heard, and much of what was reported in the press and news media concerning the accident and the investigations. When I first revealed to the Presidential Commission that Morton Thiokol initially recommended *not to launch* the *Challenger* because of the cold temperatures, right after NASA had just told the commission that Morton Thiokol *only recommended to launch*, Chairman William Rogers said to me, "Would you please come down here and repeat what you've just said, because if I just heard what I think I heard, then this may be in litigation for years to come." I took his words to heart, because I knew who would be in the hot seat for any litigation to follow: me.

Some NASA officials at the Marshall Space Flight Center in Huntsville, Alabama, and several members of Morton Thiokol senior management were in collusion and were clearly trying to cover up this bad decision to launch, and I had just pulled the cork out of the bottle. When I was essentially demoted by my company for telling the truth to the Presidential Commission, I decided at that time that I needed to document everything to protect myself from any litigation or any further retribution against me from NASA or the company.

I continued to document everything that happened after the accident, especially because I was assigned to lead the nearly impossible task of effectively redesigning the solid rocket boosters (SRBs). Only later did I learn that several members of Congress threatened to ban Morton Thiokol from receiving any NASA contracts if the company didn't reinstate me to a position equivalent to the one that I had before my testimony before the Presidential Commission. Otherwise, my company would never have given me this critical assignment.

There were many controversial issues and confrontations in the SRB redesign where I prevailed over other alternatives preferred by NASA, Morton Thiokol, and other members of the propulsion community, including the U.S. Air Force and our competitors in the solid rocket industry. I was convinced that if another accident happened on the first launch after *Challenger* that could be attributed to the solid rocket motors, I would be the focus of blame from both NASA and Morton Thiokol. Even though I considered this to be remote after completing our very thorough ground testing program, it was still a very real possibility, especially after we experienced an act of sabotage during the redesign program.

I had intentionally avoided the press and news media during their coverage of the *Challenger* accident and the ongoing investigations after the accident. I went to great lengths to avoid the news media during this time because I was worried that it would destroy my credibility with the Presidential Commission investigating the accident. I continued to avoid the press as much as possible even during the redesign and return-to-flight of the Shuttle. I only made presentations to technical societies and educational institutions. I continue to give lectures on the lessons learned from the *Challenger* accident to many universities and local sections of the American Institute of Aeronautics and Astronautics (AIAA) as a Distinguished AIAA Lecturer.

I had thought for some time that this material might be worthy of publication, but I decided not to consider doing anything with it until after I retired from Thiokol. I finally did this in the summer of 2001 after the second time that Thiokol was sold to another company within less than a year. It was a little over three years after retirement that I went up to my attic and retrieved a box full of notes that I had handwritten twenty years ago and decided to write this book. I had hoped to publish it by the twentieth anniversary of the *Challenger* accident on January 28, 2006. But given the amount of hard work it takes to complete a book (writing a book is not rocket science, it's harder!), we were lucky to finish a complete draft of the final manuscript by the twenty-second anniversary of the *Challenger* accident in January 2008.

Critically important to the completion of this book was the scholarly input and editorial expertise of Dr. James R. Hansen, one of this country's foremost aerospace historians. Having read his best-selling biography, *First*

Man: The Life of Neil A. Armstrong (Simon & Schuster, 2005), I contacted him with questions about publishing my memoir. Jim instantly recognized the significance of my story, and after meeting him at his home in Auburn, Alabama, in early February 2006, I enthusiastically solicited his assistance with improving my book. His Bibliographical Essay that appears at the back of this book is the most comprehensive review of all the major books and articles ever written about *Challenger.* Jim makes it very clear that none of these publications contain the insights that my book includes, but more important, they all suffer from numerous technical mistakes and misinformation that are revealed in my book.

Truth, Lies, and O-rings: Inside the Space Shuttle Challenger *Disaster* is the only book that has ever been published by an individual directly involved in the *Challenger* launch decision and who, then and now, is resolved to tell the truth, the whole truth, and nothing but the truth about this great national tragedy, about the effort to return the Space Shuttle to safe flight once again, and about the warnings that went unheeded in the return-to-flight of the Space Shuttle in 1988 that led to the loss of the *Columbia* and her crew in February 2003.

The *Challenger* accident was the major news story of the year in 1986 and captured the nation's and the whole world's attention. This was the first time that astronauts were killed in their journey to space in a long history of successful space flights starting with the launch of Yuri Gagarin by the Soviet Union in April 1961, some twenty-five years earlier. The Soviets had lost a cosmonaut in April 1967 when the parachute attached to the space capsule failed to properly deploy prior to touchdown in Russia. Three other cosmonauts were lost in June 1971 when their shirtsleeve oxygen system depressurized on their return to Earth; with no emergency oxygen system available, they suffocated. The U.S. space program had suffered the loss of three *Apollo* astronauts—Gus Grissom, Roger Chaffee, and Ed White—in an electrical fire in their oxygen-filled *Apollo* capsule during a routine checkout of the capsule on the launchpad in January 1967, but the United States had never lost any astronauts on their way to or home from space.

The U.S. space program had been successful in landing a dozen astronauts on the Moon and returning them home safely since Neil Armstrong first stepped on the Moon in July 1969. The miraculous rescue of the *Apollo 13* astronauts on their way to the Moon was such an extraordinary feat that it appeared that NASA could never fail or certainly could do no wrong. The *Challenger* exploding on January 28, 1986, in front of a grandstand filled

For in my father's eyes there was sorrow...

I was too young to understand
Yet I knew something was different that day.
A salty tear fell from his eyes,
His cheeks appeared bright red and flustered.
His words were soft and meek.
He talked of destruction,
He talked of tragedy.
The day seemed so gloomy
For in my father's eyes there was sorrow.

The phone rang throughout the day,
Friends and families concerned.
I felt my father was crying out of guilt,
But for what?
He had to leave late that evening,
"There's an emergency," I heard him say.
I was too young to understand
But I knew deep down that something was wrong,
For in my father's eyes there was sorrow.

Meghan McDonald
1997 - 1998

My six-year-old daughter Meghan returning home in the Morton Thiokol, Inc., company jet. (McDonald photo.)

with the astronauts' families was so shocking that it took several years for this nation to recover from it, and NASA never did recover from its badly tarnished image. This book relates why the *Challenger* so badly damaged that image and the warnings that went unheeded in the return-to-flight of the Space Shuttle in 1988 that led to the loss of the *Columbia* and crew in February 2003, more than fourteen years later.

My experiences with *Challenger* became the defining moments of my life. As such, they touched those closest to me in deeply emotional and intimate ways. My dear wife Linda, son Greg, and daughters Lisa, Lora, and Meghan have supported me throughout all the difficult and stressful years and never complained about all of the missed birthdays, weekends, anniversaries, holidays, vacations, and my long stretches away from home. While a freshman at Judge Memorial Catholic High School in Salt Lake City, my youngest daughter, Meghan, wrote a poem that captured her memories of when I first returned home from the *Challenger* accident. Meghan, now a beautiful woman of twenty-six, was three years old at the time of the *Challenger* accident. The picture of her as a six-year-old was taken inside a Morton Thiokol jet while returning home from Edwards Air Force Base in California after watching my good friend astronaut Robert "Hoot" Gibson land the Space Shuttle *Atlantis* in December 1988. *Atlantis* suffered the worst debris impact damage of any Shuttle prior to the loss of *Columbia* in 2003. It was only by the grace of God that Hoot was able to return home safely.

An accident is defined in *Webster's Dictionary* as "a happening that is not expected, foreseen, or intended." I remember being run over by my dad when I was about five years old. It was snowing, and I went outside to throw a snowball at the car just as he was turning into our driveway, and I slipped and slid under the car; my dad did not see me fall, and both the front and back tires on the passenger side ran over my legs. It was a good thing that there was lots of snow on the ground and my bones must have been like rubber then because other than some very bad bruises, I was just fine. By *Webster's* definition, that was truly an accident because it met all three criteria. But the *Challenger* may qualify as an accident by *Webster's* standards by meeting only one of the three criteria: It may have been expected and foreseen by some, but not intended by anyone.

Prologue

Minds are like parachutes. They only function when they are open.
—*Sir James Dewar*

It was a short night, and I didn't sleep well at all.

Shortly before midnight EST, my Utah-based employer, Morton Thiokol, Inc., maker of the Space Shuttle's solid rocket motors, had faxed a statement to two tension-filled National Aeronautics and Space Administration (NASA) meeting rooms, one at Kennedy Space Center (KSC), where I served as Thiokol's senior man, and the other at Marshall Space Flight Center (MSFC) in Alabama. The fax approved the next morning's launch of the Space Shuttle *Challenger*.

My company's recommendation came at the conclusion of a marathon teleconference linking Huntsville, the Cape, and our guys in Utah. Anticipated to last less than an hour, the exhausting discussion took three hours. It was interrupted only once, by an off-line "caucus" requested by my boss, Joe C. Kilminster, Vice President for Space Booster Programs back at the plant outside Brigham City.

Kilminster had asked for five minutes. The caucus took thirty. Because I was in the meeting at KSC, I heard none of it.

Prior to going off-line, Thiokol's recommendation had been *not* to launch *Challenger*. Cold weather just wouldn't allow it. The overnight forecast called for the temperature at the Cape to plummet to as low as 18°, which was extraordinarily cold for Florida, even in late January. Warming the next morning would be gradual. No way, during the scheduled launch window from 9:38 to 11:38 a.m., would the mercury rise to 53°, the coldest temperature at which a solid rocket motor on a Shuttle had ever been launched. It wouldn't even reach 40°, the lowest temperature at which we thought our solid rocket motors had been qualified to fly.

When Utah came back on the phone, *somehow* the decision had been reversed: Thiokol, NASA's sole-source contractor for the Shuttle's solid rocket motors, gave its go-ahead for the launch.

I was way beyond being perplexed and terribly upset.

NASA wanted the launch recommendation in writing. Such a request had never happened before in the Space Shuttle program. At the conclusion of every flight readiness review (FRR) within a day of the launch, Jess Moore, Associate Administrator of the Office of Space Flight at NASA headquarters and head of the NASA Mission Management Team (MMT), always polled his contractors orally, who then answered, "Yes, we're ready," or "No, we're not." This time NASA wanted our recommendation in writing, and signed by a responsible Thiokol official.

As the senior management representative from Morton Thiokol at KSC, I assumed I would be the person who would have to sign the launch recommendation. After all, that was why I was there. Each NASA contractor was required to have a senior member of management attend each launch, participate in any prelaunch problem resolution, and possess the authority to recommend launching or not.

NASA knew that I did not feel at all good about the change from our original recommendation not to launch. My concern focused on the effects of cold temperature on the performance of our O-rings, the rubber seals that our company's engineers had designed at the beginning of the Shuttle program to prevent hot gases from leaking through the joints between the segments of the Shuttle's solid rocket boosters (SRBs). If the primary and secondary O-rings did not do their job and failed to "seal" effectively because of loss of resiliency (spring-back capability) due to cold temperature, a fiery jet could escape from an SRB field-joint at ignition, impinge on the adjacent surface of the huge external tank (ET) filled with 1.6 million pounds of liquid hydrogen and oxygen that was attached to the belly of the orbiter, and cause an explosion. If that ever happened, there was likely no way for a Shuttle and its crew of seven astronauts to survive the fireball.

When Space Transportation System (STS) 51-C had launched a year earlier (January 24, 1985), with the solid rocket motors at a moderate 53°, photographs taken of the recovered boosters being disassembled at KSC indicated that a jet of black soot "blew by" the primary O-ring in one of the SRB field-joints and the nozzle joint on each booster. So why would NASA ever consider launching below 40°, the established minimum in its own specifications for the rocket motor? And what exactly had prompted my own company, whose engineers for many months had been expressing grave concerns about the effects of erosion and blowby observed on the O-rings, to reverse course and issue this perverse recommendation to launch in such cold temperatures?

NASA's Lawrence Mulloy, SRB Project Manager in Huntsville, had looked me directly in the eye from across the table at KSC when the request was made to put the launch recommendation and rationale in writing and sign it. I told Mulloy that I would not sign that recommendation and that it would have to come from the plant in Utah. It appeared to me that the need for the signature on the launch recommendation was for CYA ("Cover Your Ass") purposes only. This indicated to me that even though NASA officials got the verbal recommendation they wanted, they weren't comfortable with it either or they wouldn't have requested the signature.

Mulloy then restated to the teleconference the urgency to have the signed launch recommendation by early morning. I think Larry expected my boss, Joe Kilminster, to direct me to sign the recommendation and give it to NASA, and I was also concerned that that may happen, but Kilminster stated Morton Thiokol would provide this right away and fax it out to both KSC and MSFC.

It took a while for the fax to arrive, so all the parties in the meeting room stayed around the conference table. I told them I didn't feel very good about this launch recommendation. In fact, I soon made the direct statement, "If anything happens to this launch, I wouldn't want to be the person that has to stand in front of a Board of Inquiry to explain why we launched outside of the qualification of the solid rocket motor."

When I made that statement, the room became very silent. I was visibly upset and asked that they reconsider this decision. "If I were the Launch Director," I said, "I would cancel the launch for three reasons, not just one— the first being the concern of the cold O-rings that we have just discussed. But there are two others."

Prior to the teleconference, I had eaten dinner at Carver Kennedy's house in Titusville, where I frequently stayed while on assignment at the Cape. Carver also worked for Morton Thiokol and was responsible not only for the stacking of the SRBs (done inside KSC's cavernous Vehicle Assembly Building [VAB]) but also for booster retrieval operations. I knew that Carver had been in communication with someone at Hangar AF who had been in contact with the recovery ships at sea. Carver had been advised that the ships were in an "absolute survival mode," struggling just to stay afloat in seas swelling to thirty feet high. With sustained winds registering at fifty knots and gusting up to seventy knots, the vessels were pitching as much as 30°. Some of the retrieval equipment on the back of the main recovery ship may well have been damaged. They were steering directly into the wind, heading for shore at about three knots, and had been doing that for some time. There was no way that they would be able to support an early morning launch, because they wouldn't be in the recovery area.

Based upon the conditions at sea that I had just heard about, it appeared that it was going to be nearly impossible to recover some of the SRB hardware, either the parachutes or the frustums. I believed they were also putting the boosters at some risk as far as recovery was concerned, because the ships were heading *away* from the recovery area.

The third reason for not launching was the formation of ice. I thought that there could be an ice debris problem or even a chance that the presence of ice on the vehicle and fixed service structure could change the launch acoustics in some problematic way. I didn't know, but I didn't think it was prudent to launch under that kind of condition.

I wasn't recommending not launching because of what I knew, but because of what I didn't know, and I thought that NASA was in the same position. It just wasn't worth taking the risk with all of these unknowns.

I was told, "These really aren't your problems, Al. You really shouldn't concern yourself with them." To which I replied, "You know *all* three of these together should be more than sufficient to cancel the launch, if the matter of the O-rings that we discussed earlier isn't."

The NASA people could tell I was disturbed and tried half-heartedly to reassure me, "We will pass these on as concerns, but only in an advisory capacity."

I was then asked by Larry Mulloy where the signed fax was, because some time had transpired since the teleconference had ended, and it still wasn't there. So I said, "OK, I will go check on that." There was nothing there, and I really wondered if the fax machine was even working. It was getting very late, so I stayed there for about ten minutes before the fax finally came in. It was a single sheet of paper, and I took it to the office of my colleague, Jack Buchanan, Manager of our KSC Field Office, where we reproduced copies for everyone.

As I walked into the office of Cecil Houston, NASA/MSFC Resident Manager at KSC, Mulloy and Stan Reinartz, Manager of the Shuttle Projects Office and MSFC's representative on the Mission Management Team, were engaged in a telecom with Arnold Aldrich, Manager of the National Space

Transportation Systems Program Office at NASA Johnson Space Center and also a member of Jess Moore's Mission Management Team. They were discussing the condition of the booster recovery ships, the rough seas, and the fact the ships were, indeed, in a survival mode.

The two NASA men then briefly discussed the matter of the ice, with Aldrich reassuring Mulloy that the ice issue had been fully examined earlier in the day.

I didn't hear anything discussed about the O-ring seal problem. I presumed that topic, the focus of our marathon teleconference, had been covered while I was away waiting for the fax. Only later did I discover that I was wrong, and that the subject had not even been mentioned.

The conversation ended with Arnie Aldrich's recommendation to go on with the launch as originally planned.

I handed Mulloy and Reinartz a copy of the fax signed by Joe Kilminster. They read it and discussed whether they should notify Dr. William Lucas, Center Director of NASA Marshall. Reinartz asked, "Should we wake the old man and tell him about this?" Larry responded, "I wouldn't want to wake him. We haven't changed anything relative to launching. If we had decided to scrub, then we'd have to wake him."

I stayed around a few more minutes to talk to Jack Buchanan and then drove back to Carver Kennedy's house in Titusville. I arrived there around 1:00 a.m.

It was now January 28, 1986, the day that *Challenger* would face catastrophe.

PART I

Red Flags

A danger foreseen is half avoided.
—*German Proverb*

Nine-tenths of wisdom is being wise in time.
—*Theodore Roosevelt*

1

When It Rains It Pours

As late as 1982 very few people had heard of Thiokol Chemical Corporation, much less knew what the company did, but for me it had always been a good professional home.

I came to work for Thiokol in 1959 right after earning my chemical engineering degree from Montana State University, just a year and a half following the Soviet Union's October 1957 launch of the first artificial satellite, *Sputnik 1*. In the latter part of 1982, the Morton Norwich Company had sold Norwich Pharmaceuticals and bought Thiokol Chemical Corporation to form a new conglomerate of "Salt, Specialty Chemicals, and Aerospace." Morton brought a known identity; its little girl under the blue umbrella— "When It Rains It Pours"—was an established household trademark. Its new partner, Thiokol, was the nation's largest manufacturer of solid rockets, with roots tracing back to 1926 when a brilliant chemist, Dr. Joseph C. Patrick, accidentally invented a synthetic rubber while trying to formulate a new antifreeze. Patrick accidentally developed a polysulfide polymer from which "Thiocol" got its name. ("Thio" is from the Greek for sulfur, and "col" means glue. Thus Thiocol is sulfur glue. The "c" was eventually replaced with a "k" and the name changed to Thiokol.) This polysulfide synthetic rubber later led Thiokol to pioneer the design of case-bonded solid propellant rockets, including the one that placed the first U.S. satellite into orbit, *Explorer I*, in January 1958. The army's Alabama rocket team under Dr. Wernher von Braun got all the credit for putting up the satellite, but the Jupiter-C rocket that launched *Explorer* had three other stages besides the initial booster, which used a liquid rocket engine derived from the old German V-2s of World War II fame. The other three stages were all solids manufactured by the Jet Propulsion Laboratory (JPL) of the California Institute of Technology, with fuel using Thiokol's polysulfide rubber; JPL later became part of NASA.

Even though Morton's primary interest in Thiokol lay not in its aerospace operations producing solid rocket motors (SRMs) but in its specialty chemicals line that was complementary to Morton's own chemical operations, I and my associates enjoyed belonging to a company that was part of the blossoming "Space Age." Our aerospace operations grew in scope and magnitude, especially in the early 1980s when we became the sole-source producers for the 1.25-million-pound solid rocket motors for the Space Shuttle, as well as the contractor chosen to develop a higher-performing booster for air force launches of the Shuttle from Vandenberg Air Force Base (AFB) in California.

NASA's extremely ambitious Shuttle schedule had us all very excited. Since the flight of STS-1 *Columbia* with Commander John Young and Pilot Robert Crippen in April 1981, NASA had flown seven Shuttle missions. Its flight manifest called for flying four Shuttles in fiscal year (FY) 1984, then doubling the flight rate to eight in FY 1985 and nearly doubling the rate again to fifteen flights in FY 1986. By the end of 1988, the Shuttle was supposed to be flying at a rate of twice a month, or twenty-four launches per year, which was triple the flight rate planned for 1985. The plan was to sustain that rate into the twenty-first century. Morton Thiokol executives in Chicago, where the corporate headquarters were located, were already making fancy sales and profit projections for the Aerospace Group, which was now decisively beating both Chemicals and Salt. It was not all due to our Shuttle work. With President Reagan dramatically revitalizing the U.S. defense establishment, Aerospace soon became the major contributor to our corporate profits. We were involved in the development of solid rockets for the new air force Peacekeeper (initially known as "Missile X," or MX) and the navy's Trident II (D-5) strategic missiles, for numerous tactical air-launched and ground-launched missiles, and for nearly all of the satellite-orbit-placement motors. Along with that, we were starting to see some real promise for our fledgling automobile air-bag business.

Everything seemed to be going great for Morton Thiokol until the summer of 1983, when our major competitor, Hercules Inc., complained to NASA about our sole-source position for producing SRMs for the Space Shuttle. This pressure for competition caused NASA to solicit information from the entire solid rocket industry to determine if anyone other than Morton Thiokol was capable of producing these large SRMs at a competitive price.

It was ironic that the company stirring up the second-source issue was also a Utah-based company, one with a major subcontract from Morton Thiokol, worth over $100 million, for producing graphite-epoxy-filament-wound cases for the new generation Space Shuttles to be launched from Vandenberg AFB starting in 1986. Hercules, Inc., was also in a joint venture with Morton Thiokol to produce all three stages of the navy's Trident I (C-4) submarine-launched ballistic missile; adding to the partnership was a contract recently awarded for development and production of the first two stages of the larger Trident II (D-5) fleet ballistic missile, to be deployed on the new Trident submarines.

Even more surprising, Utah Senator Jake Garn supported Hercules in its bid to force competition for future production of the Shuttle's SRMs. Senator Garn was NASA's strongest supporter on Capitol Hill, and Hercules was one of Garn's largest supporters back home. And Garn didn't have much love for Morton Thiokol; in prior years, he had been alienated by some of Morton Thiokol's top management. Hercules was also located in the senator's residential area in Salt Lake City, where he had been the mayor earlier in his political career. Morton Thiokol was located in northern Utah, in Brigham City, where Senator Orrin Hatch was more popular. Morton Thiokol strongly supported Senator Hatch.

This talk of a second source for the Space Shuttle SRMs was cause for great concern inside Morton Thiokol top management, who had just completed sales and profit projections for the next twenty years assuming a sole-source position. The company's projections had been done to support a large capital request over the next few years from the corporate fathers in Chicago to expand the Space Shuttle production facilities so that we could meet the requirement of twenty-four "flight sets," or forty-eight of these large solid rocket motors, annually by the end of 1988.

It was this turn of events that brought me into the Space Shuttle program. As Manager of Project Engineering for Morton Thiokol's Wasatch Division in northern Utah, I had been responsible for technical management of all programs in the Wasatch Division, with the exception of the Shuttle program. I had technical responsibility for all strategic and tactical missile programs, satellite-orbit-transfer motors, automobile air-bag development, and all research and development, and had approximately seventy project engineers working for me on all of these programs. It was my responsibility to make sure that all technical requirements were being met and that all technical problems and concerns were being properly resolved.

I did not have project engineering responsibility for the Space Shuttle SRM program because Thiokol had established an autonomous Space Shuttle program office that combined the program management functions with those of project engineering for each of the major components of the SRM. These SRM component program managers all had two hats to wear, a program manager's hat and a project engineer's hat. The organizational structure was unique to the Space Shuttle program and had been originally proposed to NASA because it was the kind of organization that NASA wanted. It was a very cost-efficient organization, but in retrospect it was clearly a mistake not to have someone in our core engineering organization responsible for raising technical concerns on the program and making sure that all technical problems were being properly resolved.

Our Space Shuttle SRM component program managers were too busy trying to get hardware out the door and responding to never-ending telecoms and teleconferences with NASA to be able to provide proper technical coordination for the program. The component program managers reported directly to the Director of the Space Shuttle SRM Project in the program office, and there was no direct-line responsibility to the engineering organization. The Director of the Space Shuttle SRM Project reported to the Vice President of Space Booster Programs, who reported to the Assistant General Manager and Vice President of Program Management, who reported directly to the Senior Vice President and General Manager of the Wasatch Division.

In July 1983, I was temporarily relieved of my project engineering management responsibilities by the Senior Vice President and General Manager to develop a strategy for preventing a second source for production of the Space Shuttle's solid rocket motors. I was enjoying my job as the head of the project engineering organization and had recently been elected to chair the Solid Rocket Technical Committee for the American Institute of Aeronautics and Astronautics (AIAA). I had been a member of this technical committee for the past few years, which had provided me with an excellent opportunity to keep up with the state of the art in solid rocket propulsion and to meet with colleagues and associates from other aerospace companies, the government, and various universities. I was not seeking a new assignment, but it was to be only temporary, so I jumped in with both feet.

The first thing I did was try to become more familiar with the entire Space Shuttle system: how it operated, who the key contractors were. I also had to learn the meaning of all of the acronyms and alphabet soup used by NASA in the Shuttle program. But my focus was on what other elements of the Shuttle had been second-sourced or recompeted, and how that had worked out.

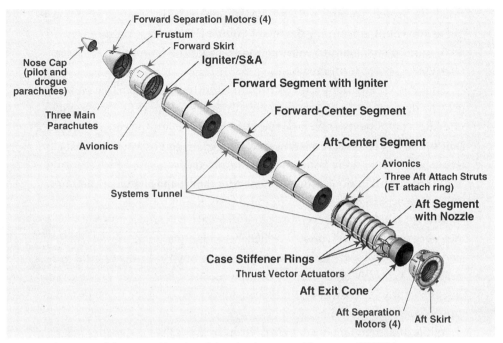

Elements of the Space Shuttle solid rocket boosters. (Courtesy of NASA.)

I was amazed to find that none of the Shuttle's hardware had ever been second-sourced or recompeted. The only Shuttle element ever reopened for competition was the Space Shuttle Processing Contract, which had been lost by a consortium of incumbents led by Rockwell International and which included Martin Marietta, United Space Boosters Inc. (USBI), and Rocketdyne. Rockwell built all of the orbiters; Rocketdyne supplied the liquid Space Shuttle main engines (SSMEs); Martin Marietta supplied the external tank containing the cryogenic liquid hydrogen and liquid oxygen that powered the Space Shuttle main engines; and USBI provided components and integration of the SRBs. The primary element of the SRBs was the solid rocket motor provided by Morton Thiokol. USBI supplied the forward and aft aluminum skirts, the external tank attach ring for attaching the SRBs to the ET, the explosive bolts for holding the SRBs on the mobile launch platform, the pyrotechnics and electronics for the SRB separation and recovery system, the hydrazine-powered hydraulic thrust vector actuation system for moving the solid rocket motor nozzles for steering the vehicle, the booster separation motors (four each on top and bottom of each solid rocket booster to separate the SRBs from the ET after motor burnout), and the nose cap, frustum, parachutes, and recovery system for the SRBs. To everyone's surprise, the winning team for the Space Shuttle Processing Contract was led by Lockheed and included Morton Thiokol, Johnson Controls, and Grumman Aerospace as team members.

This was a pretty significant contract for Morton Thiokol in that we were to provide some 500 people at the Kennedy Space Center (KSC) to assemble the SRB components and stack the SRM segments on the mobile launch platform in the Vehicle Assembly Building (VAB); receive, store, and attach the external tank to the SRBs; operate the two recovery ships for retrieving the SRBs from the ocean after each Shuttle flight; and then load them on railcars to be sent back to Utah.

Unfortunately, our timing for trying to justify why a second source was not necessary was very poor. We had been doing a good job for NASA, and we were probably the best-performing contractor in the Shuttle program. We probably could have killed the second-source issue at the time, but as Murphy's Law dictates, things can go very wrong at the worst possible time.

While disassembling the returned hardware from the maiden flight of the new high-performance motor flown on STS-8 in August 1983 (the Shuttle's first night launch), we found a major anomaly in the carbon-phenolic rings located in the forward section of the nozzle of the left-hand booster. These carbon-phenolic rings protected the metal nozzle structure from the near 6,000° F gases generated by the combustion of the solid propellant in the motor. The rings were so badly eroded that if the motor had burned another eight or nine seconds beyond its planned two minutes, the nozzle would have burned through, and a catastrophe likely would have occurred. Investigation of this serious nozzle problem severely damaged our good reputation and provided the stimulus for NASA to continue investigating the potential for another SRM source. Fortunately, I was not involved in the STS-8 nozzle-anomaly fiasco, which received near panic attention the next few months prior to the next launch, STS-9. As a result of this problem, some nozzles had to be removed and replaced at KSC to support the next couple of flights. The problem was diagnosed as a material problem in the STS-8 nozzle, and some of the same material was found in other nozzles on the motors at the Cape, requiring those nozzles to be changed-out.

After spending some time researching the second-source problem, I was requested to put together a presentation for the CEO of Morton Thiokol Inc. (MTI), Charles Locke. My presentation was to be given at Cape Canaveral while Locke was down there as a VIP witnessing the launch of STS-9 *Columbia* in November 1983.

This was my first personal contact with the CEO. My presentation pointed out certain facts related to NASA's reluctance to second-source any of the other hardware elements of its Shuttle system. One of the underlying realities working against second-sourcing was clearly evident in our own SRM contract, which involved over 400 suppliers in thirty-seven different states. Nearly half of our SRM contract dollars were outside purchases, and there was no one in the country with facilities capable of producing any of the components of the Shuttle SRMs that were currently being manufactured by Morton Thiokol. It would take a $250-million investment by another company to compete for even a portion of the business. With our planned facility expansion program set to produce the entire NASA need of twenty-four flight sets per year, it did not appear economically attractive for any other company to compete for the business. No other company had the capability of even handling the large rocket segments, which were over twelve feet in diameter and twenty-six feet long, and weighed 300,000 pounds each when loaded with propellant. We were also the only company capable of manufacturing the large movable nozzle assembly, a component that represented nearly 30 percent of the motor cost. I told CEO Locke that we should point out to Senator Garn that Hercules would have to go to California for this component. Hercules would have to use all the same qualified suppliers for basic materials and components and, as a result, 30 to 60 percent of the contract work currently being accomplished in Utah by Morton Thiokol would have to be moved to California if Hercules was successful in establishing a second source—and that was the good news for Hercules. The bad news was that it was very possible, even likely, that Hercules would not win a competitive second-source contract but that one of two more politically leveraged California companies would win, either Aerojet Solid Propulsion Company, based in Sacramento, or the Chemical Systems Division of United Technologies, Inc., in San Jose. If that happened, Utah could eventually end up losing all of its current Shuttle business, which employed more people in the private sector than any other single project in the state.

My presentation also showed how it was possible for us to significantly reduce our costs based upon increased production rates and our more recent performance based upon a demonstrated learning curve. This should make it basically impossible for any other contractor to produce motors even close to our current cost, much less what we could do in the future.

We can be so confident of our performance, I argued, that we should be willing to take a fixed-price contract for future production of the SRMs. If one added the requalification costs and facility amortization costs of the new supplier for the SRMs, it would clearly show that it was not economically beneficial for NASA to second-source the SRMs. The flight rate would have to be increased well above our capacity of twenty-four flight sets per year, and that did not appear very likely.

Locke thought we had a good story and that we needed to take that message to Senator Garn and to NASA to see if we could prevent NASA from taking some kind of official action on a second source for the SRMs.

With my presentation to the CEO, I had basically completed my assignment; however, the STS-8 nozzle problem prevented us from pursuing this plan with NASA at the time. We had to utilize all the resources we could muster to provide acceptable nozzles for the next few Shuttle flights and were in no position to argue that the second source should be dropped. I thought I was going back to my old job, but instead was asked to help assess the potential of using the Space Shuttle SRBs for application to a new expendable launch vehicle being solicited by the air force.

American space policy at the time was to discontinue further production of the nation's existing expendable launch vehicles—the Atlas, Delta, and Titan—and fly all future military, civil, and commercial payloads on the Space Shuttle. The air force wanted to maintain assured-access-to-space in the event of a Shuttle failure, so it was preparing to issue a request for proposal for a complementary expendable launch vehicle (CELV). This CELV would have the payload capability to geostationary orbit nearly equal to that of the Shuttle. It was suspected that what the air force really wanted was to upgrade its Titan 34D heavy-lift launch vehicle to increase its payload. Martin Marietta, which provided the Titan family of launch vehicles, was the odds-on favorite to win this contract, while General Dynamics (GD) was looking at similar ways to upgrade its Atlas booster so as to compete with the upgraded Titan.

To examine using Shuttle propulsion elements in an unmanned expendable launch vehicle (ELV) application, we resurrected an old ELV study done by Boeing that NASA had funded a few years earlier. In its study, Boeing designed a vehicle called the SRB-X, which used two Shuttle SRBs for the first stage and a short-length SRB (three segments rather than four) for the second stage. For the third stage, Boeing had suggested either a modified first-stage Peacekeeper (MX) or a Titan II second stage. When combined with GD's new liquid oxygen (LOX)/liquid hydrogen Centaur G Prime upper stage, the vehicle offered more than the necessary payload to geostationary orbit and met all of the air force requirements. Because the SRB-X launch vehicle would maintain the same geometric spacing of its two SRBs on the mobile launch platform as the Shuttle, the existing Shuttle launch facilities could be used without any modifications.

We decided to pursue this vehicle with Boeing as the prime contractor. Boeing told us that the air force did not want anything to do with a launch vehicle that used NASA hardware. Boeing also told us on the q.t. that the air force had sold the CELV concept to Congress basically to allow the service to further upgrade the Titan so it could eventually abandon NASA's Shuttle.

Seeking advice on how best to bid a vehicle using NASA Shuttle hardware, we went to James Beggs, the NASA Administrator. Beggs thought we had a good idea and recommended we get together with General Dynamics (Beggs was a former top management official at GD prior to becoming NASA Administrator) because the company was providing the upgraded Centaur G Prime as well as the payload carrier for all the potential CELV configurations.

We discussed the SRB-X option with General Dynamics, which appeared to be somewhat interested in bidding the concept. Our two companies worked on the concept together for a few weeks; however, when the air force announced it was about to release the official request for proposal, General Dynamics notified us that it had reconsidered and could not support us on the SRB-X vehicle. GD management thought that the company did not have the resources to submit two proposals, and their upgraded Atlas stood a better chance of winning.

We then went back to NASA and suggested that NASA bid the SRB-X for application to the air force's CELV mission. NASA would be the overall vehicle systems integrator, and we would support NASA with SRB hardware manufacturing and vehicle assembly operations in the VAB at the Cape.

Our CELV concept was very competitive with an upgraded Titan 34D and would have been a much cheaper option if launch-operations and launch-facility costs had been properly assessed by NASA and the air force. A large Shuttle infrastructure already existed, and the apparatus of that standing army could launch the SRB-X via existing mobile launch platforms from either of the two existing Space Shuttle launchpads, at Kennedy Space Center or the new launch facility planned at Vandenberg AFB in California. The

Titan, on the other hand, required extensive launchpad modifications and creation of a whole new standing army for launch operations on both coasts, a requirement complicated by the fact that the government's existing plan was to close down the Titan launch facilities.

Our SRB-X concept had three strikes against it from the beginning. It was to be a NASA-derived vehicle rather than one from the air force. It did not provide assured access to space if access was lost as a result of an SRB failure on the Shuttle. Finally, NASA's pricing policy for the SRB-X was not as competitive as it could have been.

It appeared that NASA looked at the CELV program as a way to reduce costs on the Shuttle by passing the costs off to the air force. Though most everyone thought that if the Shuttle failed, it would most likely be a result of an SSME or orbiter failure, one could not discount the possibility of an SRB failure, which, if it occurred, would stand down both the Shuttle and the SRB-X.

With our support, NASA Marshall Space Flight Center (MSFC) submitted the SRB-X proposal to the air force, but it really did not get any serious consideration. To no one's surprise, the air force selected an upgraded Titan for the CELV mission. This vehicle later was named the Titan IVA, with a new seven-segment steel-case version of the existing SRBs (produced by the Chemical Systems Division of United Technologies in San Jose) replacing the older five-and-a-half-segment SRBs. With an upgrade of the solid rocket motor, the Titan IVA developed into a Titan IVB, having a three-segment graphite-epoxy-filament-wound-case version of the rocket. The Solid Rocket Motor Upgrade would experience severe development problems at Hercules, Inc., which was awarded a fixed-price contract for the upgrade. During the development program, Hercules had to write off several hundred million dollars due to test failures, handling incidents, manufacturing errors, and propellant mixer fires. Congress later bailed out Hercules with the help of Jake Garn just before his retirement from the Senate.

On the morning of March 2, 1984, I sat in the company jet with Ed Garrison, President of our Aerospace Group, and Jerry Mason, General Manager and Senior Vice President of our Wasatch Division. We were flying from Chicago to Huntsville, Alabama, to talk with the Director of the MSFC, Dr. William Lucas, about how Morton Thiokol could team with NASA on the SRB-X proposal.

As soon as we stepped off the plane in Huntsville, the manager of our field office at MSFC, Bud Parker, was there to meet us; he came running toward the plane shouting, "A major fire has just occurred in the propellant casting area where the Space Shuttle solid rocket motors are manufactured back in Utah. The early morning radio reports indicated that several explosions had occurred and a large number of people may have been killed!"

Naturally, arriving at Lucas's office, the conversation totally bypassed the SRB-X; instead, we were trying to answer questions as to how the fire had started, how many people were killed and injured, and what impact the accident would have on future Shuttle production. After several telephone calls back to the plant, it was determined that no one had been killed and that only minor injuries had occurred but that our propellant casting house had been totally destroyed. Two Shuttle segments had been lost, two casting pits totally destroyed, and a new Shuttle casting pit building severely damaged.

The fire had started in one of four casting pits located out of doors, not in one of the four pits located inside a large new building. Our plan had been to construct a covered building for the outside row of casting pits as soon as the first new building had been checked out and declared operational. Until the exact cause of the fire was determined, the impact on Shuttle production could not be assessed, nor could casting operations be resumed.

With Jerry Mason eager to return to Utah as soon as possible to assess the damage personally and appoint an incident investigation team, our meeting with Lucas was cut short. We returned to the airport immediately and climbed aboard the company's twin-engine jet. The sun was setting by the time we arrived back at our plant site, located in a remote area at the north end of the Great Salt Lake just a few miles from Promontory Point, the historic site where the Golden Spike was driven in 1869 in commemoration of the completion of America's first transcontinental railroad. In the company jet, we made several low-level passes over the burned-out casting pit before landing on our short landing strip at the northwest end of the plant site. It was the first time in memory that the company jet had been allowed to land at the plant site, due to the short length of the runway. The company owned several smaller propeller planes that landed at the plant, but never the company jet.

A car waited by the runway to take us a few miles away to the casting pit where the fire had occurred. Being in the large solid rocket business required that our buildings be spread out over quite a large area. Over 400 buildings speckled the nearly 20,000 acres of land that made up our plant site, which made it the largest facility in the world for research and production of solid propellant rocket motors.

What we saw in our casting pit area looked like downtown Beirut after the war in Lebanon. There was evidence of fire everywhere. Pieces of large steel beams lay scattered around the area. Some were broken, some were bent, and some were partially melted from the intense heat. Against the snow-covered ground, the blackened and charred areas were clearly visible everywhere.

The fire had started in the mobile casting building, which was rolled into place on rails directly over a casting pit during propellant casting operations. The building was then removed when casting was complete. One of the Shuttle's segments was just completing its casting operation when the fire started. Propellant was cast into the steel segments from propellant mix bowls that each contained approximately 7,000 pounds of propellant. It took over forty of these individual propellant mixes to fill just one of the casting segments. Each solid rocket booster on the Shuttle was assembled from four casting segments: a forward segment containing an igniter, two identical center segments, and an aft segment containing a nozzle assembly.

After touring the accident site, Jerry Mason called a meeting in his conference room. He wanted a briefing on what was known about the accident, which people were involved, which facilities and equipment were lost, and what plans had been made to find the accident's cause, prescribe corrective action, and get back into production. Some action had been started in all these areas, and Mason appointed several people to lead these various activities and brief him on progress on a daily basis.

I thought to myself, "What a mess! I'm sure glad I'm not involved in the Shuttle program!"

I went home that night and told my wife, Linda, that things really looked bad at the plant. "I don't know what's going to happen to all those people working on the Shuttle program, but I'm sure glad I was told to keep working with NASA on the SRB-X proposal for the CELV program!"

Roughly two weeks after the accident, we submitted our technical and cost proposal on the SRB-X to the NASA Marshall Space Flight Center, to be included in its proposal to the air force. About the same time, our incident investigation team completed its analysis of the casting pit fire and was ready to report its findings to Jerry Mason. I was asked to sit in on the briefings, which concluded that the fire started in the casting pit building while a new batch of propellant was being poured into a funnel-shaped hopper located above the casting pits. A small amount of the propellant, with a consistency similar to wet cement, dripped onto the rails. When a new bowl of propellant ran over the rails to the casting pit, the propellant on the rails ignited. The ignited propellant, being propulsive, much like a firecracker without a fuse that fizzles, scooted into the hopper and down into the steel Shuttle segment that was being cast. Our people conducting this operation about two stories above ground level (there were also several people around the casting pit at ground level) immediately hit the fire alarm. All the workers were able to evacuate the building before the ignited segment caused an explosion, completely destroying the casting building, all the steel beams, and a 200-ton overhead crane assembly. Pieces of these huge steel structures flew over the heads of the workers running from the building to dirt bunkers located nearly fifty yards away.

It was an absolute miracle that no one was killed. Nearly a quarter million pounds of uncured burning propellant was ejected from the segment, starting fires all around the area. Some of the burning propellant and debris landed on an adjacent pit covered with a fiberglass lid. The propellant burned through the lid and ignited another Shuttle segment that was curing in the pit. The segment that was nearly cured—and with the consistency of a pencil eraser—ejected the casting tooling and burned like a Roman candle.

All in all, two segments with nearly 300 tons of propellant were lost. Interviews with employees in the manufacturing area revealed that ignition of propellant on the rails caused a popping sound much like a cap gun and had occurred many times before. Some employees claimed that old-timers used to put small pieces of propellant on the rails intentionally to scare some of the new hires on the crew. No one charged that this irresponsible action caused this particular fire, however.

A week after the investigation team had issued its report, Phil Dykstra, our Vice President of Program Management for all programs in the plant, including the Space Shuttle, invited me into his office for a meeting; his boss, Jerry Mason, was also there. Dykstra told me, "Al, starting next Monday, you are being reassigned to be the Director of the Space Shuttle Solid Rocket Motor Project in program management reporting to Joe Kilminster, Vice President of Space Booster Programs." Dykstra further informed me that as a result of the investigation of the nozzle near-failure on STS-8, a new Space Shuttle SRM Project Engineering organization was being formed to report directly to the Vice President of Engineering, Bob Lund. The gentleman who was currently the Director of the Space Shuttle SRM Project, Boyd Brinton, was being transferred into engineering to be the manager of the Space Shuttle SRM Project Engineering group.

I said to Dykstra and Mason, "I am happy with what I am doing and I want to know if this is one of those offers, like Marlon Brando as the God-father said, that I couldn't refuse." Dykstra responded, "Al, this is a good opportunity for you to get some program management experience." My whole career at Morton Thiokol, some twenty-five years at that point, had been spent in various areas of engineering, and this would be a promotion to a higher grade-level within program management. Mason then replied, "Al, this position carries a lot of responsibility and will have high visibility within the corporation. Both Phil and I feel that you have 'great manage-ment potential' and that this is an excellent opportunity to demonstrate it."

Talk about a challenge! I told them, "You are offering me a chance to run the Space Shuttle SRM project right after a major part of our plant had burned down, right after NASA had been given reason to be very unhappy with us for the near-failure of our nozzle on STS-8—a problem that we still did not fully understand—and right as we were in the process of expanding our production facilities to accommodate twenty-four Shuttle flights per year. Furthermore, we were in the midst of trying to stop NASA from seek-ing a second source for producing the SRMs, and we were in the middle of developing an upgraded SRM that was to be ready to fly in two years for the air force from a new launch facility at Vandenberg."

But I really had no choice other than to accept the job, not knowing that my life would never be the same again.

At home that night I told my wife that this was the first time in my career that I felt depressed after receiving a promotion. A few weeks later, after I had spent nights and weekends at the plant and my paycheck came in the mail with no increase in salary, Linda and I felt even more depressed.

When I went to my management about it, they told me that I was near the top of my old salary grade and that my promotion was up to the next grade-level but with no additional money. "If this job is as important as it has been made out to be and I have inherited all this responsibility that I've been told about," I replied, "then I feel I deserve an increase in salary as well as an increase in grade-level."

This incident should have made me more suspicious of the ways of Thiokol senior management than it did, perhaps because I was successful in embarrassing them into increasing my salary for taking on this overwhelm-ing job.

2

Tests and "No Tests"

We reconstructed our Utah casting facilities with a total change in the way we conducted the operation. We changed our propellant mix bowls to a bottom-discharge configuration like that we used in our high-energy propellant casting operations for the Trident missile, which was loaded with a propellant containing nitroglycerine. This new approach eliminated the need to tip the mix bowl to pour the propellant, which in turn eliminated the need for the propellant transfer hoppers and the rail system. At the bottom of the bottom-discharge bowl was a valve that connected the mix bowls directly to the vacuum bell so the propellant could be transferred into the motor case by simply opening the valve.

We were out of production for only three months before a number of changes had been made in the hopper-transfer process of the other casting row to allow safe casting of the solid rocket motors until we could get all our facilities modified. We also got tougher about keeping everything spick-and-span, eliminating any out-of-place propellant while still using the old casting process to allow us to restart SRM production as soon as possible. At the same time, we converted the other casting row over to the bottom-discharge approach. The minimal downtime meant there was no impact on the Shuttle flight schedule. A sufficient number of segments existed in the inventory to support what was still at the time a low Shuttle launch rate.

Meanwhile, talk of a second source for the SRMs grew louder. NASA was still very unhappy with our lack of understanding about the STS-8 nozzle-erosion anomaly. The agency didn't think too much of our initial assessment that the problem had been caused by restrained thermal growth in the carbon-phenolic nozzle parts. In our view, this caused the nozzle's carbon fibers to break, leading to pocketing—or what we called "spalling" (chunking out)—of the rocket nozzle's hot front surface.

In reviewing early static test data, prior to the first flight of the Shuttle in 1981, similar anomalous erosion had been noted in the same location of the nozzle. Why this data wasn't a cause of concern then, I do not know, but it didn't give anyone a warm feeling that it couldn't very well happen again, and it wouldn't take much more than that seen in STS-8 to be catastrophic. (If John Q. Miller from NASA Marshall Space Flight Center had not insisted that Morton Thiokol increase the design safety factor on nozzle erosion from 1.5 to 2.0, STS-8 would have been the first Space Shuttle disaster.)

Morton Thiokol engineers thought that the problem was due to the way we laid up the carbon-cloth-phenolic tape that was used to fabricate our nozzle parts rather than a materials quality problem, which NASA claimed. We felt strongly that our engineers were correct because there were other parts of the nozzle made of the same exact material that had severely eroded in the STS-8 nozzle nose inlet section but which had performed as expected in all other static tests and Shuttle flights; the only difference was the angle at which the carbon-cloth-phenolic tape had been placed on the mandrel (spindle) to make the parts. I agreed with that assessment and worked to get NASA to agree to test a new design, one with a change in the angle of how the tape was wound on the nozzle for the first graphite-epoxy-filament-wound-case (filament-wound-case, or FWC) DM-6 demonstration motor, already scheduled for testing in the fall of 1984. Meanwhile, NASA insisted that our guys at KSC remove the nozzles after each flight to inspect and measure the erosion. This was done by drilling plugs in the nozzle immediately after the boosters were brought to Hangar AF at Cape Canaveral.

It was shortly after this time that I first learned about an O-ring erosion problem in the field-joints of the SRMs. The Space Shuttle solid rocket motors were manufactured in four segments; these segments were transported to the Cape on railcars from a loading facility in Corrine, Utah, near the Thiokol plant. The segments were stacked vertically on the mobile launch platform in the Vehicle Assembly Building at KSC. There they were joined together with a tang-and-clevis joint much like the tongue-and-groove joint used in many hardwood floors. The clevis, or female part, at the bottom of the joint contained two grooves in the inner leg of the clevis; in those grooves sat two fluorocarbon O-rings designed to seal the joints from the hot propellant gases.

Since the joint is not made in the factory but is done in the field (KSC), it is called a "field-joint." The joints are pinned together with 177 high-strength nickel-alloy pins with a steel band fastened around the pins. Because the Space Shuttle was a "man-rated" system, two O-rings were used to provide redundancy in the sealing capability, because of the criticality of their function. A gas leak in the joint would most certainly be catastrophic to the vehicle and crew. The first O-ring to experience the gases was designated the

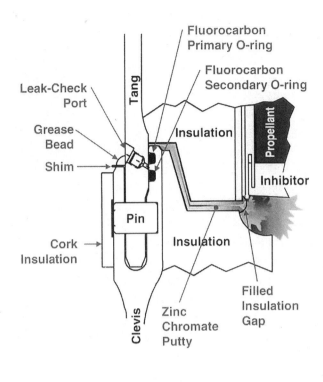

Fluorocarbon
Primary O-ring

Fluorocarbon
Secondary O-ring

Leak-Check
Port

Tang

Grease
Bead

Insulation

Propellant

Shim

Inhibitor

Pin

Cork
Insulation

Insulation

Filled
Insulation
Gap

Clevis

Zinc
Chromate
Putty

Field Joint Design

Space Shuttle
solid rocket
motor field-joint
configuration.
(Courtesy of
NASA/Morton
Thiokol.)

"primary O-ring" and the second, naturally, called the "secondary O-ring." Erosion of the primary O-ring in one of the SRB field-joints had just been observed in the last Shuttle flight, STS-41B, launched in February 1984, so it was a significant topic to be addressed at the next flight readiness review (FRR). Personally, I was very surprised to find out that such a problem existed. I was even more surprised to discover that this critical anomaly had been observed as early as the Shuttle's second flight, some two and a half years earlier in November 1981.

As if I didn't have enough problems to deal with, I was told that I needed to get involved in the flight readiness review process because both Morton Thiokol and NASA management were not happy with our previous support of this very important activity. Again, I had to jump in with both feet and get deeply involved in Morton Thiokol internal management reviews. These flight readiness reviews primarily dealt with any problems or discrepancies noted in the fabrication of the hardware sitting on the launchpad ready to fly. The FRRs also dealt with any previous problems or observations of anomalous conditions that had been noted from inspection of returned hardware from the previous flight or any other anomalous or suspect conditions noted from either fabrication or disassembly of SRM hardware since the last FRR. Because of the serious nature of the nozzle pocketing erosion problem, NASA requested that we address this particular subject on every FRR relative to nozzle materials used and inspection results of the prior flight. The O-ring erosion problem only needed to be addressed if erosion had been noted in the prior flight.

The whole FRR process seemed overwhelming to me because it involved a series of eight independent reviews prior to each Shuttle flight: an internal review with Morton Thiokol management; a review at NASA Marshall with the SRM Program Manager's Board; a meeting at Marshall with the SRB Program Manager's Board; a meeting at Marshall with the Shuttle Project Manager's Review Board; a meeting at Marshall with the MSFC Center Director's Board; a meeting at KSC with the NASA Associate Administrator for Space Flight Review Board, consisting of NASA managers and directors from MSFC, Johnson Space Center (JSC), KSC, and NASA headquarters; a prelaunch review with this same review board two days prior to the scheduled launch (launch minus two, L-2); and a final prelaunch weather briefing and assessment of any recent data or actions requiring close-out from the L-2 review board meeting one day prior to launch. Normally, we at Morton Thiokol didn't participate in the L-1 review other than to provide our verbal response that we were ready to launch. This whole review process started about six weeks prior to the scheduled launch date and culminated in the final flight readiness review two days prior to launch. Any issues still open were addressed at the L-1 FRR the day before the launch.

This flight readiness review process consumed a tremendous amount of time for all of the key NASA managers and contractor managers involved. The reviews consisted of stand-up oral presentations using overhead projection of viewgraphs with bound copies of the material distributed to all meeting attendees. Those who attended included dozens of people besides the review board members, so anywhere from fifty to seventy-five copies of the presentation material were usually available. The content of the material was continually condensed from one meeting to the next, with typically 150 to 200 charts at the initial in-house review subsequently reduced to about half a dozen charts on down to a single chart at the L-2 flight readiness review. Oral presentations were reduced accordingly from eight to ten hours for the early reviews down to five to ten minutes, especially if no serious anomaly had occurred in the previous flight that was outside of the established Shuttle flight database. By their very nature, the review boards had to be very critical of the information being presented, and board members asked numerous tough questions all along the way.

Attending my first in-house before Thiokol management in 1984, I noticed how our main presenter, John Elwell, encountered no end of difficulty in answering all the questions he was being asked about the material he was presenting. I figured that the rigors of this sort of inquisition were necessary, in that this internal review, prior to taking our information to NASA, needed to be an extremely tough and critical review by the people who actually fabricated the hardware and examined the returned hardware from prior flights. If our management seemed hypercritical, it was only to prepare Elwell for the subsequent scrutiny he would receive in the company of our customer, NASA. Still, I thought to myself, I'm sure glad I'm not standing up there making those presentations and defending all that data; I'm glad that I'm the Director of the SRM Project and one of the reviewers instead of one of the presenters. I had learned a long time ago that it was much easier to be a *critic* than the one who had to be comprehensive and *correct*.

That first internal review I sat through that day took nearly ten hours. At the merciful end of it all, our management review board gave Elwell a list of action items to complete and several recommendations on how to improve his presentation prior to going to NASA Marshall for the SRM Program Manager's review. I could tell by the verbal comments that the review board was not very happy with the way the material had been presented or explained. I learned a great deal about the Shuttle SRM from this review, and I myself offered several suggestions on how to improve the presentation. My boss, Joe Kilminster, suggested that I go to MSFC later that week to sit in on the SRM Program Manager's review and meet some of the NASA people involved in the SRM program at MSFC.

I arrived at the meeting in Alabama a little early so as to meet some of the NASA people. I also met Bud Parker, our resident office manager in Huntsville; he told me that he'd just received a telephone call from Phil Dykstra, our Assistant General Manager back at the plant, who had instructed him to tell me that I was to give the entire presentation. I was rather shocked by this last-minute decision and felt extremely uncomfortable about doing this so soon after coming into the program. I felt even worse having to tell my friend and colleague John Elwell that I had been told to replace him.

However, when I told him, John seemed more relieved than angered, and I proceeded to make the presentation.

My presentation went well, and I was able to answer all the questions that were asked. I didn't know it at the time, but I made such a good impression on the NASA people that my role as a reviewer for the flight readiness review process was very short-lived; it was, in fact, effectively over! I soon learned that, for the indefinite future, I was to be the main presenter of nearly all FRR data for all Morton Thiokol and NASA reviews.

During the next months, I worked a considerable amount of overtime, primarily to learn more about the Shuttle solid rocket motor design, manufacturing processes, and the history of the various problems that had been noted in the fabrication of the components and examination of returned flight hardware. I carpooled with three other coworkers who lived nearby. The Morton Thiokol plant site was very isolated and was approximately forty miles from Pleasant View, Utah, where I lived, ten miles north of Ogden. Most company employees carpooled or vanpooled to work. The closest residential areas to the plant site, Tremonton and Brigham City, were twenty to twenty-five miles from the plant.

One of the fellows in my carpool, Jack Hilden, worked in Strategic Operations of our Wasatch Division. At the time, Strategic Operations was manufacturing the first stage of the air force Peacekeeper (MX) intercontinental ballistic missile (ICBM) and navy Trident ballistic missiles, having had produced all the Minuteman ICBM first-stage motors and many of the third-stage motors for the Minuteman III ICBM. Strategic Operations was a separate plant on the north end of our 20,000-acre rocket factory. The other two members of my carpool worked in Space Operations of the Wasatch Division. Bob Lyles had recently moved to Utah from our Washington, D.C., office to head up business development, and Paul Ross had just been transferred from Strategic Operations to work for me as the Chief Program Manager on the Space Shuttle SRM nozzle, which was giving us so much trouble at the time. We had long discussions in the carpool about the SRM nozzle's spalling or pocketing erosion problem, occurring on STS-8, and the fact that most of the SRM fabrication problems identified in our discrepancy reports had been associated with the nozzle. Also, the major part of our FRR presentations was still dealing almost exclusively with nozzle issues. I informed Lyles and Ross about my becoming aware of the O-ring erosion problem and told them that although this, too, was a concern to everyone, it was not getting the attention that the nozzle spalling problem was receiving. I wasn't sure that the O-ring problem should be of equal concern, but I clearly recognized that it wasn't. Throughout 1984, what was clear was that nozzle erosion, not O-ring erosion, was considered by nearly everyone to be the most significant problem on the SRM.

Most of the engineering activity inside our Shuttle program was being directed toward the development of the filament-wound-case (FWC) solid rocket motor that was being developed by NASA for the air force launches from the new Shuttle launch facility at Vandenberg. Substitution of graphite fibers for the cylindrical steel sections of the SRM case reduced the case weight approximately 28,000 pounds, providing nearly 5,000 pounds more payload for the planned Shuttle launches from the West Coast.

I had been given technical responsibility for this project on top of my other duties, and it was rapidly consuming a great deal of my time. We had just received the first case segments from Hercules, Inc., which fabricated the graphite cases in their case-winding facility in Clearfield, Utah, approximately sixty miles south of our plant. The graphite cylinders were shipped to our plant, where we attached the steel end-rings to all sections, steel domes on the forward and aft segments, and a steel external tank attach section on the forward part of the aft segment. Rubber insulation was then laid up in the cases, vulcanized to the metal and graphite composite structure, lined with a rubber polymer, and cast with propellant. We were working to a very tight schedule to fabricate the first development motor (DM-6) in time for a static test in early fall of 1984.

Totally overwhelmed with my job, I frequently caused my carpool to get home quite late, and even then I carried a pile of reading material home with me every night. Any other carpool would have thrown me out, but the guys were very understanding, just first-class people.

Our lively and far-ranging discussions made the forty-mile drive to the plant nearly tolerable and always provided some comic relief to an otherwise very stressful job.

The more I seemed to learn about the Space Shuttle vehicle and launch operations, the further behind the power curve I felt. The NASA jargon and alphabet soup kept getting thicker and thicker. No one used a complete word for anything, much less a complete sentence. It was all about the STS, VAB, MLP, SRM, SRB, SSME, ET, PRCB, MECO, FRF, TVC, HPU, APU, GLS, HPOTP, HPFTP, LOX, LH2, GOX, RSS, SIR, IEA, FRR, PDR, CDR, DCR, PEB, RIDS, DR, PD, PR, PRB, OMRSD, OMI, FEC, OPT, Code M, QD, OMS, SLA, MSA, LSC, NSI, TAL, LCC, PAS, POP, RTLS, LSS, and LRU.

The phrase green SQUATCHALOIDS, which I thought were some kind of alien life, even entered our regular vernacular.

SQUATCHALOIDS are an envelope of loads occurring on the Shuttle vehicle during ascent flight. The envelope includes flight trajectory, seasonal wind loads, and three sigma dispersions in SRB thrust mismatch, thrust variations, aerodynamics, and flight control system variations; if everything is predicted to be within the envelope, then the SQUATCHALOIDS are termed "green."

My work was so all-consuming that I had virtually no time for my wife or family. I made up a report for my boss with a set of bar charts showing how I used my time, much of it in long telephone calls and teleconferences. The chart indicated that the time it took me just to be part of the teleconferences and the FRRs constituted more than a full-time job. It was not unusual to have calls or teleconferences lasting four to five hours each. Providing the required oversight to my SRM program component managers who were responsible for production of the hardware, running the FWC-SRM development program, and doing my other work forced me to work consistently 50 to 60 percent overtime, all of which was noncompensated, and I still couldn't get everything done that needed to be done. I asked my boss if a Deputy Director of the Space Shuttle SRM Project could be assigned to me to carry some of my workload, but I was turned down.

Shortly after I was appointed the SRM project director, NASA formed a Senior Materials Review Board (SMRB) at NASA Marshall and also at Morton Thiokol; this board approved or rejected all discrepancies in the critical hardware plus those that fell outside the existing SRM database. The SMRB only reviewed hardware dispositions in this category if they had been previously approved for flight use by the formal Material Review Board system, which included Morton Thiokol Engineering and Quality Assurance as well as the local NASA and local Air Force Plant Representative Office Quality Assurance organizations. Thiokol's Senior Materials Review Board included senior management representatives from Thiokol Engineering, the SRM Program Office, and Quality Assurance, and an Engineering representative from the NASA resident office at Thiokol. (NASA had only an Engineering representative, while Thiokol had reps from Engineering, Quality Assurance, and the Program Office.)

NASA appointed me to be the program office representative and SMRB chairman, and, as such, I was to personally sign my name to every discrepancy report and critical process departure that came with the delivered

SRM hardware. My signature essentially gave Thiokol management's endorsement of the SRM's flight worthiness. This didn't take much time, but it added a significant stress factor to the job.

Many times the decision to accept or reject the hardware with known defects was far from unanimous. I would have to make a final decision and defend that position with the NASA Senior Materials Review Board and FRR boards. I always felt nervous about the split decisions and lost many nights' sleep over some of them that I had accepted, wondering if I had been wrong in my assessment and whether my decision could result someday in a catastrophic failure of the Shuttle.

On the other side of the coin, when a rejection recommendation was made by the SMRB, it was equally discomforting for me to be put in the position where I had to convince Thiokol management, and sometimes NASA management, why we should scrap hardware that was worth possibly millions of dollars, equipment that previously had been determined to be perfectly acceptable by the Engineering and Quality Assurance organizations of Thiokol, NASA, and the local Air Force Plant Representative Office (AFPRO). My hair got grayer and grayer.

The O-ring erosion problem was not something that was occurring on every flight. Nor was it considered by either NASA or Morton Thiokol to be nearly as critical as the pocketing erosion of the nozzle, which was still the primary topic of discussion for every flight readiness review. Compared to the nozzle-erosion problem, and compared to the issues associated with the development of the FWC-SRM, the matter of the O-ring was still on the backburner.

Still, the issue of the O-ring design did come to the surface in conjunction with our fabrication of the first FWC-SRM development motor (DM-6). In the early fall of 1984, we were in the process of assembling this motor in our T-24 test bay for static test. The FWC motor contained two design features that we believed would improve the field-joint O-ring performance and help the nozzle-erosion problem. The FWC field-joints included a metal-capture lip on the tang side of the joint that significantly reduced the rotation, or opening, of the joint during pressurization. This would make it easier for the O-rings to maintain a seal. This capture lip slipped over the inner clevis leg of the lower segment during assembly of the joint, essentially making the joint a double-clevis arrangement rather than a tang-and-clevis joint. It was a unique arrangement that was part of the Hercules design for the FWC. The FWC motor also included a design change in the nozzle to eliminate the pocketing erosion on the carbon-phenolic nose inlet rings of the nozzle. This design change involved a different fabrication technique for this portion of the nozzle to eliminate plies of carbon-cloth-phenolic tape that were perpendicular to the flow stream; as stated earlier, this was accomplished by changing the ply angles of the carbon-cloth-phenolic tape as it was wound on a mandrel for making these parts.

I had been on the job for only a few months when I was informed that the Aerospace Safety Advisory Panel (ASAP) was going to visit our plant in early June. ASAP wanted a presentation from us concerning the design of the FWC solid rocket motor, the planned test program, and the activities that would lead up to the certification of the new FWC-SRM for flight. ASAP was an independent group of professional engineers and scientists who were responsible for periodic review of NASA's programs and making recommendations for actions that should be taken to improve safety. The panel had been established after the *Apollo* capsule fire that took the lives of three astronauts Gus Grissom, Roger Chaffee, and Ed White on January 27, 1967.

I was surprised that ASAP's agenda only included a review of the FWC-SRM program and not any discussions of the O-ring or even the pocketing erosion noted in the nozzle of the SRMs currently flying on the Shuttle. As the new kid on the block, I assumed that NASA had reviewed these potential safety problems with ASAP at meetings before I came on the program.

I prepared the presentation on the FWC-SRM and made a briefing to the Aerospace Safety Advisory Panel. Panelists had several questions about the program but seemed to be satisfied with our design and testing approach. I was surprised they were not more concerned about the limited full-scale test articles planned for prior to the first flight at Vandenberg. Two full-scale graphite cylinders were going to be hydro-burst so as to determine the ultimate case strength. In addition, a structural test article with an empty forward and aft segment was scheduled at NASA Marshall for structural loading and testing for unpressurized prelaunch bending, pressure loads, and flight loads, with the idea of confirming design structural safety factors of 1.4. The static test program involved the firing of two development motors (DM-6 and DM-7) and one qualification motor (QM-5) prior to the first flight of the Vandenberg Launch Site (VLS-1) Shuttle launch at Vandenberg. This three-motor test program seemed rather small compared to the original seven-motor test program (four development motors and three qualification motors), laid out in the original design, development, test, and evaluation program for the SRM. The graphite FWC was a major change to the design but was scheduled for only three ground tests. The upgraded version of the steel-case SRM—called the High Performance Motor—which performed its maiden flight on STS-8 in August 1983, had only minor changes from the original design and required two static tests.

The DM-6 motor test on October 25, 1984, made me really nervous. It was the first ground test of a full-scale Space Shuttle solid rocket motor that I was personally involved in—and I was responsible for the success or failure of it, the new graphite-case version. I had witnessed several of the earlier static tests of the steel-case SRM conducted between 1977 and 1983, but this was the first test of the new FWC-SRM, and it had been one and a half years since the last ground test of any Space Shuttle SRM. NASA had determined that it wasn't necessary to conduct any more ground tests after the motor was flying on the Shuttle unless a major redesign of one or more of the components had occurred, which now was the case with the FWC.

This test had great significance for the entire U.S. propulsion community because NASA, with the FWC-SRM, had clearly stepped out ahead of the Department of Defense (DOD) in advancing solid rocket motor technology, which seemed to be quite a role reversal for NASA. The space agency had always been very conservative, only adopting well-proven DOD solid rocket motor technology for application to NASA space launch vehicles and orbit-transfer motors.

Before coming to the Space Shuttle program, I had been involved in all of Thiokol's DOD production and advanced technology programs and had been very surprised, back in 1982, when NASA selected a graphite-epoxy-filament-wound case for upgrading the Shuttle for DOD flights out of Vandenberg. The air force was using this material on an advanced version of its short-range-attack-missile, the SRAM-II, to be launched from B-52 bombers, and the navy was developing graphite cases for its new Trident II D-5 submarine-launched ballistic missile. What was amazing to me was that NASA decided on a manned Space Shuttle flight of the graphite FWC-SRM before any of these weapon systems, with new high-strength rocket motor casings, had been flight tested. This proved to me that NASA wasn't as conservative as many people thought it was. NASA apparently was willing to take some significant risks to meet performance objectives.

Because our major competitor, Hercules, was fabricating the large graphite cylinders under subcontract to Morton Thiokol, the company sent a few busloads of their people up from the other end of the Great Salt Lake to witness the test. Dozens of Thiokol employees and subcontractors were there along with several busloads of students from local schools, forming a big crowd of people to witness the test. Watching from the desert highway was a host of onlookers parked in their cars from outlying neighborhoods. There must have been 3,000 people watching the test.

The morning before the test, several busloads of VIPs had been taken to the test site to have their pictures taken standing next to the huge black giant of a rocket motor lying horizontally in the test bay. The rocket motor contained 1.1 million pounds of solid propellant, was over 12 feet in diameter and 126 feet long, and produced over 3 million pounds of thrust. The huge mass of propellant in the long graphite cylinders caused the middle of the rocket casing to sag some six inches, making it look, in the cradle of its test stand, like a totally rotten banana. It was an impressive piece of hardware. In the crowd to witness the test was rookie astronaut U.S. Air Force (USAF) Major Richard Covey. Major Covey and I stood at the back end of the motor behind the nozzle—the real business-end of any rocket motor because it is where all the fire comes out. This was the first time I had met an astronaut, and I was really impressed with his knowledge, patience, and personality. I could not have guessed that I would be seeing astronaut Covey again four years later under a much different set of circumstances.

From all appearances, our motor test was an unqualified success. Its ballistic performance came very close to our predictions, the graphite case appeared to be in excellent condition and could probably be reused, and the nozzle was in better condition than any nozzle we had ever tested. What little erosion there was looked very smooth and without any pocketing. NASA was extremely pleased with meeting this major technical milestone less than two and a half years after awarding the contract.

The FWC-SRM program was under a separate contract with Morton Thiokol. It was a cost-plus-award-fee contract where NASA evaluated our performance against planned performance for the period. NASA then handed us a report card and paid a fee commensurate with the grade on the report card. The first test of the FWC-SRM was so successful that Morton Thiokol was given a 100 percent award-fee rating for its six-month performance evaluation—the first and only time we achieved this goal. It was almost unheard of in the aerospace industry. I was very proud of the achievement because not only had we conducted a very successful test, but we also had managed to recover all of the schedule time that Hercules had lost in providing late delivery of acceptable case segments; and we had done it within our planned budget.

Hercules was not so pleased because their award-fee from Morton Thiokol was not nearly as satisfactory—somewhere between 60 and 70 percent of the total award, because of case contamination problems, late case segment deliveries, and some cost overruns. In truth, the FWC Award Fee Board at Morton Thiokol, which I chaired, had given Hercules a slightly higher award-fee rating, but our company's senior management reduced the rating.

Preliminary inspection of the DM-6 motor and nozzle looked good, as did all the "quick look" data from the pressures, thermocouples, and strain gauges. I was very interested in the condition of the new nozzle parts and the improved field-joint configuration where the O-ring seals were located. I was especially anxious to see the disassembly of the motor segments when it took place several days after the test.

The nozzle appeared to be in excellent condition. Engineers who had been on the Shuttle program since it started said that the nozzle was in the best condition they had ever seen after static test or a return from flight. The field-joints were also in excellent condition: there was no O-ring erosion or blowby in any of the joints.

I was elated until I talked to my boss about the test results. He had personally gone inside the rocket motor prior to the static test. There he had applied strips of zinc chromate asbestos-filled vacuum putty on the rubber insulation in the area of the field-joints and had squeezed putty into the voids that formed between the insulation surfaces during the mating operation. We all suspected that the mating operation could trap air between the primary O-ring seal and the vacuum putty, and, if sufficient pressure developed, it could lead to blowholes through the putty during the assembly operation. Such holes had been observed during assembly of the DM-6 motor and had been repaired by my boss with a broom handle and some extra vacuum putty by packing holes and volcanoes in the putty with additional putty after assembly and motor leak-check.

I hadn't known that this putty repair operation had been done, and I was very upset with my boss, telling him that the test we just had conducted amounted to a "no-test" on the field-joint because of this makeshift operation. I asked him if we "suspend someone" inside the bore of the propellant grain during the vertical stacking operations at Kennedy Space Center to do the same thing, and he said no.

I was really floored when I found out that every field-joint of every ground test of an SRM motor had been repaired exactly the same way since the start of the program, and it had been my boss who had done them all.

What was even more surprising was that this operation never appeared in any of the assembly logs or posttest reports.

I told him that this was stupid, because we didn't fly the vehicle in that condition, and we needed to test in the same exact condition that we assembled it at KSC. I told him it was no wonder that we didn't see any O-ring erosion or blowby problems in the field-joints on static test motors. I told him that during our next test with the FWC-SRM, we should photograph the as-assembled field-joint from inside the motor, note the blowhole locations and sizes, and test the motor in the as-assembled condition.

He wasn't too happy with me after this conversation, but he agreed that I was right and that we would test the next motor without any repairs.

It was obvious to me why we had observed O-ring erosion on the nozzle-to-case joint but never in the field-joints during the static test program. This had bothered me for some time, because we had had three times as many opportunities to notice erosion in the three field-joints as compared to the single nozzle-to-case joint. The nozzle-to-case joint was inaccessible after installing the nozzle on the motor, so it was impossible to repair, and we had been observing erosion on the primary O-ring of the nozzle-to-case joint on flight motors and on several static test motors.

The most disturbing data from the FWC-SRM test came a few months later when we removed the igniter from the steel forward dome after the segments had been removed from the test bay. The igniter-to-case joint was identical to the configuration we were currently flying. The igniter was bolted to a steel adapter, which was then bolted to the forward dome. These bolted joints were sealed with a fluorocarbon elastomer gasket—a "gask-o-seal"—which was bonded to a metal retainer ring in order to prevent hot gases from escaping the joint. When we removed the igniter from the adapter, we noticed that the primary gask-o-seal had eroded completely through at a location that was in line with a blowhole in the vacuum putty that had been used to protect the seals in all the major joints. A large quantity of black soot was deposited between the burned primary gask-o-seal and the secondary seal.

This was clearly a major anomaly. We immediately established a task force under my direction to investigate the problem, because this portion of the FWC-SRM was unchanged from the design we were currently flying. We still had several igniters that had flown that had been removed from returned flight motors but were still attached to the adapters. We immediately disassembled these igniters and found soot behind the primary gask-o-seal

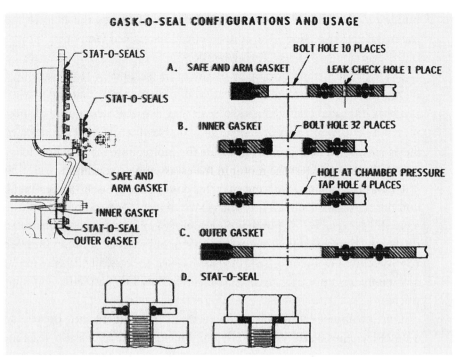

Space Shuttle solid rocket motor igniter bolted to a steel adapter and sealed with gask-o-seals. (Courtesy of NASA/Morton Thiokol.)

on the inner joint (where the igniter attached to the adapter) on four of the last eight flights flown: STS-41-C, 41-D, 51-D, and 51-G. In each case, the soot was found in a location where there was a blowhole through the vacuum putty protecting the joint; however, none of these flight motors had exhibited any erosion of the primary gask-o-seal like that we observed on DM-6.

We checked all the manufacturing assembly records and quality records on the gaskets and found that no problems had been noted. The vendor had certified the gask-o-seals prior to shipment to Morton Thiokol. We sent a quality team to the vendor and found out that, even though the firm had made all the proper inspections, a planimeter they had used in determining some of the critical dimensional features of the gaskets was out of calibration. All of the gask-o-seals showing evidence of the problem, without exception, contained more rubber in the sealing area than was allowed by the specification. This "overfill" condition of the rubber on the gaskets reduced the effective preload on the attachment bolts, which, during rapid pressure buildup in the igniter, allowed greater deflection between the adapter and the metal gasket retainer. The added deflection enabled temporary blowby of the primary gask-o-seal during ignition.

No thermal distress or blowby of any of the secondary seals was noted on any of these assemblies. Examination of the assemblies that had worked properly indicated that their gask-o-seals had all met the dimensional specifications but were in truth reused gask-o-seals that had been fabricated some time ago.

We checked the inventory of new and used gask-o-seals and found several new gask-o-seals that were out of specifications because of the overfill condition. All of the old gask-o-seals that had been used one or more times were still acceptable. We purged the inventory of all gask-o-seals fabricated during the time when the vendor inspection tool was out of calibration. We recommended flying only gask-o-seals that had successfully flown without any anomalies until proper corrective action could be verified at our supplier.

Our recommendation was accepted by the subsequent FRR boards, so Shuttle launches continued by only employing the "used" gask-o-seals for a while. This was deemed to be an acceptable solution because the gask-o-seals were considered to be one of the "reusable" components of the SRM.

Aside from the igniter gask-o-seal anomaly, everyone considered the test of the first FWC-SRM a resounding success. All of the elements new to the FWC-SRM had performed as expected. NASA was extremely happy, we were happy, and the morale of the entire Morton Thiokol company rebounded nicely from the near-disastrous fire earlier in the year.

Though continuing to work extremely long hours, I was walking on cloud nine. I had made a solid contribution to achieving a major milestone in the Shuttle program.

3

Dire Warning

Though the test of the first FWC-SRM was a success, not everything was going well. About a month before the successful test of DM-6, we removed the nozzles from the aft segment arriving back from Kennedy Space Center on STS-41D *Discovery*, which had been launched at the end of August 1984. Erosion of the primary O-ring of its nozzle-to-case joint was noted on one of the two booster nozzles. This by itself would not have been of great concern, because it had been observed before, and the amount of erosion fell within our previous experience. What was alarming was the amount of black soot between the nozzle's primary and secondary O-rings. This was the first time that this condition had been observed in the area of any O-ring seal within the solid rocket motor.

Finding soot *behind* a primary O-ring concerned me much more than any erosion of the O-ring itself, because it indicated that the primary O-ring had failed temporarily to accomplish its most essential function, which was to prevent hot gases from leaking *past* the seal. Fortunately, the secondary O-ring—a face seal rather than a bore seal like the primary O-ring—prevented any hot gas from escaping the joint. We noted no thermal distress on the secondary O-ring. This was a very disturbing condition to be noted on just-recovered hardware because we were within a few weeks of the Shuttle flying again—the flight of STS-41G *Challenger*, scheduled for early October 1984.

Appointed to head the task force investigating this problem, I needed to make a recommendation to NASA Marshall, in a week's time, whether we should fly the next Shuttle. I assembled a team consisting of our seal experts, heat transfer experts, and structures people to assess the observed condition on STS-41D. Very quickly we had to develop scenarios that explained the soot, determine the potential for this type of anomaly to lead to a catastrophic failure, and articulate a rationale for recommending whether to launch the next Shuttle or delay flight until further studies could be completed or some corrective action taken.

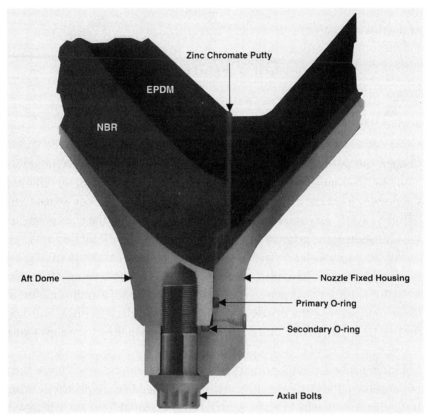

Space Shuttle solid rocket motor nozzle-to-case joint configuration. (Courtesy of NASA/Morton Thiokol.)

It was a big job to complete within a week. We worked twelve-to-sixteen hour shifts, including the weekend. Our careful examination of the recovered *Discovery* hardware indicated that the erosion of the primary O-ring seal was within our prior history, and the supporting analysis of the O-ring erosion phenomenon suggested the amount of erosion that could ever be experienced was "self-limiting." The erosion was being caused by hot gas passing through a blowhole in the vacuum putty in the joint during motor ignition and impinging on the O-ring in a localized area just downstream of one of these blowholes. The reason the process was self-limiting was that the jetting action stopped as soon as the motor reached steady-state pressure and eliminated the pressure differential during the ignition process.

This whole process occurred within the first 0.6 second after ignition.

The amount of erosion that could be experienced, we knew, was a function of the free volume between the vacuum putty and the O-ring that had to be filled, of the jet size, and of the time that it took to reach steady-state pressure in the motor. Our analysis indicated that the free volumes were small, and the times were so short, it was not possible to provide enough energy from a hot gas jet of any size through a putty blowhole to erode an O-ring sufficiently to cause it to lose its seal.

Subscale testing confirmed our analysis and also showed that over half of an O-ring had to be eroded away before it failed to function as a pressure seal. The maximum possible free volume in the nozzle joint area, combined with a maximum possible ignition time, was not enough to erode an O-ring halfway through. The maximum observed O-ring erosion on STS-41D had been 0.058 inch, which was only slightly over 20 percent of the cross section of the O-ring, which was 0.275 inch in diameter. We were confident that the O-ring provided adequate protection from jet-impingement erosion as long as the O-ring extruded into the gap between the metal parts in the joint before any significant amount of hot gas could pass underneath the O-ring. This was especially critical during the first one-quarter second of ignition, when the O-ring was moving across the O-ring groove and extruding into the gap before the metal parts started to separate.

The jet-impingement erosion on the primary O-ring did not concern us nearly as much as the evidence of soot blowing by the O-ring during the joint sealing process. What was most puzzling about the condition of this joint was that the soot between the primary and secondary O-rings was not in line with the blowholes through the vacuum putty in front of the portion of the O-ring that was eroded; the soot was several inches away. We concluded that the joint was primarily pressurized through the putty blowholes located at the point of jet-impingement erosion on the O-ring and then filled the remainder of the free volume in the joint (in front of the O-ring) circumferentially from that point. Right after assembly of the nozzle to the loaded aft segment, both O-rings of a case-to-nozzle joint were leak-checked with nitrogen through a port between the two O-rings. In the process of doing this, the primary O-ring was rolled forward in its groove, and the secondary O-ring moved back to its sealing position. The primary O-ring groove in the joint was oversized in length, allowing the O-ring to roll or slide in the groove back to its sealing position when pressurized by hot gas. We postulated that the O-ring was not moving uniformly back in the groove because of a pressure differential around the 300-inch-circumference of the joint, and that a small quantity of soot and hot gas must have blown by the O-ring (away from the point of erosion) during this O-ring movement process.

Fortunately, the quantity of gas was so low that it did not have sufficient heat content to damage the O-ring before the O-ring was fully seated and sealed. There was no evidence of any soot or hot gas ever reaching the secondary O-ring. This redundancy of two O-rings had been designed specially for the Space Shuttle. All other American solid rockets in use had only a single O-ring for sealing. With the exception of the Titan SRM field-joints, those single seals were more like the Shuttle's secondary O-ring, clamped down as it was by 100 large 1¼-inch axial bolts that attached the nozzle assembly to the aft dome of the case. This type of sealing O-ring had proven extremely reliable.

Based upon our careful examination of the returned hardware from STS-41D, and the results from analysis of the O-ring erosion during the joint pressurization process, we concluded it was still safe to fly this design. Our nozzle joint design provided redundancy to prevent hot gas leaking from the joint. Even though some erosion had occurred on the primary O-ring, and some soot had blown by the primary O-ring, our basic sealing redundancy had been maintained in the joint because both the primary and secondary O-rings maintained a pressure seal during motor operation.

My task force unanimously agreed that the case-to-nozzle joint would maintain an adequate margin of safety. The joint was safe to fly as long as it was properly assembled and passed the leak-check after assembly. Examining the manufacturing and quality records for the next flight, *Challenger's*, we could not find any discernible differences and therefore unanimously recommended flying.

As task force leader, I prepared a detailed presentation of our investigation for review in Huntsville by Jim Kingsbury, the Director of Science and Engineering (S&E) at MSFC, and his lab directors, held just prior to the flight readiness review. Halfway through my briefing I came to my discussion of the soot observed between the two O-rings. After explaining how we thought this had occurred, Kingsbury stood up and flatly stated that he didn't believe what I was saying. His Director of the Materials and Processes Laboratory, Bob Schwinghammer, stood up right behind him and said he didn't believe our version of the story either.

I did not know what else to say because we had ourselves struggled long hours over this puzzling observation and could not find any other explanation. I walked over to the NASA men and handed them the pointer: "Then you tell us what you think happened to explain this."

Kingsbury and Schwinghammer stood there for a minute looking at the viewgraph I had projecting on the wall. Finally Kingsbury looked at me and said, "I still don't believe your story, and I'm sure not going to recommend flying the next Shuttle based on what we've heard here today."

Kingsbury and his lab director walked out of the meeting. I had never been so insulted in my life. This was very unprofessional, and Kingsbury didn't even have the common courtesy to let me finish my briefing or review our conclusions and recommendations or give the rationale we had developed to indicate that there were adequate safety margins to launch the next Shuttle. We had busted our asses for the past week, working over 100 hours getting ready for this presentation, and NASA's Director of Science and Engineering walked out halfway through the presentation.

Marshall's project manager for the solid rocket boosters, Larry Mulloy, then stood up and suggested we finish our presentation for the remaining members of the S&E staff and other NASA officials who were present. After I finished the presentation, Mulloy came over to me and told me, "You really screwed up, Al." "Why? Was the presentation so bad?" I asked. "No, to the contrary," Mulloy answered, "I thought it was an excellent presentation. You guys have done an outstanding job of analyzing the problem, and I fully agree with your conclusions and recommendations.

"Where you screwed up, Al, was when you handed Kingsbury the pointer and asked him to provide his explanation of what happened. Jim is a powerful man at Marshall, and you put him into a bad position in the presence of all his subordinates."

"Well, I thought he was being very unprofessional," I responded. "The SOB got what he deserved. If he was so damned smart, he could have presented a better theory as to what happened."

It was just another example of where it is much easier to be a critic than to be correct.

I asked Mulloy, "So what do we do now?" The flight readiness review for the Level III Shuttle Project Manager, Bob Lindstrom, was coming up the next day, and I had intended to use a shortened version of this same presentation before the FRR. Mulloy suggested that I do just that and see how the FRR board reacted.

At that late date, we didn't have much choice. Even though Schwinghammer was a member of the board, and Kingsbury would be a member of the next-level FRR board (that is, the MFSC Center Director's FRR Board), Mulloy, the SRB Project Manager, didn't think that the two men would react the same way unless they had a better explanation.

As it turned out, Mulloy was exactly right. All the FRR boards agreed with our conclusions and recommendations and supported a launch recommendation based upon our analysis of the problem.

This was my first real confrontation with any of the NASA people, but it wasn't going to be the last. I figured if I could survive this one, I could survive most any situation that might develop, but I was truly worried. I had been put into this new job because NASA management had not been satisfied with my predecessor for the technical presentations he made to key management personnel. I knew that if the Director of S&E or his lab director called my boss voicing a concern about me, I wouldn't remain in this position for very long.

Apparently, that did not happen. Upon my return home, there was no confrontation by my boss about what had unfolded in Huntsville.

The pace of the Shuttle program was really starting to pick up. We had one more flight scheduled in November 1984, STS-51A, which would make a total of four flights in 1984. In fiscal year 1985, the flight rate would double to eight and then nearly double again, to fifteen, for 1986. Morton Thiokol would conduct one more development test of the FWC-SRM (DM-7) plus a single qualification motor test (QM-5), and would need to deliver the first set of FWC-SRM flight motors to Vandenberg, all in the next twelve months.

Moreover, with NASA planning to launch two dozen Shuttles a year by 1988, we were busy upgrading our facilities to produce all the required motors, with the idea of demonstrating a twenty-four-launch-rate production capacity to show it was not necessary to qualify a second source to meet future planned production of the SRMs.

The great news was that the flight of STS-51A *Discovery* in November 1984 showed no evidence of O-ring erosion or soot blowby in any of the field-joints. The nozzle exhibited minor pocketing erosion, but the depths of the pocketing were not severe and were deemed within our design margins of safety for the nozzle. Considering the fact that we had survived a near-disastrous fire in the large motor casting pits earlier in the year, we had recovered well and ended the year in good shape.

They say ignorance is bliss. In retrospect, there is no way for me to feel good about the year 1984, because flying in those four Space Shuttle flights that year were three astronauts who were scheduled to fly again in 1986, aboard STS-51L, the *Challenger*. The fact that Mission Specialist Ronald Mc-Nair (STS-41B on February 3, 1984), Commander Dick Scobee (STS-41C on April 6, 1984), and Payload Specialist Judy Resnik (STS-41D on August 30, 1984) were each scheduled for a future flight was not at all unusual. What was unusual was that all of these astronauts had completed their missions in 1984 on solid rocket boosters that had shown signs of O-ring thermal distress in two or more joints with indications known to some of us that things in that regard probably weren't going to get better.

Ron McNair's flight on STS-41B *Challenger* successfully deployed the Palapa and Weststar satellites, only to see those satellites lost in space because of failures of the carbon/carbon nozzles of their orbit-transfer motors. This event got a lot of publicity and was a personal embarrassment for Morton Thiokol because the two defective space-transfer motors had been manufactured by the Elkton Division of Morton Thiokol in Elkton, Maryland.

What wasn't well known about STS-41B was the fact that the primary O-ring in the left forward field-joint had eroded 0.040 inch and that the primary O-ring in the nozzle-to-case joint on the right-hand booster exhibited 0.039 inch of erosion.

Nor was it well known about STS-41C *Challenger*, Dick Scobee's flight, that while there was no erosion of the primary O-ring in the aft field-joint of the left-hand booster, there was a blowhole through its vacuum putty showing the effects of heat. Worse, 0.034 inch of erosion was exhibited on the primary O-ring of Scobee's right-hand SRB nozzle joint, and soot blowby had occurred in the igniter-to-case joint of the same booster.

There were bad indications also in Judy Resnik's STS-41D *Discovery* flight. The primary O-ring of the left-hand booster's nozzle joint experienced 0.046 inch of erosion, while the right-hand booster exhibited 0.028 inch of erosion on the primary O-ring in the forward field-joint, along with soot blowby on the same booster past the primary igniter gask-o-seal.

Along with all of my colleagues, I was under a mistaken impression that all of the postflight data was reviewed with each of the crews and that Mc-Nair, Scobee, Resnik, and all the other astronauts were well aware of the risks associated with their continuing to fly under these conditions. We were mistaken.

Not all of our flight readiness reviews addressed these issues, but many of them did. But even in those reviews that did, everything appeared to be "relative." Not everyone who should have heard about such problems inherent to the solid rocket motor seals actually ever heard about them.

For example, the Rocketdyne Division of Rockwell, which manufactured the Space Shuttle's main engines, was constantly finding cracked turbine blades in their engine turbo-pumps from flight operations to engine failures on the test stand. The problems were serious enough for NASA to periodically change out engines; NASA had even awarded a contract worth hundreds of millions of dollars to Rocketdyne's main competitor, Pratt and Whitney Division of United Technologies Corporation, for development of improved hydrogen and oxygen turbo-pumps for the SSMEs. I always assumed that the astronaut corps had been informed by their NASA folks about the problems with our SRB joints and that the astronauts, in conjunction with their space program officials, had determined that problems like the ones with the SSME turbo-pumps were more important to solve, with the available resources, than fixing our SRB joints.

January 1985 started off with a grave omen. We had just successfully launched STS-51C *Discovery* on January 24, 1985. Disassembly of the returned solid rocket boosters in Hangar AF at Cape Canaveral revealed a very serious problem. Two field-joints—one on the left-hand and one on the right-hand SRB—not only exhibited erosion of the primary O-rings, but large quantities of dark black soot sat between the primary and secondary O-rings in both field-joints. This was the first time we observed soot between the two O-rings in the field-joint, and it was the first time the O-ring damage displayed itself on more than one field-joint.

Both nozzles also showed evidence of soot between the primary and secondary O-rings without any erosion of the primary O-rings. This condition was not as much of a concern as the field-joint because we had observed this same condition in the nozzle earlier, and the nozzle had a good secondary seal that was truly redundant. The nozzle's primary seal was a bore seal like the field-joints, but the secondary seal was a face seal around a 90° corner from the primary seal. Deflections at this face seal were constrained, as explained earlier, by large 1¼-inch-diameter axial bolts that attached the nozzle to the aft dome of the motor. What we observed about the condition of the field-joints was much more alarming because both O-rings were bore seals in close proximity to one another, and whatever condition that could cause loss of the primary O-ring seal could also defeat the secondary seal.

Fortunately, the primary O-ring managed to maintain a seal after the initial blowby and with no blowby or serious damage of the secondary seal.

The secondary O-ring did show some heat effect, however. Nevertheless, the condition we observed in the field-joints was vastly different from anything we had seen before. Tremendously puzzled, we methodically went through all of the manufacturing and quality assurance records to determine if there might be something unique about the hardware or assembly process of STS-51C that could help to explain the observed condition. But nothing was found.

Finally, we concluded that the condition was due to the extreme cold weather that STS-51C had been exposed to in the days preceding the launch. This was the coldest launch to date in the Shuttle program: approximately 62° at launch time, which resulted in an O-ring temperature that we calculated as 53°. The three days prior to the launch, a local newspaper had noted, had been the coldest three days in Florida history, with nighttime temperatures in the teens.

The launch of STS-51C had actually been delayed a day, from January 23 to the 24th, because of the cold weather. Based on the weather forecast, NASA had grown very concerned about the potential for extreme icing conditions on the external tank and the launchpad area exacerbated by the various water systems used in the launch area. Even though the weather warmed up considerably during launch day, the extremely cold weather had kept the hardware colder than the local ambient temperature. It was clearly evident to us that the lower-than-normal temperatures had hardened the O-rings, making it more difficult for them to seal the joints.

In the next flight readiness review, we presented this very explanation as the cause of the observed condition in STS-51C, with the exact statement, "Launch of STS-51C was preceded by the coldest three days in Florida history." Recommending proceeding with the next planned launch of STS-51E (which later turned out to be STS-51D) was not difficult because we knew that, in April, such cold temperatures were virtually impossible in Florida. Reasonably comfortable that the problem we observed on STS-51C had been aggravated by record cold January temperatures, we clearly didn't expect to encounter that condition in the spring.

At that moment no one had the remotest idea that we would only a year later face a launch condition that was even worse, much worse. In retrospect, all of us should have been deeply sensitized by the implications that cold temperatures could have seriously damaging effects on the sealing capability of the field-joint; we just weren't. Ironically, Mission Specialist Ellison S. Onizuka, who flew on STS-51C *Discovery* and would also fly a year later on

STS-51L, the ill-fated *Challenger*, like his compatriots, was never told about the serious O-ring erosion and blowby condition that we certainly knew about and attributed to cold weather.

No one would have predicted that Onizuka's next flight a year later on STS-51L *Challenger* would make the weather conditions for the launch of STS-51C look like a warm day in Florida.

4

A Total O-Ring Failure

In late February 1985 I gave the keynote speech at an international aerospace conference in Tel Aviv hosted by the Israeli Institute of Technology in Haifa (known as "The Technion") on the history of the Space Shuttle's solid rocket motors. I decided to focus my talk on the development of the graphite-epoxy-filament-wound-case (FWC) solid rocket motor then currently in development.

It was a great honor for me to give this talk, especially when I discovered that Israel's Minister of Defense, Yitzhak Rabin, was to provide the conference's opening address and that Space Shuttle Commander Rick Hauck, who had just completed a successful retrieval of the Palapa and Weststar communication satellites on mission STS-51A *Discovery* in November 1984, was to be the luncheon speaker. This was only the second astronaut that I had met—the other being Major Dick Covey at the FWC-SRM test last October—but this would not be the last time that I would meet these same two astronauts. At the conclusion of his opening address, I presented Rabin with a model of the Space Shuttle. He thanked me for the gift and said he wished they could put all of Israel's problems into the cargo bay and send them into orbit. It was also quite an adventure, as I had never visited a foreign country before other than our neighbors to the north and south, Canada and Mexico.

The choice of topic for my talk was timely. The new version of the Space Shuttle SRM (DM-6) had just achieved its first successful test in October 1984. It was being developed by NASA for the air force in order to increase the payload capability of the Space Shuttle by approximately 5,000 pounds, enabling launches into polar orbit from the newly constructed Shuttle launch facility at Vandenberg AFB. The technical challenges of designing the new FWC solid rocket motor were many. Most significant, the graphite cylinders were going to be a full casting segment length—roughly double the length of the existing steel cylinders that had to be joined together to make a casting segment. The FWC's large twelve-foot-diameter graphite-epoxy cylinders contained steel rings on each end. These steel end-rings contained a large clevis that slipped over the graphite cylinders that then had to be match-machined in the factory to attach the metal ring to the composite through the machined pinholes. Large steel pins, approximately one inch in diameter, also had to be installed properly to pin the graphite sections to the steel end-rings.

The field-joints on the other end of the steel end-rings were similar to the existing tang and clevis joint that was used on the older steel-case version of the SRM but with one notable exception. The tang end of the new joint contained a metal lip, or capture-feature, to restrict the motion in the joint during pressurization at ignition so as to enhance the sealing capability of the O-rings in the field-joints. We chose to introduce this feature into the FWC because of our experience with periodic erosion of the O-rings in the field-joints and because the joint was opening wider than we had expected. We thought that the FWC-SRM design would not eliminate the O-ring erosion problem, but it would help by restricting the rotation of the joint. If we didn't make the change, the more flexible filament-wound-case would only exacerbate the problem already being experienced with the stiffer steel-case. As we figured, the capture-feature on the FWC-SRM did not eliminate joint rotation during ignition of the motor, but it did reduce the gap between the clevis and the tang at the O-ring sealing surface by as much as 60 percent.

The trip to Israel was a wonderful, eye-opening experience for me. Linda came along, and on the way home we visited Italy for a short vacation before going to Boston to visit with our oldest daughter Lisa, who was in nursing school. I knew I'd better come back refreshed and ready to go because the spring of 1985 was going to be a very busy time for the Space Shuttle program. In April 1985 the plan was for NASA to launch two Space Shuttles—and do it only a little over two weeks apart. That quick turnaround, it was hoped, would definitely demonstrate the capability of the Shuttle system to meet its projected twenty-four launches per year (two each month) by 1988.

The two April 1985 launches were of particular interest to Morton Thiokol and the people of Utah because the first launch (STS-51D *Discovery*) was scheduled to carry NASA's biggest supporter on Capitol Hill, Utah Senator Jake Garn, and the second launch (STS-51B *Challenger*) was to include Utah's first real astronaut (and NASA's oldest astronaut in training, who had never yet flown), Mission Specialist Dr. Don Lind.

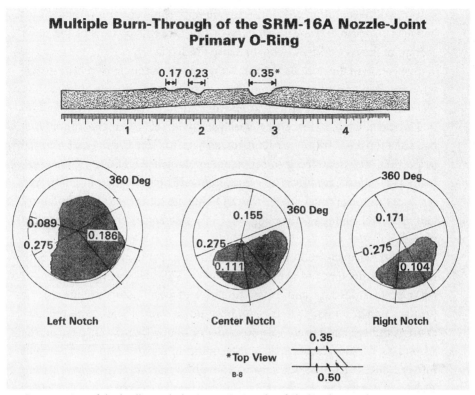

Multiple Burn-Through of the SRM-16A Nozzle-Joint Primary O-Ring

Cross section of the badly eroded primary O-ring that failed in the nozzle-to-case joint on STS-51B *Challenger* (SRM-16A). (*Report of the Presidential Commission on the Space Shuttle Challenger Accident*, Vol. 3, June 6, 1986.)

Even though the returned flight hardware from these two launches showed no evidence of any problems in the field-joints, they were far from uneventful. STS-51D (Senator Garn's flight) showed evidence of erosion of the primary O-rings in both rocket nozzles, a first-time occurrence; the magnitude of erosion in the right-hand nozzle (0.068 inch) was, in fact, one-third more than had ever been observed previously.

Neither of these two observations on STS-51D was particularly alarming at the time, but the condition observed on the second flight, STS-51B, clearly was. Don Lind's flight (STS-51B) not only showed erosion of primary O-rings on both nozzles but also that the nozzle on the left-hand side had experienced a very serious problem. The primary O-ring on this nozzle had failed to seal at all and had eroded completely through in three locations, with the worst location showing only one-third of the O-ring's original cross section still remaining. Not only was there a heavy coating of black soot between the primary and secondary O-rings, but 12 percent of the cross section of the secondary O-ring had also eroded away.

Inspection of the nozzle O-rings could only be conducted after the nozzle was unbolted and removed from the aft segment, an operation that took place several weeks after launch at Morton Thiokol's refurbishment facilities in Clearfield, Utah. Before that inspection could take place, the motors first had to be retrieved from the ocean, towed back to Cape Canaveral for initial inspection, disassembled into segments in Hangar AF at Cape Canaveral, loaded on railcars, and returned from Kennedy Space Center to Utah. As a result of this long process, this serious O-ring condition was not found until late June 1985—after another flight (STS-51G *Discovery*) had already taken place.

STS-51G also showed evidence of erosion of primary O-rings on both nozzles. Fortunately, the depth of erosion on STS-51G was well within our prior experience base and, even though soot was observed between the primary and secondary O-rings on both nozzles of STS-51G, the primary O-rings did seal, and there was no evidence of any thermal distress on the secondary O-rings. Nevertheless, the condition observed on the last flight in April 1985 (STS-51B) was serious enough to warrant a full investigation and assessment as to whether this condition could worsen to a point where the Shuttle was not safe enough to continue flying.

I was spending most of my time during this stretch on the FWC-SRM development program. Before I left for Israel, we had just delivered an important test article for the filament-wound-case solid rocket motor program, to be used in a structural test at NASA Marshall. We subjected the test article incorporating the FWC (known as Structural Test Article [STA-2]) to internal pressure loads and prelaunch bending and flight loads so as to determine if all structural safety factors had been met. STA-2 consisted of an empty forward and aft case segment with an aft skirt attached and mounted horizontally on the test stand. We applied external bending loads to the test article by connecting a series of large hydraulic actuators; engineering data came back to us through several thousand channels of information streaming from associated strain gauges, displacement transducers, and pressure gauges.

Prior to the bending loads being applied, STA-2's internal pressure and flight loads all looked good. But we knew that the prelaunch bending loads on the Space Shuttle were unique, introduced as they were into the vehicle by the start-up and check-out of the Shuttle's main engines some six to seven seconds before ignition of the SRBs. During the start-up of the SSMEs, the Space Shuttle stack literally bent over, placing very high structural loads on the aft skirt that supported the entire 4.5-million-pound vehicle. To prevent the Shuttle from tipping over, the skirt had to be bolted to the launch platform. The top of the SRBs came displaced more than two feet from vertical, and high loads got introduced into the SRB's center and aft field-joints, putting one side of the SRB in compression and the other side in tension.

During the STA-2 test, we had heard large popping noises and stopped the test at a load below the required 1.4 safety factor. Ultrasonic inspection of the FWC revealed large de-laminations in the thick graphite composite structure at the joint; these de-laminations caused the loud acoustic emissions. The joint area of the case consisted of a composite layer of graphite cloth "over-wound" with graphite filaments in the structure. Because of the concerns about high stress concentrations at the end of the graphite cloth, the engineers at Hercules were in the process of redesigning this specific area as the test program was occurring. This failure of the STA-2 test article accelerated that redesign of the joint area. Before the planned launch of the VLS-1 Space Shuttle by the air force from Vandenberg AFB in the spring of 1986, we needed to test a new test article, one incorporating a redesigned FWC.

In May 1985, we successfully conducted our last development motor static test (known as DM-7) and were preparing for the design certification review scheduled for later that summer. With a qualification motor test planned for the fall of 1985, we planned to fabricate and deliver our flight hardware to Vandenberg AFB for VLS-1 by early summer of 1985.

Everyone considered the DM-7 test a total success. We saw no problems with the FWC or any of the other design changes in the nozzle and insulation that we had introduced into our new lighter weight solid rocket motor.

The case-to-nozzle joint did exhibit, however, rather severe erosion of the primary O-ring, amounting to 0.088 inch. Since this nozzle had been disassembled before those returned from STS-51B, it represented the worst erosion we had observed on a primary pressure seal up to that point. Still, the primary O-ring seal was never compromised, and there was no evidence of any soot between the primary and secondary O-rings.

Significantly, this test was also the first and only static test in the history of the SRM program *not* to have the blowholes in the vacuum putty in the field-joints repaired *prior* to static test. Of course, we observed and photographed the blowholes and volcanoes in the putty after motor assembly in the test bay, but the motor was static tested in the "as-assembled" condition. After the static test, we noted gas paths in the vacuum putty but saw no erosion or blowby of the primary O-rings in any of the three field-joints.

Clearly, our incorporation of a capture-feature or metal lip on the tang side of the joint (restricting the motion of the joint and reducing the O-ring seal gap at ignition by 60 percent) was an improvement in the joint design. We were ecstatic over the success of DM-7; it was the first static test in which blowholes in the putty had not been repaired but were present, yet that more realistic condition did not result in any O-ring damage or primary seal blowby. True, it did not prevent blowholes through the zinc chromate vacuum putty that could allow hot gas jets to erode the primary O-ring seal. But we did not think that soot blowby or O-ring erosion was likely to occur. That type of thermal distress had been observed in only 5 to 6 percent of the field-joints that had returned from flight, equating to approximately 15 to 20 percent of the motors (three field-joints each) flown.

Being in the Shuttle program for over a year, I was absolutely overwhelmed with my job and had totally forgotten what had ever happened to the CELV that I had worked so hard on in a losing effort a year earlier, when I received a telephone call from Larry Wear, the SRM Program Manager at Marshall Space Flight Center. Wear told me that the air force had asked NASA for a briefing on the status of the Shuttle's SRM program with emphasis on "Lessons Learned" that could be applied to the new CELV program, which was starting into its development phase. The Air Force Space Systems Division in Los Angeles and its technical consultant, the Aerospace Corporation, were particularly interested in the problem areas such as O-ring erosion, pocketing erosion in the nozzle, and any other anomalies that we had seen during the history of the Shuttle SRM ground test and flight program. Wear told me that the air force had requested a tour of our Shuttle manufacturing facilities in Utah and a briefing on the SRM program. The team visiting us would consist of representatives from the air force, Aerospace Corporation, Martin Marietta, and the Chemical Systems Division (CSD) of United Technologies Corporation, the SRM contractor for the CELV.

The first thing I said to Wear was that since we were in competition with United Technologies' CSD as the potential second source for the Space Shuttle solid rocket motor, we could not allow CSD people to tour our manufacturing facilities. I also told Wear that we would host such a meeting on one condition: the air force and its contractors would have to present their own "Lessons Learned" based on their experiences in the Titan SRM development and flight program.

We were particularly interested in their experience with O-ring erosion in the field-joints and case-to-nozzle joint. Given that their five-and-a-half-segment Titan SRM motor had more opportunities for erosion than our four-segment Shuttle SRB, we figured we could learn quite a bit from one another. The Titan's field-joint design was very similar to the Shuttle's (both designs used a tang and clevis pinned joint), and our analysis indicated that the Titan SRM should be having a similar problem with O-ring erosion. The primary difference between the two joint designs was that the cases of the Titan contained the clevis on the bottom of the case (on top during joint mating) rather than on the top (on bottom during mating) as with the Shuttle SRM. It was done this way with the Titan to eliminate the potential for water being trapped in the joint while sitting on the launchpad, thus minimizing the potential for metal corrosion. For the Shuttle, we intentionally reversed the design for reasons related to assembly safety, that is, to prevent people from having to work under a suspended load during case subassemblies and during loaded case segment stacking operations.

There were other important similarities between the Titan and Shuttle SRMs. In both designs, the O-ring grooves were located in the clevis leg (Titan's SRM had only one O-ring while the Shuttle's had two, for redundancy). In both cases, the O-rings were installed just prior to mating and were carefully inspected to make sure they didn't fall out of their grooves or get contaminated prior to mating. We accomplished this most easily by putting the O-rings on the stationary bottom segments rather than hanging them from the suspended crane that lowered the case segments, where operators and inspectors would be required to work under very heavy suspended loads.

But there were important differences. The original Shuttle SRM design included a weather seal that was bonded over the joint to prevent water from entering, but this wasn't totally satisfactory in preventing seawater from entering the joint during splashdown, recovery, and tow-back operations. Since the Shuttle's cases were reused, but the Titan's were not, it was necessary to maintain a good coating of HD-2 hydrocarbon grease on all unpainted exposed surfaces to prevent saltwater corrosion. We found the grease to be more effective than the original weather seal. All that seal managed to do was trap seawater in the joint during splashdown and recovery, which made the corrosion worse, so we eliminated it. As far as splashdown and retrieval were concerned, it didn't matter which end of the case the clevis was on, because the cases were towed horizontally back to port, and the saltwater from the ocean was far more corrosive than any rainwater on the launchpad.

Also, the Titan SRM had an interference fit—a contact—at the joints of the mating insulation, while the Shuttle SRM had an intentional gap to prevent interference of the rubber insulation. The Shuttle SRM had higher structural and thermal safety factors on the insulation design; we were concerned that too much contact pressure could cause the insulation to become unbonded (or propagate an existing undetected "unbond"), which would result in insulation peeling away from the case wall at the ends of the segment joints. As has been said, the Shuttle SRM filled the gaps, during mating operation, with asbestos-filled zinc chromate vacuum putty; the Titan SRM used small quantities of this same material applied to the surfaces of the mating pieces of the rubber insulation. The putty served as a thermal barrier for the O-ring; it was used in the Titan SRM in the event that the optimum insulation interference conditions were not achieved.

As it turned out, the "Lessons Learned" meeting was very informative for the air force, but totally nonproductive for Morton Thiokol and NASA. We provided a very detailed briefing on our Shuttle SRM problems and proposed solutions, while the air force presented absolutely nothing on the Titan SRM. The Aerospace Corporation official, Joe Vanna, who had been asked to discuss the O-ring erosion experience on the Titan SRM, flatly stated they had never witnessed *any* erosion of O-rings and were confident that their interference insulation design prevented it from happening. Our people questioned whether they could actually control these huge rubber parts to such close tolerances to assure that peeling of insulation from the case wall would not take place during mating while still maintaining suitable interference during case pressurization over a wide temperature range. Our own calculations indicated hot gas should leak occasionally and impinge on the single O-ring of the Titan SRM. We had examined a similar insulation design to see how we might fix the Shuttle SRM problem and had decided that the Titan design was certainly *not* the way to do it.

Fortunately, we did not have to prepare all of this information just for the air force, given that we got nothing back in return. We prepared most of the information for a planned meeting at Morton Thiokol with Dr. William Lucas, the NASA Marshall Center Director, who had asked for a review of the progress on our understanding of the O-ring erosion problems, on the

FWC-SRM project, and on the pocketing erosion noted in the nozzles of the SRM.

They say that disasters generally come in threes. Already in 1985 we had experienced cold field-joint O-ring erosion and blowby (STS-51C) and the loss of a primary O-ring seal and erosion of the secondary O-ring in the nozzle-to-case joint (STS-51B). Early on the morning of June 3, 1985, we experienced our third disaster in the series.

It took place at our plant in Utah. A rare Wasatch thunderstorm was approaching the plant site, hard rain accompanying it. We were in the midst of mixing and casting Shuttle propellant when lightning struck one of the mixer buildings, frying the electronics that were controlling the propellant mixer. This resulted in the mix bowl dropping down while the mixer blades were still turning. The bowl-lowering device allowed the bowl to tilt, causing the outer blades to strike the wall of the mix bowl. Six hundred gallons—7,000 pounds—of Shuttle propellant ignited, completely destroying the entire building.

We had just recovered from the large motor casting pit fire of one year earlier, and the last thing we needed was another major propellant fire. Once again we were lucky that no one was killed, primarily because the mixing operations, involving such hazardous material, were conducted remotely. However, at a given moment in the mix cycle, the mixer was turned off, and the bowl was lowered to remove propellant samples manually, measure the temperature of the mix, inspect the propellant for oxidizer lumps and contamination, and scrape down the blades. Several people were inside the mixer building when this operation occurred. This time, when the lightning struck, these people had just left the building.

We had dodged another bullet, but the damage was going to make our lives more miserable for the next several months.

NASA sent in an independent investigation team that took us to task for having not conducted either a "failure modes and effects analysis" or hazards analysis on the electronic control system for the propellant mixers. Everyone at the plant had believed that the blades would just stop turning if the mix bowl started to lower—and that the bowl could not tilt—but that was not the case. We hired an outside consultant to run the hazards analysis on the mixer's electronic control system and test the various circuit boards in use. His analysis confirmed that the failure mode was, indeed, caused by the lightning strike.

The good news was, first, it was relatively easy to correct the problem in the mixer electronics, and, second, the loss of a single 600-gallon mixer did not significantly impact our propellant production. (Normally we used three 600-gallon mixers during a segment casting operation, with a fourth mixer on standby as a spare. Under normal conditions, we could successfully load the segment with only two mixers, but the third mixer provided added margin to achieve the desired casting time.) To minimize propellant mixing and casting time, Hercules Inc., one of our major competitors pushing for a second source on the SRM, had just purchased a single 1,800-gallon mixer. We could only imagine what the impact there would have been to the Shuttle program if we had been dependent on a single 1,800-gallon mixer and then lost it.

After our accident, the selling point for a larger mixer lost much of its appeal. It was clearer to NASA that the redundancy offered by our multiple 600-gallon mixers significantly reduced the vulnerability of the entire Shuttle program to a mixer fire, an event that was not that unusual in the solid propellant industry. Several years ago, CSD of United Technologies Corporation in San Jose experienced a fire and fatality while trying to repair a propellant mixer shortly after the company had just lost another mixer during operation; this series of events ended up putting CSD out of business.

The next launch of the Space Shuttle was to come at the end of July 1985. In preparation for the flight readiness review, our Morton Thiokol team assigned to investigate the dangerous condition observed in STS-51B's nozzle-to-case joint had to prepare its findings for review, first with Thiokol management and then with NASA Marshall.

Our team concluded that the primary O-ring on the left-hand nozzle of STS-51B had never sealed—and most probably leaked—during the leak check that had been conducted after initial installation of the nozzle on the motor at the factory. We knew from earlier laboratory tests that the zinc chromate vacuum putty used as a thermal barrier in the field-joints and in the case-to-nozzle joint could, at pressures below 150 pounds-per-square-inch absolute (psia), mask a leak of a bad O-ring. Below this pressure, the vacuum putty itself could act as a pressure seal. As a result of finding this out, we had changed the procedures for leak-checking from a maximum stabilization pressure of 100 psia to 200 psia.

We had then used the higher leak-check pressure on the field-joints of STS-51B. However, the nozzle for STS-51B was installed at the factory several months prior to this change being implemented. In fact, this nozzle was the very last nozzle to be leak-checked with the lower (100 psia) stabilization pressure. All subsequent assemblies, including the one that was flown in June 1985 as well as the ones on the launchpad for July, used the higher (200 psia) stabilization pressure in the leak-check process.

We introduced the higher stabilization pressure for the express purpose of making sure that the vacuum putty could not mask a leak past a bad O-ring—as well as to assure ourselves that an O-ring had been installed in the joint in the first place. If someone forgot to install the O-ring or if a bad O-ring was installed, the higher pressure would blow a hole, if it wasn't already present in that particular area, through the vacuum putty, indicating a clear leak and a bad joint. We had noticed, however, that the leak check did occasionally result in introducing additional blowholes in the vacuum putty, but this condition was much better than the possibility of not having an O-ring in the joint at all or having a bad O-ring installed. We knew from walking inside assembled horizontal static test motors that the assembly of the joint produced most of the blowholes in the vacuum putty anyway, and we had no way to prevent this from happening.

Our conclusion that the primary O-ring in the nozzle of STS-51B had never sealed in the first place seemed to be well supported by the nature of the O-ring erosion we had observed and by the analytical model developed for predicting O-ring erosion. The erosion pattern on the O-ring was distinctly different from anything we had observed before, and there was evidence of hot gas flow around the entire cross section of the O-ring's eroded area. All prior O-ring erosion had been limited to a much smaller fraction of the cross section, with anywhere from 100° to 240° unaffected. The cause behind the erosion pattern on the nozzle of STS-51B's primary O-ring appeared to be both gas flowing underneath the O-ring and jet-impingement on the body of the O-ring, something that we had not previously observed in either nozzle-joints or field-joints.

Our analysis indicated that blowby erosion alone could be more than twice as much as jet-impingement erosion and, together with jet-impingement, could result in near total erosion of the O-ring since it represented damage that was over three times as bad as what we had normally observed in prior flight motor nozzles.

Our task force made a very detailed presentation on the primary O-ring erosion failure of STS-51B. The presentation went into great detail relative to the history of O-ring erosion in the nozzle-to-case joints and the changes that had been made throughout the program in the asbestos-filled vacuum putty material and application procedures. Our report also contained detailed descriptions (with photos) of the observed condition in the left-hand nozzle of STS-51B, along with an engineering analysis explaining what we believed had happened with this particular nozzle-to-case joint from assembly through flight.

Our principal conclusion was that the severity of STS-51B's O-ring erosion could only have happened if the O-ring was not capable of sealing at all—and that we had not detected this during the leak check because the vacuum putty had managed to hold the maximum leak-check pressure of 100 psia, which had been introduced into the joint. Data from earlier tests supported this conclusion, clearly showing that the putty occasionally would hold between 100 and 150 psia for a time period longer than was used in the leak-check procedure. As this was the last nozzle assembly ever to use the lower, 100 psia maximum leak-check pressure, with all subsequent flights already increased to 200 psia, we thought that we'd eliminated this problem in all current and future assemblies.

In addition, our engineering analysis showed there was not sufficient energy in the gas jet, after eroding through the primary O-ring, to erode enough of the secondary O-ring so that it lost its sealing capability. With this conclusion in mind, we recommended that NASA should proceed with the next launch as scheduled, a recommendation that MSFC management accepted.

Apparently, though, MSFC management never then revealed to NASA headquarters just how serious this flight anomaly was, because a week or two later I received a telephone call from Paul Weitzel at NASA headquarters, asking me what I knew about the primary O-ring failure on the nozzle of STS-51B. I told Weitzel what I knew, and he suggested I come to Washington as soon as possible with my team to discuss what happened on STS-51B and explain why it was not possible for the condition to worsen to the point where a nozzle-joint failure occurred.

When I informed Larry Mulloy, the SRB Project Manager at Marshall, of Weitzel's request, he became very upset. "Headquarters people shouldn't be contacting you directly about that," Mulloy fumed. "They should be coming to Marshall for that information." Calming down, Mulloy said that we should prepare a first-class presentation for NASA headquarters but make sure that the presentation was reviewed first with MSFC before going to Washington. "I'll coordinate this with headquarters and set up the meeting date," stated Mulloy.

As it turned out, we had already planned to go to Washington the next week to present a management overview of our investigations and reports on both the status of our filament-wound-case development program and our corrective actions from the propellant mixer fire.

Without question, the loss of a primary O-ring and partial erosion of a secondary O-ring on the same Shuttle flight heightened the concern over the potential of a pressure seal failure in the case field-joint system. Those of us who were familiar with the past performance of the joints were much more concerned over the case field-joint than the nozzle-joint because the design of the case field-joint provided only *thermal* redundancy and not *seal* redundancy due to structural deflections of the case joint. In other words, in the event that jet-impingement erosion of a primary O-ring caused failure of that seal, then the secondary O-ring would provide a back-up thermal barrier that would have to be eroded through 60 percent of its diameter to lose the pressure seal in the motor case; only then could there be a catastrophic failure of the motor.

The likelihood of this happening was very remote, if not impossible, because there was insufficient energy in the gas jet during pressurization to erode both O-rings badly enough to cause both seals to fail. However, if the rapid deflection of the joint during pressurization prevented O-ring contact with the metal, then all bets were off. Then erosion could be far more severe due to convective blowby erosion of the hot gases. Add to that any significant degree of jet-impingement erosion, and the likelihood of the secondary O-ring seal (located only a fraction of an inch behind the primary seal) failing became fairly high. Furthermore, any significant amount of erosion directly on the sealing surface would probably result in a continuously leaking joint that would eventually lead to a catastrophe.

The only significant advantage of the secondary seal was that it was located in the back of the O-ring groove, where it was supposed to be during the leak check (conducted after mating the joint), while the primary O-ring was displaced to the wrong side of the groove. This resulted in some O-ring motion plus the time required for the primary O-ring to move to the back of the groove during the critical rapid pressurization of the joint occurring in the first 0.6 second after ignition. The nozzle-to-case joint did not suffer this same inadequacy: it had a genuine redundancy in that the secondary seal worked as a "face seal" around a corner. Unlike the primary seal, which was a bore seal, the secondary seal surface was restrained by 100 large 1¼-inch bolts, which prevented deflection of this portion of the joint even though the nozzle primary seal suffered deflections similar to the field-joint O-ring seal.

As a result of this significant difference between the two joints, I asked my program manager, Howard McIntosh, if the unmachined case forgings at our vendor, the Ladish Company in Milwaukee, Wisconsin, had sufficient material to machine a capture-feature on the tang side of the joint (similar to one that we were using on the steel end-rings of the filament-wound-case) that would reduce the motion in the joint. McIntosh looked into it and reported back that we needed to notify Ladish right away because they were scheduled to rough-machine some of the new forgings to the existing design fairly soon. Howard told me that we could meet our production schedule for the next six to twelve months with our existing case inventory, but that we would definitely need the new cases to meet our projected flight rate of twenty-four flights per year by 1988. I told him to tell Ladish to put the new forgings on hold until we could furnish them new engineering drawings to machine these cases with a capture-feature in the joint.

This was in late July 1985. I then went to my boss, Joe Kilminster, to tell him what I had done. He agreed with the need to do it, but said we couldn't do it without the approval of Walt Rupinski, Vice President of Material and Contracts, because this change had not been approved by NASA, and Morton Thiokol would be exposing itself to some considerable cost risk if we took this action on our own. Kilminster suggested we both go to Rupinski and explain our actions; he'd want to know exactly what the cost risks were, both in terms of the impact on case segment deliveries from our vendor and of the higher machining costs with the new design.

Rupinski was understanding and told us to proceed. We notified Ladish to hold the case forgings until we received approval from NASA to implement our recommended change to the case field-joint design. This was an important first step because we didn't have many more new case segments to be purchased under our existing contract, and if we had to start with new steel billets and forgings, it could be anywhere from a year and a half to two years before we got the change incorporated into a Shuttle flight.

5

An Impotent Task Force

In spite of the O-ring erosion problems, which got progressively worse, the Shuttle continued to fly. The loss of the primary O-ring with significant erosion of the secondary O-ring in the nozzle joint on the *Challenger* flight (STS-51B) on April 29, 1985, raised the level of concern, so much so that Morton Thiokol would form an in-house O-ring Seal Task Force in July and stimulate a NASA headquarters review of the problem in August 1985. Despite the growing concern, however, Shuttle flights continued at a very accelerated and ambitious pace. We had just completed two launches in April 1985 when we had noted the severe nozzle O-ring erosion problem. Nonetheless, the Shuttle launched again in June (STS-51G), July (STS-51F), August (STS-51I), twice in October (STS-51J and STS-61A), and again in November (STS-61B). The launch of STS-61C *Columbia*, with Congressman Bill Nelson from Florida aboard, was scrubbed during the terminal countdown in December 1985 but then launched successfully on January 12, 1986. In other words, from April 12, 1985, to January 12, 1986, we successfully launched nine Shuttle flights in exactly nine months—an average of one per month. This was more than double the rate that had occurred in any previous nine-month period.

As ambitious as this seemed, we would need to double that flight rate again within the next two years to meet the planned launch rate of twenty-four flights per year by 1988. The thing was, postflight examination of the returned hardware from the flights indicated that the nozzle-joint O-ring erosion was not an occasional but a *routine* occurrence: eight of the last nine flights revealed the problem. O-ring erosion in the case field-joints was far less severe and much more infrequent, but the truth was we saw minor erosion in a field-joint on STS-61C (January 1986) and soot blowby (without erosion of the primary O-ring) in two joints on STS-61A (October 1985).

Many of us felt the stress. At the end of a long meeting at NASA Marshall in late July 1985, I asked Larry Wear, NASA's SRM Program Manager at MSFC, to join me for dinner because we still had some unfinished business to take care of before I returned home to Utah early the next morning. During our chat, a deeply worried Wear said to me: "Al, you know there is a real possibility of a Shuttle failure. It's going to happen sooner or later. I think it's inevitable. Most likely the failure will be due to a problem with one of the high-speed turbo-pumps in the Space Shuttle's main engines."

It wasn't surprising to me that Larry Wear was focusing his worry on the SSMEs. Larry had been involved in the development of these very high-performance cryogenic liquid oxygen and liquid hydrogen engines, even being sent by NASA as a resident representative at the Rocketdyne Division of Rockwell International in California where the Shuttle's main engines were made. Larry had seen several catastrophic engine failures during the development program and knew that Rocketdyne was still experiencing an occasional engine failure on the test stand. "NASA is just not prepared to deal with a Shuttle failure," Wear lamented, "and neither am I, no matter what the cause. If a disaster comes as a result of a main engine, I'll feel some responsibility for that; if it comes because of an SRM failure, now that I'm program manager for the SRM, it'll be even more difficult to deal with."

"I feel exactly the same way, Larry," I answered. "I don't think I could deal, either, with the loss of the Shuttle due to a failure in the SRM. In every flight readiness review, I'm the one who always has to defend the flight worthiness of the motors. It really makes me uncomfortable to be the person who has to approve and sign every report for every discrepancy and process departure that is outside our experience base—discrepancies that could lead to a catastrophic failure.

"There are times, Larry, when I lose sleep over my job worrying about making a mistake or accepting a piece of discrepant hardware that could result someday in a failure. It'd be much easier to decide to scrap every piece of questionable hardware, but I know it's not fiscally responsible to do that if the engineering assessment indicates it's safe to fly. These components can range in cost from several hundred thousand dollars each to several millions. As much as all this *that I know about* bothers me, it's *what I don't know* that scares me the most.

"Larry, I've been in the solid rocket business for twenty-six years now, and I've seen a number of catastrophic failures. The Shuttle boosters release a tremendous amount of energy in a very short period of time, and there's no way we can make them 100 percent reliable. I just hope I'm not associated with the Shuttle program when one of these motors finally fails. I feel exactly like you do, Larry. I'm not prepared psychologically or emotionally to deal with an event that would be so traumatic—a national tragedy. I can't even comprehend the impact it would have on my personal life and on my company.

"But I agree with you, Larry, that the most likely source of a catastrophic failure rests in the SSMEs, not the SRBs."

We ended this philosophical conversation with the stark reality that all we could do was to try to do our best to prevent a failure in the Shuttle system from ever happening. Larry Wear was only a few years older than me, and we both needed to get by the next seven to eight years without a Shuttle accident to make it to an early retirement successfully.

Back at the plant, my bosses wanted me to prepare the briefing to NASA on what had happened on the nozzle-to-case joint on STS-51B and our rationale for continuing Shuttle launches on the planned schedule. The first thing I did was convene the team that had just completed its investigation. Every member felt comfortable that we had properly diagnosed the problem and believed, even though we couldn't guarantee the problem would never happen again, that even under the worst conditions imaginable, the secondary O-ring would always seal and never be compromised. If a similar condition occurred in the case field-joint, however, we all thought that under certain conditions, the secondary O-ring might *not* seal if the primary O-ring failed, and that this would result in a catastrophic failure of the Shuttle on the launchpad. In fact, one of our team members, Roger Boisjoly, had just written a memorandum to his supervisor stating just that.

Because of our deep concern, I decided to change the scope and content of the presentation being prepared for NASA headquarters. It was clear that these NASA officials were not well informed about the problems with the O-ring seals and that the problem occurring on STS-51B was not even the most critical problem we were facing with the solid rocket motors. At the time, NASA management in Washington, Huntsville, and Houston were far more concerned about the pocketing erosion observed in the nozzles and about qualification of the FWC-SRM than they were about the erosion or blowby observed in the pressure seals.

My new presentation addressed not just the STS-51B nozzle-to-case joint problem but all of the SRM pressure seal problems. My talk included the performance history of all SRM pressure seals, including the igniter-to-case joint, the case field-joint, and the nozzle-to-case joint for all returned flight hardware and static test motors. I included details of the various joint designs, the history of changes made in the asbestos-filled vacuum putty material used in the joints as a thermal barrier, the application of the putty material, the changes made to the joint leak-check procedures, and the correlation of these parameters to observed erosion or seal blowby in the various SRM joints. At the request of SRB Project Manager Larry Mulloy, my draft was faxed to Marshall for review. The men in Huntsville agreed it was important to emphasize for NASA officials our conclusion that the field-joint was the primary concern due to joint deflection and secondary O-ring resiliency. The only addition Huntsville wanted was for me to add a Thiokol recommendation that it was safe to continue flying the SRMs if certain inspections and leak-check procedures were verified at the time of assembly of the various joints.

But Mulloy and his colleagues in Huntsville also wanted something *eliminated* from my presentation. They asked me to cut a statement that said that *data obtained on resiliency of the O-rings indicate that lower temperatures aggravated this problem*, but that we could leave in the statement about our concern about joint deflection and secondary O-ring resiliency. In retrospect, it was a serious mistake for that point to be taken as minor, but I agreed to make this change.

I made the presentation at NASA headquarters on August 19, 1985. Since it was my first full-fledged presentation to NASA management in Washington, and because the subject matter was so important, I prepared a number of fancy, multicolored viewgraphs. Copies of the viewgraph material were neatly bound into a package and distributed to all those attending the meeting. The Associate Administrator for Space Flight, Jess Moore, was supposed to attend the meeting but was unable to attend; he sent his deputy, Mike Weeks, along with several key members of the propulsion organization at NASA headquarters.

The focus of my presentation was on the SRM pressure seal problems. I aimed to raise to a higher level of concern the question whether the field-joint would be able to seal properly under all potential operating conditions and environments.

One of my viewgraphs, entitled "Primary Concerns" was "Field-Joint—Joint Deflection and Secondary O-ring Resiliency." The next viewgraph addressed the "Field-Joint—Highest Concern" and made clear that the SRM field-joint had been reclassified by NASA (Change No. 23, SRB Critical Items List, February 3, 1983) from Criticality 1R (meaning "redundant") to Criticality 1. Criticality 1 meant a failure of a single component such as the primary O-ring would result in loss of mission, vehicle, and crew. The chart also showed that the secondary seal only acted as a redundant seal in the early phases of the ignition transient of the motor (before the pressure could deflect the joint); in the latter half of the ignition transient, there was a high probability the secondary O-ring would *not* seal if required to do so. All of this would occur in the first 0.6 second after the electrical signal was sent to ignite the SRBs on the launchpad.

My "Conclusions" chart stated that "the primary O-ring should not erode through, but, if it leaks due to erosion or lack of sealing, the secondary seal may not seal the motor."

The first bullet point on my "Recommendations" chart underscored that "the lack of a good secondary seal in the field-joint is most critical, and ways to reduce joint rotation should be incorporated as soon as possible to reduce criticality." This was clearly the most important improvement that could have been made, because if we could eliminate joint rotation entirely, both O-rings would seal under all operating and environmental conditions, *independent of O-ring resiliency.*

Another recommendation that I made in the presentation was to include a capture-feature (a metal lip on the tang side) of the field-joint to restrict the motion in the joint during case pressurization at ignition. No one in Washington seemed to disagree with that recommendation, even though they did not come out and say, "Go do it."

I thought my presentation went very well; in fact, one member of the propulsion staff asked me if he could keep the viewgraphs as a model for future briefings. Larry Mulloy subsequently used my presentation in its entirety to close out an action item he had received from Level I back in April 1984, "To perform a formal review of SRM case-to-case and case-to-nozzle joint sealing procedures to ensure satisfactory and consistent closeouts."

A day after the meeting with NASA headquarters, we formed a full-time O-ring Seal Task Force back at our Utah plant; the group had, in fact, been meeting unofficially for about a month already. I directed the task force to examine our joint design, conduct tests and analysis to better understand the design's limitations, and recommend changes in the design to improve the reliability of the joints. Specifically, I told them to use the next FWC-SRM static test, scheduled for November 1985, to evaluate an improved joint seal design.

I returned to my primary job, which at the time was to complete the FWC-SRM development and qualification program and get these motors ready for the first flight from the new Shuttle complex at Vandenberg

(Launch Complex-6), planned for April or May 1986. I had plenty on my plate to accomplish this, but if this wasn't enough, I also had to prepare for all of the flight readiness reviews for the ongoing Shuttle flight program. This meant I alternated with my boss, Joe Kilminster, at our console at Kennedy Space Center as Thiokol's senior management representative supporting prelaunch countdown and launch decisions related to flight readiness and acceptability of our SRMs for launch.

For the important final static test of the FWC-SRM (QM-5), we recommended to NASA the installation of a Teflon spacer ring in the O-ring groove just forward of the primary O-ring in the nozzle-to-case joint. This spacer ring would provide added thermal protection from any hot gas jets impinging on the primary O-ring. We thought the spacer would also enhance dynamic sealing of the primary O-ring, because the current width of the O-ring groove was resulting in excessive travel of the O-ring during pressurization of the motor. We also recommended installing a larger cross-section O-ring in one of the case field-joints to provide added thermal margin and a larger O-ring "footprint" during joint motion.

NASA accepted our recommendations, but Jim Kingsbury, Director of Science and Engineering at Marshall, wasn't convinced that we had provided adequate seal redundancy at the attachment of the metal end-rings to the composite case in the FWC-SRM design. These metal end-rings contained two O-rings that sealed on a polyurethane surface coating that we applied to the inner surface of the composite case just prior to installing the end-rings on the composite cylinders. We then leak-checked these O-rings twice, after assembly and prior to committing the case for insulating. Kingsbury was concerned that the polyurethane surface might not be an adequate sealing surface for the fluorocarbon O-rings; if not, hot gas could permeate through the polyurethane coating and possibly even through the composite case. We told Kingsbury we felt very comfortable with the sealing design of the metal ring to the composite case, because this whole area was covered with a rubber insulation that was vulcanized to the metal attach-rings as well as to the composite case. In the exact area Kingsbury was most worried about, our insulation provided added seal redundancy, and, in the general areas of the case, it provided the entire pressure seal for the motor. Composite cases, we knew, were pervious to gases, but our experience indicated that our rubber internal insulation provided an adequate seal for the entire motor. In our view, the rubber insulation over the metal attach-ring added seal redundancy over and above the two O-rings.

Kingsbury didn't agree with our conclusions and recommended that we obtain more data to convince him that our current design for the FWC-SRM was adequate. He did recommend that we not change the QM-5 configuration but that we delay the static test of the motor from November 1985 to February 1986 in order to allow sufficient time to incorporate the recommendations that we'd made relative to the improvement in the nozzle-to-case joint and the case field-joint O-ring size. Both the Shuttle program office and technical people at NASA Marshall seemed to be very wary about accepting any recommendation that they considered might add risk to the program.

While taking care of my primary responsibilities, I also stayed abreast of what was going on inside our O-ring task force. The main engineer attached to it, Roger Boisjoly, remained deeply concerned over the loss of the primary O-ring in the nozzle-joint of STS-51B in late April 1985. Only the successful operation of the secondary O-ring had saved that flight, and even this O-ring had suffered significant damage. Boisjoly was more familiar with the SRM seal design than anyone at Thiokol or NASA, and he knew that had we suffered a similar situation in the field-joint, there was a good chance that the secondary O-ring would not have sealed, resulting in a catastrophic failure of the Shuttle during liftoff. From the first that I had heard about Roger's concern, I had agreed with it, making it a part of my August 1985 presentation to NASA headquarters by concluding that, even though the nozzle O-ring failure on STS-51B was very serious, it was the case field-joint sealing system that was the most critical concern.

Boisjoly had felt strongly enough about the problem that he had written a memorandum in July 1985 to Bob Lund, our Vice President of Engineering, explaining in detail what the consequences would be if we encountered this kind of situation in the case field-joint. The memo read: "This would result in loss of the Shuttle vehicle and its crew." This letter was largely responsible for getting a formal task force assigned to the O-ring seal problem, with Roger as a key member of the team. Even though the task force was started in July, a task force leader, Don Ketner, was not selected until the task force was formalized in August 1985 after the meeting with NASA headquarters.

In September 1985, the task force submitted its first recommendation to NASA Marshall for solving the O-ring seal problem: to incorporate the capture-feature in the field-joint. After considerable discussion, MSFC formally rejected the proposed change, saying that we hadn't shown how the change would increase the service life of the SRB case joint—the net result being that our change would increase the cost of the Shuttle program.

NASA instructed us to resubmit our engineering change proposal (ECP) with a cost-benefit analysis. The agency recognized that it was clearly going to cost more money for Ladish to rough-machine the forgings with the capture feature we wanted and that more work would also have to be done at the Rohr Corporation, where the final machining operations and inspection occurred. Howard McIntosh, our SRM Case Program Manager, and I wanted to get this capture feature added to the field-joint, but we couldn't come up with a cost-effectiveness justification, so we decided to resubmit the ECP with a cost rationale that implied that we could improve our reuse potential with the cases using our revised design, thereby offsetting the added cost. Marshall actually didn't respond to this until April 1986, nearly three months after the *Challenger* accident. Meanwhile, we instructed Ladish to keep the forgings on hold—at Morton Thiokol company risk for the added costs.

A few months into their work, several members of our O-ring task force became very frustrated at the lack of progress. They were finding it increasingly more difficult to get tests run, instrumentation installed, and support from those in Thiokol's manufacturing area responsible for the full-scale Shuttle SRM hardware that was needed for some of the testing. Production of flight segments to be shipped to Kennedy Space Center clearly had priority in our plant, and, since most of the hardware planned for use by the O-ring task force was eventually going to be returned to the inventory for flight use, all of the mountains of paperwork, procedures, and seemingly endless signatures of approval had to be obtained before any testing could be accomplished. Equipment and support from our development laboratories were also difficult to obtain since their workload was also dramatically ramping up with the increased Shuttle flight rates.

Boisjoly was not the only one who was frustrated. One of his colleagues on the O-ring Seal Task Force, Scott Stein, wrote a memo to his boss, Jack Kapp, on October 1, 1985, complaining of lack of support for the O-ring Seal Task Force in critical areas of the plant. That same day, Bob Ebeling, Manager of Ignition System and Final Assembly in program management, wrote to me in a weekly activity report, pleading for "Help!" from top management to improve the ability of the O-ring task force to accomplish its jobs and obtain better cooperation from other areas of the plant. Ebeling was the senior manager on the task force and was responsible for coordinating all activities of the task force with other organizations of the plant.

I called Ebeling into my office. I was very sympathetic with his concerns and told him that I would set up a meeting for him with my boss, Joe Kilminster, to discuss the problem. I knew that Kilminster believed that we simply couldn't take any shortcuts in obtaining test data when we were using flight hardware that would be returned to the inventory for later use. This was a philosophy with which I wholeheartedly agreed and thought that Kilminster could provide some real help.

I also told Bob that we should ask for some external help in solving the O-ring sealing problem. Ebeling knew that Boisjoly was preparing a presentation for a National Society of Automotive Engineers conference in about a week, and at that conference he could solicit help from the entire U.S. seal industry.

The meeting with Joe Kilminster took place four days later, on October 5. I wasn't able to make it because I was preoccupied with flight readiness reviews for upcoming Shuttle flights and with preparing for launching the FWC-SRM from Vandenberg, but I told Ebeling, "This is an important meeting, Bob. We need to get this issue resolved as soon as possible."

Regrettably, all the meeting with Kilminster managed to do was make members of the task force even more frustrated. Rather than a meeting for problem resolution, the task force members concluded that it degenerated into a "bull session," one in which Kilminster did not provide the support and priority to the task force that I anticipated him giving.

In early December 1985, Brian Russell, a program manager working for me in the ignition system and final assembly area, received a telephone call from Jim Thomas in Larry Wear's SRM Program Manager's office at NASA Marshall. The call concerned the status of a Problem Assessment System (PAS) report. This was a reporting system used by NASA to document every anomaly noted on a Shuttle flight, in a postflight inspection, or in a ground test of any Shuttle propulsion component. It amounted to a large IBM printout that over the years had gotten so thick and so overwhelming that nobody paid any attention to it.

Jim Thomas told Brian Russell that his boss, Larry Wear, had just left a meeting with Jim Kingsbury, Marshall's Director of Science and Engineering, who complained that both MSFC and its Shuttle contractors were being delinquent in closing out items on the PAS report and that we needed to immediately reduce the size of our anomaly list. How this was done was by conducting a test or analysis and taking some corrective action to show that the particular anomaly wouldn't occur in the future or be significant enough to endanger flight safety. In the next few weeks, memos arrived from NASA specifically requesting "action."

On the PAS report were several items involving our solid rocket motors— many of them concerning our observation of individual putty blowholes or O-ring erosion and blowby. Consulting with Brian Russell, we decided the best way to reduce the list was to combine all the observations under the single heading "pressure seal anomalies," referencing applicable flights and tests and noting the range of conditions representing our experience base to date. I suspected this probably wouldn't work because some time back I had recommended doing this same thing, and NASA wouldn't accept it. Anyway, lumping things together like this wouldn't remove or close out anything on the PAS report, which is what the S&E director wanted.

In my mind what we were already doing to focus on the O-ring problem by appointing our O-ring task force was a much more meaningful and visible means for attacking the problem than a PAS report, which was just being thrown in the trash can without it being read by anyone. Without our task force, the PAS report would only get thicker, because the O-ring erosion and blowby problems we had been seeing would only continue until the problem was fixed. What made it even more absurd was that MSFC did not even prepare or maintain the PAS reports itself; it contracted out the activity to a support engineering firm, which meant that no one in NASA had any real ownership of the anomaly tracking system.

My way through this bureaucratic tangle was to have Brian Russell draft a letter to NASA Marshall suggesting that we remove the O-ring problem from the PAS report with a recommendation that we address the issue *at each FRR, reporting the progress being made by the O-ring Seal Task Force.*

I hoped this approach would satisfy Jim Kingsbury's desire to reduce the size of the PAS report but also satisfy our objective of making the O-ring problem and task force activity more visible.

6

"In the Interest of Avoiding Pain"

The last half of 1985 brought new faces into the Shuttle program. The SRB chief engineer at Marshall, Bill Horton, retired and was replaced by a very competent engineer, Keith Coates. Bob Lindstrom, the manager of the Space Shuttle Projects Office at MSFC, also retired; Lindstrom had been responsible for the management of all of the Shuttle's propulsion elements furnished by MSFC: the SRBs, ETs, and SSMEs. Larry Mulloy, manager of the SRB Project, had reported to Lindstrom.

Bob Lindstrom was highly regarded by all of us for his thorough flight readiness reviews. Bob had a great talent for asking the right questions. However, his replacement, Stan Reinartz, was a total stranger to the Shuttle program and never did become comfortable in his position. I believe Reinartz's prior assignment at Marshall had been associated with advanced technology, upper stages, and satellite systems. I can recall making a comment at Lindstrom's retirement roast in Huntsville's Von Braun Center that "Bob Lindstrom's job couldn't have been that tough because they replaced him with somebody who had absolutely no Shuttle experience." In fact, the joke circulated later that in the first Shuttle program meeting that Reinartz attended, when somebody made a comment about "the ET," Reinartz immediately ran out of the room and into the hallway to "phone home," and when they told him that the ET was to hold "the LOX," Reinartz went out and bought some bagels!

The Shuttle Projects Office Manager's position was the most important management position relative to the Shuttle program at Marshall. This position represented Huntsville's only official member of the overall Mission Management Team supporting Shuttle launches and decisions to launch or scrub at Kennedy Space Center. In all the flight readiness reviews Reinartz attended, his lack of participation was always very noticeable. He seemed to be particularly uncomfortable in FRR board reviews chaired by Dr. William Lucas, the MSFC Director. Reinartz clearly relied on decisions and recommendations from his various project managers; this contrasted dramatically to Bob Lindstrom, who frequently challenged his project managers' recommendations.

In this same time period, NASA Administrator James Beggs was brought under investigation by the government relative to some charges filed against his former employer, General Dynamics. The charges concerned the improper use of independent research and development funds by General Dynamics in support of a research and development contract on a tactical weapon system for which General Dynamics was in competitive development. Jim Beggs was a key General Dynamics executive responsible for this area at the time and was being investigated for his possible criminal involvement. As a result, Beggs temporarily resigned his position as NASA Administrator and was replaced by a man who was allegedly a good friend of the Reagans, Dr. William Graham.

Dr. Graham had no Shuttle or space launch systems background at all; his background was in the "black" (secret) weapon systems development area for the Department of Defense. It quickly became well known that Jim Beggs was not at all happy with the naming of Graham to this position. Beggs thought that Graham was not qualified to lead NASA and that his background, coming from a very closed and controlled type of organization to a very open organization so visible to the public's eye, was more of a detriment than an asset. As it turned out, the charges against Beggs were eventually dropped, but the damage to his reputation was clearly done. The charges against him ruined his career, and he left NASA forever.

Thus, as its darkest hour approached, the space agency was without a competent leader at headquarters in Washington, D.C., or at the Marshall Space Flight Center in Huntsville, Alabama.

At the end of October 1985, we launched STS-61A *Challenger*. Postflight examination of the recovered case hardware showed some soot between the primary and secondary O-rings in two field-joints: the aft joint on the left-hand motor and the center joint on the same motor. The quantity of soot was not nearly as large as when STS-51C launched on that cold day in January 1985, and the soot deposits did not penetrate as far around the joints. STS-61A showed no evidence of soot between the seals on the nozzles, whereas soot had been observed on both nozzles of STS-51C. Also, the soot between the O-rings in the field-joints of STS-61A was thin and light gray in color rather than the thick, jet-black soot we observed on STS-51C, and there was no O-ring erosion at either of the two joints on STS-61A, in contrast to the severe erosion on both primary O-rings on STS-51C.

The puzzling part of the data was that the O-ring temperatures for STS-61A were one of the warmest on record: 75° as contrasted to 53° for STS-51C. With no visible thermal distress to the joint or O-ring seals, added to the fact that what we observed was far less severe than on STS-51C—meaning the data was well within our prior experience and database—we did not believe it needed to be a topic of discussion in the approval process for the subsequent launch, STS-61B, scheduled for November 1985.

Meanwhile, our O-ring task force was having trouble in getting priority in the company laboratory and manufacturing area for tests that would provide data to better understand the limitations and safety margins of the field-joint design. Roger Boisjoly, a member of the task force and lead engineer for the improved joint design, became very frustrated over the lack of progress.

The truth was that Morton Thiokol was coming very close to overextending its resources. We were at the peak for producing SRM segments; we were barely supporting the rapidly increasing flight rate; and we had made a conscious decision to demonstrate that we could load segments with propellant at a rate of sixteen segments per month—four per week—to show we were capable of meeting the planned twenty-four flights per year by 1988. This was considered extremely important because NASA was currently evaluating proposals from our competitors for establishing a second source for SRM production to meet the increased flight rates and reduce production costs.

The company had also submitted a "Buy III proposal" to NASA for production of sixty-six flight sets of motors, with an option for as many as ninety flight sets. NASA was going to use our proposal to compare costs with those submitted by our competitors in their second-source proposals so as to determine the economic viability of a second production source.

We knew exactly what our motors cost, and we knew how far up the learning curve we had climbed to get to the reduced price of our 1985 motors over those that we delivered the previous year. We felt very comfortable about achieving further cost reductions based upon our favorable cost reductions to date, and with the new facilities and equipment we were incorporating to meet the twenty-four-flights-per-year rate, we thought we could improve our cost position even further. As a result, we decided to submit a firm-fixed-price bid to NASA for the motor production part of the proposal rather than a cost-plus-incentive-fee bid, which had been requested from the potential second-source suppliers. We knew they could not possibly meet our fixed-price offer and that if they lied and bid a lower number under a cost-plus type of contract, NASA would have to justify why it accepted such an offer, because the cost risk to the government would be extremely high. By accepting our fixed-price, on the other hand, NASA's cost for the SRMs would be significantly lower than the budget plan and would bring no cost risk to the government.

In addition to these production costs, NASA would have to pay for qualification of a new supplier to produce our currently qualified design and then amortize these costs over the production program. We estimated that this qualification program would cost $60 to $100 million, depending on who won the second-source contract.

Each of the eligible bidders would also have to invest in considerable facility upgrades to enable them to manufacture and test these large solid rocket motors; amortization of these facility investments would run from $200 to $300 million and further increase a company's overhead and production costs.

We felt fairly comfortable that it was not economically feasible for NASA to go with a second source for the SRMs—and that economic feasibility was always the total basis for any second-source decision. Conversely, our competitors thought they had convinced NASA that a second source would *save* the agency money, and they knew they could bid lower artificially without any financial risk to themselves under a cost-plus contract. Our strategy was that by submitting a fixed-price bid, we'd make it very difficult for NASA to justify giving our competitors a cost-plus contract, when none of our competitors could afford to accept a fixed-price contract to do this work. It was simply too risky if one didn't already have experience comparable to ours.

I felt confident we could do the job for what we bid. Jerry Mason, our General Manager, agreed with this approach and directed us to submit our Buy III proposal with the fixed-price option for production and delivery of sixty-six flight sets of SRMs.

However, there was a second portion of the contract that NASA wanted to maintain under the structure of a cost-plus-award-fee. Morton Thiokol was currently working under a cost-plus-award-fee for work involving flight support and special study activities. We were also doing the development work on the FWC-SRM on a cost-plus-award-fee basis. NASA management preferred this type of contract because it gave them much more control over the program. They could make us respond to their wishes because the amount of fee we earned was totally dependent on their assessment of our performance. NASA basically gave us a report card covering activities over a certain period of time comparing our performance against goals that were agreed upon to be accomplished during that specific award-fee period. We were required to submit a self-evaluation report on how well we achieved these goals as well as a recommended score or grade. NASA then reviewed our reports, made its own evaluation, and agreed with or changed the scores accordingly, either up or down—actually very seldom, if ever, up.

In late 1985, we knew that we were not performing as well as we should under the flight support portion of the contracts; in fact, we were "underrunning" this portion of the program because, as noted earlier, we didn't have enough people working in this area. I recommended that we make a significant increase in this portion of our proposal to provide the necessary increase in engineering and laboratory support to accommodate the increased flight rates and resolve the various anomalies that had come along, such as the nozzle-pocketing erosion and the O-ring task force. Initially, management was reluctant to include this request, because, in the midst of preparing for negotiations on our proposal, they did not want to increase the proposed cost to NASA. Finally, however, Thiokol management agreed with my assessment and asked me to prepare a new estimate with a full justification for the added cost. I did this, and they agreed to submit this change to NASA.

In early December 1985, at roughly the same time we were being told to act to clear up our PAS anomalies, two serious problems developed at Kennedy Space Center during the assembly of the hardware for the *Challenger* flight designated STS-51L, scheduled for January 1986. The first problem involved the aft exit cone of the nozzle on the right-hand SRB. Visual inspection of that aft exit cone revealed massive unbonds of the phenolic liner from its aluminum shell. Because we did not have a replacement exit cone immediately available, we sent some of our engineering and quality control people to Cape Canaveral to inspect the cone ultrasonically, quantifying the amount of unbonded area and determining if the cone could be repaired. Our people concluded that the unbonds could be repaired by injecting epoxy adhesive in the unbond area—a fix quickly accomplished by trained technicians sent to Kennedy from our plant in Utah.

The second problem resulted from an improper segment handling operation at KSC. The mishap occurred when the handling rings were being removed from one of the left-hand-center segments just prior to stacking. In Shuttle assembly, the handling rings are connected to a segment by the same high-strength metal pins that connect the segments together. In this case, all of the pins were not removed when the crane was lifting off the handling ring, causing a high load on those left-in pins, resulting in a slight bending of the clevis leg and some minor deformation of the pinholes. Structural analysis of the SRB segment indicated that required safety margins could be met, and the segment could be successfully mated with the proper O-ring squeeze in the joint.

Our engineers were happy with the fix and recommended using the segment, but the MSFC program office, at the recommendation of S&E Director Jim Kingsbury, decided to substitute a new segment instead. This was accomplished by using one of the other center segments already in storage at KSC in the Remote Processing and Surge Facility behind the Vehicle Assembly Building.

The segment with the bent clevis leg was not a major issue at the subsequent flight readiness review because it had been replaced with a perfectly acceptable segment; however, the decision to use the repaired aft exit cone became a big issue in Huntsville. Jim Kingsbury did not agree with the recommendation to use the repaired aft exit cone on the right-hand solid rocket booster for STS-51L, even though the cone had been 100 percent ultrasonically inspected before and after the repair, with the data indicating the repair had been successful. Engineering analysis also showed that aluminum pins used to pin the cone's phenolic liner to its aluminum shell were more than sufficient to hold the liner even if it became totally unbonded from the epoxy adhesive.

I presented all this data to a pre-FRR board meeting that Kingsbury chaired. Kingsbury stood up and said, "I do not agree with Thiokol's recommendation to use the right-hand aft exit cone in its repaired condition." I immediately responded, "As a result of all of our testing, we actually know more about the right-hand aft exit cone than the left-hand exit cone, and I personally feel more comfortable flying it than the other, which, though manufactured at about the same time, has never been ultrasonically inspected and has been accepted for flight purely on a visual inspection prior to assembly."

"Mr. Kingsbury," I said, "visual inspection can only detect unbonds on the very edge if they are sufficiently separated from the aluminum shell. The right-hand aft exit cone unbonds were clearly visible to us, but the major area of unbond detected by the ultrasonic inspection was *deep* inside the part. The cause of the unbond in the right aft exit cone is not known, but since both exit cones were processed the same way at approximately the same time by the same people, it is reasonable to assume that the left-hand aft exit cone may have some unbond areas as well."

Kingsbury answered, "Since the engineering analysis indicates that the mechanical pins will retain the exit cone if it becomes totally unbonded, hot gas can't ever reach this area unless it breaches an O-ring seal. Since no unbonds were detected during visual inspection, I consider the left-hand exit cone to be acceptable."

"But what you're saying," I replied, "is also true relative to the repaired aft exit cone, Mr. Kingsbury, but, in addition, the right exit cone was 100 percent ultrasonically inspected, and we *know* it has no significant unbonded areas. It then has to be that the repaired right-hand aft exit cone must also be acceptable, based upon your logic for accepting the left exit cone."

Kingsbury wouldn't agree and said that he would not support Morton Thiokol's engineering recommendation to use the repaired aft exit cone for STS-51L.

After nearly everyone left the room, I asked Larry Mulloy, SRB manager for MSFC, "What do we do now? As far as I am concerned, logic and good judgment support the use of the repaired aft exit cone. How could anyone feel more comfortable with a cone that they knew less about, even if it was readily available?"

"Al, here at the Marshall Space Flight Center," Mulloy said, "logic and good judgment are not always enough for making a decision. Many times logic and good judgment yield to fear and superstition in the interest of avoiding pain."

"So what do you suggest we do, Larry?"

"Well, your engineering people have put together some very strong technical rationale for using the repaired exit cone, and I'm not sure Jim Kingsbury will have a good basis for refuting your rationale when you present it to Dr. Lucas at the next FRR."

As it turned out, Mulloy was correct. When I presented the same information to the Level III Center Directors FRR, it was accepted, even by Kingsbury. There were no dissenting votes from any board member relative to proceeding with the launch of STS-51L as planned. Apparently, Dr. Lucas knew about Kingsbury's previous concern, so Lucas asked him privately after the vote if he was comfortable with the decision. Kingsbury answered that he really wasn't. Lucas then asked if his degree of concern was sufficient to recommend not using the exit cone and delaying the launch. Kingsbury replied that he was concerned but that he was not necessarily recommending against the launching with the repaired exit cone.

Every year, just before Christmas, it was customary to have a year-end review at MSFC; we highlighted the past year's accomplishments in the SRB program and made forecasts for the coming year. This meeting was a fun occasion, a good time for socializing with all the key members of the SRB team.

As usual, Larry Mulloy gave a humorous award to one of his contractors for the most memorable remark made during the past year. This time, Jerry Mason, Senior Vice President and General Manager of our Wasatch Operations, got the prize for a comment he made to Mulloy after our propellant mixer fire in June, which occurred just one year after we experienced a casting pit fire. Mason's comment was, "I can understand how someone could shoot himself in the foot once in awhile, but in our case it's absolutely amazing how quickly we're able to reload and keep firing."

Mulloy made only a brief cameo appearance at the party because he was occupied with problems that had developed with STS-61C, whose launch had been scrubbed just the day before. This scrub, the first ever due to an SRB problem, had occurred on the first attempt to launch STS-61C *Columbia*, a flight with Florida Congressman Bill Nelson on board. The problem involved an "over-speed indication" on the turbines used in the auxiliary power unit that was part of the steering system for the SRB nozzle. This unit started up during the terminal count just seconds before launch and, when the over-speed indication came on, it violated a launch commit criteria that caused the system to be shut down. Only a day after the launch was scrubbed, Mulloy and the USBI people responsible for this component were running from one meeting or teleconference to another and did not have much interest, much less time, in joining the Christmas festivities, especially when they knew they were probably going to have to tell people they would be working over the Christmas holidays to fix the problem.

For this launch delay, Larry Mulloy took a lot of heat from NASA Level I and II Shuttle program people; worse, he was thoroughly chastised by the MSFC Center Director, Dr. William Lucas. In the end, it turned out that the over-speed indication had been due to a very inexpensive electronic part that had been procured from a new vendor. This was the first flight use of that item.

With all of the heat that Larry Mulloy and NASA Marshall took for the scrub, I figured that Mulloy was not going to have the SRB blamed for any launch delay in the future. It was also clear that everybody else in the Shuttle program office at Marshall would think twice before they caused a launch delay that could be blamed on the MSFC and then have to answer to Dr. Lucas.

I was particularly disappointed that Mulloy could not stay for many of the Christmas party activities, because I had a special plaque for him. As a result of the frustrations we had both gone through at the flight readiness review for the planned launch of STS-51L *Challenger*, with its repaired aft exit cone, I had a plaque made with the inscription: "Logic and good judgment will always yield to fear, superstition, or local politics in the interest of avoiding pain!"

Originally I had entitled the plaque "The SRB Decision Process," but I decided the phenomenon was much more generic, so I just called it, "The Decision Process." I signed the plaque "*A.J. L.B. McLoy*," which is a combination of Allan J. McDonald and Larry B. Mulloy. Since Larry was not at the party at the time, I presented the plaque to Larry Wear, the Solid Rocket Motor Project Program Manager at MSFC, who passed it on to Mulloy, who hung it on his office wall.

In December 1985, NASA completed its evaluation of proposals from several of our competitors in the solid rocket business for establishing a second source for producing the Shuttle's solid rocket motors. We had submitted our sole-source Buy III proposal for the next "buy" of sixty-six flight-sets of motors several months earlier, a contract worth well over a billion dollars, and NASA was using our proposal as a baseline to compare the proposed costs from our competitors.

It was not NASA's plan to second-source the entire rocket motor, as all the metal parts were reusable, and long-lead forgings were already on order. The second-source contracts were primarily for case segment preparation, insulating, lining, propellant loading, final assembly, nozzle fabrication, and shipping to KSC. We knew we had submitted a very competitive fixed-price bid for this portion of the Buy III program based upon our current experience base and established learning curve, and, as we suspected, none of our competitors came close to meeting our bid. This should have made it clear that establishing a second production source for the SRMs was not going to be economically viable.

NASA issued a press release in mid-December 1985 that basically said just that, and we were ecstatic. However, much to the surprise of our senior management, a few weeks later NASA announced it was going to continue its assessment of second-source production for the SRMs "for national se-curity reasons."

As serious a blow as this was to Morton Thiokol, our management grew even more concerned when NASA chose not to sign a contract with us for the Buy III program activities based upon the proposal we had submitted several months earlier. We had been in various stages of contract negotia-tions for several months, and we had expected to have a signed contract once NASA made the decision relative to potential second-source activities. The flight set deliveries in the proposed Buy III program were not part of any second-source production plans.

It appeared to us that NASA was intentionally dragging its feet on our sole-source Buy III contract in order to maintain some leverage with Morton Thiokol until firm plans for second-sourcing could be made. This was par-ticularly disturbing to us because we were one of NASA's best-performing contractors, recently selected as a finalist for the 1985 NASA Productivity Improvement and Quality Enhancement Award. One of the other finalists, Martin Marietta (Michoud Operations), which manufactured the Shuttle's external tank, had recently been successful in avoiding any second-source competition for its element of the Shuttle. Martin Marietta was doing good work for NASA and was reducing the costs of the ET in a government-owned facility.

All of this seemed rather unfair to us because one of the reasons NASA reconsidered second-sourcing of the SRMs was the complaint from our competitors that Morton Thiokol had a cost advantage because we already had a facility capable of doing the work. But in our case, unlike with Mar-tin Marietta, our facility had been financed by our own company capital, whereas Martin Marietta had been producing ETs in a totally government-financed and government-owned facility, which should have made it easier for the government to second-source the ET than the SRB.

One week before the scheduled launch of the STS-51L *Challenger*, NASA formally announced that it was proceeding with plans to seek bids for a second-source production of the SRBs for the Shuttle. If this news wasn't bad enough for Morton Thiokol, Larry Mulloy notified Morton Thiokol management that he would be chairing a meeting at Kennedy Space Center to discuss the possibilities of transferring total responsibility for postflight SRB disassembly in Hangar AF from Morton Thiokol to USBI. It was well known that earlier in the Shuttle program Larry Mulloy and George Mur-phy, then USBI President, were the best of friends and that Mulloy spent a considerable amount of time at Murphy's comfortable Florida home during Shuttle launch activities. Murphy was a very aggressive marketing man. He was one of the most personable people I've ever met, and he had a great sense of humor. Mulloy enjoyed being around him.

Furthermore, USBI was brimming with former NASA employees and was always receiving, it seemed to us, favorable treatment over Morton Thiokol. For example, USBI had just been successful in competing for the Solid Rocket Booster Assembly Contract over a team of Lockheed Space Operations and Morton Thiokol. USBI had submitted a very low bid for this work and was having great difficulty in meeting its proposed cost; it was clearly looking for more work to get healthy. It certainly seemed to us that NASA Marshall, through Larry Mulloy, was trying to help out USBI by transferring the Hangar AF Booster disassembly work from its current processing contractor—Lockheed and Morton Thiokol—to USBI.

We were concerned that this might be just the tip of the iceberg. If NASA was successful in taking the disassembly work away from us, the next step might be to replace us with USBI personnel in the SRB stacking operations inside the VAB.

It was not the perfect situation for levelheaded engineering thinking to trump vital business concerns inside Morton Thiokol management, not with billions of dollars at stake. But the sad truth was, if Morton Thiokol management had not feared that SRB activities might be second-sourced to our competitors, a key domino leading to the *Challenger* disaster of January 28, 1986, might never have fallen.

Above: The solid rocket booster floated vertically (spar mode) in the ocean while divers prepared to retrieve the boosters and recovery hardware for reuse. (Courtesy of NASA.)

Left: Divers 100 feet below the surface prepared to insert a plug into the nozzle to remove water in the rocket motor cavity, enabling the spent rocket motor casing to float horizontally. (Courtesy of NASA.)

After water removal through the plug, the booster was towed back horizontally (log mode) to Port Canaveral. (Courtesy of NASA.)

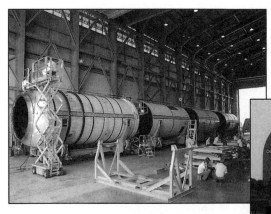

Before the SRB's four segments could be prepared for shipment in railcars back to the Morton Thiokol plant site in Utah for reloading and reuse, technicians had to disassemble and inspect them while stored in Hangar AF at Cape Canaveral. (Courtesy of NASA.)

This aerial view shows a snow-covered Space Shuttle solid rocket motor propellant casting area, which was destroyed by a propellant fire in March 1984. The total destruction (blackened area) included two outside casting pits containing Shuttle segments and mobile casting building as well as collateral damage to an adjacent newly constructed (white) indoor casting building. (Courtesy of NASA/Morton Thiokol.)

Uncured solid propellant was poured from a 600-gallon mixing bowl into a funnel-shaped hopper connected to a Space Shuttle solid rocket motor segment located in a pit below the ground level. (Courtesy of NASA/Morton Thiokol.)

USAF Major Richard Covey posed with Allan McDonald behind the nozzle of the first graphite-epoxy-filament-wound-case (FWC) solid rocket motor (DM-6) prior to a static test in October 1984. (Courtesy of NASA/Morton Thiokol.)

A lighting strike on a Space Shuttle solid rocket motor propellant mixer in June 1985 caused a catastrophic fire, with far-flung debris. (Courtesy of NASA/Morton Thiokol.)

At an international aerospace conference in Tel Aviv, Israel, in February 1985, I presented a Space Shuttle model to Yitzhak Rabin, then Israeli Minister of Defense (later prime minister). (McDonald photo.)

PART II

Misdiagnosis

It is not good to stop knowledge from going forward. Ignorance is never better than knowledge.

—Enrico Fermi

Regret for the things we did can be tempered by time; it is regret for the things we did not do that is inconsolable.

—Sydney J. Harris

7

The Teleconference

The start of 1986 was a very hectic time for Morton Thiokol and the Space Shuttle program. We were in the process of expanding our facilities and increasing our production rate to support twenty-four Shuttle flights per year in just two years. We were also completing the development and qualification of the FWC-SRM for its first use in a scheduled launch from the new Shuttle launch facilities at Vandenberg Air Force Base, scheduled for the spring. We had just completed the assembly of the final test and only FWC-SRM qualification motor (QM-5) was in the test bay for a February firing. The first flight set of the FWC-SRMs were already stacked on the launchpad at Vandenberg AFB.

Many at NASA and in Congress were questioning whether the space agency could accomplish all of this at a time when it was also developing a new and more powerful liquid oxygen/liquid hydrogen orbital transfer stage to be deployed from the cargo bay of the Shuttle. This new stage, called the Centaur G Prime, was to be used by NASA for some of its planetary missions and was scheduled for its initial two flights in May 1986. But before any of this could happen, it required modification of two orbiters for carrying the cryogenic propellant upper stage in the cargo bay—one being the *Challenger*, which was scheduled to fly the STS-51L mission in January.

Under the Reagan administration's "Shuttle Only" policy, the Shuttle was to be the only U.S. launch vehicle. All the existing expendable launch vehicle production lines were in the process of shutting down. Both NASA and Morton Thiokol had a lot on their plates, and it was important to demonstrate again that the Shuttle could, indeed, routinely launch twice in one month. That had been accomplished first in April 1985, again in October 1985, and was now planned for January 1986. If all went as planned, we might have *three* Shuttle flights in May 1986: two from Kennedy Space Center and one from Vandenberg.

Albeit with many challenges, the space agency was riding high, enjoying strong support in the White House and in Congress and a high degree of confidence from the American people. The only threat to NASA was the development of the complementary expendable launch vehicle that had been pushed through Congress by the air force to provide the DOD assured-access-to-space in the event of a Shuttle failure. NASA was concerned that once the CELV became operational, the air force would abandon the Shuttle, jeopardizing the twenty-four-launches-per-year rate on which NASA's budget for the late 1980s was predicated—and further having its budget threatened by the needs of a multibillion dollar West Coast Shuttle launch complex at Vandenberg.

Still, the CELV threat to NASA's budget future and NASA's preeminent position as the sole provider of American space launches was too far in the future to cause much real concern at NASA in January 1986. What disquieted agency leaders the most was the knowledge that they had clearly bitten off more than they could chew in meeting their immediate goals for 1986.

On January 12, 1986, NASA was finally successful in launching STS-61C *Columbia* with Congressman Bill Nelson on board. In order to accommodate the congressman, Payload Specialist Gregory Jarvis moved from his slot on the crew of STS-61C to that of the next flight, STS-51L *Challenger.* Jarvis had been bumped once before (STS-51D *Discovery* in April 1985) to accommodate Utah Senator Jake Garn.

Originally, STS-61C was to launch on December 18, 1985. Before finally lifting off Pad 39A, it suffered a total of seven launch delays. Two of the delays were due to bad weather at the transatlantic landing abort sites and at the Cape, where visibility fell below the minimums acceptable for an emergency landing of the orbiter in the event of an abort. The press put a lot of heat on NASA for the seven delays, continually repeating that it was being delayed more than any previous Shuttle. In fact, the press was merciless, ridiculing NASA's plan to begin launching two Shuttles per month when it was taking nearly a month to launch just one after it had been declared ready to go.

Joe Kilminster sat in the Launch Control Center at KSC for the launch of STS-61C as our senior management representative, which meant that it would be my turn to do the same for the next launch, that of STS-51L *Challenger.* Prior to my going to the Cape, I attended a meeting in Huntsville. It concerned a revision to the Buy III proposal that I had submitted, requesting additional engineering support hours for more detailed examination of returned SRB hardware and for problem solving. We were very anxious to clear all the hurdles we could to break loose the Buy III contract and welcomed the opportunity to clarify any questions Marshall might have about our proposal.

Larry Wear, of the MSFC program office, agreed to the need for increased engineering activity for flight support, but Marshall's cost people took the position that our design was mature and should not be encountering so many new problems to warrant added effort over what we were currently expending. "We still have several unresolved problems," I countered, "and we're about to *triple* the launch rate!" The cost people then wanted to know if I could distinguish between flight rate–related increases and those associated with problem resolution, because the latter, they said, should show some increased learning and reductions in the "out" years. I had data that showed just that. The reduction in problem resolution would be more than offset by the increase in flight readiness activities and in the postflight inspections needed to support twenty-four launches per year.

All present agreed that the proposed changes were appropriate. I felt good about the meeting and was eager to jump aboard an airplane to Orlando to support the L-2 and L-1 flight readiness reviews for STS-51L *Challenger*.

At the L-2 FRR at KSC on January 24, 1986, Larry Mulloy presented the flight readiness position for the SRBs. No significant issues came up other than concern about the weather. A rainstorm was on the way that could affect the launch. Prior to the L-1 review the next day, we met with Huntsville's Dr. William Lucas and Jim Kingsbury to review the results of the flight hardware from the just-returned STS-61C. Though there was really nothing unusual to report, Larry Mulloy did mention that we had seen 0.004 inch of erosion on the left aft field-joint O-ring. This amount of erosion was well within our historical database and much smaller than we had observed on prior flights.

The formal L-1 review went very smoothly. No significant technical issues had developed since the previous day's L-2 review. All outstanding actions were closed out, and the STS-51L mission was given the go-ahead for launch the next morning, January 26, pending the weather forecast.

The weather forecast from Patrick Air Force Base was very bleak. The probability of rain showers during the launch window was so high that the decision was made to delay the launch one day. Most everybody was happy over this decision because January 26, the scheduled launch date, was "Super Bowl Sunday." Unfortunately, the weathermen were wrong again. Sunday morning dawned beautiful; it would have been a perfect day for the launch.

I watched the Super Bowl at Carver Kennedy's house, a game in which the Chicago Bears absolutely annihilated the New England Patriots. Carver was the Vice President in charge of our KSC Space Services organization in support of the Shuttle processing contract that assembled the Shuttle's segments and installed the ET in the Vehicle Assembly Building. Carver was known as the "Czar of the VAB." He was also responsible for the SRB retrieval ships and recovery and disassembly of the segments in Hangar AF to return to Utah. In my opinion, Carver was one of the best technical managers in the aerospace industry; in fact, Carver had just been selected as the outstanding manager of Lockheed at the Kennedy Space Center. This was very unusual because the honor was supposed to be reserved for Lockheed employees only. Carver earned it because, as a member of the Lockheed team, he was clearly the best manager on the entire Space Shuttle Processing Contractor team. He knew rocket propulsion very well and knew how to effectively manage people. For most of his professional career, he had been with Morton Thiokol.

As Thiokol's senior management representative in the Launch Control Center at Kennedy, it was my job to attend all the flight readiness reviews and report into firing room No. 2 at the SRB computer console at least three hours before the opening of the launch window. This was the time in the launch countdown sequence that the Shuttle came out of a planned two-hour hold and entered the terminal countdown phase. The two-hour hold was planned to resolve any problems that may have developed during the countdown that might affect the flight readiness of the vehicle before entering the terminal phase of the countdown. The two-hour hold could be extended if required, with the duration depending upon the length of the launch window. For me, that meant I had to get up before 5:00 a.m. to be on station at the SRB console in the Launch Control Center by 6:00, three hours before the scheduled launch time a little after 9:00.

That Monday morning I got up earlier than usual, around four o'clock, anxious to get this launch off and go back home. I had always done a considerable amount of traveling for the company since coming to work for Thiokol in 1959, but never so much as in the past months—and, with the drastic increase in the Shuttle launch rate, it was only going to get worse. Seldom was I gone more than two to three days, but the flight between Utah and Huntsville or Cape Canaveral was long and tiring. This trip had already been tough, away from my loved ones almost a week. Right after the launch, I was headed to the airport in Orlando.

When I arrived at the Launch Control Center early that morning, Carver Kennedy was sitting at one of the consoles operated by Lockheed Space Operations. I took a seat right across the aisle at an SRB console operated by Ike Rigdell from USBI and Charlie Mauldin from NASA Marshall. Putting on my headset, I connected to a communications network that included NASA's Mission Management Team at KSC and the MSFC launch support team at the Huntsville Operations Support Center.

For the most part, USBI ran the show when it came to the operation of the SRBs; its staff had responsibility for all of the active systems on the SRBs, including the power supply and the electrical and pressurization systems active during the terminal countdown. The only Morton Thiokol systems active during the countdown were the electromechanical "safe-and-arm device" for arming the firing circuits to the SRBs five minutes before launch and the electrical calibration checks for the operational pressure transducers on the SRMs that were used for separation of the motors from the external tank after motor burnout. Because USBI controlled most of the active systems, USBI always had an operator sitting at the SRB console with a representative of NASA Marshall. Both USBI and Morton Thiokol were required to position a senior management representative sitting behind the SRB console operators to monitor all the data and participate in prelaunch communications in the event that a problem occurred. This launch was unusual in that USBI had several extra people supporting the console operator, necessitated by the problem of the SRB auxiliary power unit turbine overspeed indication that had occurred on the previous Shuttle launch. With USBI people needing all the chairs around the SRB monitor, I moved over to a vacant seat near Carver Kennedy to monitor the launch operations.

The countdown went fairly smoothly until the time came to close the door to the orbiter's crew compartment. Upon latching it, a pad technician mechanically removed the handle from the door. On this occasion, however, the handle could not be removed. No amount of pulling and pushing could unstick the handle, so Lockheed Space Operations engineers requested power tools to help them remove it. It took a long time to locate battery-operated tools and, when they arrived, the batteries were almost dead. The whole affair grew embarrassing, because it was holding up the launch, and it wasn't a very high-tech item, just a door handle. The affair started to look like a rerun of *The Three Stooges*. Eventually, the technicians decided to employ a sophisticated aerospace tool called a "hacksaw." With it, they cut off the handle and declared the Shuttle ready for launch.

But now the weather was turning bad. We had used up most of our launch window dealing with the door handle, and, by the time the problem was solved, a front had approached—one with strong enough ground winds to violate the Shuttle return-to-launch-site landing constraints.

So, we scrubbed to the next day. That afternoon walking to my car in the lot outside of the Launch Control Center, I felt a wind stronger than I had ever experienced at KSC. I wore contact lenses, and sand blew into my eyes so badly that I could hardly see. I had to lean forward to keep from being blown over, and ever few steps I got hit by a gust that stopped me in my tracks.

Shortly after I arrived at Carver Kennedy's home in Titusville, I received a telephone call from Bob Ebeling at our plant in Utah. Bob told me that a local Florida weatherman in Orlando was predicting an "extreme" cold front that would drop temperatures by early the next morning to possibly as low as 18°. "Bob, that really concerns me for the O-ring seals," I said. "Engineering here is already working that problem," Ebeling told me. "What we need are the expected ambient temperatures at the launchpad in the hours prior to the scheduled time of launch," which was 9:38 a.m. the next morning. "I'll call KSC and see if I can get an hour-by-hour forecast of temperature," I told Bob.

Turning to Carver Kennedy, I asked, "Who should I call to get this data?" Carver knew exactly whom to call inside the Launch Operations Center. What he learned was that freezing was expected to occur before midnight, with a minimum temperature of 22° expected by 6:00 a.m. rising only to around 26° at launch time.

I immediately called Ebeling with this information and with instructions that our engineering people use this data to predict actual O-ring seal temperatures at the opening of the launch window and assess the impact of such low temperatures on the sealing capability of the field-joints. I told Bob I'd arrange a teleconference for later in the day that would include the NASA/MSFC management people at KSC, the NASA engineering people at MSFC, and our engineering people back in Utah. "Engineering needs to analyze the situation and make a recommendation as to what temperature it is safe to launch," I told him. "Also, make sure you get Bob Lund [Thiokol's Vice President of Engineering] involved, and get him to make an engineering recommendation during the teleconference, because this has to be an engineering decision, *not* a program management decision." I knew that program managers were always under severe pressure to maintain schedules and control costs but that this was a thorny technical problem whose resolution—in terms of whether to launch—should be based upon technical rationale only. "I'm not in a position to make that assessment here by myself, Bob, so get engineering to prepare some charts and fax them to both KSC and MSFC."

I tried to reach Larry Mulloy at his hotel but failed, so I called Cecil Houston, the NASA/MSFC Resident Manager at KSC. I alerted him of our concerns about the sealing capability of the field-joint O-rings at the predicted cold temperatures and asked him to set up the teleconference. Cecil said he would set up a direct connection linking Morton Thiokol engineering in Utah, MSFC engineering in Huntsville, and the conference room next to his office at KSC. He said he'd also track down Mulloy and Stan Reinartz, NASA/MSFC Shuttle Project Manager, and have them come to his office. He'd also notify Jack Buchanan, manager of our local Launch Support Services, to join us. Cecil suggested we set the meeting for 8:15 p.m. EST so a decision could be made before the fueling of the external tank began, scheduled for midnight. I agreed with the plan and told Cecil I'd meet him in the conference room at a quarter past eight.

Carver Kennedy and his wife, Martha, could tell I was nervous and rather uptight about the situation. They suggested that we not go out for dinner as planned and just have a bite to eat at home, because it would take me half an hour to get to KSC. I agreed, but really had no appetite.

It was the first time in my two-year career in the Shuttle program that we were experiencing a last-minute crisis of this magnitude just before a launch.

I drove out by myself to KSC. Cecil Houston's office was in a trailer complex across the road and just south of the Vehicle Assembly Building. When I got there, Larry Mulloy, Stan Reinartz, Jack Buchanan, and Cecil were in the conference room waiting for our charts from Thiokol to arrive by fax. The charts didn't start coming in until around 8:30 p.m. and it was close to 8:45 before the teleconference began—and even then we had not received any charts on the conclusions and recommendations. Because this was such a last-minute affair, the engineers in Utah didn't have sufficient time to even type any of their charts. All of them were handwritten, with the exception of a few that had been taken directly out of the August 19 "Erosion of SRM Pressure Seals" presentation that I'd made to NASA headquarters. It was 9:15 before the charts with the conclusions and recommendations finally showed up.

Thiokol's engineering presentation consisted of about a dozen charts summarizing the history of the performance of the field-joints, some engineering analysis on the operation of the joints, and some laboratory and full-scale static test data relative to the performance of the O-rings at various temperatures. About half the charts had been prepared by Roger Boisjoly, our chief seal expert on the O-ring Seal Task Force and staff engineer to Jack Kapp, Manager of Applied Mechanics. The remainder were presented by Arnie Thompson, the supervisor of our Structures Section under Jack Kapp, and by Brian Russell, a program manager working for Bob Ebeling.

The first few charts were basically mine from August with updated data obtained from flights flown since August. The first contained a history of O-ring erosion and blowby observed in the field-joints of prior flights; the second reiterated the concerns we presented in the August 19 meeting about the inability of the secondary seal to maintain pressure integrity during most of the motor operation if the primary seal should ever fail due to erosion and/or blowby. Boisjoly spent a considerable amount of time on this chart, pointing out that the timing function for the primary O-ring to seal was extremely important during the ignition phase.

"With colder temperatures, we're moving away from the direction of 'goodness,'" Boisjoly explained, "to where the sealing capability of the primary O-ring is going to be reduced." This was extremely important because the chart clearly showed that there was a high probability that the secondary seal would not be capable of sealing the joint if the primary O-ring failed during the last half of the ignition transient—a period lasting from 0.330 to 0.600 second. On the chart, Boisjoly drew an "exaggerated" sketch of the pressurized field-joint that clearly showed how the secondary seal would not be in contact with the tang sealing surface while the primary O-ring was extruded by the motor pressure into the gap between the two seals.

Emphasizing another point made back on August 19 to NASA headquarters, Boisjoly discussed again how the loss of seal contact resulted from joint rotation. "The point I'm making," Roger underscored, "is that if the primary O-ring fails to seal before the joint rotates, then, yes, the secondary seal has a high probability of sealing the joint. But if the joint rotates and the primary seal is lost, then the secondary seal will be unable to seal the joints. That's why the 'timing function' for the dynamic seal is so important, gentlemen." Roger felt very strongly that the cold temperature would slow down the timing function, making it more difficult for the primary O-ring to seal the joints.

Primary Concerns (Cont)

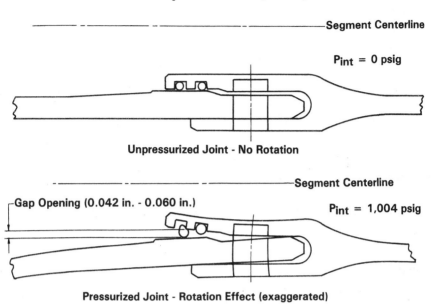

Exaggerated sketch of opening field-joint. (*Report of the Presidential Commission on the Space Shuttle Challenger Accident*, Vol. 3, June 6, 1986.)

Boisjoly's next chart showed how cold temperature would reduce all the factors that helped maintain a good seal in the joint: lower O-ring squeeze due to thermal shrinkage of the O-ring; thicker and more viscous grease around the O-ring, making it slower to move across the O-ring groove; and higher O-ring hardness due to low temperature, making it more difficult for the O-ring to extrude dynamically into the gap for proper sealing. All of these things increased the dynamic actuation time, or timing function, of the O-ring, when at the very same time the O-ring could be eroding, creating a situation where the secondary seal might not be able to seal the motor, not if the primary O-ring was sufficiently eroded to prevent sealing in the joint.

Roger then showed a chart with data from the flight of STS-51C (which used the SRM-15 motors) exhibiting large areas of jet-black grease from soot between the O-rings in one field-joint on each booster, indicating a temporary loss of the primary seal at ignition. STS-51C, in January 1985, had been the coldest launch to date, and Roger was absolutely sure that the condition observed on STS-51C was "a direct function of the cold temperature" on that launch.

Boisjoly also presented some laboratory data on what temperature did to the hardness of the fluorocarbon O-ring material called Viton—data that showed that Viton got much harder at low temperatures, making it more difficult to pressure actuate and extrude the O-rings into the seal gap. Roger said, "The colder temperatures predicted can change the O-ring material from a hard sponge to something more like a brick."

When Roger finished, Arnie Thompson came on the network with test data on the resiliency of the O-ring, on dynamic blowby, and on compression set of the O-ring. Resiliency—the ability of a material to spring back to its original shape and dimensions after being compressed or squeezed—was very important for an O-ring that was trying to seal a joint that was opening. Thompson focused on some "static" data we had obtained from compressing an O-ring between two plates and then rapidly moving away one of the plates and observing how long it took for the O-ring to spring back and contact the other plate without any pressure assistance. The data showed that at 100°, the O-ring would never lose contact with the plate, but at 50°, it lost contact and took ten minutes to recover fully. Everyone agreed this was an extreme case and not representative of the SRB joint because the motor pressure worked to extrude the O-ring into the gap. Nonetheless, the point was made: cold temperatures can dramatically reduce the ability of O-rings to seal on surfaces that were moving away from each other, as in our field-joint.

Arnie Thompson also presented data on cold gas blowby tests that showed no leakage at 75° or at 30°. The test hardware (which was not full scale) actuated the O-ring but was not simultaneously capable of opening the joint. Arnie explained that the subscale tests had been conducted with argon and that argon, being inert gas, was very difficult to detect for leakage. The tests were being rerun with Freon to improve detection, but the results were not yet available.

Following Thompson, Brian Russell went over a chart documenting the history of blowby in the field-joints and nozzle-to-case joints. His chart clearly showed that the worst-case condition ever observed occurred on the SRM-15 hardware that had flown on STS-51C the previous January. Both the case and nozzle-joints had been affected. Russell's chart also showed that we had seen blowby in the field-joints of only one other flight—in the two joints of the left-hand booster of STS-61A, during one of the warmest launches to date, in October 1985. When this chart was presented, Larry Mulloy immediately challenged, "How then can you conclude that temperature has anything to do with blowby?"

Roger Boisjoly answered, "We need to explain the differences between the data obtained on STS-61A in October and that observed on STS-51C in January. I personally inspected the SRM-15 motors on STS-51C when they were disassembled in Hangar AF at KSC, and a colleague of mine, Jerry Burn, inspected the disassembly of the SRM-22 motors used on STS-61A." Roger observed that both joints of the SRM-15 motors had primary O-ring erosion; he also saw a large quantity of jet-black soot over a large portion of the circumference between the primary and secondary O-rings—from 80 to 110°. It was Roger's understanding that the SRM-22 joints, on the warmer launch of STS-61A, had *no* primary O-ring erosion and only a very light gray coating of soot over a small area of the circumference between the O-rings—30 to 40°. "I feel strongly," Roger continued, "that the observed difference between these two flights is most probably due *only to temperature*—that STS-51C had O-ring temperatures of 53° and STS-61A had O-ring temperatures of 75°."

Mulloy wouldn't budge. Since NASA had launched the Shuttle several times at temperatures between 53° and 75° without observing any blowby on field-joints, Larry could not believe that we could conclude that temperature had anything to do with this problem. He thought the data was inconclusive.

Boisjoly disagreed and went on to his last chart, which compared the O-ring squeeze of the primary seals used in STS-51C (SRM-15) with those currently existing in the STS-51L (SRM-25). The chart focused on the magnitude of primary O-ring erosion on the joints of STS-51C and where the blowby had been located. Finding no correlation between O-ring squeeze and the observed problems in the joint, Boisjoly's chart showed the configuration of the joints in STS-51L was very similar to the configuration of the joints in STS-51C.

The next chart that was presented included a limited history of O-ring temperatures from some of the static tests and flight motors as compared to the planned launch of STS-51L (SRM-25). The chart clearly showed that STS-51C was the coldest flight to date, with an estimated O-ring temperature of 53°. The coldest static test occurred on development motor No. 4 (DM-4), which had an estimated O-ring temperature of 47° and was static tested at an ambient temperature of 36° with no observed effects on the joints or O-rings. It was at this time that I made my first comment on the four-wire network telecom, and that was, "I don't believe that the static test history is a valid test for the O-rings because the blowholes observed in the zinc chromate putty after assembly and leak-check were manually filled prior to static test." Someone on the network back in the plant in Utah made a similar comment. The chart did show that the planned launch of STS-51L (SRM-25) at 26° would result in predicted O-ring temperatures of 27–29°, which was far away from our prior experience base. The chart also included the SRM-22 motors used on STS-61A that was launched at 78° with an estimated O-ring temperature of 75°. It was this data again that NASA, namely Larry Mulloy, took issue with in regards to any conclusions relative to temperature affecting the ability of the O-rings to properly perform their function. There was a lively exchange between several of the parties on the teleconference relative to what all this data meant. Roger Boisjoly stood fast, saying he believed the data indicated that the cold temperatures were going away from the direction of goodness. He was asked to quantify his concerns, and he could not do that. He could only relate what he observed on STS-51C a year earlier, which resulted in a "qualitative" conclusion that the cold temperature of that launch was a major contributor to the heavy soot blowby that was observed.

The meeting had been going on for over an hour now, and it was shortly after 10:00 p.m. before we discussed the final conclusions and recommendations. All of the previous charts had been presented by the engineers who had prepared them, and now Bob Lund, Vice President of Engineering, was about to present Morton Thiokol engineering's conclusions and recommendations. So far, the meeting was conducted just as I had requested, and that was to have the engineers present the technical data and have the Vice President of Engineering present Morton Thiokol's conclusions and recommendations relative to the planned launch. I was very surprised at NASA's challenging of the data that had been presented, because this action was totally out of character with any past flight readiness review meeting that I had ever attended. In all of these other meetings, NASA always challenged the data relative to whether it was accurate enough or well enough understood to recommend a launch, and it was data far more insignificant and much better understood than we were discussing this evening. At the time, I thought that NASA was just testing us to see how strongly we believed the data that we were presenting and whether we really thought that there was a real technical concern about the ability of the field-joints to properly seal at low temperatures.

The "Conclusions" chart said that temperature of the O-ring was not the only parameter that controlled the blowby that was observed on SRM-15 at an O-ring temperature of 53°. It also stated that development motors had been tested at lower temperatures than this with no observed blowby but that the packing of the putty in these motors resulted in better performance. The conclusion was that at about 50°, blowby could be experienced in the case joints and that the temperature for SRM-25 (STS-51L) on January 28, 1986, would be 29° at 9:00 a.m. and 38° at 2:00 p.m. The conclusions also stated that we had no other data that would indicate that SRM-25 was any different than SRM-15 other than temperature. There wasn't much discussion or comment by the participants on the network relative to the "Conclusions" chart.

The final chart included the recommendations, which resulted in several strong comments and many very surprising reactions from the NASA participants in the teleconference. The first statement on the "Recommendations" chart stated that the O-ring temperature must be equal to or greater than 53° at launch, and this was primarily based upon the fact that SRM-15,

CONCLUSIONS:

o   TEMPERATURE OF O-RING IS NOT ONLY PARAMETER CONTROLLING BLOW·BY

SRM 15 WITH BLOW·BY HAD AN O-RING TEMP AT 53°F

FOUR DEVELOPMENT MOTORS WITH NO BLOW·BY WERE TESTED AT O-RING TEMP OF 47° TO 52°F

DEVELOPMENT MOTORS HAD PUTTY PACKING WHICH RESULTED IN BETTER PERFORMANCE

o   AT ABOUT 50°F BLOW·BY COULD BE EXPERIENCED IN CASE JOINTS

o   TEMP FOR SRM 25 ON 1-28-86 LAUNCH WILL BE 29°F 9 AM
         38°F 2 PM

o   HAVE NO DATA THAT WOULD INDICATE SRM 25 IS DIFFERENT THAN SRM 15 OTHER THAN TEMP

The handwritten "Conclusions" chart presented by the Vice President of Engineering of Morton Thiokol to NASA. (*Report of the Presidential Commission on the Space Shuttle Challenger Accident*, Vol. 4, June 6, 1986.)

RECOMMENDATIONS :

○ O-RING TEMP MUST BE ≥ 53 °F AT LAUNCH

DEVELOPMENT MOTORS AT 47° TO 52°F WITH
PUTTY PACKING HAD NO BLOW-BY
SRM 15 (THE BEST SIMULATION) WORKED AT 53 °F

○ PROJECT AMBIENT CONDITIONS (TEMP & WIND)
TO DETERMINE LAUNCH TIME

The "Recommendations" chart restricting the launch temperature for *Challenger* to
53° F or higher. (*Report of the Presidential Commission on the Space Shuttle Challenger
Accident*, Vol. 1, June 6, 1986.)

which was the best simulation of this condition, worked at 53°. The chart
ended with a statement that we should project the ambient conditions (temperature and wind) to determine the launch time.

Mulloy immediately said he could not accept the rationale that was
used in arriving at that recommendation. Stan Reinartz then asked George
Hardy, Deputy Director of Science and Engineering at NASA/MSFC, for his
opinion. Hardy said he was "appalled" that we could make such a recommendation, but that he wouldn't fly without Morton Thiokol's concurrence.
Hardy also stated that we had only addressed the primary O-ring, and did
not address the secondary O-ring, which was in a better position to seal because of the leak-check. Mulloy then shouted, "My God, Thiokol, when do
you want me to launch, next April?" He also stated that "the eve of a launch
is a helluva time to be generating new launch commit criteria!" Stan Reinartz entered the conversation by saying that he was under the impression
that the solid rocket motors were qualified from 40° to 90° and that the 53°
recommendation certainly was not consistent with that.

Larry Mulloy then asked Joe Kilminster, Vice President of Space Booster
Programs at MTI, what the program office recommendation was, and Kilminster said that he would not recommend launch based upon the engineering position that was just presented. Mulloy then challenged the engineering position based upon his own assessment that the engineering data
was inconclusive. He mentioned that we had presented data that observed
blowby on cold motors and warm motors, and he wanted more quantita-
tive data that the temperature really affected the ability of the joint to seal. I

could see the fire in his eyes, and that he was in no way going to accept the recommendation that was just made based upon the data presented.

Stan Reinartz then asked Joe Kilminster to respond to Mulloy's remarks, and he couldn't. Joe then requested a five-minute off-line caucus to reevaluate the data, and Reinartz accepted that request.

It was at that time that I decided if we were going to reevaluate the data, we needed to consider a comment made earlier by George Hardy relative to the ability of the secondary O-ring to seal, because nobody responded to his comment. I stated on the network that lower temperatures are in the direction of badness for both O-rings, because they slow down the timing function, but that the effect is much worse for the primary O-ring compared to the secondary O-ring because the leak-check forces the primary O-ring into the wrong side of the groove, while the secondary O-ring goes in the right direction, and that this condition should be evaluated in making the final decision for recommending the lowest acceptable temperature for launch. Based upon the data presented in the second chart, I considered this very important because how much delay there was in getting a good reliable primary seal affected the capability of the secondary O-ring to seal. This chart was the only chart presented that attempted to quantify the behavior of the joint that Larry Mulloy was asking about. When MTI agreed to caucus to reevaluate the data, I thought it was important to consider this issue because that was really the crux of the matter, and it appeared to be an opportunity to "quantify" the engineering assessment to satisfy Mulloy. I did not believe I was capable of doing this, and it should be an engineering assessment and an engineering recommendation to launch or not. It should not be a program management decision.

It was perfectly clear that MTI engineering's original recommendation of 53° was based solely on the experience of STS-51C a year earlier. It was evident from the STS-51C data that the primary O-ring temporarily lost its seal during the early phase of the ignition transient, and that is why the secondary O-ring worked perfectly as a redundant seal with very minor thermal distress and no visible blowby of hot combustion gases. At the time, this was not considered a desirable condition, but it was considered an acceptable condition. The secondary O-ring was in the proper sealing position as Hardy had earlier indicated, which made it a truly redundant seal on STS-51C. The real question to answer was how temperatures much colder than 53° affect this. What is the lowest temperature at which we can no longer depend on the *secondary* seal because the timing function was slowed down sufficiently that we were in an area of a truly Criticality 1 joint, where we could no longer depend upon the secondary O-ring to be capable of sealing the joints. It was this latter condition that had caused the field-joints to be downgraded by NASA from a Criticality 1 redundant (Criticality 1R) to a Criticality 1 in late 1982. I thought George Hardy's comment concerning the secondary O-ring was important, and should be considered if we were to change our recommendation from 53°. The question in everybody's mind was, how close are we to a cliff?

I was personally very shocked at what I heard from the NASA people participating in the teleconference. I had never heard George Hardy speak so harshly. I had heard Larry Mulloy talk that way before, but it was always in situations where he was challenging our rationale on why it was *safe* to fly. This was the first time that NASA personnel ever challenged a recommendation that was made that said it was *unsafe* to fly. The flight readiness review process was always structured around the contractors having to prove beyond a shadow of a doubt that their hardware was safe to fly.

For some strange reason, we found ourselves being challenged to prove quantitatively that it would definitely fail, and we couldn't do that. I was probably more shocked by this dramatic change in philosophy than anyone else, because of the prior FRRs that I had participated in concerning the repaired nozzle aft exit cone and the damaged center segment on this particular flight set of solid rocket boosters (STS-51L) that were initially rejected by NASA/MSFC personnel, even though MTI engineering said they were safe to fly. The Director of S&E at MSFC, Jim Kingsbury, George Hardy's boss, initially rejected both of these components as not being safe to fly. He changed his mind on the aft exit cone because no one else agreed with him, but he did not change his mind on the center segment with a slightly bent (0.050 inch) outer clevis leg. Mulloy also agreed with rejecting the center segments as not being safe to fly. In both cases, I presented a very detailed "quantified conclusive engineering analysis" that showed that these two components were perfectly safe to fly and did not compromise the required safety factors contained in the contractor-end-item (CEI) specification for the motor furnished by NASA.

NASA managers' reason that evening for not accepting the no-launch recommendation was that they thought the data presented was "inconclusive" and not "quantifiable." In the past, data that was inconclusive or not quantifiable was automatically considered as being unsatisfactory to support a recommendation *to* launch.

8

The Caucus

It was a unique turn of events. Shuttle program managers from NASA Marshall were applying a tremendous amount of pressure on Morton Thiokol—namely, on Joe Kilminster to change his mind to support a decision to proceed with a Shuttle launch. Larry Mulloy and others had basically ignored our engineering recommendation put forward by Bob Lund. Mulloy had then asked Kilminster what his recommendation was. That was exactly what I did not want to happen. Kilminster had agreed with the recommendation of his engineers, but Mulloy knew that he had leverage with Kilminster, because Kilminster essentially worked for him and was his program management executive counterpart at MTI. I wanted this to be an engineering recommendation only, because I knew all about the schedule pressures and other pressures weighing down on our management team as a result of NASA's continued interest in possibly second-sourcing some of the SRB production. Our behind-schedule position on the current program and lack of a signed contract from NASA for the next sole-source procurement for sixty-six more flight sets of motors were tremendous leverage for NASA.

While the MTI personnel were off-line—on mute—in Utah for their caucus, Mulloy, Reinartz, and I engaged in a discussion concerning the 40° to 90° qualification temperature of the motor, which had been raised by Reinartz in the teleconference. "I wasn't involved in the qualification of the steel-case SRM," I told them, "but based upon my experience with the FWC-SRM, I don't believe that every component and element of the SRM is qualified to 40°. My interpretation of the specification is that 40° to 90° is the operating temperature range and includes *all* components of the SRM."

"The 40° requirement applies to PMBT [propellant mean bulk temperature] only," Mulloy retorted, "and STS-51L, at time of launch, will be at 55°. That means other components can be below 40° as long as the PMBT never drops below this value."

"That's ridiculous, Larry," I said, "because the propellant is such a massive insulator [over 1.1 million pounds] that it never changes the PMBT in a few hours, even with tremendous external temperature extremes. I'm *sure* the specification didn't mean that! It could be 100° below zero outside for several hours, and it wouldn't change the PMBT by more than a few degrees."

"NASA should consider delaying the launch until late afternoon," I told them, "when temperatures are expected to reach 48 to 50°. That's an acceptable launch window because the launch time was set originally at 3:45 in the afternoon."

"We've considered this and rejected it," Mulloy said, "because of problems with weather and maybe visibility at one of the TAL [transatlantic abort landing] sites."

Overhearing our conversation, Cecil Houston, MSFC's Resident Manager at KSC, commented, "It will be Thursday before early morning temperatures are expected to be in the fifties." However, Cecil didn't seem to be too alarmed at that, and seemed to think a delay was not an unreasonable solution.

By this time, the five-minute off-line caucus requested by MTI had gone on for nearly twenty minutes. Since NASA had concluded that the data presented earlier on the charts to be inconclusive relative to temperature, I expected MTI either to find more supporting information for the 53° launch temperature recommendation or evaluate lower temperatures and make a new recommendation.

I suspected that was why the off-line caucus was taking so long.

I suspected also that if a new recommendation could be supported, it might be for as low as 40° because of the question Reinartz had raised on the qualification requirements for the motor, unless some new calculations supported a better number. I knew that our engineers in Utah were under a great deal of pressure to "quantify" the basis for any recommendation and that that was probably the reason the off-line caucus was taking so long. Boisjoly and the gang were probably making some quick calculations to support the original launch recommendation or support a different launch temperature.

The caucus had lasted almost a half an hour when Kilminster finally came back on the line. "We've reassessed the data," he said, "and concluded that, even though the lower temperatures are a concern, the temperature effects are inconclusive. Therefore, MTI recommends launching."

Kilminster then read some rationale supporting the new recommendation. Much to my regret, the Morton Thiokol caucus had reversed our engineering's original no-launch recommendation and supported a launch decision. The rationale for supporting a launch was not much different than the previous rationale for not launching below 53°. The only *new data* was that an O-ring could tolerate three times the erosion that had been seen on last year's STS-51C (SRM-15) and still maintain a seal, but the erosion behavior of an O-ring had nothing to do with temperature. Everything else about the rationale, including that which dealt with the timing function for sealing (which could be affected by temperature), was still negative. The new rationale ended with the statement that the solid rocket motor for STS-51L *Challenger* would not be significantly different than that for STS-51C twelve months earlier.

With the exception of the erosion data, the statements were basically the same ones as those Mulloy had found unacceptable and unquantified before the caucus.

Reinartz asked if there were any disagreements with the Thiokol recommendation. Hearing none, he considered the matter "properly dispositioned." No one from NASA had asked any questions about the validity of the data used to change the recommendation from "don't launch" to "launch," whereas they had challenged nearly *all* the data Thiokol had presented before the caucus.

It was clear that NASA had finally gotten the answer it wanted.

George Hardy, sitting in Huntsville, said that KSC needed both the launch recommendation and rationale in writing and signed by a responsible Thiokol official. He also stated that Thiokol needed to fax a copy of the launch recommendation to the Huntsville Operations Support Center.

At both the L-2 and L-1 flight readiness reviews, it had been up to me to provide the go-ahead for MTI. I had done it in response to an oral poll taken at the conclusion of each FRR meeting conducted by Jess Moore, NASA Associate Administrator for Space Flight and head of the Mission Management Team. As was customary, the senior management representative at KSC from each Space Shuttle element contractor was polled at the end of these meetings to support a launch recommendation. As I was that representative from Morton Thiokol, I assumed I would be the person who would also have to sign this launch recommendation following the teleconference. After all, that was why I was there. Each contractor was required to have a senior member of management attend each launch and participate in any prelaunch problem resolution with authority to recommend launching.

A request to provide a signed recommendation to launch after an L-1 FRR had *never* happened before in the history of the Shuttle program. I didn't feel good about any recommendation that could lead to a launch at a temperature below 53° just because our engineers felt forced to concede after a three-hour discussion that their data was "inconclusive."

From across the table, Mulloy looked me directly in the eye when George Hardy made the request that our launch recommendation and rationale be put in writing and signed. "I won't sign that recommendation, Larry. It will have to come from the plant in Utah."

It was clear to me what was going on. Even though the NASA people finally got the recommendation they wanted, they weren't comfortable with it either, or they wouldn't have requested the signature.

"It's imperative that we get this signed launch recommendation by early morning," Mulloy declared.

I'm sure Larry expected Joe Kilminster to direct me to sign the recommendation and give it to NASA, and I was concerned that in fact it might happen. But Kilminster told the teleconference he'd provide it right away and fax it to both Kennedy and Marshall. "Al, you stay and receive the fax," my boss told me, "and deliver copies to Larry, Stan, and Cecil still this evening." Turning to Cecil, I asked, "Where is the fax machine?" "Down at the other end of the building," he said.

I felt extremely puzzled by the drastic change in the launch recommendation, but I didn't voice any objection at the time, not even when Reinartz asked over the network if there were any disagreements with Thiokol's recommendation. I'm sure I would have disagreed if I hadn't thought that he was directing that question to the people at MSFC. As far as I knew, the launch recommendation had been made by MTI engineering for Joe Kilminster to deliver, just as with the original recommendation. I'd taken the position from the outset that any launch recommendation needed to be an engineering decision and not a program management decision. I was the Director of the Space Shuttle SRM Project, responsible for program management, not engineering.

What I didn't realize at the time was that the revised launch decision was *not* an engineering recommendation. I had thought that Kilminster was relaying the engineering recommendation after engineering had conducted a reevaluation of all the previous data, considered some additional data, and conducted further engineering analysis. Joe had been very emphatic prior to the caucus that he supported the engineering recommendation not to launch. I thought that our two senior people at MTI, Kilminster and Lund, had come to full agreement on the new recommendation in conjunction with the engineering people, who had provided all the information for the teleconference.

What I did not know was that Kilminster's boss, Cal Wiggins, Vice President and General Manager of the Space Division, and Jerry Mason, Senior Vice President of Wasatch Operations, Cal Wiggins's boss, had been in charge of the off-line caucus. Neither of these men had made his presence known to anyone at the teleconference—something that was contrary to the established protocol of acknowledging all senior management people participating in a NASA/contractor teleconference.

What really happened at MTI during that five-minute-caucus-that-took-half-an-hour? As I later learned, there was a conflict between our management, under great pressure (as they were) from their most important customer, and our engineers, who were responsible for making the technical recommendations. It was a classic case where good technical judgment and common sense seem to have yielded to the philosophy that "the customer is always right." It was just not good politics—and not good business—to go against your most important customer's wishes when you are as vulnerable as Morton Thiokol thought it was relative to second-source issues—especially with an unsigned contract from the customer for the next, and probably last, noncompetitive buy of solid rocket motors.

The frustrations of our engineers in Utah had been heightened by the fact that they could not *prove* that the O-rings would fail to seal at lower temperatures. Jerry Mason had started the caucus by saying, "We need to make a management decision." Arnie Thompson, one of the engineers strongly opposed to the launch, then walked over to the table where the senior managers were sitting and laid out sketches of the joint design along with copies of the data that had been presented originally. Surely they could see that cold temperatures were detrimental to obtaining a good seal, Arnie thought. What he got from Mason and Wiggins was a cold stare.

Roger Boisjoly then went to the table and laid down the photographs of the jet-black soot that had been observed between the primary and secondary O-rings on STS-51C. Roger was visibly upset: "Look *carefully* at these photographs! Don't ignore what they are telling us, namely, that low temperature causes more blowby in the joint!"

All he got, too, were stares.

Feeling totally helpless and that there was no point in pursuing the matter any further, Boisjoly also sat down.

None of the other MTI engineers who were present in that room spoke up against management. But *none* of the engineers in that room took a position in *favor* of launching, either. Jerry Mason, in fact, made the statement, "Am I the only one who wants to fly?"

No one answered him.

He then directed his question specifically to his three senior managers: Wiggins, Kilminster, and Lund. Mason received a supporting response from Wiggins and Kilminster but no response one way or the other from Lund.

"We're just going over and over the same information," Mason directed, "and it's time for a decision."

Mason knew that Lund's position as Vice President of Engineering was key, because our original recommendation had been an engineering recommendation presented by Lund to NASA during the teleconference.

When Lund still gave no response, Mason said, "It's time for you, Bob, to take off your engineering hat and put on your management hat."

It was an intimidating statement, sufficient to make Lund agree with his boss that the launch should proceed as scheduled, with none of the constraints that engineering had originally imposed.

The decision had been made. The caucus ended.

During the caucus, NASA management at the Huntsville Operations Support Center had discussed the "inconsistency" in the engineering data that we'd presented as the basis for not recommending a launch below 53°. Morton Thiokol's chief project engineer on the SRM program, Boyd Brinton, had participated in that meeting but took no position; in fact, there was only one person there who vocally supported our original MTI recommendation not to launch. That was MSFC's Ben Powers, who expressed his opinion against the launch to his boss, Dr. John McCarty, Deputy Director of the Structures and Propulsion Laboratory at Marshall. Apparently, McCarty didn't consider Ben Powers's opinion sufficiently important to relay it to George Hardy, the senior NASA manager at the Huntsville Operations and Support Center (HOSC).

Finding this out some weeks later, I was not surprised that Ben Powers's opinion was not held in high regard. Ben had a reputation for not supporting most any launch relative to any concerns being raised. He always had something on file supporting why he could not be unequivocal about a launch. He had disagreed many times with the information that I myself had presented recommending prior launches. I always believed that some day he was going to be the only one correct. When that day came, he would be the boy crying wolf, and people would discount his opinion precisely when it counted the most.

Again, it took some time for the signed fax with the new launch recommendation to get to us, while all of the parties waited around the conference table at KSC. "I don't feel very good about this launch recommendation," I told them. "It's very difficult to quantify at which temperature the seals may be acceptable and where they aren't, based upon this data. Some of the data is inconclusive and difficult to quantify."

I then stated, "Though I don't agree with the 40° temperature qualification of the motor—that is, that *all* elements are qualified to that—it is my understanding there were a lot of people at both NASA and Thiokol that had signed up to that, through the design certification process and critical design reviews. I don't understand how NASA can accept any recommendation below 40°, especially when the expected temperature will be 26° or even lower."

"I may be naive about what generates launch commit criteria," I said at the conference table, "but I'm under the impression that the 40° qualification was generated based upon the qualification of *all* elements or subsystems of the Shuttle. Anything outside that qualification is a launch commit criteria, and we've never gone outside that envelope. I don't understand why NASA would ever want to launch below 40° if that is what the SRM is qualified to."

Getting no response, I made the direct statement, "If anything happens to this launch, I wouldn't want to be the person that has to stand in front of a Board of Inquiry to explain why we launched this outside of the qualification of the solid rocket motor or any Shuttle system."

The room remained silent. I was still very upset, and so I asked that they reconsider the decision: "If I were the Launch Director, I would *cancel* the launch for three reasons, not just one. The first reason being the concern of the cold O-rings that we have just discussed, but there are two others. I've just left Carver Kennedy's house in Titusville, and, as you know, Carver's responsible not only for the stacking of the SRBs but also for the retrieval operations. Carver's been in communication with someone at Hangar AF who contacts the booster recovery ships at sea. That person has told him that the recovery ships are in an 'absolute survival mode.' They're in seas as high as thirty feet. The winds, which are at fifty knots sustained and gusting to seventy, are pitching the boats as much as 30°. The seas are so rough that they may very well have damaged some of our retrieval equipment. They're steering directly into the wind, heading for shore at about three knots, and have been doing that for some time. There's no way they'll be able to support an early morning launch, because they won't be in the recovery area.

"There are also some 'firsts' on this launch we better consider. This is the first time we're going to use a new electronic control system for separating the nozzle extension cone on the SRMs at apogee, rather than just before water impact while it's under full main parachute. We're also going to separate the parachutes at water impact for the very first time. Based upon the conditions at sea, it's going to be nearly impossible to recover that hardware, either the parachutes or the frustums. We're going to be putting the boosters at some risk as far as recovery is concerned, because the ships are right now heading *away* from the recovery area!

"The third reason for not launching is the formation of ice. With our sound suppression system being a water system, there's probably going to be a lot of ice on and around the pad. I'm not an expert on these matters, but I feel there could be an ice debris problem or possible chance that some abnormal acoustics could be a problem. Noise could have an effect on structures or electronics.

"I don't know for sure, but I don't think it's prudent to launch under this kind of condition.

"I'm recommending against launching not because of what I know but because of what I don't know—and NASA should feel that it's in the same position. It just isn't worth taking the risk with all of these unknowns!"

"But, Al, these really aren't your problems," I was told. "You really shouldn't concern yourself with them."

I answered, "You know all three of these together should be more than sufficient to cancel the launch, if the one we discussed earlier wasn't."

Knowing that I was very disturbed, all the NASA people could say was, "Al, we'll pass these on as concerns, but we'll do it in an advisory capacity only."

Larry Mulloy wanted to know where the signed fax was, because some time had transpired since the end of the teleconference. Going down to the other end of the building to check, I found nothing, and really wondered if the fax machine was even working. Finally it came in about ten minutes later; I waited for it the whole time. When it came, it was just a single sheet of paper. I took it to Jack Buchanan's office, where we reproduced copies for everyone.

Returning to Cecil Houston's office, a new telecom was under way. Talking together were Mulloy and Reinartz with Arnie Aldrich, Manager of the National Space Transportation Systems Program Office from NASA Johnson. Aldrich was a member of the overall NASA Mission Management Team at KSC. The three of them were in the middle of discussing the condi-

MTI Assessment of Temperature Concern on SRM-25 (51L) Launch

O   Calculations show that SRM-25 o-rings will be 20° colder than SRM-15 o-rings

O   Temperature data not conclusive on predicting primary o-ring blow-by

O   Engineering assessment is that:

   O   Colder o-rings will have increased effective durometer ("harder")

   O   "Harder" o-rings will take longer to "seat"

      O   More gas may pass primary o-ring before the primary seal seats (relative to SRM-15)

         O   Demonstrated sealing threshold is 3 times greater than 0.038" erosion experienced on SRM-15

   O   If the primary seal does not seat, the secondary seal will seat

      O   Pressure will get to secondary seal before the metal parts rotate

         O   O-ring pressure leak check places secondary seal in outboard position which minimizes sealing time

O   MTI recommends STS-51L launch proceed on 28 January 1986
   O   SRM-25 will not be significantly different from SRM-15

Joe C. Kilminster [signature]

JOE C. KILMINSTER, VICE PRESIDENT
SPACE BOOSTER PROGRAMS

MORTON THIOKOL. INC.
Wasatch Division

INFORMATION ON THIS PAGE WAS PREPARED TO SUPPORT AN ORAL PRESENTATION

FAXED TO:

MSFC   # 205-453-5725
KSC   # 305-867-7103

9:45 PM MST
27 JAN 1986

Allan McDonald's boss, Joe Kilminster, signed and faxed this typewritten chart to NASA on the eve of the launch recommending proceeding with the launch; this was the document that McDonald refused to sign. This chart will forever remain one of the most historically significant documents in America's aerospace history. (*Report of the Presidential Commission on the Space Shuttle Challenger Accident*, Vol. 1, June 6, 1986.)

tions of the booster recovery ships and the high seas conditions and the fact that the ships were in survival mode. Aldrich asked Mulloy, "Is there any risk to recovering the boosters?" Larry said, "No, I don't think there's any significant risk, because the boosters have beacons and monitors, and we're going to have aircraft in the area. There is, though, a high probability that we won't recover the parachutes or frustums. We'll have to make a decision to launch on the basis of a high probability of not recovering that auxiliary hardware, but I don't think we're compromising the recovery of the boosters themselves, not significantly."

"What's the value of the parachutes and frustums?" Aldrich asked Mulloy. "Right under a million dollars," Mulloy estimated. "Can you afford to lose this hardware and still support the program schedule?" Aldrich wanted to know. "I think we have sufficient inventory to do that," said Larry. "Well," Aldrich said, "in no way should you have those ships attempt to turn around in these conditions, Larry, because the safety of the ships is very important. Tell them to continue on towards shore until it is really safe to turn around."

Aldrich and Mulloy also spoke briefly about the ice issue, with both men agreeing that the subject had been addressed adequately earlier in the day.

I didn't hear anything discussed about the O-ring seal problem. I presumed, incorrectly, that it had been done while I was waiting for the fax, because that was the first concern we had addressed.

The conversation with Arnie Aldrich concluded with a recommendation to proceed with the launch as originally planned. I handed Mulloy and Reinartz a copy of the fax signed by Joe Kilminster recommending proceeding with the launch. They read the fax, then discussed whether they should notify Marshall's Center Director, Dr. Lucas, who was at the Cape for the launch.

"Should we wake the old man and tell him about this?" Reinartz asked Mulloy. "I wouldn't want to wake him," Mulloy responded. "We haven't changed anything relative to launching. If we had decided to scrub, then we'd have to wake him."

I stayed around a few more minutes to talk to Jack Buchanan, Morton Thiokol's Resident Manager at KSC, and then drove back to Carver Kennedy's house in Titusville, where I arrived around 1:00 a.m. on January 28, 1986.

9

"Obviously, a Major Malfunction"

I had to get up at 4:30 a.m. in order to be at the Launch Control Center to take my place at the SRB console by 6:00. I remember walking in from the parking lot in the dark, carrying my headset in one hand and briefcase in the other. Driving in, I heard on the radio the temperature was 22°.

Inside the Launch Control Center, I went to the SRB console and, just like the day before when the launch had been scrubbed because of high winds, all of the available support personnel seats at the SRB console were taken by people from United Space Boosters Inc. So I sat down at the adjacent console with Carver Kennedy just as I had on the previous launch attempt.

At the console, I quickly surveyed several of the on-pad video cameras. Numerous cameras supported a Shuttle launch by looking at several areas of the vehicle and launch support structures and equipment at various levels. I saw huge icicles hanging everywhere and said to myself, "They aren't really going to launch this thing today."

During a planned hold at T minus three hours, the countdown was stopped for two hours to survey the ice on the vehicle and launchpad. The Ice Team went out to launchpad 39B to inspect the vehicle, the mobile launch platform, and the launch support structures. I was sure they would cancel the launch right then and there.

This was the second visit by the Ice Team, as it had surveyed the pad between 1:30 and 3:00 in the morning. Even then, when the ambient temperature was 29°, the inspectors found large quantities of ice on the fixed service structure (FSS), mobile launch platform, and launchpad apron. Water troughs below the nozzles of the solid rocket boosters had been filled with an antifreeze solution good to 16°, but sheets of ice were beginning to form nonetheless. Several of the water lines on the FSS were left running overnight to prevent the lines from freezing, but large areas of ice three inches thick were forming as the water dripped and accumulated. Even worse, a light wind was blowing the water everywhere.

On the second visit to the pad, when ambient temperatures ranged between 26° and 30°, the Ice Team noticed increased thickness of ice everywhere. The water troughs containing the antifreeze were frozen solid. Directly under the nozzles of the SRB, a large quantity of ice was removed from the troughs. Icicles were hanging from the stiffener rings on the aft segment of the left-hand SRB and on the aft skirt. Sheets of ice had formed on the left side of the aft booster case and aft skirt as well. The icicles were removed, but the sheets of ice were left intact. Charlie Stevenson, head of the Ice Team, told Horace Lamberth, NASA Director of Engineering at KSC, that based upon what he could see, "The only choice you got today is *not* to go!"

With a hand-held infrared pyrometer, the Ice Team measured the surface temperature of various elements of the Shuttle vehicle and launchpad area. Measurements on the left-hand SRB segments and on the aft skirt were within a few degrees of the measured local ambient temperature of 26°, between 23° and 27°. However, for some strange reason, the temperature that was measured on the right-hand solid rocket booster near the area of the aft field-joint was only 9°. An even lower temperature was measured on the aft skirt of the same booster—a mere 7°—and a reading taken on the mobile launch platform below the boosters read 12°. In the water troughs below the SRBs, the ice and unfrozen antifreeze solution was determined to be between 8° and 10°. The Ice Team recorded these measurements in a notebook, along with other temperature readings on the external tank, orbiter, and FSS.

Completing the inspection of the vehicle on the launchpad, the Ice Team briefed the Mission Management Team shortly after 9:00 a.m. The team didn't discuss the temperature measurements they had taken on the SRBs because those measurements were recorded for engineering information only and were not part of the formal launch commit criteria (LCC). Discussion was limited to ice observations only, which were part of the formal LCC.

Unbeknownst to me at the time, neither the Ice Team nor the Mission Management Team had been informed about Morton Thiokol's earlier concerns about the sealing capability of the field-joint O-rings at temperatures as low as 29°, much less 9°.

During the 9:00 a.m. meeting, Bob Glaysher, Vice President of Rockwell and Program Manager for Orbiter Operations Support at KSC, told the Mission Management Team that based upon the Ice Team's report, "Rockwell cannot assure that it is safe to fly." This conclusion was based on concern that ice debris from launch might impact and damage the sensitive heat-protecting tiles on the underside of the orbiter. Glaysher had been in a telephone conversation with Dr. Rocco Petrone, President of Rockwell's Space Transportation Division in Downey, California. An ex-NASA Director of Launch Operations at KSC and a former Director of MSFC, Petrone told Glaysher, "Make sure NASA understands that Rockwell feels it is not safe to launch."

Jim Beggs, the NASA Administrator, at the time on leave from the space agency, was watching TV coverage of the launch in Washington, D.C., and he, too, was concerned about all the ice. I learned later that Beggs called NASA officials at KSC, urging them to wait for warmer weather.

The Mission Management Team decided to remove as much ice as possible and conduct an analysis of the potential trajectory of the remaining ice debris to determine the probability of impact with the orbiter. At the same time, it chose to continue on with the countdown, knowing that a hold was planned for later on, during which a last-minute inspection of all the ice conditions could be made.

During the course of its inspection, the Ice Team had removed 95 percent of the ice from the frozen antifreeze solution in the troughs below the SRB nozzles as well as several icicles hanging from the left-hand SRB skirts. A smaller crew was sent out to the pad a third time at 10:30 a.m., about one hour before the launch, when the ambient temperature stood at approximately 35°. By this time the ice was starting to melt from the upper levels of the fixed service structure in the areas hit by direct sunlight. The troughs were still experiencing freezing, so a shrimp net was used to remove 95 percent of the ice once again. Only the two south water troughs remained ice free, as had been the case in the first two inspections. Icicles on the left-hand SRB aft skirt were also removed, and any loose ice on the deck of the mobile launch platform was removed. The crew returned to the Launch Control Center and reported that all ice that could be removed had been removed, but ice remained on the lower segment of the left-hand aft SRB and in several areas of the FSS. Ice was still on the external tank, but it was determined to be tolerable and within previous experience with icing on the ET thanks to the cryogenic fuels in the tank. The Mission Management Team at KSC heard a report from Mission Control in Houston that their trajectory analysis of the ice remaining on the FSS indicated that it would not impact the vehicle during launch. The countdown resumed with a final go-ahead for launch.

With the exception of Stan Reinartz, neither the Mission Management Team nor the astronauts on board *Challenger* were ever made aware of the concerns raised by Morton Thiokol the night before concerning the ability of the O-rings to seal in such cold temperatures. Furthermore, the four space veterans in the *Challenger* crew—Commander Francis Scobee, Mission Specialist Judith Resnik, Mission Specialist Ronald McNair, and Mission Specialist Ellison Onizuka—had all flown on a previous flight that exhibited thermal distress or erosion of O-rings and/or blowby of primary pressure seals in two or more joints of the solid rocket boosters. Onizuka had flown on STS-51C just a year earlier, in January 1985. His flight exhibited O-ring erosion in two field-joints, with heavy black soot blowby in both of these joints, along with soot blowby in both booster nozzle joints. That Shuttle, STS-51C (*Discovery*), was the coldest launch prior to STS-51L and was the primary basis of concern by the Morton Thiokol engineers for launching the *Challenger* in even colder weather.

None of the *Challenger* crew was ever made aware of the concerns of the O-rings sealing properly on this cold launch *or* the grave observations concerning their own previous Shuttle flights.

STS-51L was the first flight for the other three crew members: Pilot Michael Smith, Payload Specialist Gregory Jarvis, and schoolteacher Christa McAuliffe. This was navy Captain Smith's first opportunity to pilot the Shuttle. Jarvis had been bumped off two previous flights basically for reasons related to Washington politics. McAuliffe, from Concord, New Hampshire, had just been selected over thousands of candidate schoolteachers to be the first ordinary citizen to fly in space; McAuliffe had referred to her flight as the "ultimate field trip." Her proud parents, Ed and Grace Corrigan, with their younger daughter Lisa were anxiously watching the launch from a special viewing area. Christa's backup and close friend, schoolteacher Barbara Morgan from McCall, Idaho, was also at KSC to witness the launch that very cold morning.

Some say "ignorance is bliss," but in this case it contributed to a national disaster.

I remember switching channels to monitor the terminal phase of the countdown through my headset at the console. At 11:38 a.m., the countdown progressed through ignition of the Space Shuttle main engines; this came at between six and seven seconds before SRB ignition.

I broke out into a cold sweat.

As the countdown proceeded from six down to zero, I had one eye glued on the digital readout of the solid rocket motor chamber pressure data that was displayed on the console's screen and the other eye on an adjacent TV screen monitoring the launch. The SRM chamber pressure was extremely important because both SRBs must ignite simultaneously and in 0.6 second come up to a maximum pressure of around 920 psia, with over three million pounds of thrust each. If one SRB lags the other SRB by as little as one-tenth of a second, the whole vehicle could explode by the overturning moment produced by the differential thrust of the two boosters on opposite sides of the external tank.

Ignition always makes me nervous, but this time I was far more concerned because of the effects that the extremely cold temperatures could have on the sealing capability of the O-rings in the SRB field-joints. I breathed a sigh of relief as *Challenger* lifted off the pad. I really believed that if the O-rings were to fail, it would be at ignition during the critical 0.6-second time period. The whole thing would explode on the launchpad: 2.2 million pounds of solid propellant in the SRMs and 1.6 million pounds of liquid hydrogen and oxygen in the ET. I continued to watch the SRB chamber pressures and the TV monitor of the flight until Mission Control in Houston gave the *Challenger* crew the "Go for throttle up!" Commander Scobee responded, "Roger, go at throttle up!" The very next instant, the vehicle exploded. *Challenger* was at an altitude of nearly 50,000 feet traveling at a velocity just under Mach 2. My initial reaction was that the SRBs had separated from the ET prematurely, when they should have continued to burn for nearly another minute before separation.

I kept looking for the orbiter to come out of the fireball shown so clearly on the TV monitor. Everyone in the Launch Control Center was shocked. It was silent, except for several people sobbing in the background. I kept hearing a voice over the network saying, "RTLS! RTLS!" ("Return to Launch Site"), but I could see nothing coming out of that huge fireball except the two solid rocket boosters, which were still flying.

As shocking as the explosion was to those inside the Launch Control Center, it was absolutely terrifying to the astronauts' families witnessing the launch outside in a special viewing area, as was indicated by the expressions of pain and anguish on the faces of the Corrigans, Christa McAuliffe's parents and younger sister Lisa, that appeared in a photo in the *New York Times* the day after the accident, January 29, 1986. The *New York Times* later reported that their review of the launch tape "from liftoff to the announcement of the explosion" showed that the photo of the Corrigan family witnessing the launch "was in fact made slightly before the explosion and that the suggestion that the family was reacting to the explosion was mistaken." I do not believe this was the case since the announcement of the explosion by the NASA commentator was not made until nearly forty-five seconds after the explosion. The expressions on the faces of the Corrigan family are clearly those of terrified family members; I believe they knew that something had gone horribly wrong at the time the photo was taken. The actual photo was not reprinted at the request of the Corrigan family.

For nearly forty-five seconds after the explosion, the commentator on NASA Select TV continued to provide an update on the *Challenger's* altitude, downrange position, and velocity, as if nothing had gone wrong. When the debris from the explosion started falling toward the ground, he finally exclaimed dryly, "Obviously a major malfunction. . . . We have no downlink. . . . We have a report from the flight dynamics officer that the vehicle has exploded."

One of the solid rocket boosters did a full 360° flip in the air and continued on its trajectory before the air force range safety officer finally—but not until some 110 seconds into the flight, or 37 seconds after the explosion—pushed the button to fire a linear-shaped charge attached to the upper three segments of the SRB, blow open the casing, and terminate its thrust. The SRBs were just starting into "tail-off" and would have burned out on their own in another 10 seconds.

All the doors in the Launch Control Center were immediately locked, the telephone lines all disconnected, and all data associated with the flight frozen on the screen and immediately impounded. At seventy-three seconds, I remember looking at the last recorded chamber pressure data and noting that the right-hand SRB was approximately 20 psia lower than the left-hand SRB. Even though this represented less than 3 percent of the pressure at the time, it was clearly outside of our normal experience. I also noted that the thrust vector actuation system, which moved the movable nozzle to steer the vehicle, was more active than normal prior to failure, but was far from exceeding demonstrated operating limits.

Naturally, the initial question in my own mind was whether this failure could have been caused by an O-ring leaking in an SRB due to the cold temperature that we had agonized over the night before. I really thought that this was not the case, however, because the SRBs were the only things that continued to fly, and everyone knew that solid rocket motors don't continue flying with hot gas leaks or holes in their sides. They explode!

In spite of my underlying worries, I felt fairly confident that the accident was either a result of an ET structural failure or more likely due to a failure of a cracked high-speed turbine blade in the Space Shuttle main engine; after all, those blades operated at fantastic 30,000 to 40,000 revolutions per minute.

With my cohorts, I was locked in the Launch Control Center for a couple hours before being assigned to one of the failure teams. We were also informed that Vice President George Bush was flying down from Washington to speak to the personnel who were in the Launch Control Center at the time of the accident.

Everywhere at Kennedy Space Center there was mass confusion. NASA was clearly not prepared to handle such an accident. Even though NASA managers had some kind of disaster plan in the books, they really never paid any attention to it, and the first thing they did was to make wholesale changes to the names of those who were designated to lead various failure activities.

Since I had to stay for a failure team formation meeting, Carver Kennedy said that he would leave the Launch Control Center as soon as they unlocked the door and call Ed Garrison, President of the Aerospace Group of Morton Thiokol in Chicago, and let him know what happened. Carver asked me if I had anything to pass on to Garrison. I told Carver, "Tell Ed and Charles Locke, our CEO in Chicago, to both get their asses down here as soon as possible, because there was nothing more important in the corporation deserving their attention at this time." Carver thought that was a good idea and told me he would pass that on, but probably not in those same words.

Carver called me in the Launch Control Center just before Vice President Bush was to arrive. He told me that both Locke and Garrison were flying in from our Chicago headquarters on the corporate jet and would land at the Titusville airport. They wanted me to meet them at the airport and take them to one of the condominiums that Thiokol leased on the beach at Cape Canaveral. I told Carver that I couldn't do that; I needed to stay for Bush's visit to the Launch Control Center. I asked Carver if he could pick them up and take them to the condo and said I would meet them there later that evening.

It was late in the afternoon when Vice President Bush addressed those involved in the launch of the ill-fated *Challenger*. He reiterated what a terrible tragedy it was, but that we must learn from this experience and get America back into space safely again as soon as possible. He did not want to assign blame to anyone and wanted to make everyone aware that the country possessed the resolve to continue our space program in spite of this terrible accident. He was trying to provide some comfort and encouragement to a group of people who were still in shock.

After the vice president's visit, I met with a group of NASA and support contractors to discuss how the investigation was going to be conducted and who would be assigned to the various teams. The principal failure analysis team would be at KSC and headed up by Jess Moore, NASA Associate Administrator for Space Flight and head of the Mission Management Team. Moore would have several high-level NASA officials working with him, including NASA Marshall's Director, Dr. William Lucas. There would be several supporting teams covering all of the primary elements of the Space Shuttle and launch complex. The primary analysis of the propulsion system elements would be headquartered in Alabama at the Huntsville Operations and Support Center, under the direction of Jack Lee, Marshall's Deputy Director. Bob Schwinghammer, Director of the Materials and Processes Laboratories at MSFC, would head up the team that examined the solid rocket motors.

Moore reiterated that all of the flight data had been impounded and that no one was authorized to copy or leave the room with any of the flight data. All of the data would be treated as "Secret," and a special badge, card, and code number would be required for anyone wanting to gain access to the data at the Huntsville Operations and Support Center.

It all sounded legitimate. I did not suspect at the time that a cover-up by some people in NASA was already under way.

10

"Don't Blame Yourself"

It was nearly 11 p.m. before I left the meeting laying out NASA's initial approach to investigating *Challenger*. I immediately drove to the condominium where Thiokol's executives Charles Locke and Ed Garrison were staying in Cape Canaveral north of Cocoa Beach. As I was explaining to them what all had transpired at Kennedy Space Center since the accident, the phone rang. Jerry Mason, Senior Vice President and General Manager of Morton Thiokol's Wasatch Operations, was on the line to tell Ed that he was forming a failure analysis team at the plant in northern Utah. They were in the process of impounding all the manufacturing records and preflight information on the solid rocket motor hardware that had been used in this flight, and he was assigning Joe Kilminster, our Vice President of Space Booster Programs, to head up the Thiokol team that would support Bob Schwinghammer's SRM Failure Analysis Team at NASA Marshall in Huntsville.

While talking with Garrison, Mason asked him if he had seen me. Ed related that I had just gotten in from KSC and was there, at the condo, briefing him and Charles Locke on what was going on at Kennedy relative to the formation of a failure analysis team.

Mason asked to speak to me. After summarily repeating what he had told Ed, Jerry told me to catch the first flight out in the morning and come back to the plant. I asked if he really didn't want me to join the SRM Failure Analysis Team at Huntsville. "No," he said, "I already have people assigned to that team, and you are to return home immediately."

So surprised was I by this decision that, after hanging up, I just stood there in amazement for a few minutes.

While I was on the phone, Locke and Garrison asked Carver Kennedy to arrange an early morning meeting with Dr. William Lucas. They wanted to tell Dr. Lucas that the resources of our corporation were available to help find the cause of this failure and that Morton Thiokol was willing to help in any way we could.

Carver really did not know Lucas, so he asked if I could locate him and arrange a meeting first thing in the morning. It took several calls, but I finally located NASA Marshall's Director, and as one would suspect, under the circumstances, he was extremely busy and didn't think he had any time available for a meeting. Garrison quietly cued me to suggest an early breakfast meeting, and Lucas agreed to that.

After watching several instant replays of the disaster on several different TV stations, we called it a night. Carver and I both told Ed and Charles that we didn't think the accident was caused by a failure of the SRBs, because they appeared to have kept flying after the accident. It was probably a Space Shuttle main engine or external tank failure that caused the accident.

As I drove away from the condo around 1 a.m. and headed back to Carver Kennedy's Titusville house for the night, I searched my mind as to whether there was any possibility that this type of failure could have resulted from a leaking O-ring in a field-joint. It really bothered me, but I couldn't come up with a scenario that supported a failure like that happening some seventy-three seconds after SRB ignition.

What bothered me even more were the orders I had received from Jerry Mason, our General Manager, to return to the plant immediately and not participate in the SRM Failure Analysis Team. That bothered me so much that, when I arrived at Carver's home, I told him I was going to call Mason right then and there and tell him that I was going to Huntsville tomorrow to review all the flight data, because I just had to assure myself that this failure was in no way caused by the SRB. I wouldn't be useful to anyone for anything until I convinced myself that the O-ring concerns had nothing to do with the catastrophe.

Carver understood but advised against calling the General Manager at two o'clock in the morning. He knew I was very tired and upset. Both of us had been up for nearly twenty-two hours and had gotten, at best, a disturbed three-and-a-half-hours' sleep the night before. He suggested I wait until morning to call.

I couldn't understand why I was left off the SRM Failure Analysis Team; I knew more about the operation and design of the solid rocket motor than anyone. I could understand why he put Joe Kilminster, my boss, in charge of the SRM Failure Analysis Team, but to leave me off the team entirely when I presented all of the flight readiness reviews on this flight and was the senior Thiokol manager at KSC really blew my mind. It just didn't make sense.

I was so upset, I ignored Carver's good advice and called Mason, getting him out of bed. I told him I really needed to go to Huntsville and review all the data. I told him I knew he had instructed me to come home, but I needed to convince myself that this accident was not our fault before I could do anything else. I was sorry I had awakened him, but I had to make him aware of my feelings and needed to know why I shouldn't do this. Specifically, I would review the "quick look" telemetry data from the orbiter and review some of the flight films. I would not stay at Huntsville with the SRM Failure Analysis Team and would return to the plant the following day.

Mason replied that he needed me back at the plant to keep the program running and formulate some plans on how we would get back into production after shutting down for a period of time. Reluctantly, he told me it was OK to go to Huntsville, but he expected me to return to the plant the following day.

It turned out to be another very short night. About 6:00 a.m., Carver received a telephone call from Ed Garrison saying that Lucas had been called into a very early morning meeting and was unable to make the breakfast with Ed and Charles Locke. Lucas appreciated their offer to help and suggested that we all go to Huntsville and talk to Jack Lee, his Deputy Director and head of the MSFC Failure Analysis Team, because that's where all the data would be and where the SRM Failure Analysis Team would be located.

Garrison and Locke had never met Jack Lee, nor had Carver Kennedy, so Ed asked if I could arrange for a meeting with Jack later in the day. I called the new manager of our Huntsville field office at MSFC, Ivan Adams, and he arranged a meeting with Lee. Garrison asked me to meet him and Locke at the Titusville airport in a couple hours to fly with them to Huntsville.

Still very upset, I continued to fret that maybe, just maybe, the cold O-rings had something to do with the accident. As I was telling this to Carver, my voice started to quiver, and Carver's wife, Martha, came over to me and gave me a big hug. She could tell I was stressed out and told me she was sure everything was going to turn out all right. I could only think of the horror that those seven astronauts must have gone through as they were engulfed in the huge fireball.

As a tear came to my eye, I said, "It may have been all my fault." "Don't blame yourself," Martha countered. "You did all that you could have done, and I'm sure that it is really not our problem."

Carver agreed and displayed the morning edition of the *Orlando Sentinel*. On the front page was a huge color photograph of the *Challenger* disaster clearly showing both solid rocket boosters flying normally at a considerable distance away from the huge fireball caused by the explosion of the liquid oxygen and liquid hydrogen from the external tank. Based upon the picture, Carver declared: "The SRBs are probably the only system that is *not* a major suspect in the failure."

I was so impressed by the photo that I asked if I could keep the paper to show to Garrison and Locke on the airplane. "Sure," Carver said, "I'll get another one."

After downing a few more cups of coffee and packing my bags, I jumped into my rental car and headed to the Titusville airport. Carver told me to leave the car keys at the General Aviation lobby because he had made arrangements for someone to pick up my rental and turn it in to the agency in Cocoa Beach, because they did not have a rental counter at the Titusville airport. I was running a bit late and drove pretty fast to get to the airport, because I knew that both Garrison and Locke did not look kindly on anyone who kept them waiting.

Fortunately, I arrived at the airport at the same time they did. Dropping my car keys off at the desk, I handed my baggage to one of the pilots waiting for us in the terminal. I climbed aboard with the newspaper under my arm and briefcase in hand.

It was the first time I had ridden on the corporate jet, and I was impressed. The flight attendant brought us coffee and a tray of donuts, rolls, and fruit juices. As we sat around the table to talk, Ed and Charlie asked me if I had any new information. I told them I didn't, but based upon the picture on the front page of the morning's *Orlando Sentinel*, it appeared that the SRBs were in good shape.

I showed the newspaper and could see them breathing a sigh of relief. Charlie asked me if he could have the paper, and even though I really wanted to keep it, I felt like a 900-pound gorilla was asking me to give him something. I told him he could have anything he wanted.

Even though I felt reasonably comfortable that the SRBs didn't cause the disaster, I thought that I should make both men aware of the teleconference that had occurred on the night before the launch. I told them that Morton Thiokol *engineering* was concerned about the cold temperature and the effects it may have on the O-rings, and I told them about our original recommendation not to launch at any temperature below 53°. However, I continued, NASA thought that the data used for making that recommendation was inconclusive so we had decided to reevaluate that data. After reevaluating, the guys in Utah deemed the data inconclusive and recommended to proceed with the launch. I told Ed and Charlie that I hadn't fully understood why they changed their minds, but that Joe Kilminster had faxed to KSC and MSFC a signed recommendation to launch. I had a copy of the fax with me, which I gave to them. I told them I didn't believe the accident was our problem, but they should be aware of all this information just in case.

Waiting for us at the General Aviation terminal in Huntsville was our field office manager, Ivan Adams, who gathered up Ed and Charlie to take them out to Marshall for a meeting with Jack Lee. Garrison and Locke were taking the company jet back to Chicago right after the meeting, and, as I planned to fly to Utah on a commercial airliner the next day, I went over to the main passenger terminal to rent a car.

I went from one counter to the next, with no luck. The *Challenger* accident had brought such an influx of people participating in the accident investigation, along with the news media and press from all over the world, that not a single car was available in Huntsville. I told all the rental companies that I was also involved with *Challenger* and needed a car to get to Marshall as soon as possible.

Finally, one of the agents said there was an old delivery van he could rent out, but that it was in town and would take a while to deliver. Desperate for transportation, I said OK and waited. Twenty minutes later, it arrived—a beat-up old white panel truck with a floor shift that really didn't want to shift. I drove the wreck out to the Redstone Arsenal, where I received a pass to get onto the military property, and from there to the NASA security office, where I obtained a car pass, a badge with a secret clearance, a magnetic card to get into the Operations and Support Center, and a secret code to punch in at the center after inserting the card.

Immediately upon arrival at the Operations and Support Center, I ran into Larry Mulloy in the hallway. I asked him what he was doing there, and he said he was doing what I should be doing and that was going through his files.

"Remember that discussion we had the other night about what temperature the SRBs are supposed to be qualified to?" Mulloy asked.

"Sure, why?"

"Because I've just finished going through reviewing the contractor-end-item specification for the solid rocket motors, and you should do the same. Our specification refers to *Volume X, 'Natural and Induced Environments,' Johnson Space Center Document 07700 for the Shuttle Vehicle*, which is supposed to be qualified to launch at 31°. So we were well within that requirement at time of launch. It was 36°."

"I'm unfamiliar with that document," I told Mulloy, "but I'll check it out when I get home."

I asked Larry if they had any new data that might indicate what caused the accident. He said he wasn't aware of any, but that I should sign in with Bob Schwinghammer's SRM Failure Analysis Team, which was getting organized in one of the rooms just down the hall.

I went into that room and sat down. NASA engineers were listing all of the recorded SRB data on a blackboard so as to determine if there was any abnormal data. All of it appeared to be normal. Even the chamber pressure showing a 20 psia lower pressure in the right-hand SRB was within specification at seventy-three seconds after launch when the data stopped coming. There was also no data indicating a problem with the SRBs.

That fact—that the SRBs appeared to continue operating normally long after the explosion—made everyone feel fairly confident, and myself fairly comfortable—that the accident was not caused by a malfunction of the SRBs.

After reviewing all this data, I was sufficiently convinced and decided to go to the airport and see if I could get a flight home.

Walking out the door of the Operations and Support Center, Larry Mulloy yelled at me down the hall. "I need to get on the phone with Jim Kingsbury at KSC right away." Kingsbury, Director of Science and Engineering at Marshall Space Flight Center, was at the Cape reviewing the film coverage of the launch.

"Some of the films indicate that hot gas is coming out of the side of one of the segments on one of the solid rocket boosters," Larry declared.

"That's absurd," I shot back. "Jim Kingsbury doesn't know what the hell he is looking at, because solid rocket motors don't continue flying around with holes burned through the side of them. They explode!"

"Well, Jim's on a teleconference right now, so you should come into the room and talk to him."

Quickly getting inside that conference room, I could hardly believe what I was hearing, so I asked Kingsbury, "Are you sure that the hot gas is coming from the side of the SRB and not the ET?"

"We're sure that's the case," Jim answered.

"Can you tell whether the hot gas is coming from an area in the vicinity of one of the field-joints?"

"It's hard to tell," said Jim, "but it appears to be coming from the aft segment of the right-hand SRB near the inboard side toward the ET, in the vicinity of the ET attachment ring, which is in the general area of the aft field-joint." Jim went on to say that he was bringing a copy of the film back with him to Huntsville that night, to review it with Marshall management and with the chairmen of the failure analysis teams.

The moment he spoke of hot gas possibly coming from the aft field-joint, I felt my heart would stop beating.

11

"When Is the Space Shuttle Going Up, Daddy?"

I left the conference room to find a room at one of the local motels, which were pretty much booked solid. I finally found a room, and I called the plant back home to relay what I had heard and that I was staying overnight in Huntsville to review the films that evening. I grabbed a bite to eat and returned to the HOSC, where everyone was anxiously waiting to see the films. Jim Kingsbury came in shortly after I arrived, and reviewed the films, which clearly showed hot gas leaking from the aft field-joint of the right-hand SRB. The leak appeared to start at T+59 seconds—the exact time of maximum dynamic pressure (Max Q) on the Shuttle vehicle. This is a time of maximum aerodynamic loading on the vehicle that causes it to vibrate and flex. The leak appeared to grow, and the plume finally became large enough to impinge on the ET, burning a hole through the external tank, which caused the explosion at approximately T+73 seconds. The films also clearly showed the right-hand solid rocket booster with hot gas leaking from the aft field-joint, with the hot gas plume continuing to grow in size after the explosion. After reviewing these films, my worst fears had been realized. The *Challenger* accident was indeed caused by a failure of one of our SRBs. Worse yet, it may have been caused by the very problem that we had tried to avoid that cold winter night before the launch—the O-rings had failed to seal in one of the field-joints due to the extremely cold temperatures.

It was very late that night when I finally returned to my motel. I was mentally exhausted and depressed. It was one of the saddest times in my life. I remember calling home that night and my three-year-old daughter, Meghan, answered the telephone and said, "When are you coming home, Daddy?" "Pretty soon, Meghan, pretty soon." "When is the Space Shuttle going up, Daddy?" I could not answer; I had a big lump in my throat. She knew her daddy always came home after the Space Shuttle went up. I did not know what to say, so I finally uttered, "Pretty soon, Meghan, pretty soon." I was thankful Meghan didn't see the horror of the fireball that engulfed *Challenger*, which must have appeared on television a hundred times during the past two days.

Linda got on the phone and asked me when I was coming home. I told her I didn't know, because things had dramatically changed for me now. I told her about the films I had just seen and about the teleconference we'd had the night before the launch. Choking up, I couldn't finish my conversation with her. "Are you OK, Al?" "I'll be all right. I just wish I were home. But I know I really can't leave now." She told me that our oldest daughter, Lisa, had called her earlier in the day asking if I had been involved in any way with the accident. Linda had told her that I had been in Florida for the launch and the last she'd heard was that I was going on to Alabama, but she expected me back either today or tomorrow. She suggested I call Lisa in Boston, where she was a student in the College of Nursing at Boston College, because she was really worried about me.

On the phone with Lisa, I couldn't tell her what had happened without crying. "I feel like it's my fault that seven American astronauts have just been killed, because I was unsuccessful in stopping the launch and my gut feeling had been that it was too risky to fly in such cold temperatures." Beginning to cry herself, Lisa said, "Don't blame yourself, Dad." Before hanging up, she asked if I was going to be OK, and I told her I would be. This crying conversation was very good therapy for me. I was able to tell my daughter everything that happened, which had been locked up in me over the past few days. Even though I was extremely sad thinking about the seven heroic astronauts who did not have to die, I felt much better after talking to my family. I was homesick as well, having been gone for eight days.

I laid awake most of the night replaying the scene of *Challenger* exploding and of the teleconference the night before. Suddenly it occurred to me that maybe, just maybe, there might be other contributing causes to the accident. We had to make sure that nothing was overlooked. I carried a slim hope that we'd find some other cause for the accident or, if we didn't, we'd find some other reason for the field-joint leaking other than the cold temperatures about which we had been so concerned. There was a reasonable chance that this could be the case, because I couldn't figure why the joint would start leaking so late in the motor operation and not leak at ignition.

Normally benefiting from seven to eight hours of sleep a night, I hadn't had eight in the past three nights combined. Eating a quick breakfast, I went out to the HOSC, where the hallways were buzzing about what was in the

Challenger films. I went into the room for the SRM Failure Analysis Team; very few members were there yet. I went down the hallway, got a cup of coffee, and then went into the "War Room," where the entire Shuttle flight data was recorded and additional data was being faxed from KSC. This room also housed the management team in charge of the failure analysis at MSFC. It took a special code to get into the War Room. The door had a special lock with a keyboard panel on the door for punching in a series of secret code numbers. Not all of the people in the HOSC were given the access code. I went inside, and several members of MSFC management were already there discussing the films. They had requested copies of the films from other camera locations, including those around the launchpad. Unfortunately, the on-pad camera that was in direct line with the suspect area indicating a hot gas leak during the flight was inoperable; it had frozen up prior to liftoff. Murphy's Law strikes again. They were also reviewing some flight trajectory information to correlate with the thrust vector actuation data on the SRB nozzles to see if the steering control was correcting for the side forces induced by the leaking SRB field-joint prior to the explosion. I finished my cup of coffee and returned to the SRM Failure Analysis Team room.

Bob Schwinghammer was listing all the data that might reflect the operation of the SRBs, including thrust vector control data. Several of the team members were being appointed to examine certain data sources that were to be reviewed with the total SRM Failure Analysis Team during the next few days. It was now January 30, two days after the accident, and it was important that all the data be reviewed carefully before any conclusions were drawn. Schwinghammer was very adamant about that, and it was truly the right thing to do; however, members of the press were becoming unbearable, as they were so overly anxious to hear from NASA as to what was the cause of the accident. Every day NASA held a press conference saying it had a huge pile of data to review and that it was going to take some time to sort through it all and identify the cause of the accident. I was personally relieved that NASA had not released to the press that a failure in one of the SRBs was the principal cause of the accident. I wondered how long it was going to keep that fact under wraps with all of the different people who had now seen the films. I had called the plant and asked our engineering people to run an analysis on how large a hole would have had to develop in the casing to match the 20 psia drop in pressure noted in the right-hand SRB prior to the explosion. I also asked our heat-transfer and thermodynamics people to run some calculations to determine how large the plume would be coming out of the aft field-joint at approximately the 31° location: Could it have impacted the struts attaching the SRB to the ET, and would the plume have impinged directly on the ET?

I worked long hours again that day, not getting back to my motel until around 10:00 p.m. I had a message from Bob Lindstrom, ex-manager of the Shuttle Projects Office at MSFC. Stan Reinartz had taken Bob's place in the Shuttle program. Bob had retired from NASA at the end of the summer of 1985 and was being employed by Morton Thiokol as a consultant during the accident investigation. He lived in Huntsville and wanted to come over and talk to me. I told him what I knew. He seemed shocked when I told him about the teleconference the night before the launch. He had apparently not heard anything about it, and he couldn't believe that NASA/MSFC management had reacted the way they did over the original Morton Thiokol recommendation not to fly at any temperature below 53°. Lindstrom personally knew all of the NASA participants in the teleconference very well, and he just couldn't believe what I was telling him. Notifying me that he would be working under a contract with Morton Thiokol, he suggested I converse with him each night to review any new data or findings. He also wanted to involve some of the other people from Thiokol who were supporting the SRM Failure Analysis Team.

What was really ironic was that Thiokol had more people in Huntsville at the time negotiating the Buy III contract than were working on the SRM Failure Analysis Team. I couldn't believe it. Before the *Challenger* accident, the Buy III negotiations were scheduled to take place the current week, and both NASA and Morton Thiokol management decided to go on with the negotiations and fact-finding as scheduled in spite of *Challenger*. I asked the leader of our team: "Why in the hell are we negotiating the production of a solid rocket motor that doesn't work right? We don't know yet what it will take to fix the problem, but it's clear it will take some time, and it will be a much different design than we are currently manufacturing!" I was told they had their instructions, and those instructions were to negotiate the Buy III production proposal that we had submitted to NASA nine months earlier, and that was exactly what they were going to do.

But the very next day we literally found the smoking gun. I had been puzzled from the start over why the joint would start leaking so late in the firing and not at ignition. Well, everyone learned the answer when we finally reviewed a copy of a film that had been taken from a second launchpad camera viewing the area of the right-hand SRB's aft field-joint. During ignition of the motor and liftoff, a large puff of black smoke emanating from the aft field-joint of that SRB was clearly evident. The smoke was being directed upward along the side of the vehicle just like it should if it were coming from the joint, because the clevis part of the joint faced up.

If the gas had come from a burn-through in the case or from a structural failure of the case, it would have come straight out perpendicularly to the case rather than being directed upward. The black puff of smoke was visible on the film from 0.678 to 2.5 seconds after SRB ignition, and then it disappeared. There was no visible gas leak after T+2.5 seconds or any drop in chamber pressure in the motor until T+59 seconds, when a luminous plume was first observed coming from the same area as the black smoke exiting at liftoff. At exactly the same time (T+59 seconds), the vehicle went through some severe vibration, bending, and structural loading due to maximum dynamic pressure and high upper atmospheric wind shears, as shown in the rocket plume. At this time, the chamber pressure in the right-hand SRB also started to drop below the predicted pressure.

It was now evident that the joint failure had been initiated at ignition by a failure of the O-rings to seal off in the aft right-hand solid rocket booster field-joint.

I had already seen another flight film that appeared to show some unusual burning in the area of the orbital maneuvering system (OMS) on the orbiter. These engines were not supposed to be operating until the Shuttle had achieved orbit and, for a brief period, I thought we might have found another contributing factor to the accident; the SRB failure may not have been the sole cause. However, shortly after seeing the black smoke in the launchpad film, it was determined that what appeared to be some kind of malfunction of the OMS engines on the orbiter was really a reflection on the film. It now seemed like an airtight case that a leak in the field-joint of the SRB was the sole cause of the accident, and that NASA was going to have to release that information to the hungry press and news media.

In its wisdom, NASA chose to release to the media that afternoon the information about the hot gas plume seen coming from the SRB, but it said nothing about the puff of smoke at launch. Concerned that reporters would soon find out I was the Director of the Space Shuttle SRM Project at Morton Thiokol, I quickly called my wife to warn her. "If they call," I told her, "just tell them I'm out of town and can't be reached for comment." NASA's press statement emphasized that it was far from completing the investigation, but that the primary cause of the *Challenger* accident did appear to be due to a hot gas leak in one of the field-joints of the SRB. In their normal way of trying to assign blame, the press asked, "Who makes these solid rocket motors, and where are they built?" NASA responded by telling them that Morton Thiokol supplied the SRMs from its plant in Utah.

Shortly after that news was released to the press, the switchboard at our plant in Utah lit up like a Christmas tree. Most of the calls came from reporters trying to get exclusives, but several calls were threatening against Morton Thiokol and those at the company responsible for managing the Space Shuttle program. Some of the callers referred to Thiokol as "murderers" for killing the astronauts aboard *Challenger*. Thiokol took these calls very seriously by putting extra guards around the plant and informing all key Shuttle managers, including myself, that local police would be monitoring and patrolling our homes closely for the next few days. I was told to call my wife and let her know that she might see police cars driving frequently by our house in Pleasant View and not to be alarmed. The company didn't think the threats were real, but extra precautions were being taken just in case.

I called Linda and told her the situation and not to be alarmed if she saw police patrolling the neighborhood. She was home with my three-year-old daughter, Meghan, and our sixteen-year-old daughter, Lora, a junior in high school. Our twenty-two-year-old son, Greg, was taking private pilot lessons and also living at home at the time. Linda was very nervous, so I told her to have Greg stay around home for the next couple days until the reaction to *Challenger* subsided. "When are you coming home, Al?" "I don't know, but I'll call you every day." The next morning a friend at the Utah plant told me that someone had spray-painted the words "Thiokol—Murderers" on the side of one of the overpasses on the main road to the plant. That made everybody very nervous.

The next morning I went into the HOSC again very early. A cup of coffee in hand, I went into the War Room to see if any new data of interest had come in. Walking around the room, I came up behind Dr. Judson Lovingood, Deputy Director of the Space Shuttle Projects Office at MSFC. Lovingood was sitting in front of a fax machine that was transmitting some data from the Ice Team's report about conditions the morning of the launch. Numerous pages of data were flowing out, and Lovingood suddenly picked up a handwritten page containing temperature measurements that had been measured by an infrared pyrometer (IR) gun on the SRBs the fateful morning. On that faxed page were two numbers that had been circled, 7° and 9°.

These were the two temperature measurements that had been made just below the field-joint of the right-hand SRB. Similar temperatures recorded for the left-hand SRB were shown to between 23° and 27°, close to the local ambient temperature at the time. Lovingood picked up the sheet containing this data and, looking over his shoulder, exclaimed to me, "Boy, I'll bet you could really make something out of this!" Looking at it, I answered, "Yes, I think I really could."

Lovingood went over to the copy machine and made me a copy of the data. I was really excited because we'd found the silver bullet to go with the smoking gun—the last piece of the puzzle, the answer to my question, "Why was it that only the aft field-joint on the right-hand SRB leaked hot gas and the other five field-joints didn't?" They should all have been at about the same temperature.

I immediately took the data to Bob Schwinghammer, head of the SRM Failure Analysis Team. "This may well explain why the failure occurred in that particular joint," I explained. "Maybe there was some kind of cryogenic fuel leak from the ET valves or the fuel lines to the SSMEs that had caused the low temperatures in the area of the aft field-joint. That might also explain why the joint had started leaking again some fifty-nine seconds later, if it was being continuously cooled by a cryogenic fuel leak, causing the O-ring finally to become glassy and eventually fail." It was clear we had a lot of work to do, but this gave me some hope that maybe we weren't the entire cause of the accident; maybe we were the cause of the second failure, not the first. Schwinghammer immediately passed on this information to the SSME and ET Failure Analysis teams to see if they could find any unusual data in their area that might indicate a possible fuel leak. Meanwhile, I called back to the plant to our gas dynamics and heat transfer people to try and model how much liquid oxygen or liquid hydrogen would have to leak to destroy the sealing integrity of the O-rings in fifty-nine seconds.

On February 3, nearly a week after the accident, President Reagan announced that he was forming a Presidential Commission to investigate the *Challenger* accident. At about the same time, Schwinghammer, as head of the SRM Failure Analysis Team, announced that we were to prepare some charts for a briefing to Dr. Lucas, Director of the MSFC, on the status of the failure analysis. As a member of Jess Moore's Failure Analysis Team at KSC, Lucas was arriving at the HOSC during the afternoon of February 4, and had requested a briefing from all of the failure analysis teams at MSFC.

Schwinghammer wanted to put together a very thorough briefing covering all of the possible failure modes of the SRBs that could have contributed to the failure and insisted that we not eliminate any possible causes prematurely. The SRM Failure Analysis Team was to list every possible way the SRB could have failed and then find any and all data that might support that failure mode or at least that couldn't be ruled out by that data. Schwinghammer compiled a long list of SRB failure modes, including igniter failures, lightning strikes, various types of nozzle failures, propellant failures, case failures, insulation failures, joints and O-ring failures, and on and on. "We're going to go down the list, one by one," he declared, "and record any SRB or vehicle data that just might support that particular failure mode or that cannot be ruled out."

"I don't think that's necessary," I said, rising to my feet. "We can eliminate 90 percent of that long list of failure modes immediately just from what we have seen in the films. The photographic coverage of the launch is real data, too, and it's probably the best and most important data we have. Based upon the film data, we should focus our investigation in the area of the aft field-joint. Even though it appears from the films that the joint is leaking hot gas past the O-rings, we cannot totally eliminate the possibility of an insulation or structural failure in this area. We should continue to evaluate those possibilities."

Much to my surprise, Schwinghammer said he wasn't going to make an assessment at this time based upon the film data at all, only on the recorded data and other observations. What surprised me even more was when he stated that "most of the people in this room have not seen any of those films. As far as I know, only you and I have seen those films."

"Then everybody in this room needs to see those films right away," I replied, "otherwise, we will be here, pointlessly, all night. I recommend that we approach this problem in a manner just opposite to what you have just suggested. Instead of searching for any piece of data that might support one of the myriad of failure modes you have listed on the blackboard, let's list *any* piece of data that absolutely refutes or provides concrete evidence that the accident could not have been caused by that particular failure mode. That way we'll get to an answer a whole lot sooner. Using the films as the primary source of data, we can eliminate over 90 percent of possible failure modes in two minutes."

Schwinghammer snapped back at me: "I am in charge of this investigation team, not you, McDonald, and if you aren't going to support the team as I direct, you should get the hell out of the room!"

"I'll do just that," I retorted. "I'll take my chair out into the hall where I could do something productive. I expect Dr. Lucas and Jack Lee will want to have a pretty good idea as what really caused the failure, and I don't think your approach to the problem is going to get there in the very short time we have left to prepare." With that, I walked out of the room and took up residency in the hallway.

A little later I walked down to the War Room to make sure I had access to all of the data available. I reviewed everything that had come up recently from KSC and JSC to see if any new data might support (or cast doubt on) what appeared to be an obvious failure of the SRB field-joint to seal at ignition. In particular, I noticed some written information that had come up from our Morton Thiokol people at the Cape who were in charge of mating the SRB segments in the VAB about a previous disassembly of an SRB field-joint that had been exposed to a rainstorm outside of the VAB before being de-mated. The technicians who had de-mated the field-joint noticed that the clevis part of the joint was filled with water. When they pulled the joint apart, water flowed out of the pinholes in the clevis. Since the clevis faced up, it tended to collect water during a rainstorm. The outer leg of the clevis was open to the atmosphere and only sealed off from the weather with a bead of grease, which did not make for a good weather seal.

Our people at the Cape also noted that *Challenger* had sat on the launch-pad during STS-51L through several rainstorms; in fact, over seven inches of rain had fallen at the pad while *Challenger* was there. There was a good chance, if not a sure thing, that the clevis joint had filled with water prior to the freezing temperatures. Even though this wasn't unique to the aft field-joint of the right-hand SRB, it could have contributed to the failure.

I made a copy of this information and went back to my new office in the hallway. I sat in one of those student chairs with the folding arm to write on.

Pondering all the data, I worked to organize it in support of the single failure scenario of a leaking SRB field-joint. Reflecting on the probability that the clevis had been full of rainwater, I thought to myself that if the water froze in the clevis, it would expand and try to push the clevis farther apart, resulting in reduced squeeze on both the primary and secondary O-rings. Furthermore, maybe the expansion from the ice could have forced the secondary seal off its seat and filled the secondary O-ring extrusion gap with ice. This could have further increased the timing function for obtaining a good secondary seal in the joint and made the secondary seal act more like the primary seal, which we knew had been on the wrong side of the seal groove as a result of the leak-check. If all this were true, our rationale that the secondary seal would be in a better position to seal than the primary might have been wrong—maybe it was in a worse position because of the ice in the extrusion gap. I decided to spend the rest of the day piecing this puzzle together, developing some kind of logical format to present all this data in the event that the Schwinghammer team didn't come to the same conclusions.

I spent most of that day summarizing all the data that I had, making sure I couldn't be accused of ignoring something important. It was another long day, and I was getting very tired. Finishing a thorough review, I came to the conclusion that I needed to spend the rest of the time constructing a logical way to interpret and present the data in a manner that would support a single cause of the accident. Wrestling with this problem for a while, I decided to assemble the data into various phases of Shuttle operations: from the on-pad environment prior to launch, prelaunch loading prior to SRB ignition during SSME start-up, SRB ignition and liftoff, SRB flight up to the time of Max Q, data from Max Q to the explosion, and data and observations after the explosion. I wanted to highlight first-time occurrences and any data clearly out of our prior experience base or that indicated a worst-case condition for the vehicle even if it was within specification or design limits. Having outlined this approach, I slept for a few hours, ready to get up early in the morning to prepare charts summarizing the results. Before hitting the pillow, I discussed the new data with Bob Lindstrom. I didn't tell him that Schwinghammer had thrown me out of the SRM Failure Analysis Team room. Lindstrom and Schwinghammer were very good friends, and I may have gotten into a lot of trouble with Lindstrom if he knew what had occurred.

Returning to the HOSC early the next morning (February 4, exactly one week since the accident), again I sat out in the hallway, where I prepared some handwritten charts. In my view, I was constructing a very methodical approach to assessing the failure, one that included examination of all prelaunch, launch, and postlaunch data. Based upon that examination, I felt sure of my conclusion: that the accident was most likely caused by a leaking aft field-joint in the right-hand solid rocket booster. The failure of the joint to seal was most likely the result of the cold ambient temperatures and local cooling of that particular joint from cryogenic fluids from the ET and/or SSME. The joint was most likely filled with water that had frozen, further exacerbating the sealing problem, causing the joint to leak at ignition. The joint resealed itself with molten aluminum oxide in the propellant combustion gas that solidified when it came in contact with the cold steel in the joint and then leaked again at Max Q during a time of high wind shears aloft, causing the vehicle to vibrate and flex. It was the vehicle response to these severe aerodynamic loads that caused sufficient vibration to break loose the fragile ceramic seal made by the aluminum oxide around the eroded portion of the O-ring that initially leaked at liftoff. There was sufficient hot gas flow from the joints to penetrate the ET, which ultimately failed, causing the explosion. A suspected cryogenic leak may have contributed to both the initial leak and the final joint failure, resulting in the ET exploding.

There was *no* data refuting the conclusion that the joint failed because of the adverse weather conditions and the extremely cold temperatures that had been observed in the area of that particular joint the morning of the launch. I made a few copies of my charts and viewgraphs just in case the SRM Failure Analysis Team didn't present a similar story with the same basic conclusion.

I listened with great interest to Bob Schwinghammer's SRM Failure Analysis Team presentation. He started with a series of charts listing all the possible failure modes, with the field-joint failure being way down in the pack. He was just beginning to discuss all the data that could support each of these potential failure modes when one of the members of the Shuttle Failure Team interjected that he didn't understand why Schwinghammer was wasting everyone's time going through all these different modes when the films clearly showed that most of these possibilities were not credible. The films clearly showed hot gas coming from a field-joint in one of the SRBs. "What we want to know," the man said, "is where is the data that explains what may have caused *that!*"

Sitting in the front row, I probably shouldn't have been grinning from ear to ear, as I was, with Schwinghammer trying to explain why he hadn't had enough time to put all of that together yet. "Bear with me," he asked the audience, "and I'll address that particular failure mode in just a minute." But the failure teams were restless and told him to skip most of the material he was going to present and address that specific subject, because it was the primary suspect at this point, and everyone needed to focus their energies on the most likely causes at this time. As for the completeness Schwinghammer wanted, everyone believed they could color in those squares later.

Flustered, Bob skipped to the few charts he had addressing a potential joint leak. His presentation was terribly disappointing to the audience, because it didn't address the body of data for the failure mode that was far and away the prime suspect for the accident.

So many having voiced their disappointment with Schwinghammer's presentation, I stood up and asked if they would allow me to present an SRB field-joint failure scenario that I had worked on for the past couple of days and found very logical. Given the go-ahead, I showed that all of the data, including the films, were entirely consistent with the O-ring failure scenario. I told them that I had not found *any* data that refuted it as the primary failure mode. "I still have some questions relative to the source of some of the contributing factors to the failure of the field-joint," I explained, "but there is no doubt in my mind that this failure was the main cause of the accident."

The reaction of the audience surprised me; they were all basically silent. There were no supporting statements, but also no one who challenged the failure scenario I presented. I left several copies of the presentation with the Accident Investigation Team and left the meeting. As events unfolded later, it was clear that NASA didn't want to believe what I had just told them.

Later the next day Larry Mulloy tracked me down. He wanted to tell me that one of the astronauts assigned to the Accident Investigation Team was going through the files containing documents and presentations on SRB problems, issues, and flight readiness reviews. This unnamed astronaut had found a copy of the August 19, 1985, presentation that I had made to NASA headquarters in Washington, D.C., on "Erosion of the SRM Pressure Seals." The astronaut was shocked by the fact that my presentation on the SRM pressure seals had clearly identified the SRM field-joint as the highest concern—and that the joint opening and O-ring resiliency were reasons for that concern. Mulloy told me that we needed to update my presentation with observations and results from all the Shuttle flights since August 1985, and we also needed to show what progress had been made through the O-ring Seal Task Force to better understand the O-ring problems and provide corrective action. Larry suggested that we "status" the milepost schedule chart that I had included in my presentation summarizing the analysis and testing that had been planned.

Badly in need of a little R&R, I decided this was a good opportunity to get back home. I had been working eighteen to twenty hours per day ever since the accident and was totally exhausted. I made airline reservations for

Salt Lake the next day, February 6, on a flight that left out of Huntsville in the late afternoon.

I ate dinner the night before leaving with Boyd Brinton, who had taken my place on Schwinghammer's SRM Failure Analysis Team. Brinton was the Chief Project Engineer for the SRM Program at Morton Thiokol, and he had been at the HOSC during the launch and during the teleconference the night before the launch. Boyd told me he had talked with several of the engineers at Thiokol who had participated in that teleconference, and that the original decision not to launch below 53° had been the engineering recommendation, but that Thiokol management, not engineering, changed the recommendation during the caucus, after NASA management challenged the original no-launch recommendation. Both Roger Boisjoly and Arnie Thompson, the engineers who had prepared most of the presentation material, initially recommended not to launch and had not changed their minds. They continued to argue for a launch delay, but they were overruled by management. Brinton said that the decision to proceed with the launch was made by our Senior Vice President and General Manager, Jerry Mason, and had been fully supported by Calvin Wiggins, Vice President and General Manager of the Space Division. Bob Lund, Vice President of Engineering, and Joe Kilminster, Vice President of Space Booster Programs, had been reluctant to change the original no-launch recommendation that they had supported, but were persuaded to do so by Mason, who had stated that he accepted Larry Mulloy's rationale that the data was not conclusive enough to impose the 53° launch temperature requirement. Brinton also told me that Roger Boisjoly had told him that Mason asked Bob Lund to take off his engineering hat and put on his management hat: "That it's time to make a management decision." The four vice presidents in the meeting in Utah then unanimously voted to proceed with the launch in spite of the objections from the engineering people.

This was the first time that I was aware that Mason and Wiggins were even involved in that meeting, much less responsible for changing the original no-launch recommendation. I had thought that the final recommendation, just like the original recommendation, was an engineering decision just like I had requested. Brinton also told me that he was very surprised that the NASA Marshall people reacted the way they did to the no-launch recommendation. He was even more surprised that George Hardy reacted the way he did, as it was totally out of character for him. Hardy was normally a very thoughtful, considerate, and rather quiet individual. For him to say he was "appalled" at Morton Thiokol's recommendation took everyone by surprise.

The next morning I returned to the HOSC to pack up my things and check to see if any more data had come in that I should be aware of before heading home to Utah. Walking down the hall, I noticed a number of people in a large conference room watching a television monitor. The NASA Select TV channel was broadcasting live the first meeting of the Presidential Commission on the Space Shuttle *Challenger* Accident. The chairman of the Presidential Commission was William P. Rogers, former Secretary of State for President Nixon. Other members were Neil Armstrong, Vice-Chairman and former astronaut and first man to walk on the Moon; Dr. Richard Feynman, a Nobel Prize winner in physics from California Institute of Technology; General Donald J. Kutyna, Commander of the U.S. Air Force Space Systems Command in Los Angeles; Dr. Sally Ride, NASA astronaut and first American woman in space; Dr. Eugene Covert, head of the Department of Aeronautics and Astronautics at the Massachusetts Institute of Technology (MIT); Robert Hotz, former publisher of *Aviation Week and Space Technology* magazine; Dr. Arthur Walker, a physicist from the University of California at Berkeley; David Acheson, an attorney from Washington, D.C.; Robert W. Rummel, former Vice President of Trans World Airlines; Joseph Sutter, Executive Vice President of Boeing; and Dr. Albert Wheelon, physicist and Executive Vice President of Hughes Aircraft Company. This certainly was a mighty impressive group of people. The commissioners asked NASA to brief them on the entire Space Shuttle system and its operations in order to provide them with the necessary context and background for them successfully to investigate the cause of the accident. NASA sent several of its management people from the various NASA centers to conduct this briefing.

I walked into the back of the conference room and stood up against the wall to see what was going on. On the TV, Dr. Judson Lovingood, Deputy Director of the Space Shuttle Projects Office at MSFC, was briefing the Commission on Marshall's responsibility in the Space Shuttle program— that being all the propulsion elements, notably the SRBs, ET, and Space Shuttle main engines. I was anxious to hear what Lovingood had to say, and was surprised how many questions he was asked about the SRBs. I was even more surprised by how many times he didn't know the answer—or at least didn't give them one. At the close of his part of the briefing on the booster, Lovingood offered: "There was a question earlier about . . . a concern by Thiokol on low temperatures. We did have a meeting with Thiokol. We had a telecom and discussion with people in Huntsville, people at the Wasatch Division, and people at KSC. The discussion centered on the integrity of the O-rings under lower temperature. We had the project managers from both Marshall and Thiokol in the discussion. We had the chief engineers from both places in the discussion. And Thiokol recommended proceeding on with the launch. So they did recommend a launch."

"When was that meeting?" Chairman Rogers asked. "That was the 27th," replied Lovingood. It started around quarter to 5 p.m. Central Time." Lovingood then quickly asked, "Is there anything else on the booster?" With no one responding, Chairman Rogers uttered, "I guess not." Lovingood then proceeded to talk about the external tank.

"Boy, did he just dodge a bullet!" I thought to myself, standing against the wall in the back of the conference room. "Well, this is obviously a crash course on the Space Shuttle system for the Presidential Commission, and I'm sure NASA will go into more detail about the prelaunch activities at some later date." I watched the briefing for a while longer, picked up my overstuffed briefcase, and headed out the door to catch my plane. I had been away for two weeks, and I felt like a college kid going home for Christmas right after final exams. I was so exhausted that I slept all the way home, when I usually cannot sleep a wink on airplanes.

After welcoming kisses and hugs with my family, Linda said, "You need to get some rest, and you are not going into work tomorrow, are you?" She knew I had been working eighteen to twenty hours a day and that ten to twelve of those hours every single day for the past nine days since the accident (including an additional eighteen to twenty hours on both Saturdays and Sundays) were noncompensated overtime. I had totally forgotten about my son Greg's birthday on February 3 and my wife's on February 5. I apologized for forgetting their birthdays, but told her that I needed to go in to work tomorrow, to update the presentation that I had made in August for Larry Mulloy, but that I would take the weekend off to catch up on my sleep (on Saturday) and go skiing with my son Greg and daughter Lora (on Sunday). "You're crazy, Al," Linda declared. "You should take a day of management-granted-time-off for all of the overtime you're donating to the company. You better get some rest before you end up sick in bed. You need to let someone else out there take over for a while. As far as I'm concerned, you've already done more than your share." "Linda, it's real important for the company, and for me, to do everything I am doing. I don't know what's going to happen. They just might decide to stop flying the Space Shuttle altogether. Such a decision would have a major impact on the company and on me, personally." I really shouldn't have told her this, because now she really had something to worry about.

Entering my office the next morning, I noticed a print sitting on my desk taken from the *Orange County (Calif.) Register* depicting the *Challenger* accident; it was called "The Final Frontier" and showed the *Challenger* astronauts climbing the fingers of God. Sent to me by a friend in California, I liked the sketch so well that I asked one of the fellows in our publications department to have it enlarged and framed for me, so I could hang it on the wall of my office. He brought it back framed with photos of the *Challenger* astronauts superimposed on the picture along with my gold Space Shuttle–embossed business card.

That day I received a phone call from the American Institute of Aeronautics and Astronautics relative to the nomination of Thiokol's Ed Dorsey for the AIAA's Wyld Propulsion Award, the highest award given by the AIAA for rocket propulsion. As chairman of the AIAA Solid Rocket Technical Committee, I had supported Dorsey's winning the award because of his leadership in the successful design, development, and test and evaluation program for the solid rocket motors for the Space Shuttle. In light of the evidence that the *Challenger* accident was caused by a failure in one of these solid rocket motors, the AIAA was now concerned about giving this award to Dorsey. Instead, AIAA suggested to me that we offer the award to Dominick Sanchini, of Rocketdyne, for his contributions to the successful development of the Space Shuttle main engines. I agreed that it would probably not be appropriate to give the award to Dorsey at this time, and that I was OK with giving the award to Sanchini at the AIAA Joint Propulsion Conference, to be held in the summer of 1986.

I spent the rest of the day updating the August 1985 presentation on SRM pressure seals that I had given to NASA headquarters, using all of the new information and knowledge that we had accumulated since last summer. I updated the presentation by inserting a blue colored sheet behind each page of the presentation that was updated and with a dash "a" ("-a") placed as the page number for reference. I did this with all of the pages having data as well as with all the "Conclusions" and "Recommendations" charts at the end of the presentation.

One of the items listed on my general "Conclusions" chart stated: "The primary O-ring in the field-joint should not erode through, but if it leaks due to erosion or lack of seating, the secondary seal may not seal the motor."

To that statement I now added the following statement in bold print: "Data obtained on resiliency of the O-rings indicate that lower temperatures aggravate this problem."

Another one of my August 1985 statements, this one on the "Recommendations" chart, had read: "Analysis of existing data indicates that it is safe to continue flying existing design as long as all joints are leak-checked with a 200 psia stabilization pressure, are free of contamination in the seal areas, and meet O-ring squeeze requirements." To this statement, I now added in bold print: "and environmental conditions stay within our acceptable experience database."

Everyone at our plant was demoralized by the bad press that the company and its employees were receiving as a result of a leak in the SRBs being identified as the cause of the *Challenger* accident. A good many employees in the operations area voluntarily reduced their workweek and pay period to thirty-two hours in order to reduce operating costs during the accident investigation. Other areas in engineering and some in the quality control area worked large quantities of overtime in supporting the failure investigation. All of the prior building records for the hardware flown on STS-51L were carefully reviewed to determine if any anomalies or inconsistencies in the records could contribute to the understanding of what had happened. All the prior engineering analysis, presentations, reports, and documents that said it was safe to fly were cataloged and reviewed. We knew that this failure was going to require a total reassessment of the SRM design. I went home the night after my first day back at the plant more depressed than ever.

I did just what I had planned to do. I updated the August presentation on Friday at the plant, slept most of the next day, and went skiing with my son and daughter on Sunday. It felt great standing at the top of the mountain at Snowbasin, which was less than twenty miles from my home in Pleasant View, just north of Ogden. I was finally able to get my mind off *Challenger* and just enjoy the great outdoors. I had a great time skiing with my family, but the feeling was short-lived. Getting home, my wife greeted me at the door with word that I was supposed to call my boss immediately and be prepared to go right to Washington, that night, on the company jet. I called my boss, Joe Kilminster, who told me Larry Mulloy had called him and that we were being asked to support an "emergency" meeting of the Presidential Commission the next day. I needed to bring a copy of the updated version of the August 1985 SRM Pressure Seals presentation that he had asked me to prepare.

On Sunday, February 9, 1986, the *New York Times* published an article concerning some NASA correspondence on the problems with the O-rings in the SRM. The newspaper had received some internal NASA headquarters memoranda, written by an employee by the name of Richard Cook, which identified the O-rings as a serious problem needing to be fixed, but that had not been funded by NASA. Richard Cook was a budget analyst at NASA headquarters, and he had talked to several engineers about the problem and written several memos concerning how serious the problem was, but it wasn't being funded because it would take too long and too much money to fix it.

When this hit the press, the Presidential Commission received numerous telephone calls from the media asking what it was all about. NASA had told the Commission nothing about these memos, and the panel was embarrassed that it didn't know anything about them. Rogers said his commissioners would get to the bottom of it, and asked NASA to brief them on what they had known about problems with the O-rings before the accident—and specifically respond to the information and allegations made in Cook's memos, memos that were in some way supplied to the *New York Times* but not to the Presidential Commission.

The meeting with the Presidential Commission was scheduled for Monday afternoon, February 10, so NASA wanted Morton Thiokol to help it prepare a presentation at the NASA headquarters building at 7:30 a.m. that morning. That meant that I couldn't leave Utah any later than 11:00 p.m. MST to arrive at a meeting at NASA by 7:30 a.m. EST. We left from Ogden at 9:00 p.m. our time, flew all the way to Washington, and checked into a motel around 4:00 a.m. local time for a few hours' sleep before going over to NASA. I was sure glad I had slept most of the previous day, because here I was back on a twenty-hour-a-day schedule.

Inside NASA headquarters, it was mass confusion. No one seemed to know what they were doing and seemed to be more preoccupied by how such information was being leaked to the *New York Times* than the gravity of the issue. Larry Mulloy immediately asked me for a copy of my "updated" August 1985 presentation. He had also requested that I bring two other people with me to answer some of the anticipated questions relative to the O-ring problems. One was Dr. Mark Salita, a senior scientist at Morton Thiokol in the Gas Dynamics and Heat Transfer Section. Dr. Salita had developed the computer program that we were using to predict the effects of hot gas impingement on the O-ring seals. The other fellow from Morton

Thiokol was Don Ketner, Salita's boss and Supervisor of the Gas Dynamics and Heat Transfer Section. Ketner was also the head of our O-ring Seal Task Force, and he had been acting in that capacity since last August.

Mulloy quickly thumbed through the updated briefing I had given him and pulled out several pages to add to the presentation material he had already selected for himself. Nowhere in NASA was there anything but real confusion and lack of direction as to what material should be presented to the Presidential Commission, who should present the material, and just how the meeting was going to be conducted. NASA officials knew that the agency had been placed into a very embarrassing position by the *New York Times* article. Their strategy appeared to be more focused on trying to discredit the source of the information, Richard Cook, than on being up front and open with all of the data and information that they had available to them on this issue.

Eventually, NASA decided that it should address the *New York Times* article (and its source of information) first. The majority of the presentation to the Presidential Commission would be in the hands of Mulloy. Larry would discuss what was known about the O-ring seal problems, the technical issues being addressed, and what NASA and Thiokol had been planning to do to resolve these problems. Mulloy would answer all the questions but could ask for help from specific others in the room, if necessary.

After a quick lunch, our threesome from Thiokol headed over to the Old Executive Office Building next to the White House, where the first closed session of the Presidential Commission on the *Challenger* accident was to be held.

Photos taken the morning of the *Challenger* launch showed large icicles on the Shuttle vehicle and launch support structure. (*Report of the Presidential Commission on the Space Shuttle Challenger Accident*, Vol. 1, June 6, 1986.)

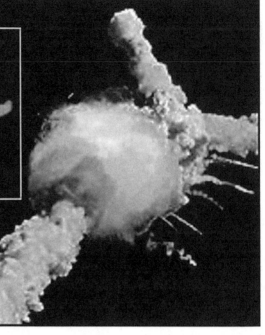

At T+76 seconds, the two solid rocket boosters exited the explosion of the *Challenger*. (*Report of the Presidential Commission on the Space Shuttle Challenger Accident*, Vol. 1, June 6, 1986.)

**The Explosion
(T Plus 76 Seconds)**

7 Minutes Later

Seven minutes after the launch, people could see from the highly distorted rocket plume that there were some very high wind shears aloft (Courtesy of NASA.)

MAXIMUM DYNAMIC PRESSURE AND WIND SHEAR

**Joint Leak Restarts
T Plus 59 Seconds**

**Joint Plume Visible
T Plus 60 Seconds**

By T+59 seconds, the leak in the aft field-joint of the right-hand solid rocket booster was clearly apparent; a second later, there appeared a fully developed plume.(*Report of the Presidential Commission on the Space Shuttle Challenger Accident*, Vol. 1, June 6, 1986.)

LIFTOFF

**Leak at
T Plus 0.678 Seconds**

At T+0.678 second, momentarily after liftoff, the first evidence of a leak surfaced, when a puff of smoke traveled up the side of the booster from the right-hand solid rocket motor aft field-joint. (*Report of the Presidential Commission on the Space Shuttle Challenger Accident*, Vol. 1, June 6, 1986.)

PART III

Search for the Truth

All truth passes through three stages. First, it is ridiculed. Second, it is violently opposed. Third, it is accepted as being self-evident.

—*Arthur Schopenhauer*

If what we see is doubtful, how can we believe what is spoken behind the back?

—*Chinese Proverb*

A word once let out of the cage cannot be whistled back.

—*Horace*

12

Shocking the
Presidential Commission

I was very nervous going into this closed meeting and sensed that NASA was nervous, too. In fact, I wasn't even sure I was supposed to be in the meeting room at all.

We were all standing in the hall as the members of the Presidential Commission entered the meeting room. NASA then instructed its people on exactly where they were to sit in the room yet didn't say anything to Dr. Mark Salita, Don Ketner, or myself as we were about to walk into the room. I turned to Don and Mark and told them to just go in and sit down in the back of the room, which we did.

Chairman William Rogers called the meeting to order and introduced General Chuck Yeager as a member of the Presidential Commission who had been unable to attend earlier meetings because he was off breaking another flight record. Yeager then informed the Commission that he had prior commitments and would not be able to attend any future meetings, either, at least not until sometime in March.

Chairman Rogers then stated, "The *New York Times* and other newspaper articles have created an unpleasant and unfortunate situation. We would hope that NASA and NASA officials will volunteer any information in a frank and forthright manner. This is not an adversarial procedure."

Dr. William Graham, acting NASA Administrator, responded that NASA would cooperate fully with the Commission, but added what I thought was a very odd comment, one that could have led someone to believe that maybe everything that NASA had presented to the Commission or, for that matter, would present in the future, might not always be correct:

"Should any error, partial, incomplete, or potentially misleading statements be found, an amendment to the testimony should be filed in order to clarify the issue of concern. We realize that it is possible for NASA to occasionally misspeak or delete something inadvertently."

I found these comments by the acting head of NASA more than peculiar. They seemed to me to be apologizing in advance for untrue or misleading testimony coming from NASA.

It was readily apparent to everyone there that Commission members were deeply upset by the fact that reports of O-ring failure as the suspected primary cause of the *Challenger* accident had appeared in the press before NASA had shared the first thing about the subject with the Commission. To read about it first in the *New York Times* was more than embarrassing. It made them downright angry with NASA for not being more open with its information.

By now, it was abundantly clear to me that NASA, for some time, had been trying to keep as much information under wraps as possible. Starting to feel the same way, William Rogers severely scolded NASA for not being more open with the Commission about these matters. The chairman acknowledged that it was going to be difficult for his investigation to keep totally ahead of the press, but his panel needed to know about a problem or potential problem as soon as possible from NASA, so it could at least demonstrate an awareness of all the critical issues and tell the press it was looking into them.

Rogers declared that it was his intention, throughout the investigation, to conduct all closed meetings with sworn testimony and that, eventually, all of this information would be released to the public. Furthermore, many of the future meetings of the Commission would be open to the news media and the public. In fact, he was planning the first open meeting for tomorrow, Tuesday, February 11, at which time the SRB O-ring problems addressed in the *New York Times* would be discussed.

I was impressed by his remarks. It was very clear to me that Chairman Rogers was admonishing everyone to tell everything they knew, because it would eventually come out.

After those opening remarks, the meeting started with a discussion by NASA headquarters personnel on why Richard Cook's memos, the primary source for the *New York Times*'s story on the O-ring problem, were not accurate. The immediate strategy of the NASA spokesmen was to cast doubts on the credibility of the information in those memos by labeling Cook as a man with no technical training.

After minimizing Cook's credentials, responsibilities within the agency, and source of the information in his memos, Larry Mulloy offered to explain what was known about the technical problems associated with the O-rings.

With the Commission's consent, he discussed how the O-rings operated and reviewed what was known about problems involving the SRB field-joint opening during pressurization.

Summing up, Mulloy stated that the current design of the solid rocket booster's field-joint was "an acceptable situation" and that he had "no data today to change that," a statement that totally flabbergasted Chairman Rogers.

"Not even today!?" Rogers shot back.

Mulloy held his ground: "No sir, not even today."

Bill Rogers couldn't believe what he had just heard and quickly responded, "Could I go back to your last answer? Are you suggesting that you never came to the conclusion that these things did not cause the accident?"

"Sir, I'm not aware of anything that has caused the accident yet."

What a crock! I couldn't believe what I had just heard, because Larry Mulloy had been exposed to all of the data that I was, including the presentation that I had given to the MSFC Accident Investigation Team just a week ago, which clearly showed what caused the accident. At that point, I began to fathom why Dr. Graham said such odd things in his opening statement relative to the fact that the Commission might hear some "misleading testimony" that may need correcting.

It was amazing to me that NASA was still not being up front and open with the Commission even after Chairman Rogers's strong opening statements on this very subject. It was quite clear to me that NASA was intentionally withholding information from the Commission.

Mulloy then started through a large stack of viewgraphs. He talked for a considerable length of time before the commissioners asked for a short break. When they returned, Dr. Sally Ride had a stack of pink telephone slips in her hand. Speaking first, she commented that during the break she had received a telephone call from a reporter from the *Washington Times* wanting to verify some information he had just received. The reporter had asked her if the Presidential Commission was aware that one of the Shuttle contractors had told NASA the night before the launch that it was concerned about the cold temperatures.

So Dr. Ride turned to Mulloy and asked him, "Was there any internal correspondence related to this potential concern on the operation of the O-ring or joint?" "There was no correspondence," Mulloy answered, "but there was concern initially for the cold temperatures relative to the SRB recovery batteries and later relative to the O-rings." Larry then referred to the first hearing of the Presidential Commission, conducted on February 6, when, concerning the evening teleconference of January 27, "Dr. Lovingood had mentioned that."

Larry mentioned the names of many of the people who had been involved in the teleconference and closed with a statement, "Thiokol presented to us the fact that the lowest temperature that we had flown an O-ring or a case joint was 53°, and they wanted to point out that we would be outside of that experience base."

Chairman Rogers asked Mulloy, "Who did that for Thiokol?"

"It was the Director of Engineering for Thiokol, a gentleman named Bob Lund."

Quickly Mulloy added the names of those present in Utah during the teleconference, notably Joe Kilminster, the Vice President of our Space Booster Program and our Project Manager. Larry concluded by saying, "After hearing the discussion, *we all* concluded that there was no problem with the predicted temperatures for the SRM, and I received a document from the solid rocket motor project management at Thiokol to that effect, that there was no adverse consequence expected due to the temperatures on the night of the 27th."

Sally Ride then asked Mulloy: "I guess maybe what I'm asking is, we read in the *New York Times* about NASA internal memos where people within NASA were suggesting problems with the erosion before, and I guess I am wondering whether similar memos exist relating to problems of launching with the O-rings at low temperatures."

"I'm not aware of any such documents at Marshall," Mulloy responded, "but that isn't to say there aren't any. I need to go research my files to make sure."

It was becoming more and more apparent to me that NASA management had no intention of informing the Presidential Commission about Morton Thiokol's initial recommendation not to launch the *Challenger* and that, if at all possible, they were going to *cover this matter up*. Mulloy's testimony clearly indicated that a party line had been established on the subject and that Mulloy and other NASA officials were never to say anything other than the fact that Morton Thiokol recommended to proceed with a launch.

Even though what Mulloy said was literally true, his comments were misleading at best and, at worst, a flat-out lie.

When Mulloy was asked whether NASA may have some documents in its files relating to problems of launching with the O-rings at low temperatures, he and I both knew that was indeed the case. The original Thiokol recommendation not to launch because of cold temperatures was in a document in NASA's files. I was flabbergasted by his comment, "I am not aware of any such documents."

Every one of the hearings of the Presidential Commission started with the swearing in of the participants on the basis that they were supplying information and answering questions under oath. Even though this was a fact-finding Commission and not a jury sitting in judgment of those participating, individuals could be accused of perjury after the hearings were over. This was highly significant because there was a high probability that the *Challenger* accident and the loss of seven American heroes' lives were likely to result in some litigation against both NASA and Morton Thiokol after the hearings were completed.

I thought that the truth about the decision-making process leading to launch must also be exposed to the Commission, so that the process, too, like the poor joint design, would be fixed before another Shuttle flew.

Listening to Mulloy, I came to a resolute personal decision. Somehow, I had to tell the Presidential Commission the whole truth about the meeting the night before the launch and the fact that Morton Thiokol did initially recommend against launching *Challenger*. Thiokol had been *very concerned* about the effects that cold temperatures might have on the O-rings in spite of what Mulloy had just said. Furthermore, our recommendation was given to NASA in writing, and I just happened to have a copy of that document with me.

I raised my hand to make a comment, but no one acknowledged me because I was sitting in the back of the audience where all the "nonparticipants" were sitting. Chairman Rogers asked Jess Moore about a comment that the Commission had heard about, not about Thiokol, but relative to a telephone call from Rockwell expressing concern about the icing conditions just twenty minutes before the launch. Moore told Rogers that Arnie Aldrich did have a meeting before launch on that subject where the Rockwell representative did express some concern, but he wasn't aware of any phone call. Rogers then asked whether someone from Rockwell would be present tomorrow at the open hearing, and Moore said he would get somebody from Rockwell in tonight.

Rogers then made a very cogent point: "I'm not sure it is essential, but if Rockwell was the one that raised the concern, then we want somebody from Rockwell to say, 'I raised a concern, we talked it over, and our concern was satisfied and we said go-ahead.' As long as we still have that concern on the part of Rockwell, if you testified or someone testifies from NASA that there was the meeting and everybody was reasonably satisfied then someone from Rockwell comes along and says that's not so, 'we told you *not* to go ahead and you went ahead anyway,' that is the kind of thing we want to try to deal with at these meetings."

Of course, as far as I was concerned, that was extremely close to our case with the O-ring situation—not exactly, but very close. I put my hand down because Larry Mulloy stepped back on the floor. I thought, "Finally, he's going to tell the Presidential Commission the details of exactly what happened on the night of the 27th."

But astonishingly, Mulloy went right back to the viewgraph machine and his canned presentation. After responding to a few questions on the data he had presented earlier, Larry very casually continued on with the remainder of his viewgraphs, stopping for questions from the Commission members. Again I raised my hand, but no one acknowledged me. I then stood up to try to get someone's attention, but I was ignored. Mulloy kept right on going.

I stood there in frustration waving my hand for what seemed like several minutes. When I started walking toward the floor, Mulloy finally acknowledged me. As I had originally prepared many of the charts that he was using, I think he thought that I wanted to comment on some of the data that he was presenting, which was generating several questions.

Larry said, "Mr. Chairman, Allan McDonald from Morton Thiokol wants to make a point."

"I wanted to make a point about the meeting," I said. "That meeting was set up at the Cape, and we tied Marshall in and Thiokol back in Utah about the concerns of the lower temperatures. The meeting was set up to send material on the fax so that people could review data and concerns and our basis for the concerns. That data was transmitted to Kennedy and also to Marshall from our office at Thiokol and was presented by our Vice President of Engineering. I asked him to give that briefing. He did that.

"The recommendation at that time from the data that was sent out from Thiokol was *not* to launch below 53° Fahrenheit because that was our lowest acceptable experience base and had demonstrated some blowby from a year ago. We also had some data that indicated the poor resiliency of response of the Viton seal to low temperatures.

"That was the first transmittal of information, and you should be aware of that and where the data was discussed."

"Who in the hell are you?" Chairman Rogers said to me.

"I'm the Director of the Solid Rocket Motor Project at Morton Thiokol."

"Would you please come down here and repeat what you've just said," Rogers then declared, "because if I just heard what I think I heard, then this may be in litigation for years to come."

I didn't realize it at the time, but those statements would have a major impact on me for the rest of my life. The members of the Presidential Commission couldn't believe what they had just heard; they were shocked.

13

Cover-up

Chairman Rogers had chastised NASA officials at the beginning of the meeting for not being up front with the Presidential Commission relative to controversial documents like those printed in the *New York Times*, and he had just told them to make sure that someone from Rockwell didn't object to the launch. Sally Ride had just asked Larry Mulloy if he was aware of any documents that related to any problems of launching with the O-rings at low temperatures, and Larry had told her he wasn't aware of any.

Following my statement, Commissioner Donald Kutyna, a USAF general, immediately fired a question in my direction: "You said not to launch below 53°, and what was the actual temperature?" I wasn't able to respond fast enough because Larry Mulloy instantly replied, "The actual temperature predicted at that time, based upon Thiokol's calculations, was 29°."

I figured that Mulloy was now going to answer all the questions and finally provide all the details relative to the Morton Thiokol–NASA meeting the night before the launch, so I sat down on a bench. No sooner had I sat down than Chairman Rogers, with a very puzzled look on his face, set his gaze squarely back on me and said, "Could you stand up again and say that a lot louder so we could all hear it?" "What I said," came my response, "was when the concern of the predicted cold temperatures at the Cape was transmitted to our plant, our engineers were asked to examine that and see if there was any concern about launching the SRM or any concern with any component. The people who were working the O-ring seal problem were concerned. They'd called me at the Cape and said, 'We've looked at the predicted temperatures, and it looks like the O-rings are going to be very cold. We've run some tests in the past few months that show the resiliency of the O-ring to be very sluggish at low temperatures. The O-rings become very hard, and we would like to review that information.' I answered back to them, 'Yes, I'm concerned about that also. I'll set up a meeting to review the information and recommend whether we want to launch or what temperature we're willing to launch.'

"I called Mr. Mulloy at the Merritt Island Holiday Inn but didn't reach him; I had the wrong number or something. I then called Mr. Cecil Houston, Resident Manager of MSFC at Kennedy, and told him the concern. He said, 'Fine, I'll get everyone on the network. I have a four-wire system right off my office in a conference room and we'll get all the proper parties involved. You tell your plant to be available at 8:15 p.m. and transmit the charts and the data that you have both to Kennedy and to Marshall, and I'll contact Marshall to have the program people there.'

"Cecil Houston set that meeting up, and I went on over to KSC. There was a group of people at Marshall and a group of people at Thiokol. The material was a little bit late getting there. We waited for about a half-hour and then finally the material was transmitted from Thiokol. In that material was the data that we had on our erosion history and the fact that, a year ago, we did see some blowby of the primary seals of the case joint. That was the lowest temperature we had calculated the O-rings to be in the flight vehicle, which caused us some concern. We also had run some resiliency tests on the O-rings where essentially they were squeezed in between two plates that were removed very rapidly, and they would not respond. We were concerned about what impact that might have for lower temperatures. As a result of that, our recommendation at the time was to not launch below 53°, because we didn't know how much farther below that we could go and be in the acceptable range."

General Kutyna then asked me, "And on this launch it was 29°?" Mulloy interjected, "No. That was the single dimension analysis that Thiokol had run during the discussion we had on the 27th. Since then we've run a multinode thermal model and I believe it is 25°."

Chairman Rogers interrupted, "Before we come to that, I'm not sure I understand." Turning to me directly again, he said, "Am I hearing you say that you recommended against launch and you never changed your mind?" "No, I did not say that," came my response. "We did change our mind afterwards."

Rogers asked, "What brought you to that decision?"

"Well, NASA concluded that the temperature data we had presented was inconclusive and, indeed, a lot of the data *was* inconclusive because the next worst blowby we had ever seen in a primary seal in a case-to-case field-joint was about the highest temperature we had launched at. And that was true—the next worst blowby."

Sally Ride wanted to know "Which one was that?"

"I can't remember exactly. I have it in my notes."

"I have it here," Mulloy offered.

"We didn't calculate the effects of all that from the data that we had," I continued, "but we did have some data that indicated that the timing function of the O-ring seal was going in the wrong direction, in the direction of badness. The O-ring was getting harder. The grease in there was getting more viscous. The time to seat the O-ring took longer, and it would be more difficult to extrude because of the hardened O-ring. We didn't know exactly where the right temperatures were that would make it so it could not seal, but it was in the wrong direction and the temperatures that were being reported for the 51-L were so far away from our experience base that we didn't feel comfortable operating that far away."

Giving up searching through his paperwork for the temperature blowby data, Mulloy interjected, "I don't remember which one it is. We could get it, but on one of those it was 75°."

"That was 22-A," I said. "If you'll look at the double asterisk [on the chart Mulloy was showing] we saw a certain amount of soot behind the primary O-ring even though we didn't see any erosion. Apparently some gas got past the primary O-ring between the two O-ring seals."

"I still don't understand your explanation," Rogers lamented. "Did you change your mind?"

"Yes. The assessment was that the data was not totally conclusive, that temperature could affect everything relative to the seal. But there was data that indicated that there were things going in the wrong direction and this was far from our experience base. NASA was asking us for a reassessment and for some more data to show that the temperature in itself can cause this to be such a serious concern that we had said it would be."

"At that time," I continued, "Thiokol in Utah said that they would like to go off-line and caucus for about five minutes and reassess what data they had there or any other additional data. That caucus lasted for, I think, a half-hour. When they came back on they said they'd reassessed all the data and had come to the conclusion that the temperature influence, based on the data they had available to them, was inconclusive; they, therefore, recommended a launch."

Chairman Rogers then queried, "When you say it's inconclusive, what does that mean? You told them the day before *not* to do it, and now you got some more data and you say it's inconclusive and so you changed your mind?!"

"I was not at Wasatch in that discussion. I was at Kennedy and do not know what other data they were looking at other than the charts that I had in front of me and that others had in front of them at both KSC and Marshall. I *do not* know. I *do* know that when they came back on they said they had reassessed and concluded it was OK to launch. Then Thiokol was requested to put it in writing."

Understandably, Rogers wanted greater clarity: "I think, in view of the very serious nature of this and the fact that it will be scrutinized for years, that we should have precisely what the data was before we present it."

"I have that in my notes, sir."

"But you're just conveying information that pertains to the decision somebody else made."

"I have the faxes that were distributed at both of those meetings—all of the charts from the original meeting and from the one afterwards."

"Who made the decision from Thiokol?" Rogers wanted to know.

"I do not know who made the final decision. I do know that the fax was signed by Joe Kilminster, my boss, our Vice President."

A question came from Commissioner Arthur Walker, a professor of applied physics at Stanford University: "So there's no evidence that the evidence was actually looked at in this caucus that persuaded Thiokol that your first view was incorrect, that perhaps the first view wasn't based upon solid evidence?"

"I cannot say specifically because I was not there to see it. I think you need to get the people that were at that meeting to discuss that."

Rogers was clearly disturbed by what he was hearing: "I'm sure you can see the logic of what you're saying. You recommended against a flight on one night and then have meetings with NASA people and they seem anxious to go ahead, or at least they were asking questions about it, and they gave you some data and you checked back to your home office and you got word back from home office to go-ahead because the evidence is inconclusive!"

"He said something about *having* something," Commission member Richard Feynman noted. "He does have some data there from the people who made the decision, I guess," Rogers allowed.

"I have some of that, but I also have the material that was reviewed at the meeting by all three parties, because I *was* party to that."

"Which meeting now are we talking about?" Rogers asked.

"This was the first meeting scheduled to review why we had any concerns about low temperature, and I have that as a matter of record."

Rogers had every reason to demand precise information: "Let's be sure that we make it clear if it's one meeting, two meetings, who was there, when it was held, because otherwise it gets all blurred in the minds of the listener, and I gather you had several meetings on the subject—two meetings, three meetings."

"Well, there were probably three meetings, yes."

"Well, make that clear or have your people make that clear," Chairman Rogers ordered.

General Kutyna then asked me, "In your opinion, what is the greatest indication of the problem—the amount of erosion or the fact that you have some soot, i.e., something that has a lot of erosion like this? Is that more of a problem than something that has a little soot and no erosion?"

"I think soot was more of a problem than the erosion. That shows that you had violated the primary seal in some way, to get gas between the two seals. Erosion itself, as long as you don't violate the seal and it still has integrity, is not a problem."

Rogers cut off the conversation and asked for a short break. During the recess, I overheard the only comment made by Chuck Yeager during the entire affair, probably because it was the rare occasion the general attended a meeting. A few Commission members were already commenting on the need to redesign the SRB joint, to prevent this from ever happening in the future, and Yeager declared, "Hell, give me a warm day and I'll go fly that son of a bitch!"

Apparently General Yeager didn't think it was necessary to stand down the Space Shuttle program until a totally new SRM design was qualified to fly. He was absolutely right. We should have entered the bore of the boosters at KSC after stacking and filled the holes in the vacuum putty in the joints just like we did on our static test motors, and put on electrical tape heaters on the field-joints and kept flying the Shuttle while the redesign program was being conducted. We should have limited the Shuttle flights to military crews for high-priority classified DOD missions only during this time frame. After the accident, we successfully static tested eleven SRMs (no failures) of the *Challenger* design that had completed fabrication before the accident; we tested these motors with the putty packing and joint heaters. Some of these motors were more than seven years old by the time they were tested. One of these motors (Technical Evaluation Motor-10) that was successfully static tested in January 1993 contained the damaged center segment *rejected* by NASA for use on STS-51L *Challenger*. Ironically, Thiokol recommended using that segment on *Challenger* and NASA said no, but when Thiokol said "don't launch" because of cold temperatures, NASA was appalled with that recommendation!

Recess over, the Rogers Commission discussed how it would handle the next morning's public hearing, its very first. Larry Mulloy wanted to get to the last three charts of his presentation, however, which explained what NASA had been doing to better understand the O-ring problem. His charts had been taken directly from the August 19, 1985, presentation that I myself had given to NASA headquarters.

"I think we can wind up with this one," Mulloy suggested. "These are general conclusions about what happens with O-ring erosion. That data was good back on August 19th, and it is still good today."

Again, I was shocked. I had given Mulloy, at his request, an "updated" presentation that modified these conclusions, and he was not using or even referring to it! Directly behind the critical page of the August 19 presentation I had placed a blue page with a dash a (-a) label on it; it contained the new information in bold print showing how the presentation would change based upon our newest information. But here was Mulloy doing everything he could to cover up the real situation.

I couldn't help but become very suspicious, not just of Mulloy, but of all NASA for its lack of candidness.

Near the end of the session, Sally Ride asked if there were any documents that were given to the Commission that had not been made public in the *New York Times*. NASA's Jess Moore responded, "Yes, there was a document by Russ Bardos to Mr. Winterhalter, Director of Propulsion at NASA headquarters, suggesting that the SRM system should continue to fly and that we should not make any quick fixes to the SRM in preparation for the QM-5 test." Ride wondered whether all the documents should be given to the press, but Phil Culbertson, NASA Associate Administrator, felt uncomfortable with that, "because there are probably some cases where everyone but one person said yes and that person may have been right, and that needs evaluation."

Chairman Rogers quickly interceded: "Anybody that has lived in government in Washington knows all of those things come out, whether it is the president of the United States or anyone else." It was clearly evident that Rogers was irritated with NASA management about their lack of forthrightness and reluctance to be up front with all the information, so his tone grew even stronger: "Let me tell you—and I have already told some of the members of the Commission—I have two letters from Congress, one from the Senate, one from the House, saying they are not going to do anything while we are going through this process. But as soon as it is over, they are going to analyze the report we make, and they are going to have an overview with us. At that time, we will be in the same position you fellows are in now. We will be spending three or four or five days up there and trying to say why we said what we said in the report, and that is just the way it operates."

Phil Culbertson continued to fret: "You must expect that you probably haven't seen the last of these memos where somebody says, 'Hey, don't do it.' You'll probably come across many. We are searching our files to find those things."

"Well, that's fine," Rogers stated.

If Rogers had only known with crystal clarity how dishonest NASA was being about those "don't do it" memos, he would have been much more distressed than he was already becoming. The most important of those memos was the original decision by Thiokol not to launch *Challenger*! It should have been on the top of the pile! Yet it wasn't anywhere to be found, and no one from NASA dared to even mention it! If I hadn't done so, I don't believe NASA or Morton Thiokol management would have revealed the information. It would have eventually leaked out, causing great embarrassment to everyone.

With the meeting breaking up, I asked Chairman Rogers if he wanted me to have the people at Thiokol who were involved in changing the launch decision to be available tomorrow. He said no. His plan was to address that subject later in the week when the Commission convened at KSC. He didn't think the subject would come up so soon, but it would be a good idea if I were there just in case. I said I would attend the meeting.

Shortly after 6:00 p.m., four hours after it started, the Monday, February 10, 1986, meeting of the Rogers Commission ended. No meeting, no four hours, would ever have more impact on my professional life.

The next morning, the Presidential Commission held its first public hearing. The meeting was structured around presentations by NASA personnel, primarily by Larry Mulloy, to explain what NASA knew about the O-rings, SRB joints, and O-ring erosion problems.

Right off the bat, Mulloy made a totally false statement: "The last time we saw any erosion on O-rings was the January launch the year before [1985]."

It was patently untrue. We had observed O-ring erosion in the field-joint of the very last launch, STS-61C, earlier in January 1986, as well as in numerous nozzle joints over the past year.

Adeptly, Mulloy evaded all questions from the Commission concerning cold temperatures and the possible impact that weather may have had on the performance of the O-rings in the field-joints. Chairman Rogers helped him in this regard, explaining that the Commission was scheduled to review all weather-related problems later in the week and preferred to avoid serious discussion of this sensitive issue, the very one that I had shined a light on the day before.

In the meantime, General Kutyna had shown Feynman his old Opel GT in his garage and told him that the carburetor leaked when it was cold because the seals got stiff and lost their resiliency. This likely gave Feynman an idea, and that was to obtain an O-ring and submerge it in ice water for a good length of time to show how it would lose its ability to spring back to its original shape. Feynman ran around Washington, D.C., in a taxi all morning looking for a hardware store to get a C-clamp and some pliers so he could do the experiment on stage for the public meeting. Mulloy's pompous attitude during the previous day's closed hearings and his adamant position that he still wasn't convinced that cold temperature could affect the performance of the joint had prompted Feynman to prepare for a little live demonstration in front of the TV cameras.

NASA had brought a model of the SRB joint to pass around to the commissioners, so when it came to Feynman, he pulled out one of the two O-rings with the pliers he had in his pocket to do his experiment. He then realized he did not have a glass of ice water like he had the day before, so he got up and asked one of the attendants if he could have a glass of ice water. Shortly thereafter they brought ice water to all of the Commission members, and Feynman dropped the squeezed O-ring in the C-clamp into his glass of ice water.

In the very first meeting of the Commission the previous week, Kutyna kidded Feynman about his messy hair and teasingly advised that he should comb it if he wanted to make a good impression on TV; Kutyna said, "Co-pilot to pilot: comb your hair." But Feynman was a real maverick and didn't care about how he looked on TV or any other time for that matter. The physicist depended only on his cleverness. He knew how to reduce the most complex problem into a very simple explanation or analogy that even the news media and average person on the street could understand. Those fortunate to have been a Feynman student related that one of his greatest attributes was to be able to reduce a problem as complex as one in quantum physics to something everyone can relate to.

With Mulloy belaboring his briefing and skillfully avoiding even the possibility that cold O-rings may not perform properly, Feynman itched to steal the show. Wanting the timing to be just right, Kutyna, who sat next to him on stage, whispered, "Copilot to pilot: not now, not now."

Following Mulloy's lengthy briefing, Chairman Rogers called for a short break. When the Commission reconvened, Feynman could wait no longer.

Before testimony from the day's featured witness, Richard Cook, the author of the memos published in the *New York Times,* began, Rogers announced as soon as the Commission came to order, "Dr. Feynman has one or two comments that he would like to make."

Immediately, Feynman pulled out his glass of ice water with the cold O-ring in it. "This is a comment for Mr. Mulloy," he said while conducting the experiment. "I took this stuff that I got out of your seal and I put it in ice water, and I discovered that when you put some pressure on it for a while and then undo it, it doesn't stretch back. It stops at the same dimension. In other words, for a few seconds at least—and more seconds than that—there is no resilience in this particular material when it is at a temperature of 32°. I believe that has some significance for our problem."

The blood rushed out of Larry Mulloy's face. Quickly, Chairman Rogers responded to prevent, at that early moment in the hearings, further embarrassment to Mulloy and NASA: "That is a matter we will consider, of course, at length, in the session that we will hold on the weather. It is an important point, which I'm sure Mr. Mulloy acknowledges and will comment on in a further session." Rogers then moved right to Richard Cook's swearing in.

Richard Cook was a budget analyst for NASA, and commissioners were tough on him relative to his credentials and motives for writing the "poison pen" memos about the Shuttle that he did. There was concern about his genuine understanding of the technical problems, and concern about his motive for writing memos pertaining to technical problems (with the O-ring seals) that weren't in his purview at NASA.

It took a lot of effort on his part for Cook to convince the panel that he was only relating the seriousness of a problem that he had been told about by the technical people at NASA; his responsibility involved "budgetary impacts" of such problems. That was his motive for writing the memos.

Feeling personally embarrassed by the fact that they were not aware of Cook's—or anybody else's concerns—about the SRB O-rings until they had read about it in the *New York Times,* some members of the Commission were trying to discredit Cook so that the significance of his memos would be reduced until they had time to really get down to the bottom of all this.

Battering away at the issue of his expertise, which was in budgetary matters only, Cook weathered the assault fairly well until just before the lunch break, when Rogers asked Cook about a very recent memo about the *Challenger* accident, which he had written on February 3, 1986. "Why did you write that memo?" the chairman asked. "It was in the heat of the moment," Cook nervously replied.

From this moment on, Cook had a very hard time keeping his composure. He tried to justify his actions by saying he had done it again from a budgetary concern only. However, this recent memo was quite technical and appeared to be very critical of NASA headquarters management, the NASA Failure Analysis Team, and MSFC management, in particular. Even though he denied it was the purpose of this or any other of his memos, it became clear that Cook had been frustrated with NASA management before the accident for not providing adequate funds to fix the O-ring problems that were known, and he seemed even more concerned that a thorough analysis might not be done even now by the NASA Failure Analysis Team. He commented that he was glad that there was a presidential-level commission involved in the investigation, but thought that the truly knowledgeable engineers at NASA headquarters, MSFC, and Thiokol were being left out of the accident investigation. Curiously, Rogers asked him if it were true that he was in the process of changing jobs in the government. Cook acknowledged that he was returning to the Treasury Department next week but that it had nothing to do with this matter. It was a total coincidence.

The Commission then heard testimony from several members of NASA management concerning Cook's memorandums. All the while, Dr. Feynman wanted to get back to Larry Mulloy, his ice water experiment, the impact of cold temperatures on the performance of the O-rings, and, most important, what was known before the launch. "There were data presented, as we have discussed," Mulloy acknowledged, "by some—by Thiokol engineering—that there was a suggestion that possibly the seals shouldn't be operated below any temperature that it had been operated on previous flights. . . ." But again, Rogers cut off the conversation, saying the Commission was going to Kennedy Space Center on Thursday and would address the weather and temperature effects with the Thiokol people there. Clear to all that he didn't want to get into that discussion just yet, the hearing was adjourned.

14

"God, That Took a Lot of Guts"

It was shortly after 4:00 p.m. when I left the building and immediately rushed to National Airport to catch a plane to Huntsville so that I could join the SRM Failure Analysis Team. When I went into the HOSC the next morning, I had a message telling me to be at KSC on Thursday, the next day (February 13, 1986), to provide support for the testimony that would be given to the Presidential Commission by Carver Kennedy, Vice President of Morton Thiokol Space Services. Carver had been asked to discuss the stacking and assembly of the STS-51L hardware, and he knew that we had experienced several problems in that area. He wanted me there to support his testimony. The message further stated that the Commission, on Friday, February 14, would be hearing testimony on the decision to launch *Challenger*. To prepare for that, Ed Garrison, President of the Aerospace Group of Morton Thiokol, wanted a meeting the night before the hearings with all Morton Thiokol employees who would be involved with that testimony.

Quickly, I assembled all of the information I had concerning the assembly of the STS-51L hardware. I also had some notes that I had made concerning the January 27 discussions prior to the launch; I had written them down a week ago, on February 6. After hearing Mulloy's testimony and growing concerned that if I myself didn't have an opportunity to provide this information to the commissioners, they might never know about the original recommendation by Morton Thiokol engineering not to launch. I was so concerned that if the company airplane crashed that this information would be lost forever, I gave a sealed copy of my notes to our Huntsville Field Office Manager, Ivan Adams. I told him to give the envelope directly to the Presidential Commission if anything happened to me.

The company plane was late getting to Huntsville from Ogden, and as a result we were late getting to KSC. The plane landed at the executive airport near Titusville, the closest one can land to KSC unless one lands on the Shuttle landing strip. Since we were late, and I was the only one who had been requested to support the afternoon testimony of the KSC stacking operations, our local Launch Support Services office manager, Jack Buchanan, had a car available for me to rush over to the hearings. I was about half an hour late, but the hearings were running a few hours behind schedule. There was no way they would end by five o'clock, as planned. Everyone was told to stand by in the hallway because they were going to continue the session into the early evening.

All day long, NASA personnel had been briefing the Commission in a closed meeting concerning the results of their current failure analysis, film review, flight data, and potential failure causes. Each investigation team from KSC, MSFC, and JSC briefed the Commission. Though I was not present at these briefings, I reviewed the written testimony after it was released by the Presidential Commission in the summer of 1986. Regrettably, but not surprisingly, not a single person with NASA mentioned Thiokol's original recommendation not to launch *Challenger*. The only information presented that was controversial was the data concerning the extremely cold temperature measurements that were made on the SRBs on the morning of the launch.

Giving that part of the briefing was Rick Bachtel, a thermal analyst from MSFC. Bachtel told the Commission that, based upon the ambient temperature early on the morning of the launch, they had predicted a minimum temperature of 21° on the SRB case, but on the pad that morning it had actually measured 9° on the right-hand SRB, at a time when the left-hand SRB was very close to ambient. He tried to explain the difference by some error of 4° or 5° in the measurement and the remainder by the fact that the right-hand SRB was more exposed to the cold night sky. Commissioner Dr. Eugene Covert from MIT stated that NASA seemed to be putting too much faith in its model; he wanted to know the temperature difference between the two SRBs early on the morning of the launch that was scrubbed the day before. Bachtel answered that the SRB temperatures were in the 30s and that both SRBs were within 2° of each other. Former astronaut Captain Bob Crippen added, "We're still trying to understand the temperature measurements, but this was the first time that we've ever seen that kind of a difference between the left- and right-hand SRBs."

The meeting continued with discussions on the films and the reconstruction of various accident scenarios, starting with a leak in the joint of the right-hand SRB. Interpretation of the films initially indicated that the leak in the joint not only burned through the external tank, causing a leak in the hydrogen tank, but also burned through the struts that attached the SRB to the ET, causing the right-hand SRB to rotate into the top of the ET containing the liquid oxygen. It was later determined that the hydrogen tank aft dome also failed about this same time, which resulted in a very high thrust load propelling the hydrogen tank into the LOX tank above. The breaking of the LOX tank ultimately caused the explosion when it came into contact with the liquid hydrogen. This event blew the orbiter away from the ET; the force of the explosion combined with the aerodynamic forces on the vehicle literally tore the orbiter apart, separating the crew compartment from the rest of the orbiter.

I stood in the hall until after 7:30 p.m., when the Commission finally decided to adjourn, having not gotten to Carver Kennedy's testimony on the assembly of the SRBs. Carver and I would have to wait until the next morning, following testimony covering the Thiokol temperature discussion, which was to be limited to ninety minutes to get back on schedule.

Ed Garrison's confab with all of the Morton Thiokol employees who were going to testify was being held in the Titusville Ramada Inn starting at 8:00 p.m., so I rushed over there to try to get some dinner before the meeting. Everyone was already in the conference room, so I asked the dining room to bring some dinner over, as everyone else had eaten. Next to Garrison at the table sat one of our corporate attorneys from Chicago plus representatives from McKenna, Connors & Cuneo, a Washington, D.C., law firm just being brought in to represent our company. The rest of the seats were taken by Morton Thiokol employees Jerry Mason, Senior Vice President and General Manager of Wasatch Operations; Calvin Wiggins, Vice President and General Manager of the Space Division; Joe Kilminster, Vice President of Space Booster Programs; Robert Lund, Vice President of Engineering; Don Ketner, Supervisor of the Gas Dynamics and Heat Transfer Section; Roger Boisjoly, Senior Engineer on the O-ring Seal Task Force; Arnold Thompson, Supervisor of the Structures Section; Boyd Brinton, Manager of the SRM Project Engineering Department; and myself, Director of the Space Shuttle SRM Project—all of whom could be testifying tomorrow. Garrison opened the meeting with a few comments about the need to listen to the attorneys very carefully and do exactly what they told us.

The attorneys took over, wanting to make sure that all of us understood that although this was neither a trial nor a judicial hearing, all of us who were to testify had some basic rights. First of all, no one should *volunteer* any information. When a question was asked, we should respond with a simple "yes" or "no" and not elaborate on the answer. We were not obligated to provide any more information than that. The attorneys made it clear that they were not telling anyone to lie: "You must tell the truth, but don't elaborate on the answer, as it may be misconstrued later, which could cause you some problems."

Ironically, I had been involved in a civil suit filed by Thiokol several years earlier against the U.S. Navy Joint Cruise Missile Project Office disputing the award of the booster for the Tomahawk cruise missile, where I acted as chief witness for Thiokol, the plaintiff. The case was filed in the federal district court in Salt Lake City by the same law firm now representing Thiokol for *Challenger*. I had received the same instructions from one of these same attorneys concerning my cross-examination by U.S. government defense attorneys, so I had heard these words before.

In my own mind, I knew this was a totally different situation. When you are being cross-examined in the courtroom, it is the intent of the questioning attorneys to discredit you and your previous testimony. Their objective is to impeach your testimony, so it is generally to your advantage to answer a question only with a simple "yes" or "no." After seeing how NASA had intentionally avoided telling the Presidential Commission about the prelaunch discussions with Morton Thiokol, I knew if we didn't tell them all the facts, they would probably never be revealed.

Furthermore, I had been living with this bitter feeling now for over two weeks since the accident. It had been eating away at me, and I had lost considerable sleep. I hadn't told my wife about it because I didn't want her to worry any more than she already was. At this point in time, it wasn't clear whether the Shuttle program would ever continue and, if it did, whether Morton Thiokol would be allowed to continue producing the solid rocket motors. I shared some of her concerns about Thiokol losing the Shuttle contract, but thought it was unlikely I would lose my job.

I had been hoping that NASA would voluntarily provide all information to the Commission. Then I wouldn't have to do anything but verify what they had already reported. After listening to NASA management mislead the Commission earlier in the week, and recognizing that NASA in five separate hearings before the Commission (February 6, 7, 10, 11, and 13) had failed every time to address the real circumstances surrounding the January 27 teleconference, I was convinced that Morton Thiokol employees needed to provide every last detail about the prelaunch meeting.

Listening to the instructions from the corporate attorneys, it was clear that my company was about to follow the same trail as NASA, which I believed was a real mistake. I had detected that Chairman Rogers and his colleagues did not think that NASA was being up front and totally honest with the Commission—maybe the NASA people had listened to similar instructions from the government attorneys.

I could not restrain myself anymore, so I stood up: "I do not agree that it would be in the best interest of the company or the employees testifying tomorrow to respond to the Commission as instructed."

Ed Garrison quickly responded: "Many of us in this room may be experts in technical matters, but these people are the experts in legal matters, and we need to listen to what they have to say."

"I wasn't sure that they were legal experts," was my response, "because a few years back I developed a case for these same lawyers that no one could lose, but they managed to lose it! Furthermore, I would like to ask them one simple question and that is, 'How many times have any of them appeared before a Presidential Commission?'"

"Never," was their response. "Well, I have, and therefore I believe I'm in a much better position to advise those that are testifying than our attorneys. I don't believe it would be a good idea to create an adversarial relationship with this Commission, and if we don't voluntarily tell them everything we know, we'll be doing just that. This Commission hearing is not like a court trial. The Commission is chartered to investigate the causes of the *Challenger* accident so that the proper corrective actions can be taken to get the Space Shuttle flying again safely as soon as possible. The Commission is a fact-finding body, and unlike a jury in a trial, Rogers's panel was not sitting in judgment of those testifying."

"It's my intent tomorrow," I advised the group, "to tell the Commission everything I know about the circumstances leading up to the accident in as much detail as I can remember. By doing so, I hope that I'll never have to appear before them again and respond to the question, 'Why didn't you tell us about that when you were here before, Mr. McDonald?'"

The attorneys were shocked, and Garrison had such a scowl on his face that if looks could kill, I would've been a dead man. For a few seconds, you could hear a pin drop—no one dared say anything. Finally, Roger Boisjoly broke the silence: "I agree with Al, and I'm going to do the same thing." All the attorneys could do was reiterate that we were simply not obligated to tell the Commission everything we knew and that it was best to keep our testimony as short as possible.

The meeting broke up at a very late hour on this very tense and somber note. What I did not know at the time was that Morton Thiokol had coordinated all of its planned testimony with the Marshall Space Flight Center while I was on my way to KSC to support Carver Kennedy's testimony. Larry Mulloy had structured the format for the presentation to the Presidential Commission so that he would give a timeline overview, Jerry Mason would provide an overview of the decision process, Bob Lund would address the engineering recommendations that had been made, and then Joe Kilminster would present the final recommendations that came following the caucus in Utah. During the strategizing at Marshall for these presentations, Mulloy allegedly remarked, "Looks good, if McDonald doesn't throw another turd on the table."

The Commission's closed hearing on the circumstances surrounding the decision to launch *Challenger* started at KSC at 8:00 a.m. on February 14, 1986. Chairman Rogers opened the meeting by stating, "The first item on the agenda is the Thiokol discussion referring to comments made in the Commission's Executive Session by Mr. McDonald. . . . This meeting today is in the nature of an investigation and not really a hearing, so it is not necessary to swear anybody in. So, although it is going to be recorded, it is not going to be sworn testimony. What I have in mind particularly . . . is to be sure we know precisely what the facts were, who was involved, and . . . if there were any documents relating to those discussions."

Mulloy was the first to speak. He began by indicating that it was not until 5:15 p.m. on January 27, the day before the launch, that he was informed that Thiokol engineering had some concerns regarding the function of the O-rings at the predicted temperatures. Chairman Rogers immediately zeroed in, wanting to know if the first discussions concerning the weather that Mulloy remembered were at 5:15 p.m. "Yes sir," Mulloy answered, too quickly.

Arnie Aldrich, Manager of the National Space Transportation System at Johnson Space Center, quickly came in to correct him: "Excuse me, Larry. To be consistent with the discussion we had yesterday, there was a Mission Management Team that met earlier discussing the weather."

Hearing a discrepancy between the two NASA men, both of whom had been sworn in the previous day, Rogers interrupted: "Excuse me. I think I've changed my mind. I think all the people from Thiokol that are going to talk today, let's swear them all in at one time, because, if we don't, and we have sworn in the other witnesses, it may in retrospect look a little odd."

All of us from Morton Thiokol were then sworn in at the same time.

Mulloy continued with his testimony relative to the events that transpired on the eve of the launch. He provided a chronology of discussions on the O-ring seals along with a summary of both the original Morton Thiokol recommendation not to launch below 53° and the final Morton Thiokol recommendation (signed by Joe Kilminster) to proceed with the launch. Larry didn't mention anything about my conversation with him and Stanley Reinartz during the caucus or with them after the final Morton Thiokol recommendation had been made. He only said that Morton Thiokol would discuss its deliberations in more detail, beginning with the testimony of Jerry Mason, Senior Vice President and General Manager of the Wasatch Division of Morton Thiokol, whom he introduced.

As Mason stood up to testify, Chairman Rogers said, "Mr. Mason, might I suggest in your discussion with us today that you please disclose anything that you know about that *may turn up*. If you have documents that we don't know about that would be embarrassing to you, tell us about them *now*. We don't want to have to pry information out of you. *You know what's there.* Tell us the whole story, if you will."

You can imagine how I felt. These were basically the same remarks that I had made to all of those from Morton Thiokol in front of Ed Garrison the night before—and it was a far cry from the instructions given by our attorneys, to volunteer nothing and answer each question with a simple "yes" or "no."

Commission members had numerous questions for Jerry Mason, and he had great difficulty in providing good answers to many of them. Rogers asked, "How many people at Morton Thiokol in Utah were against the launch versus those supporting it?"

"About half a dozen—five or six in engineering."

"How many felt it was OK to launch?" Rogers queried.

Mason deflected that question to Bob Lund, Vice President of Engineering, who responded, "I guess there were five or six"—the same number reported to be against the launch.

Ex-astronaut Bob Crippen, who was in the audience, was uncomfortable with what he was hearing: "Mr. Chairman, if I may make an observation. Since the earliest days of the manned space flight program that I've been associated with, and Mr. Neil Armstrong [the Commission Vice-Chairman] has been associated with, our basic philosophy has been: 'Prove to me we're ready to fly.' Somehow, it seems, in this particular instance, we have switched it around to, 'Prove to me we are *not* able to fly.' I think that was a serious mistake on NASA's part, if that was the case."

In my view, Crippen—the pilot for STS-1, the very first Shuttle test flight, in 1981—had hit the nail right on the head, because that is exactly what had happened. And not even the numbers of five or six in favor provided by Mason and Lund were correct, as I had not been able to find a single engineer who recommended to Morton Thiokol management to proceed with a launch when that decision was made. Even stranger to me was Arnie Aldrich's comment that neither he nor Jess Moore, NASA's Associate Administrator for Space Flight, was aware that such a meeting occurred or that Morton Thiokol had expressed any concern about the low temperatures. Aldrich stated that Dr. Lucas and Stan Reinartz were physically adjacent to him and to Jess Moore in the KSC firing room for several hours before the launch, but not one word had been spoken to them about an SRB concern with the temperatures and seals. It was obvious from Chairman Rogers's sour facial expression that he was as surprised by Aldrich's comment as I was.

Dr. Feynman then asked Lund, "What are the names of your *four* best seal experts, ranked in order of their ability?" "Roger Boisjoly, Arnie Thompson, Jack Kapp, and Jerry Burn," Lund replied. "What was the opinion of your top *two* experts?" Feynman wanted to know. "Did they agree that it was OK to fly?"

Roger Boisjoly and Arnie Thompson took it upon themselves to respond that they were not in agreement to fly. "And we never changed our minds," they declared.

Earlier, Jerry Mason testified that he had only polled his senior management team—Lund, Kilminster, and Wiggins—who unanimously approved the launch; he had not polled his engineers. But he had to have known very well that Arnie Thompson, Supervisor of the Structure and Seals Design Section, had been very adamant about not launching.

Chairman Rogers now suspected as much and turned directly to Thompson: "Were there other engineers that would take issue with you on this matter, when it came to the launch question of whether to 'go' or 'no go'? Did you have others who opposed your point of view?"

"I have twenty-four or twenty-five gals and guys working for me," Arnie responded, "and I know none of them that would have opposed this viewpoint that are involved in the case/nozzle joints!"

Rogers wanted absolute clarification: "All of those people would have said no to the launch?"

"My judgment is yes, that is true," Thompson replied.

Later on, Mason tried to rationalize why it was appropriate to proceed with a launch at a temperature well below the experience base of 53°. "If you ask anybody if 50° would be OK," Jerry postulated, "I think everyone would sign up at 50°. So every flight in this program has had to break some frontiers. When we went from STS-1 to STS-2 we went into a temperature regime that we hadn't been in. As we moved further into the program, we were working, to some degree, in an extrapolated area."

Very disturbed, and rightfully, by this rationalization, Sally Ride quickly interjected: "The time you go through frontiers is during testing, not during the flights. That's the way it's supposed to work!"

"That was a poor choice of words on my part," Mason apologized. "What I was trying to convey was that we are always working somewhat in the extrapolated area."

"But that is not this case here," Chairman Rogers countered. "Here you had a lot of warning. You've all been discussing the O-rings and seals. You've all had concerns. It's all over the papers. All you have to do is look at it, as we have, and you can see that's been a major concern. And now you have a situation where there is a lot of concern and a lot of your people expressed a negative vote, and our people in NASA didn't even know about it! I mean, that is unbelievable to me! At least, if there had been a calculated decision and everybody said yes, 'we all know about it and we're willing to take the risk,' and there is some risk, and I suppose you have to do that. Every flight has some risk. But this one seems to be so difficult to explain."

As confusing as the decision to launch was to me, its details were tremendously more confusing to many members of the Presidential Commission, the press, and the American public—and the technicalities only grew more confusing. Earlier in the hearings, NASA had tried to explain why the SRB field-joint had been changed from a "Criticality 1R"—"R" meaning "redundant"—to a Criticality 1, which meant the loss of vehicle and crew if a single component, such as the primary seal, failed. NASA had made this change to the critical items list because it had been found during testing that the secondary seal might not, in some circumstances, be able to seal the joint because of the deflection of the case, which could cause the joint to open at full pressure. Because effective sealing of the joint could be beyond the capability of the design, at the worst acceptable case and O-ring dimensions, NASA deemed it necessary to change the Criticality level of the SRB field-joint from 1R to 1.

Mulloy presented many of his explanations to the Commission almost as if his objective was obfuscation. For example, in explaining why, on the night of the teleconference, he had challenged Morton Thiokol's original engineering recommendation not to launch below 53°, he presented his rationale in writing to the Commission. He had listed ten conditions involving NASA's cold O-ring assessment, with four of them relating to the secondary O-ring, which was in direct violation with the critical items list that listed the field-joint as a Criticality 1 component with *no secondary* seal capability. The four conditions were: (1) no *secondary* O-ring erosion or blowby appeared to date in the field-joints; (2) colder temperatures might result in greater primary O-ring erosion and some heat-affected *secondary* erosion because of the increased hardness of O-ring, resulting in some slow seating; (3) the *secondary* seal was in position to seat (200 psia/50 psia leak-check); and (4) the primary seal may not seat due to reduced resiliency; however, during periods of flow past primary, the *secondary* seal will be seated and seal before significant joint rotation occurs. Mulloy's conclusion had been that "risk recognized at all levels of NASA management is applicable to STS-51L."

When Mulloy finished presenting his self-serving version of the substance of the teleconference to the Commission, Stan Reinartz asked George Hardy from NASA Marshall to comment on Larry's assessment. Hardy stated, "I basically agree with it, but I would not recommend launch against Thiokol's recommendation." Mulloy jumped back in: "It was at that time that Mr. Kilminster requested an off-line caucus."

Both Roger Boisjoly and Arnie Thompson then told the Commission that their most serious concern was that the cold temperature might slow down the timing function for the primary O-ring, preventing a good seal. A delay of only 0.170 second due to the cold temperatures could put the Shuttle into a time regime wherein the secondary O-ring might not be capable of sealing the joint. "We couldn't depend upon the secondary seal," Roger and Arnie explained.

After hearing all this discussion, the Commission's Joseph Sutter exclaimed, "I just don't understand it. I have listened for two hours, and I still don't understand it! I seem to hear two different things here. It's a helluva way to run a railroad on a critical item like this!"

Seemingly exasperated like the rest of the commissioners, Chairman Rogers called an end to the day's business.

I had sat there all morning waiting to hear Mulloy and/or Reinartz discuss the conversations that I had had with them during the Morton Thiokol caucus and after the final recommendation by Morton Thiokol to proceed with the launch. I was genuinely hoping that they'd reveal these discussions to the Commission so I wouldn't have to testify. Since I had been with Mulloy and Reinartz at KSC, I fully expected the Commission to ask me if I had anything more to add, especially since I had been the one who caused this particular hearing to be held in the first place.

But apparently Rogers had heard all he wanted to hear on the subject for the moment and brought the discussion to an end. He wanted to get on to the next item of the agenda.

Once again I had to stand up in front of this very hostile audience and volunteer to testify: "Since I caused this meeting to come about, I would like to testify, I guess."

Vice-Chairman of the Commission Neil Armstrong said, "I was just going to say that other Thiokol people might want to testify and should be given the opportunity."

Rogers apologized for the oversight. "The schedule is tight, and we're already lagging behind, but please, come up and say what you have to say."

As deliberately as I could manage, I told the Commission about my discussions with the NASA management on the evening before the launch about why they could not accept a recommendation to fly below the qualification temperature of the SRBs. I explained that I had given them three good reasons not to fly: the cold temperature effects on the O-rings, the heavy seas jeopardizing the retrieval of the SRBs, and the ice on the launch platform. I told the Commission of my strong suggestion to delay the launch until the afternoon, when it would be warmer, around 50°, and how NASA had rejected that. I recalled I had told them very explicitly, "I think these three reasons combined should be absolute criteria not to launch this thing, and if I were the Launch Director I wouldn't do it!" I told the Commission of my warning to NASA, "That if we're wrong and something goes wrong on this flight, I wouldn't want to have to be the person to stand up in front of a Board of Inquiry and say that I went ahead and told them to go ahead and fly this thing outside what the motor was qualified to!"

In response to my testimony, Arnie Aldrich affirmed that he had, indeed, received a telephone call that evening from Mulloy and Reinartz, but they had only discussed the concerns relative to the SRB retrieval operations and the ice on the launchpad. They had never mentioned then, or at any time the next morning, Aldrich said, that Morton Thiokol had a concern about cold temperatures affecting the O-rings in the joints.

Sally Ride asked a critical question: "I would also like to ask the people from Marshall why they decided not to advise Mr. Aldrich of Thiokol's concerns?"

Stan Reinartz replied, "I would be glad to go through it again, Sally, but I think I indicated earlier the basis for the decision not being brought up as a Level III item, one violating waivers or constraints to the launch."

Minutes before the session adjourned, Dr. Alton Keel, the Commission's Executive Director, said to Bill Rogers, "Mr. Chairman, could I just ask one question—not to embarrass anyone or put anyone on the spot, but just for the sake of establishing the record so it is not left to inference? The inference from your testimony, Mr. McDonald, is that you were under pressure, perhaps *unusual* pressure from NASA officials, to go ahead with the launch. Is that an accurate inference or not?"

"That is an accurate inference, yes."

"And did I understand, too," Chairman Rogers added, "that you did not sign off on this one?"

"No, I did not."

"Was that unusual?" Rogers queried.

"I believe it was, yes." Because that's why I was there! Every Shuttle contractor is required to have a senior management representative present at Kennedy Space Center for every launch from the L-2 [launch minus two days] flight readiness review right up to launch time. We were there to approve or not approve all the decisions relative to flight readiness and launch safety issues that came up right to the last minute.

Mentally exhausted, it was difficult for me to respond to these tough questions at the end of the meeting. There I was, standing alone in front of all of my own management: in front of Ed Garrison, President of our Aerospace Group, who was there as an observer; in front of the entire NASA hierarchy, our company's most important customer. In front of the whole world, it seemed, I was responding to a question with my very honest opinion that we had, indeed, been under unusual pressure from NASA officials to go ahead with the launch. No one else from NASA or Morton Thiokol dared imply, much less state publicly, that NASA had applied pressure to Morton Thiokol to change its recommendation from "do not launch" to "launch." Adding fuel to the fire was my answer to Chairman Rogers's question that I had *not* signed off on this launch, even though it was my reason and responsibility for being at KSC.

I could see the distress—in some cases, anger—in the eyes of many who were there in the room directed at me. My testimony made it very clear that NASA had not been, and was still not being, forthright with the Presidential Commission. It also indicated that Morton Thiokol senior management was trying to protect NASA by making it appear that nothing that had happened had been at all unusual, process-wise. I had worked with all of these people from Morton Thiokol for many years, and I had worked closely with many of those in NASA management for the past couple years. I had gotten to know them very well, and I considered them all as personal friends.

I was deeply saddened by having to give the testimony I did, but I knew I had to do it, to clear my own conscience and provide the truth to the Commission. Again, I had hoped that all of this information would come out before I had to testify; there had been plenty of opportunities for that to happen.

Then I hadn't even been asked to testify.

I knew what I had just said could ruin the careers of several very good engineers and managers. I knew it could be devastating to NASA, which was a revered and respected government agency. Worse for me, I knew it would be very damaging to Morton Thiokol senior management, my bosses who had reversed the company's original engineering recommendation not to launch. At the same time, I really thought that I had just ruined my own career and could well lose my job.

My voice cracked as I ended my testimony. As I walked away from the speaker's stand in front of the Commission, I could not hold back my tears. I was embarrassed for my company and for NASA, but overpowering all of my other emotions was the knowledge that seven American heroes had lost their lives in a tragic accident that really shouldn't have happened.

I was hoping no one would notice my tears. I had been raised to believe that real men don't cry, but as much as I tried to hold back the tears, I couldn't.

As I walked to the door, Sally Ride got up from her seat at the Commission's table, came over to me, put her arm around me, and gave me a big hug. "God, that took a lot of guts," Ride told me. "I'm glad someone finally leveled with this Commission."

Sally Ride then walked over to Roger Boisjoly and gave him a hug, too.

It took a lot of guts and genuine compassion for Sally Ride to do what she did. Her appreciation for what I had done put blood back into my face, air into my lungs, and steel back into my backbone. I knew I could not become a quitter. I would not run off and hide. I must continue to pursue the truth until all the important problems were identified and until all that really needed fixing got fixed.

Earlier in the session, Chairman Rogers had requested that if anyone had made any notes concerning the meeting between Morton Thiokol and NASA, these notes should be turned over to the Commission. "One other thing I would like to comment on, and I hope everybody is listening to this," Rogers declared, "I would like everybody to consider that the Commission has requested from each of you all of your private notes—not only your official notes and your official files but any private notes you have in your own handwriting, or any other notes you have, particularly as they relate to these conversations we've been talking about this morning."

Our company's Jerry Mason immediately asked, "Is it OK if we gather those up and deliver them on, like, Monday?" "Yes, that would be fine," Rogers replied, "but anybody who wants to give us documents should feel free to give them to the Commission directly, if they would like to."

As I had written down a week ago everything that I could recall about the launch decision, I immediately turned my notes over to the Commission, which later would be published as part of Volume IV of the Commission's final report. Roger Boisjoly had kept a daily log of his work, which contained many notes about the teleconference, and he, too, turned them right over to the Commission, along with copies of memos he had written in 1985 about his serious concerns with the O-rings. Mulloy and Reinartz apparently provided written details of their actions as well as notes they had made as a result of the Commission's request.

As I was leaving the room, I was chastised by our corporate attorneys for not first giving them a copy of my notes and letting them supply these notes to the Commission at a later date. After all, the attorneys complained, Chairman Rogers had said that if we didn't have our notes readily available, it was OK to supply them by the first of the week.

It was later reported by the press that Chairman Rogers immediately called President Reagan and told him that the decision to launch *Challenger* was "flawed" based upon the testimony that he had heard that day. I don't know if Rogers used my name when he gave President Reagan the news, but it was later reported by the media that it was my testimony he was talking about.

15

A Leper in the Limelight

Leaving the meeting room, I ran into Carver Kennedy in the hallway. He was waiting outside to testify before the Commission on the SRB assembly operations for STS-51L, a subject postponed and rescheduled from the previous day's meetings. Carver could tell I was very disturbed but asked how things went. "Very badly," I said. "I just want to get out of here. I'm sorry but I'm not going to be able to support you in your testimony as we had planned."

On the ride to TYCO Executive Airport with Jack Buchanan, manager of Launch Support Services for Morton Thiokol, neither of us said a word. Inside the company jet was like riding in a flying funeral parlor. I could feel the cold stares that Roger Boisjoly and I were receiving from our senior management; we both felt like lepers. The weather was terrible, and the airplane ride was a rough one as it was raining all the way to Huntsville, where we dropped someone off and took on extra fuel. Still very upset, I was on the verge of getting airsick. Unwilling to tolerate the icy silence on the long ride home all the way to Utah, I got off the airplane, saying I didn't feel well. My plan was to take a commercial flight home the next day, Saturday, February 15, 1986.

I hadn't been in my motel room thirty minutes when I started receiving telephone calls from various reporters. Not wanting to talk to any of them, I couldn't fathom how they knew to call me much less where I was. The meeting I had just left was a "closed meeting" of the Presidential Commission. Not wanting to talk to any news media, I packed my belongings and drove over to John Thirkill's house. Thirkill was the general manager of Morton Thiokol's Huntsville Division, and both he and his wife, Mary Lou, were good friends of mine. John had been Deputy Director of the Space Shuttle SRM program during the development and early flights of the Space Shuttle and later became Vice President of Engineering for the Wasatch Division. I told John what had happened before the Commission, and he reacted almost as if he didn't believe what I was telling him. I could tell he had some real reservations about what I had done; he was very defensive about Morton Thiokol management. Two of John and Mary Lou's best friends were Joe and Shirley Kilminster. I could tell John was very uncomfortable with my refusing to sign the final go-ahead for launch recommendation, which meant that Joe Kilminster had to sign it. "It could be the smartest thing I've ever done in my life," I told Thirkill, "but I'm very sorry I had to put Joe in that position."

I called Linda and told her the bad news and why I wasn't coming home on the company plane that night and why I was spending the night at the Thirkill home. She was also very upset about how the hearing came out: that I didn't do what Morton Thiokol senior management had told me to do and that my testimony had no doubt angered NASA management even more than my own management.

When I got home the next day, I received a phone call from Tom Davidson, a good friend of mine who was the Technical Director in our corporate headquarters working directly for Ed Garrison, our president. Davidson wanted to get together with me the next day, even though it was a Sunday. One of the kindest and most considerate people I knew, Tom played the role of father confessor at Thiokol. Whenever anyone in the corporation was frustrated with their job for any reason whatsoever, they called Tom.

Tom and I talked for nearly three hours that Sunday. I told him what I knew and what had happened at KSC, including the meeting with Garrison the night before the hearing and what had transpired during the hearing. I could tell that he felt very uncomfortable about what I was telling him. I can't say that Tom didn't believe me, but he did comment that some of the things I was telling him didn't seem to be in total agreement with what he had heard that Jerry Mason had told Garrison, but Tom wouldn't elaborate any further.

I had a difficult time going to sleep that night, because I was starting to get the feeling that the whole world was against me. I dreaded going out to the plant the next day and facing all of those people. I was also totally exhausted to a point that I was so emotionally drained and tired that I couldn't relax. I had been working for twenty-six straight days, since January 22, when I'd left home for Huntsville to negotiate a portion of the Buy III SRM production program for Morton Thiokol. During that stretch, I had only been home for one day to work and a couple weekends. Since the *Challenger* accident on January 28, many of those workdays were sixteen to twenty hour days. No wonder I was exhausted.

My time in the plant the next morning wasn't as bad as I had anticipated. I had expected the rumor mill to be buzzing about my testimony that was given at the closed hearing of the Presidential Commission the previous week, but no such conversation seemed to be occurring in the hallways or cafeteria. Everyone was still in a very somber mood while doing what they could to support the SRM Failure Analysis Team by providing documents, drawings, test results, and analysis of the joints and O-rings for the data file that was being impounded at the request of NASA. Everyone was concerned about their job and the future of the company. My secretary, Larene Thompson, asked me how the hearings went. "Not pleasant," I replied, "but as well as I could expect under the circumstances." I spent the next couple days helping to locate all the records concerning the SRM field-joints and O-rings so that these records could be inventoried and copied for the impounded data files.

My two days back in Utah constituted a lull before the storm.

On Tuesday evening, February 18, 1986, I came home from work and turned on the TV in the kitchen to watch the evening news. My family was just starting our dinner when Peter Jennings of *ABC News* in New York interrupted his news broadcast to report there had been a break in the *Challenger* story and was going live to Lynn Sherr with *ABC News* in Washington, D.C.

Sherr reported that information had just come out concerning a closed hearing of the Presidential Commission occurring last week at Kennedy Space Center: "Chairman William Rogers, based upon testimony heard from a Mr. Allan McDonald of Morton Thiokol who was at KSC during the launch of the *Challenger*, stated that the decision to launch *Challenger* was 'flawed.'" According to this new information, "Mr. McDonald had warned NASA the night before not to launch the *Challenger* because of concerns about the O-rings at the cold temperatures, but he was overruled by his superiors." Correspondent Sherr concluded her report, "ABC will provide further information on this late-breaking news as it develops," then turned the broadcast back over to Peter Jennings for the rest of the evening news.

My whole family was shocked. I had discussed what had happened in the Presidential Commission hearings with my wife, but not with my children. I had told Linda that this was very confidential information, and I didn't want the rest of the family to know how deeply I was involved until it was released to the public by the Presidential Commission.

It didn't take five minutes before the telephone started ringing off the hook. Calls came in from TV stations, radio stations, newspapers, magazines, neighbors, fellow employees, and God-only-knows-who-else wanting to get the scoop. I refused to answer the phone, telling my wife to inform all callers that I wasn't home right now and she didn't know when I would return.

For the next few hours, the phone rang constantly. Surprisingly, the story was not picked up by the local TV stations for their late-night news. Either they missed the story or we missed hearing it as we switched from channel to channel.

Just starting to fall asleep around 11:00 p.m., the telephone rang again. I heard my wife say that I wasn't interested in talking to any reporters. Linda was about to hang up when the caller told her that maybe I'd be interested in the story he was writing about what I had said to the Presidential Commission for the early morning edition of the *New York Times*. The journalist said he had obtained his information from a source that was at the Commission's meeting. This got my wife's attention and she asked me if I wanted to hear what the *New York Times* was about to print. I told her to give me the telephone.

The caller identified himself as David Sanger, a science editor for the *New York Times*. Hearing that I wasn't interested in talking with him about the Presidential Commission, Sanger said he wasn't asking me to do that. "I just thought you'd be interested in the story I've already written involving your testimony to the Commission. I just want to make sure what I have is accurate, and I am giving you the opportunity to verify or correct what I have before we go to press, because, either way, the story will appear in the morning paper."

How could I not listen to what the reporter had to say? I wanted to make sure it was accurate. Sanger read me what he had written. Most of his story was accurate, and it was clear that he had, indeed, obtained the information from someone who had been in the meeting room while I was testifying. There were a few minor errors in the story, however, and I corrected them for him. He thanked me for my help and told me he'd make the corrections and send me a copy of the paper. I spent another sleepless night wondering what kind of grief this was going to bring me.

Rather than waiting for my carpool, I decided to go to work very early by myself, leaving home at 6:00 a.m. rather than 6:45. Boy, was I surprised! Sitting on my front doorstep was a TV newsman from the local NBC affiliate,

KUTV, in Salt Lake City. He asked if I would speak to him on camera about the news story that had appeared in the early morning edition of the *New York Times*. "I have to get to work and don't have time for an interview," I told him, getting into my car for the forty-mile drive to work.

I had been at my desk for about an hour when I began to hear people walking down the hallway; I could hear them talking about the news story on ABC the night before. As anxious as that made me, I was even more nervous thinking about the *New York Times* story that surely would find its way to the plant.

I didn't have to wait long. Around 8:00 a.m., Jerry Mason, Senior Vice President and General Manager, called me into his office. "Al, why in the world did you release to the press all this information about the decision to launch the *Challenger*?" a very perturbed Mason asked. "I didn't give *any* information to the press," I answered. "Then where did all this information come from that's in the morning *New York Times*? It's a very detailed story about what was said in last week's closed hearing of the Presidential Commission and the whole story is attributed to you, Al!"

"I haven't seen the article, Jerry, but I did not give any information to the *New York Times*." I told him that a reporter from the *Times* had already written a story and that he had called me last night to verify the accuracy of the story but that he didn't get any of the information from me.

Scowling, Mason handed me a copy of a fax containing the *New York Times* article. Our marketing office in Washington, D.C., had faxed the article to him.

Reading the article, I became furious. Sanger had rewritten the article as if it were based on an interview that he had had with me! He attributed the whole story to me! Furthermore, he hadn't changed any of the information I had corrected! I told Mason what had really happened, but it was clear to me that he didn't believe me. He thought I had intentionally leaked this information to the press just to get into the limelight.

I stormed out of Jerry Mason's office with a copy of the article and returned to my office. I told my secretary to find a number for David Sanger at the *New York Times* because I needed to talk to him right away. When Larene finally located him, she told him that Allan McDonald was on the line and wanted to talk to him.

"You son of a bitch," I blasted, "you wrote that article as if I had given it to you in an interview, and you know that's not true! I never did that! All I did was verify what you had already written. Furthermore, you ignored the corrections that I told you about; you printed the article just like you read it to me without the corrections that you said you would incorporate!"

"Mr. McDonald, I apologize for the corrections not being made, but that happened because my article was already typeset when I talked to you last night. As for my attributing the information to you, you don't understand how this works, Mr. McDonald. Once you had verified the story, then it *was* your story."

"I understand that now," I shot back. "I also understand that if you could change the story to say that I was the source for your information, you could have also changed it to make the corrections! Please understand me, Mr. Sanger! Never again will I talk to you or to your newspaper!" A year later, this story helped Sanger win a Pulitzer Prize.

No sooner had I hung up the telephone when my secretary told me that I had another call on the other line. She told me it was from some U.S. senator's office in Washington and that she had put them on hold, checking to see if I was in. I told her to put the call through.

The call was actually from two senators: Don Riegle (D-Mich.) and Ernest "Fritz" Hollings (D-S.C.), both of whom were members of the U.S. Senate's Subcommittee on Space, Science and Technology. The senators said they were calling on behalf of the subcommittee concerning the article that had appeared in the *New York Times* that morning. Their subcommittee, they emphasized, needed to know what I knew about the circumstances that led up to the decision to launch *Challenger*; it was important for them to get to the bottom of this. Senator Hollings was particularly concerned that President Reagan may have been at the root of the problem, by insisting on launching *Challenger* because he was supposed to give his State of the Union address later that same day and had planned on mentioning Christa McAuliffe, the schoolteacher in space. Hollings asked if I knew anything about the Reagan involvement, and I told him I did not. He said that I needed to get back to Washington right away so I could tell them what I knew. Senator Riegle told me to catch an airplane right away, so that I could be in the nation's capital to talk to them that very afternoon. He would arrange for "a private, secret, off-the-record meeting" with the four ranking senators of the subcommittee: Slade Gorton (R-Wash.), William Danforth (R-Mo.), plus Hollings and Riegle. I told them it was impossible for me to get to a meeting in Washington that afternoon. It was two hours later there than it was in Utah; I was two hours away from the Salt Lake City airport;

and it would take me at least five hours to fly to D.C. "Fine," they said, "then we'll set up the meeting up for 3:00 p.m. [EST] tomorrow." I was to report to the chief counsel's office for the subcommittee in the Russell Senate Building, and he would make sure I got to the right place.

This sounded awfully serious, so I told them I would do that. Hanging up the phone, I immediately walked down to the general manager's office, told him about the call, and asked if I could use the company jet so I could get in and out of Washington, D.C., as fast as possible. He said he would see what he could do.

My anxiety heightening, I returned to my office and started worrying about what I was going to say to the Senate subcommittee and how it would be perceived by the Presidential Commission. The best thing to do, I decided, was to call the Executive Director of the Presidential Commission, Dr. Alton Keel, and let him know how the story got into the *Times* so that the Commission knew I wasn't out disclosing what was going on in their closed sessions. I also needed to tell them about the Senate subcommittee asking me to come back to Washington to tell them what I knew.

Surprisingly, Dr. Keel wasn't that upset about the *Times* story; he was more concerned about me talking to the Senate subcommittee. He asked if I was being subpoenaed by the committee, and I said I thought not. "If you're not being subpoenaed, and if you feel uncomfortable about the situation," Keel said to me, "you really don't have to go or tell them anything. It's really up to you. I really appreciate that you called me about this matter. My personal advice is that you come to Washington and talk to the Senate subcommittee since it's going to be a secret, off-the-record discussion with just the four ranking senators on the committee. Our Presidential Commission is planning to hold public hearings in Washington next week on the decision to launch *Challenger*."

I told Keel that I would talk to the senators and would call him afterward and let him know what transpired. "I'd appreciate that," he replied, "because the Presidential Commission wants to be responsive to the congressional committees. However, it is the Presidential Commission that has the charter and responsibility for the investigation of the *Challenger* accident, not Congress."

Hanging up the telephone, I started to organize some of the data I had stacked up on my desk. I might need to bring some schematics on how the field-joints and O-rings operated and information on the problems involving them. I also took copies of all of the charts that had been presented on the evening before the launch, including those that recommended not launching below 53°. I also grabbed the single chart signed by my boss, Joe Kilminster, to proceed with a launch after the caucus was held in Utah with Morton Thiokol senior management.

It wasn't very long before my secretary received a phone call from the general manager's secretary, saying the company airplane would be available to take me to Washington tomorrow.

I was sure glad when the day was over, because I was dodging everyone who had heard about the story on TV or what was written in the *New York Times*. Leaving my office, I told Larene to tell everyone I was busy preparing for a meeting in Washington the next day.

As I got close to my house in Pleasant View, I saw vans with satellite dishes mounted on top from every major television station in Salt Lake City on both sides of my street. Slowing down, I noted even more TV vans in the circular driveway in front of my house. Putting my foot on the gas, I drove by my house to the top of the street and turned around. Passing my house once again, I headed down the street to my friend Jack Hilden's house, only a few blocks away. Jack's wife, Jeri, came to the door and asked me to come in. I told her about the TV vans at my house and that I really didn't want to talk to any of them.

Inside the Hilden home, I called Linda, telling her what had happened at work and that I had to go to Washington tomorrow on the company plane. I was already in deep trouble with the senior management at Morton Thiokol over the *Times* article, which wasn't my fault. I certainly didn't need to show up on television again this evening. I asked her if she knew when they planned to leave, and she said some of them had been there since I left this morning; she had no idea when they might leave. They had all come to the door several times waiting for me to return. Jeri Hilden overheard the conversation and offered to let me to stay overnight in their guest bedroom. "That would be great," I said, "but I don't even have a toothbrush and razor much less clean clothes to wear to Washington." Linda suggested that she put together my shaving kit, some clean underwear, socks, T-shirts, and some pressed white shirts in a grocery bag and have one of the kids bring it down to me. "Good idea," I answered, "but you should have Greg (our son) bring it down in his Volkswagen, because if one of the kids comes out of the house and starts walking down the street, it won't take a rocket scientist to figure out where they are probably going."

16

Walking Out

The next morning as I arrived at the hangar at the Ogden airport where the Morton Thiokol jet sat ready, I was surprised to see two other gentlemen from Morton Thiokol besides the pilot and copilot. Standing inside the office at the hangar were Calvin Wiggins, Vice President and General Manager of the Space Division, and Lee Dribin, one of our corporate attorneys. I said nothing to either of these gentlemen; I just climbed aboard the plane with my briefcase and sat down. I brought no luggage because I planned to return home from Washington that night.

Wiggins and Dribin also boarded the plane. After we had been in the air for about half an hour, Wiggins broke the silence: "I'll bet you're wondering why we're going to Washington with you."

"Well, I am a bit surprised to see you and Mr. Dribin on the plane. I thought I was the only one that the Senate committee had asked to meet with them. Maybe they also requested that someone who participated in the caucus at Utah be available. Is that not the case?"

"No, we were not specifically requested to participate," Wiggins replied, "but Jerry Mason feels that someone from our senior management who was present at the caucus does need to be there in case some questions are asked that do not involve you directly. What's more, Mason has talked to corporate about the Senate's request for a meeting, and corporate suggests that Mr. Dribin needs to be present."

Our company jet landed at Dulles International Airport's General Aviation terminal. A member of our Washington office, John Haddow, met us there and drove us to our corporate marketing office at Crystal City. Already at the office was a local Washington, D.C., attorney hired by Morton Thiokol especially for the *Challenger* investigation, who was to accompany us to the meeting in the Russell Senate Building.

A former staffer for Utah Senator Orrin Hatch, Haddow took us to the Russell Senate Building. I noticed a man with a video camera parked in front of the building photographing everyone entering the front door. That seemed strange, but I presumed it was being done for security reasons. Wrong! As soon as we stepped out of the car, the cameraman quickly came walking toward us. "Are you Allan McDonald?" he asked the attorney, who was leading our way. "No, I am not," he declared. "Are you from Morton Thiokol?" Again, "No." We all kept on walking.

It was evident that something had changed concerning our "secret," "off-the-record" meeting.

Inside the office of the Chief Counsel for the Senate Subcommittee on Space, Science and Technology, we were told that there had been some changes relative to the meeting. First of all, it was not going to be a meeting with just the four ranking members of the subcommittee; the full subcommittee was waiting for us in the upstairs conference room. Second, what I had to say would be sworn testimony for the record. Finally, the meeting would not be closed or secret; it would be open to the press and news media. "I realize that these are not the conditions you agreed to on the telephone, Mr. McDonald, but it's the way the senators want it."

I was furious: "Okay, fine. Then I do not plan to attend this meeting. I'm leaving right this minute and returning home to Utah."

"Since you volunteered to discuss this matter with selected members of the subcommittee and was not subpoenaed to do so," responded the Senate lawyer, "it is certainly within your prerogative to do just that. But let's see what the senators might want to say to you at this point."

Ringing the upstairs conference room, he doubtlessly told some staff member that I had arrived but was "reluctant" to come upstairs and talk with them given the changed conditions of the meeting. Apparently, a few members of the subcommittee wanted to know my specific objections that were preventing me from coming upstairs to talk to them. Quickly it became a three-way conversation with the committee staff counsel relaying my responses to the questions. Finally, I asked the lawyer if I could speak directly to the Committee over the phone. "Certainly," he said, and gave me the phone.

"Mr. McDonald," said a voice without identifying himself, "would it be OK if we put you on the speakerphone in our conference room, because there are several members of the committee who want to talk to you about your concerns about appearing before this committee?"

"Fine, I don't have any problems with that. However, I want all of you to know that I did not agree to such a meeting and, as far as I am concerned, I am in no way obligated to speak with you since you have changed the conditions of the meeting without my consent."

I recognized the voice of Senator Don Riegle: "Since you have made the long trip from Utah all the way to Washington, D.C., shouldn't you at least take the time to come upstairs and tell the senators what you know?"

"Since the nature of the meeting has changed so much, Senator, what I think I should do is just turn around and go back home."

Next came the unmistakable drawl of Senator Fritz Hollings: "It is very important that this committee get down to the bottom of this matter, and your testimony before this committee, Mr. McDonald, is critically important. There is evidence that President Reagan planned to mention that we had a schoolteacher in space during his scheduled State of the Union address the night of the launch, and this committee needs to find out if it was this pressure from the White House that caused the *Challenger* to launch when it was so darned cold."

"I don't know a thing about that allegation," I countered. "All I know is what I heard during a meeting the night before the launch concerning the cold temperatures."

Hollings pressed on: "This is extremely important. You need to come forth and tell us exactly what you do know."

"I was willing to do that based upon the original ground rules for this meeting, but those have changed so much that I don't feel comfortable participating in this meeting."

"Please reconsider," urged Senator John Rockefeller. "You are already here, and it is important that this committee hear your story."

"Under the circumstances, I don't have a good feeling about the whole situation, and I really do think I will return home."

Senators Danforth and Riegle then suggested that the meeting be temporarily delayed and reconstituted more like that originally planned. "That would be better," I conceded, "but even that would not be the same."

One of the other senators then suggested that the subcommittee revert to an off-the-record meeting with just the four ranking senators. He said the meeting could take place somewhere else where the press would not be present.

"At this point, I just don't want to participate."

"I can't understand why you won't talk to us under the same exact conditions you agreed to before," Senator Hollings scolded.

"Well, the fact is that this meeting has already been disclosed to the public. And now you're willing to go somewhere else in Washington to have a secret meeting. That will only make it appear that there's something mysterious or sinister going on, and I don't want to be associated with such a meeting."

Hollings then complained, "I don't understand that! These are the very same conditions for the meeting that you agreed to in the first place!"

"I don't see it the same way, Senator. If we had had the meeting as originally planned, the news media would not have known anything about it. But now they'll think that we're holding some kind of clandestine meeting behind closed doors, and I will not be party to that. As far as I am concerned, this whole affair appears to me to have been orchestrated for someone's political benefit—and certainly not my benefit. I will have no part of it. I'm returning home and will talk to those members of the Presidential Commission who are scheduled to be at our plant site in Utah tomorrow."

With that, I hung up the telephone and turned around to John Haddow: "How in the hell do I get out of here?"

Haddow, who had previously worked for Senator Hatch in this very building, knew a way out of the back. We literally ran out and around to our car so we could avoid the press. Driving off, we saw a herd of press and news cameras focused on the front door waiting for us to exit the building. Back at Morton Thiokol's office in Crystal City, we told Jim Robertson, our Washington office manager, what had happened. Robertson immediately called our CEO, Charlie Locke, in Chicago, and told him I had just walked out of the Senate subcommittee hearing. Hanging up, Jim related that our CEO had said, "Boy, has he got leather balls!"

I called Alton Keel to tell him what had happened. "You did what?!" he exclaimed. "I had to do it, Dr. Keel," I responded. "It was the only thing I felt I could do."

"Well, it appears to me," Keel offered, "that the Senate subcommittee is trying to upstage our Presidential Commission, because we plan to hold public hearings on the decision to launch *Challenger* next week."

We soon learned that it was Senator Hollings who was the one responsible for changing the conditions of the meeting. Hollings was up for re-election and had held a press conference earlier in the day stating that he had located the mysterious Allan McDonald from Morton Thiokol and had personally arranged for him to come to Washington for a hearing with his Senate subcommittee. Allegedly, after thinking about it overnight, Hollings had decided that the public had a right to know what went on concerning the decision to launch *Challenger* and opened the meeting to the press. Hollings's statement had even aired that morning on ABC's *Good Morning America*.

I had thought that I smelled a rat and was glad I had pulled the rug out from under the senior senator from South Carolina. I don't think Hollings gave a damn about me, the *Challenger* accident, or anyone else. He was only interested in finding a way to use this very emotionally charged subject to benefit himself in his bid for reelection.

We jumped into John Haddow's car and headed for Dulles. About fifteen minutes from the airport, Haddow called our pilots at the general aviation terminal to see if the airplane was ready to go. The pilots said the terminal was loaded with media folks and TV cameras. It would be better, they said, if we passed by the executive terminal to a locked gate that led directly to where our jet was parked. A pickup truck with its lights on would be waiting to show us the way and take us through the locked gate.

The plan worked but, as we pulled up next to the airplane, a whole bank of TV cameras that had assembled outside the chain-link fence surrounding the airport activated their telescopic lenses. Cal Wiggins got out first. As he started up the stairs into the airplane, he threw both hands up in front of his face to mask it from the cameras. I followed a short distance behind, briefcase in hand, trying to act very nonchalantly as if nothing had happened. The TV cameras immediately focused on Wiggins.

Once in the air, seeing how tenacious the media was, I asked the pilots to check with the Ogden airport and our hangar to see if any reporters were waiting for us to land. The answer came back that TV trucks were everywhere, including next to the hangar and even along the taxi strip to the main runway.

"Maybe we should ask for clearance at Hill Air Force Base," I proposed. One of our pilots, Shelby Pierce, had flown jets for the air force out of Hill Air Force Base. He said we could do that if it was a real emergency, but this probably wouldn't qualify. Pierce suggested we land the airplane at the Brigham City Municipal Airport about twenty miles north of Ogden; he would arrange for a car to pick us up at the airport.

Calling ahead, we learned that a couple of TV trucks were waiting for us to land there as well. Dick Benson, our chief pilot, then asked if anyone had left their cars unlocked in the hangar with the keys in the ignition. My wife had dropped me off, but Cal Wiggins had parked his Chevy Suburban in the hangar. Benson thought that he could get approval from the tower to have one of the aircraft maintenance guys drive the Suburban out of the hangar directly out to the end of the runway after we landed.

It was getting to be more like a James Bond movie all the time. As soon as our plane came to a complete stop on the runway in Ogden, Wiggins's Suburban came squealing right up next to our jet. As we piled out of the airplane into the Suburban, a whole fleet of trucks with TV cameras and floodlights started their motors and raced after us. We laid rubber and came to a screeching halt in front of the fence on the backside of the hangar. Unlocking the gate, our accomplice on the ground waved us through and locked the gate behind us so no one else could get out.

Wiggins drove me to Pleasant View. Instead of dropping me off at home, I had him stop at the bottom of a hill at the local corner grocery and gas station, a place called Teddy Bears. Sure enough, when I called home, my wife said several television news trucks were still parked outside. I called Jack Hilden and had him pick me up for another overnight at his house.

On the 10:00 p.m. news, I saw a short clip where Senator Hollings, in his southern drawl, told the press, "I brought this fellow Allan McDonald from Morton Thiokol in Utah all the way here to Washington, D.C., but the man didn't even have the courtesy to talk face-to-face with a United States senator! All he did was talk to us over a squawk box in our Senate conference room." One of the news channels also had the picture of Cal Wiggins walking up the steps of the airplane at Dulles with his hands in front of his face, like he was Al Capone being arrested. You can imagine how I felt when the TV commentator said, "Here you see Mr. Allan McDonald boarding his company jet to leave Washington after stonewalling the Senate."

Work the next day was buzzing about how I had stood up the Senate and embarrassed the good-ol'-boy senator from South Carolina. But some of us had more important matters to think about. Three members of the Rogers Commission—Dr. Eugene Covert, David Acheson, and Robert Rummel— were going to be at the plant to interview individuals who had participated in the infamous January 27 NASA/Morton Thiokol teleconference. I was on their list as well.

In our nicest conference room, the trio conducted confidential interviews with selected members of the staff. Although the interviews took place individually, the company had two of our corporate attorneys present during each individual discussion. None of us knew exactly what to expect. Basically, I thought I'd repeat the information I had shared at the closed hearing of the Presidential Commission at the Kennedy Space Center on February 14. In support, I gathered up copies of the history of O-ring problems and all other relevant data.

When I was finally called in, Covert, Acheson, and Rummel were very cordial. "We just have a few questions for you concerning your testimony at KSC. We just need some clarification and amplification of some of the testimony you've already given. None of our questions are particularly earth-shaking."

From their very first question, it was obvious they wanted more details about the January 27 teleconference: "What is your opinion, Mr. McDonald, as to why NASA appeared to be so reluctant to cancel the launch?"

"Well, the pressure on Larry Mulloy had to be very high," I began, "because Larry was the one who had to cancel the previous launch originally planned for December. You might remember there was a card failure in an integrated electronics assembly of the hydraulic power unit on one of the SRBs. This was the first time that an SRB had officially held up a launch, and Mulloy was reluctant to let that ever happen again."

"Do you think Mulloy came on too strong in this teleconference, or was he just making good technical points relevant to the data being presented?"

"I thought Larry came on very strong, especially when he asked, 'When will we ever be able to launch if we're to hold to 53°?' and when he made a sarcastic comment concerning what he called 'a new set of launch commit criteria' surfacing at the last minute."

The men from the Presidential Commission asked me how I felt about the other people from the Marshall Space Flight Center. "I admired Dr. William Lucas very much," I told them, "but it bothers me to see his program managers so reluctant to raise difficult issues with him." "What about Jim Kingsbury?" they asked. "To be honest," I answered, "I've had some bitter disputes with Jim. At one meeting where I was reviewing a planned Level III FRR presentation, Jim even walked out on me. Jim Kingsbury is a smart guy also, but his people are also afraid to talk to him or go against his wishes."

"Marshall has this pecking order," I explained. "When one of their strong bosses speaks, everybody moves their chairs back to the wall. Sometimes that happens here at Morton Thiokol when Jerry Mason speaks, and I worry about that also."

"Stan Reinartz is too inexperienced," I continued. "He doesn't have the confidence to do the job he found himself in; he's far from being a Bob Lindstrom."

It took them roughly twenty minutes to complete their questions. I was just about to get up to leave, when Dr. Covert asked, "Mr. McDonald, are you actively participating in the SRM Failure Analysis Team at Marshall?"

"I'm a member of the team headed up by Marshall's Bob Schwinghammer," I replied, "but I don't necessarily agree with either the approach or the findings reported by that team to date."

"What do you mean by that, Mr. McDonald?"

"Gentlemen, I've been diagnosing this accident nonstop ever since it happened, and I believe I have come up with a rational explanation for the accident that is compatible with all of the data obtained to date. This includes films as well as measured data."

"Have you given that data to the SRM Failure Analysis Team?"

"I have, but they don't seem too interested in what I've done. I made viewgraphs of a failure scenario and presented them on 4 February to Jack Lee at Marshall and to his *Challenger* Accident Failure Board."

"Do you have a copy of this material?" Covert inquired.

"I do. It's somewhere in the stack of information I've brought along to this meeting."

Shuffling through the pile of papers stacked in front of me, I found a copy of the February 4, 1986, presentation, which I had handwritten, and gave it to Covert. "I've made two copies, so you can have this one."

"Mr. McDonald," interjected one of our corporate attorneys, "please remember that all of our documents that are to be turned over to the Presidential Commission need to be logged in and sent first to our legal department. From there, they will be formally transmitted on to the Presidential Commission."

I could tell that our lawyers weren't very happy with me, but I told Covert that he could keep that copy, and that I would make another for our corporate attorneys.

Back in my office, my boss, Joe Kilminster, told me I needed to attend an afternoon meeting in another one of our conference rooms in preparation for the public hearing of the Presidential Commission scheduled in Washington, D.C., for the following Tuesday, February 25. The afternoon meeting was to involve all of us from Morton Thiokol who were scheduled to testify. Several of the people who were on the list to testify had been contacted by Jerry Mason and Joe Kilminster while I was in Washington for the Senate subcommittee hearing. They were told to prepare viewgraphs on what they planned to present to the Presidential Commission.

"This hearing of the Presidential Commission is extremely important," we were told by our bosses at the start of our afternoon conference, "and everyone participating from MTI must present information that is very concise and accurate. We must make sure that we don't contradict one another or confuse members of the Commission."

One by one, people presented drafts of their viewgraphs for review by our management and corporate attorneys in accordance with an agenda Jerry Mason had prepared the previous day. My turn came last. "Where are your viewgraphs, Al?" Mason asked.

"I don't have any. Furthermore, I'm not sure I'll prepare any charts, because I don't think the Presidential Commission wants to hear a canned presentation. I think they want to ask questions based upon what they have heard to date."

Bob Lund spoke up, stressing that he absolutely needed charts and schematics in order to explain to Commission members the configuration of the field-joint and how it operated, so that the commissioners got a better understanding of the issues pertaining to the January 27 teleconference.

"That would probably be worthwhile, Bob," I ventured, "but to construct a whole viewgraph presentation on what all transpired during the teleconference that evening I think is unnecessary—and unwanted."

Jerry Mason and Cal Wiggins thought it was better to have the charts and told everyone to work over the weekend, if necessary, to get them done. "We'll have a final review of all of the charts on Monday afternoon in our Washington, D.C., office, prior to the Tuesday hearing," Mason concluded.

I left the meeting puzzled and confused as to what was going on and whether I should be spending the weekend preparing viewgraphs. Driving home from work, I noticed that most, but not all, of the TV trucks had left. So it was time to head to the Hildens again. On the phone, my wife admitted she had invited several of the news reporters in for a cup of coffee, because it was so cold outside. Most of them were quite nice, she said, but a couple of them were downright rude. Two or three had come to the door and insisted on an interview with me. When she told them I was not home, they then said they wanted to talk to other members of the family. Linda replied that no one in the family was talking with the press, at which time the reporters became very angry and insulting to my wife. One fellow from an NBC television crew in California, Phil Clapick, who had been invited inside for a cup of coffee, came to Linda's rescue. He told these jerks that they needed to respect my wife's desire for privacy. He shut the door in their faces and told my wife that sometimes you need to do that, because that's the only message some people understand. When I called my wife again the next morning, she said all of the television news trucks and reporters had finally left, and it was safe to come home.

All that the reporters had gotten about me was that I was a fellow who hadn't returned home for a week. This fellow Allan McDonald was still a mystery to them. One of the TV stations did manage to obtain a photograph of me that had appeared in the *Brigham City Box Elder News and Journal* a year earlier, when I had been invited to Israel to participate in an International Aeronautical and Astronautics Conference. The photo subsequently appeared on a number of TV network broadcasts.

I relaxed the best I could that weekend. When Monday morning came, I climbed aboard the company jet in Ogden with the other Morton Thiokol people, all of whom had been busy all weekend preparing charts for their testimony. On the plane, they exchanged copies of their charts with one another to make sure everyone's information was consistent and accurate. No one asked me to look at their charts.

Jerry Mason had constructed a detailed agenda for our participation in the hearings. He assigned an approximate time allocation for each person involved to make sure we could cover all the material in the time allocated by the Commission. I didn't have any charts, but I did bring copies of the notes that I had previously given to the Commission. I also brought along a copy of the February 4 failure scenario that I had given to NASA and to the Commission; a copy of the August 19, 1985, SRM Pressure Seals presentation I had made to NASA headquarters; a copy of the AIAA technical paper I had written in the summer of 1985 on the development of the Space Shuttle SRMs; and a chronology of presentations and engineering reports we had given to NASA over the years on problems related to the SRM O-rings and joint design. I felt nicely rested after my weekend at home and laid back in my seat and dozed for most of the trip to Washington.

By the time we checked into our rooms at the Crystal City Marriott, it was nearly time for dinner. The plan was for us to eat and then go over to our Washington office, located in an adjacent building connected by the Crystal Underground. What Mason wanted was a "dry run," a full dress rehearsal of what all of us would say and do before the Presidential Commission the next day. Each of us was to present his charts to the audience in exactly the same manner we would do for the Commission. Each of us was to stay within our allocated time, and any question or concern with our charts needed to be corrected that night.

I was the last one on the agenda, so it was going to be a very long evening before they got to me.

The plan was for Jerry Mason, our Senior Vice President and General Manager, to speak first. He would explain the Morton Thiokol management structure and where each of the subsequent participants was located in the organizational chain. Mason would then discuss the circumstances that led up to the teleconference with NASA on the eve of the *Challenger* launch. He would lay out where each of the participants was physically during the teleconference and summarize their particular area of responsibility.

In the dress rehearsal, Mason told us he did not plan to get into the details of any of the charts that had actually been presented at the January 27 meeting or into any interpretation of any of the data that had been discussed that night. He would introduce Bob Lund, Vice President of Engineering, and Joe Kilminster, Vice President of Space Booster Programs, who would then provide those details. Lund had prepared several viewgraphs showing the detailed configuration of the SRB field-joint and how it operated when pressurized. He was to go through all the data presented at the teleconference, while Kilminster covered the caucus held in Utah and how the final recommendation to proceed to launch was made.

Mason thought it was important to provide this information to the Commission before anyone from Morton Thiokol said anything else. Engineers Roger Boisjoly, Arnie Thompson, and Brian Russell would then follow, making any comments they cared to make; Roger and Arnie had served on the O-ring Seal Task Force, and Brian had supported the task force work from his post in the program office. The three of them had prepared most of the charts that had been used during the January 27 teleconference.

Mason wanted the company to provide the full story of what had happened in Utah before involving me, because I had been off-site at Kennedy Space Center. This seemed very logical, but I knew it would be late in the evening before they got to me.

Before starting the dry run, Mason requested that everybody turn in their viewgraphs to the overhead projector operator so he could put them in the proper order as designated by Mason's agenda. Roger Boisjoly pointed out that he had not made any new viewgraphs and would need to use some of the charts that were to be presented by Bob Lund. Roger planned to bring along to the hearings some of the notes he had made in his daily log as well as some of the background material that he had already given to the Rogers Commission. Arnie Thompson said he was in a similar position, expecting to use some of the charts that he had personally constructed that were now in Bob Lund's presentation.

I was the only one who did not plan to use any viewgraphs at all, which greatly bothered Cal Wiggins. "Al," he declared, "you really need to prepare some charts to present concerning what transpired at KSC."

"I don't know what I need to present, Cal, because I don't think the Presidential Commission wants to look at viewgraphs and prepared statements from canned presentations. My guess is, they really only want to ask questions based upon what they have already heard. If it's necessary to use viewgraphs to clarify answers to questions, so be it, but I don't think we should plan on making this grandiose, orchestrated presentation to the Commission. I don't have any viewgraphs, because I don't know what they plan to ask me. I think we'd be better off if we *all* responded that way."

Both Jerry Mason and Cal Wiggins made it emphatically clear that we were all going to proceed as planned, because that was the best approach. They told everyone to go ahead and make their viewgraph presentation just as they would do it during the hearings tomorrow. The marching order was to include me.

I grew increasingly uncomfortable with what I was hearing as the night wore on. Jerry Mason, in his opening presentation, did not tell the whole story as to how Morton Thiokol had come to make its final recommendation for NASA to proceed with the launch. "Mr. Mason," I stated, "I do not agree with your explanation as to why our management here in Utah wanted to caucus off-line. You are implying to the Commission that one of the things that triggered the caucus was the comment *I made* in Florida about assessing the ability of the secondary O-ring to seal, because it was placed in the proper sealing location as a result of the leak-check, and this was considered a supportive comment for launch.

"First of all, Mr. Mason, I did *not* make that comment *before* it was decided to have a caucus, so it couldn't have *triggered* the caucus. I made the statement concerning the secondary O-ring *after* Joe Kilminster said they wanted to take five minutes off-line to reevaluate the data based upon the comments that were made in the teleconference.

"Secondly, the statement concerning the improved ability of the secondary O-ring to seal was originally made by George Hardy at Marshall. What

I said was that we should *assess* the comments made by Hardy because they were important.

"Lastly, Mr. Mason, I didn't make that comment to reflect a position either to support a launch or not to support launch. I made it strictly in terms of being something that should be evaluated during the caucus, because we had not addressed it earlier, as indicated by George Hardy's comments. I had indicated earlier that night that there was some good news and some bad news associated with the secondary O-ring's ability to provide a seal at low temperatures."

Mason acknowledged my concern with how he was interpreting the chart, but said that it was too late to change it; what he promised to do was clarify the situation in his actual testimony. "But I have no problem with what is stated on the chart," I clarified. "What I am objecting to is which chart your information is on. The summarized statement that you are attributing to me, even if you attribute it correctly, should show up on the *next* chart discussing what happened *after* Joe Kilminster asked for the off-line caucus to consider comments, because it will be obvious to everyone that Joe asked for the caucus to respond to NASA's comments only, primarily those made by George Hardy and Larry Mulloy. Without those comments, no caucus would have been held, and no change in the launch recommendation would have occurred.

"I think that is pretty darn significant, Mr. Mason, and I think the record should reflect that."

Even though Mason indicated that he could not correct the charts at this late hour to support his testimony tomorrow, I thought that Mason, surely, would make all of this clear orally before the Commission.

I was stunned when he didn't.

Halfway through the dry run, someone came into the room with a press release from the Presidential Commission concerning the planned public hearing, which was to start at 9:30 a.m. the next morning in the auditorium of the U.S. State Department building. The agenda listed the names of those who would be testifying in the order they would be doing it. To everyone's surprise, the press release stated that the hearings would begin with testimony from me.

Mason was furious. He turned to the corporate attorneys, "It's been my understanding that you all had precoordinated our agenda with the legal counsel for the Presidential Commission and that they had accepted our plan, which has me testifying first and McDonald testifying last! Get on the telephone right now and get this whole thing straightened out, because this is the way we're going to proceed tomorrow!"

It took a few minutes for our lawyers to get a hold of the legal counsel for the Commission, and a few minutes more than that before the Commission's counsel could get back to them with a reply from Chairman William Rogers. In the meantime, Mason had everyone continue with the dry run. When the answer finally came from the Commission's counsel that "Chairman Rogers plans to proceed on the schedule that was released to the press" and "if Morton Thiokol management has some overriding reason why this isn't appropriate, Chairman Rogers would be more than happy to meet with them in closed chambers tomorrow morning a half-hour before the scheduled hearings," Mason's face turned from red to purple. He told the lawyers to arrange for such a meeting, because he was not going to accept the revised agenda.

"That might be a mistake, Mr. Mason," I ventured, "because we don't want to foster an adversarial relationship with the Presidential Commission going into these public hearings. If we insist on our own agenda over what the Chairman has already decided, we may just do that. I would strongly suggest that we cooperate fully with the Commission and not appear to be confrontational or combative."

Mason snapped back at me: "It's the Commission that has gone back on its agreement! We're not asking for anything that they haven't already agreed to, and we need to get that straightened out.

"I and a few others will go ahead and meet with Chairman Rogers before the hearing. It isn't necessary for you to attend, Al. You just be there about fifteen minutes before the scheduled hearings like everybody else."

When it finally came my turn for the dry run, Cal Wiggins asked me for my viewgraphs, knowing full well that I had none. Again, I told him and everyone else that I didn't see the need to prepare any. Wiggins then asked me what I was going to say, and I repeated that I wouldn't know until the commissioners asked their questions.

On that dissonant note, the dress rehearsal ended. Everybody went back to their hotel rooms without anyone saying a word to me.

17

"The Extremes of Allowable Engineering"

I didn't sleep very well that night, so nervous was I about testifying in public. I felt even more uncomfortable about testifying first, against the wishes of my management, who planned that I would testify last. Taking a taxi to the State Department, I arrived shortly after 9:00 a.m., not wanting to be late. Walking into the auditorium where the hearings were being conducted, I was completely overwhelmed by the huge bank of floodlights focused on the stage and the multitude of TV cameras focused on the Presidential Commission's seating area and the witness table. I sat down all by myself in an empty row of seats directly in front of the stage. It wasn't long before the auditorium filled to the brim with reporters and interested bystanders from NASA, Congress, and the general public.

The other Morton Thiokol people who were scheduled to testify were still not there, and it was just five minutes before the hearings were to begin. Nervously, I got out of my seat and walked into the hallway adjacent to the auditorium, where I noticed several Morton Thiokol people, including Roger Boisjoly, Arnie Thompson, and Brian Russell, standing outside of the Commission's chambers. They told me that the others were inside the Commission's chambers with Jerry Mason and the corporate attorneys discussing the order of witnesses.

My watch said 9:30 sharp. "The hearings are supposed to be starting," I remarked, "and neither the commissioners nor most of those testifying are even in the auditorium."

I decided to stay in the hall rather than walk back into the auditorium, where all the floodlights were now on and all the cameras running. It seemed like hours before the chamber doors opened and the Morton Thiokol people walked out, scowls on their faces. Jerry Mason walked up to me, "You will be testifying first." Chairman Rogers would not compromise. He would allow Mason to testify right after me, but I would be the first person called to the witness stand.

I headed toward the auditorium. Just as I was walking through the door, I felt a hand on the back of my shoulder. Turning around, I saw Chairman Rogers motion to me to step back into the hallway. The rest of the Commission members walked past me to their seats on the stage. Rogers could tell I was very nervous. He put his hand on my shoulder and told me that I was going to be the first witness. He told me to relax, and if at any time I felt too uncomfortable to proceed, to just tell him that I would like a short break, and he would immediately call a recess.

Inside the auditorium, I sat down in the only open seat in the front row. Rogers then entered the auditorium and took his place on the stage.

"Today the Commission will focus on the events leading up to the decision to launch the *Challenger*. The Commission wants to be fair in the presentation of the facts, because the subject matter may involve possible human error as distinguished from equipment failure. The Commission will attempt to give a right of reply as soon as possible to any person who believes he has been unfairly criticized or whose actions may have been inaccurately portrayed.

"Before beginning today's hearings, I want to note a change in the order of testimony. The first witness this morning will be Allan McDonald. Following Mr. McDonald, and at the request of Morton Thiokol, the Commission will next hear Jerry Mason, who is Executive Vice President of Morton Thiokol."

Asked to come on stage and sit at the witness table, I was sworn in. After directing me to provide some background on my education and professional career at Morton Thiokol, Chairman Rogers asked me to tell the Commission what I knew about the circumstances that led to the decision to launch *Challenger* and, in particular, what my involvement had been in the discussions held the night before the launch.

I proceeded to tell them everything I knew. Periodically, members of the Commission interrupted my testimony to ask a question. I answered every question the best that I could. Some of the questions concerned the original design and qualification test program for the Space Shuttle solid rocket motors, and I told them I had not been involved in the program during that time and they should address those questions to those who had been directly involved. They also asked me several questions about the management caucus held in Utah during the teleconference on the eve of the *Challenger* launch, and I also told them that I wasn't there and that they needed to address that issue with those who were.

The answers I gave basically provided the same information as I had volunteered in the closed hearing of the Presidential Commission held at Kennedy Space Center on February 14. The major difference resulted from the Commission's request that I tell them everything I knew about the circumstances leading up to the January 27 teleconference and what happened during that teleconference to produce Morton Thiokol's original recommendation not to launch followed by the request for an off-line caucus to reevaluate the data. The Commission was particularly interested in my personal involvement in these activities and my recollection of what others participating in the teleconference had said.

At the Commission's closed hearing at KSC, my testimony was basically limited to the conversation I had had with the NASA people in Florida during and after the Thiokol caucus in Utah. But now, in the public hearing, Chairman Rogers was trying to find out if there was anyone at Morton Thiokol, or at NASA, who had supported a position to launch prior to the off-line caucus in Utah. This didn't surprise me. What did greatly surprise me was the suggestion, which had apparently been passed on to the Commission by some individuals at NASA and even within Morton Thiokol, that the statement I had made (*after* Morton Thiokol requested the off-line caucus to reevaluate the data) about examining the ability of the cold secondary O-ring to seal was a statement in some way in support of the launch. In my view, I never ever made *any* such statement. What I did was restate a comment that had been made earlier by Marshall's George Hardy, a statement that we at Morton Thiokol had not answered. Now the very point I had taken issue with in Jerry Mason's dry run the night before about this statement became a central element in the Commission's line of questioning.

"Could I ask you to stop there just for a moment, Mr. McDonald, and go back? We will come to the caucus in a minute, but go back and explain what was said to convey the decision of no-launch. I gather at that point the decision by Thiokol was to recommend against a launch?"

"That is correct. At that point, the recommendation was not to launch below 53°."

"Who did the talking on that subject?"

"It was Bob Lund, Vice President of Engineering, who presented that position."

"Were you able to ascertain from that conversation how the engineers as a group stood on that?"

"Well, the engineers as a group, I can't speak for the group. I was not there, but I did hear the engineers who presented the charts, that they actually presented as part of that, the support of the 53° recommendation. And I thought they were very strong in their convictions as to why they felt uncomfortable to go outside that experience base."

"And who were they?"

"Roger Boisjoly, I think, was one of the strongest ones, and Arnie Thompson, who presented those positions and presented the charts."

"And Mr. Lund himself at the time?"

"Mr. Lund himself at the time didn't present the detailed technical charts, but he did present the conclusions and recommendations, and that was his recommendation as Vice President of Engineering."

"Was anyone who was on the telecom from Thiokol's side recommending launch?"

"At this time, no, there was no one that recommended launch. I don't recall there was anyone at either Marshall or Thiokol or from NASA who didn't agree that cold temperatures were going in the wrong direction and didn't help anything. No one from Thiokol at that time recommended launch."

"What were the comments by NASA officials about the recommendation, as you recall?"

"As I recall, there were some fairly strong comments about being appalled by the recommendation, about trying to institute new launch commit criteria at the last minute, about when we will ever get this launch off. I thought those were fairly strong comments."

"And those are comments, according to your recollection, were made by whom?"

"The comments relative to the launch commit criteria and 'when we will ever get this off' were made by Larry Mulloy. The comment about being 'appalled' was made by George Hardy, who was at Marshall. I think I recognized his voice, but that was by voice recognition."

"And so it was decided. . . . Are there any other questions that any other member of the Commission has up to this point?"

"Mr. McDonald, I have one question," said USAF General Donald Kutyna. "Before you went off the net, did you ask or make a comment about the secondary O-ring seal seating?"

"Was I *asked* to make one?"

"No, did you *make* a comment?"

"I did make a comment, yes."

"Could you recall that comment, please?"

"Yes. I think I read it to you, and I would be glad to do that again. I made the comment that lower temperatures are in the direction of badness for both O-rings, because it slows down the timing function, but the effect is much worse for the primary O-ring compared to the secondary O-ring because the leak-check forces the primary O-ring into the wrong side of the groove, while the secondary O-ring goes in the right direction, and this condition should be evaluated in making the final decision for recommending the lowest acceptable temperature for launch. That was a comment I made."

"That confused some people didn't it, Mr. McDonald? Some of the witnesses I spoke to thought that was a comment in *support* of the launch—the fact the secondary O-ring seal *would* seat."

"Well, my comment was a 'good news/bad news' comment. There was good news and there was bad news. The good news was that the secondary seal was in the right position, but that's not unique to temperature. It's always that way. The bad news was that the primary seal was the one we were depending on for the full-ignition transient, and it was going to be a lot worse than it was. But even the secondary, as I mentioned, wouldn't be as good cold as it would be normally."

"But by this comment you were not supporting the launch?"

"No, by that comment I was not supporting the launch. I was just saying it was an important consideration and I thought that if we could run some calculations to show that the temperature did not affect the timing, then *that* would be supportive of the launch. If it did, and that was a concern, if we pushed that timing out long enough. . . . We had a chart in there that said if that timing went beyond the 200 psia regime, while we were still eroding the primary O-ring—that for whatever reason if we ever bypass it at that time, we can't depend on the secondary. That is what was important."

"So at the time of the caucus, then, you never favored launch?" Chairman Rogers asked for clarification.

"No."

"And you made it clear that you were opposed to launch?"

"Well, I never said I was opposed to launch. I just made a few comments about why I thought some of the data was not appropriate, like the static test, for saying the O-rings were good to 36°. I made the comment about the lower temperatures affecting both O-rings but that it affected them a little bit differently because one of them, a dynamic O-ring, it moved. One of them had a lot more movement, which affected it more than the other one did, because it has to move across the groove. I made these comments."

"But you accepted the recommendation?" Rogers asked.

"I accepted the recommendation, yes. I thought it was the appropriate recommendation."

"In other words, you accepted the recommendation of no-launch that was made by your company?"

"That is correct."

Vice Chairman Neil Armstrong followed up. "Would you be surprised, Mr. McDonald, if your comments were interpreted by both your own company personnel and by Marshall personnel as being supportive of the Marshall position?"

"Yes, I would be surprised at that. I wouldn't be surprised that that would be evaluated as the effect of that, but I would be surprised that it was interpreted as supporting a launch."

What I couldn't tell Neil Armstrong was that I wasn't at all surprised that the Commission had heard this interpretation from some of its witnesses, because it was becoming clearer and clearer to me that both Morton Thiokol and NASA management were trying to find any shred of evidence they could to support a launch decision and discredit my earlier testimony before the Commission. Sometime later, I received a telephone call and a note in the mail from Wiley Bunn, the Director of Reliability and Quality Assurance Office at NASA Marshall, supporting my actions and testimony and warning me that Marshall management was plotting against me and that I should watch my backside.

As my testimony proceeded, I became more and more relaxed. The stage lights remained very bright, hot, and uncomfortable, but I managed to stay relatively cool and composed. My wife had warned me before I left home not to talk with my hands, which I have a habit of doing. She said I should leave them folded on the table. I started my testimony that way, but it wasn't long before I was using my hands to explain how the joint was assembled; once I almost knocked a glass of water off the table.

The next part of my testimony, out of necessity, became very technical. The Commission had already heard a great deal from NASA about the different "Criticality" levels associated with Shuttle design components and was particularly distressed that our SRB field-joint design had been designated by NASA in 1983 as a "Criticality 1" component rather than a "Criticality 1R."

Criticality 1 meant that a single point failure in that element could result in a catastrophic failure of the Space Shuttle vehicle, with loss of the orbiter and crew. Criticality 1R (R stood for redundant) indicated that it would take two failures in that particular component before a catastrophic failure could occur.

In the case of the SRB, the original design intent was to provide pressure-seal redundancy in the field-joint by incorporating two O-ring seals—a primary and a secondary O-ring that could be pressure-tested and checked for leaks after assembly. This was accomplished by pressurizing the cavity between the two O-rings through a leak-check port in the hardware and determining the pressure-seal integrity by monitoring any pressure decay in the area. During motor operation, the joint had to contain an average pressure of approximately 660 psia for approximately two minutes. The leak-check was conducted for ten minutes at 50 psia and only allowed a 1 psia drop in pressure over that time period. The stabilization pressure was increased to 200 psia when it was discovered that the vacuum putty in the joint could possibly mask a leak below 150 psia. The pressure was then dropped to 50 psia to conduct the leak test.

The original analysis of the joint, which had been conducted back in the mid-1970s, indicated that seal redundancy was maintained, because the joint closed during pressurization. It wasn't until after considerable hardware had been fabricated and tested that it was realized that the joint didn't close during pressurization; it actually opened. Some measurements of the joint opening had been made through the leak-test ports during "hydro-proof" of some of the rocket cases at Morton Thiokol and also during the pressurized structural tests conducted at Marshall, but the data from these various tests were not in agreement. The tests conducted at Marshall involved a horizontal assembly while the tests at Morton Thiokol were vertical assemblies, just as the launch vehicle itself was. The data from the horizontal structural tests at Marshall indicated the joint could open as much as 0.060 inch, while Morton Thiokol's hydro-proof data indicated only 0.030–0.040 inch of opening. Based upon Marshall's higher number, it was determined that it was possible that the secondary O-ring might not have sufficient squeeze to maintain a seal if it were called upon to do so after the joint had fully opened if the O-ring was contained in hardware that represented the combination of "extremes of allowable engineering." These extremes included the maximum clevis opening, the minimum thickness tang, the maximum O-ring groove, the minimum cross-sectional diameter O-ring, and the maximum expected operating pressure of the motor. Morton Thiokol engineering and NASA Marshall engineering could not agree on what data was correct. Our company's Roger Boisjoly thought the NASA data might be misleading because of its horizontal test assembly, while engineer Leon Ray from Marshall insisted that the NASA data was real and that the wider gaps, not Thiokol's narrower gaps, should be used in assessing the SRB joint for redundancy. Because of this disagreement between Boisjoly and Ray, Larry Wear, the SRM Program Manager at MSFC, had approved a series of tests to determine the real joint opening.

At the time of the *Challenger* accident, this "referee test" was still being prepared. At Leon Ray's insistence, the SRB joint had been downgraded by NASA in late 1982 from Criticality 1R to Criticality 1, because Morton Thiokol could not show that the secondary O-ring would maintain contact on both sides of the joint with such a large gap.

Thiokol never agreed to the change, but we did calculate actual O-ring squeezes using "as-built" hardware dimensions. We presented these at every flight readiness review to assure ourselves that we would never fly a set of hardware that didn't meet the requirements for seal redundancy. We never assembled any hardware with all of the worst-case dimensions, and we never experienced a (3-sigma) maximum-expected-operating pressure after the Criticality was changed from 1R to 1. However, in going back into the records on the earlier Shuttle flights before this change occurred, we noted that a field-joint on the fourth flight of the Shuttle (STS-4) did not meet the criteria. Ironically, this particular set of boosters was never retrieved because of parachute failures, meaning that the remains of the boosters rested on the bottom of the Atlantic Ocean.

After my lengthy explanation of how the field-joint had been changed from Criticality 1R to Criticality 1, several Commission members, notably Dr. Sally Ride and Joseph Sutter, expressed stern criticism of how all of this had been managed by NASA and Morton Thiokol. Ride believed that once the joint had been redesignated as a Criticality 1, NASA's system would not allow us to develop a set of flight rationale that depended on the ability of the secondary O-ring to seal. The critical items list for the SRB should have kept the joint as a Criticality 1, because of the inability of the secondary O-ring to *always* provide a seal.

Not only was Dr. Ride absolutely correct technically, she was correct in the broader view of what made sense within the entire system of flight safety and risk assessment. Since the recommendation to proceed to launch provided to NASA by Morton Thiokol management included rationale related to the ability of the secondary O-ring to seal, it should have required a real-time launch waiver, processed by Marshall through NASA's Mission Management Team, because that flight rationale was in direct conflict with the established critical items list for the SRB. As far as the Mission Management Team knew, we were flying the SRBs on the basis of a Criticality 1 field-joint—and doing so on the premise that the primary O-ring would *always* seal, but could not depend on the secondary O-ring to seal, therefore seal redundancy in the joint was lost. The flight rationale for STS-51L stated that the ability of the primary O-ring seal could be seriously degraded by the cold temperatures, but that the secondary O-ring *would* seal, which was in total violation of the established critical items list and should have required a waiver from the Mission Management Team prior to launch. The critical items list did not address O-ring resiliency even though it was an even more important issue relative to the ability of the O-rings to seal at the cold temperatures forecast for this launch—and this issue should also have been addressed in the waiver.

Physicist Richard Feynman was very interested in the dynamics of the field-joint, particularly the pressure actuation and extrusion of the O-rings into the gap created by the motor pressure. I explained to Dr. Feynman how this occurred and why it was so important to understand the timing function in the early phases of ignition for assessing the ability of the O-rings to seal properly.

"You suggested, Mr. McDonald, that the secondary seal would not be much affected by the temperature," Feynman stated, "but now you are telling us that because of the complete or nearly complete loss of resilience—that is, the tendency to spring back—the secondary seal would require very little rotation to open. Do I understand that correctly?"

"I said it wouldn't be as affected as much as the primary seal," I replied, "because it doesn't have to move from one end of the slot to the other. As to the effect of resilience, you're absolutely correct. It still has the same problem. As far as extruding into the gap, it has still got the same problem."

"We were talking about a discussion about the secondary seal being redundant until the metal parts rotate," Feynman noted. "As the pressure starts to build up, it can't move the primary seal until there is some pressure, and then there's a very small pressure, perhaps, and a very small rotation. Isn't it true that the rotation is more or less proportional to the pressure, or is there a delay of some sort? Why is there a time delay between the two?"

"There is some delay," I said, "because the joints are stiff enough that, under certain pressures, they just don't move at all, until sufficient pressure has built up to produce a rotation. So there is a delay in that time, but at some point in time it becomes a direct function of the pressure."

"Aren't the laws of elasticity such that everything is proportional to force and all of the spaces are proportional to the force?" Feynman asked. "Wouldn't it be true that at every joint there is some rotation and the rotation is proportional to the pressure?"

"Well, I think when you are down to a few psia or 50 psia, I don't think you are rotating anything," I answered.

"You are rotating it one-tenth as much as you rotate it at 500 psia, is that right?" he asked.

"Well, the structure is so stiff that I can't believe you rotate it at all at 50 psia," I responded. "You are moving the O-ring seal back into the groove, however."

I was so involved in the issues at hand that I did not give a thought to the fact that I was arguing a scientific principle with a Nobel Prize–winning professor in nuclear physics from the California Institute of Technology on national television, but my wife did. Watching the hearings on CNN at a friend's house (we didn't have cable TV), Linda wondered how I dare do such a thing. Later I told her, "I have great respect for Dr. Feynman. He is an extremely intelligent man. But I know a lot more about solid rockets than he does."

At the close of my testimony, the questions returned to the possibility that management pressures had prompted Morton Thiokol engineers to change their recommendation not to launch.

"At the conclusion of your testimony in executive session," Chairman Rogers said to me, "Dr. [Alton] Keel asked whether you felt that you were under pressure, or had been under pressure, or the company had been under pressure, to reverse its decision, and I think your answer was yes. Do you remember that?"

"Yes, definitely. There was no doubt in my mind, I felt some pressure. I feel that I have a responsible management position, and I felt pressure."

"Would you explain the reasons for feeling pressure?"

"Well, I have been in many flight readiness reviews, probably as many as anyone, in the past year and a half at Thiokol, and I have to get up and stand before, I think, a very critical audience at Marshall, and a very good one, justifying why our hardware was ready to fly. I have to get up and explain every major defect and why we can fly with that defect. And for the most part, they are very minor—very, very minor. I've been hassled about how I'm sure that it is OK to fly with these defects—you know such things as losing vacuum on a carbon-cloth part in the nozzle while the part is basically cured. It is a critical process. There are a lot of critical processes, and I have to address every one of those in great detail as to why I am sure that part has not been compromised. It has been that way through all of the reviews I've ever had, and that's the way it should be. It is not pleasant, but that's the way it should be."

"But I was surprised at this particular meeting," I continued, "that the tone of the meeting was just the opposite of that. I didn't have to prove that I was ready to fly. In fact, I think Bob Crippen made the most accurate statement I have ever heard. His conclusion from that meeting was that the philosophy seemed to have changed. He had the same impression I did, that the contractor always had to stand up and prove that his hardware was ready to fly. In this case, though, we had to prove it *wasn't* ready, and that is a big difference.

"Yes, I felt that was pressure."

"Can you explain a little more," Rogers coaxed, "what source the pressure came from in your mind?"

"Well, I think the strongest statements were made by Mulloy, and even some of the people from Marshall that were on there, Hardy, made what I think were fairly strong statements that I took as pressure about 'when we will ever fly this thing,' and the launch commit criteria that we shouldn't be generating at the last minute, and being 'appalled' by our recommendation to fly at temperatures only as low as 53°. That, to me, was pressure. It may not have been interpreted that way by others, but it was pressure to me."

"Any other pressure that you want to refer to from this time or at any time?"

"No. I just felt that the way the comments were made, as strong as they were made, and the fact that the conditions in justifying this launch were so much different than anything I'd been involved with before."

"As far as the telefax was concerned, were you normally to sign that telefax, or was the procedure that was followed the normal procedure?"

"Well, I'm not sure what the normal procedure is. I felt that since I was there as the senior official at the Cape that I'm the one that usually was responsible for that. I know at the L-1, when we have a normal launch, there is a poll that is conducted by Jess Moore of all the contractors, whether we are ready to fly, and I am the guy there that has to get up and say, 'Thiokol's hardware is ready to fly.' I felt that it was my responsibility. That's why I'm there. I can't recall whether we were asked to sign anything like that before. We have a presentation that is given at every flight readiness review that is signed off by the principal parties. Joe Kilminster usually signs all of those on the formal review, but if anything comes up afterwards it has been done on an oral basis. But I don't know. I never was put in that position, and I don't know if Joe has. I don't believe he has."

"So, the fact that a written decision was requested, as far as you were concerned, was not a normal way to do it?"

"It was not normal as far as I was concerned."

"And, as I recall your earlier testimony, you testified that you made it clear that you would not sign a written statement approving the launch. Am I correct?"

"Yes, that is correct. And I think that has been misinterpreted, at least by the press. They said that I was overruled by my supervisor. That is not true at all. I chose not to sign that. He didn't overrule me. I felt that that decision, when I started, was an engineering decision by the people that understood the problem the best, that had all of the data and facts, and they were the ones that should recommend it. And that is why I said that. It wasn't that I was overruled."

"Mr. McDonald," spoke up Commissioner David Acheson, a former U.S. Attorney for the District of Columbia, "did you consider bringing your concerns about the final recommendation to the personal attention of Mr. Moore, Mr. Aldrich, or Dr. Lucas?"

"Well, I'm very familiar with the process by which these things are reviewed, and I was absolutely positive and sure that they were brought to their attention, because that is the way things go. I talked to the SRB Project Manager. I talked to his boss, the Shuttle Project Manager in the Shuttle Project Office, and I assumed that those all went through those reviews. I had no doubt in my mind that they had."

"You believe that NASA folk indicated they would pass your concerns along," Richard Feynman commented, "and I presume that you thought there were three concerns that you were talking about: the O-rings, the ocean sea, and the ice. But could they have meant only the last two concerns, the sea and the ice? That is one question for you. And the other is, who were the NASA folks that promised to pass your concerns along?"

"I guess people could have interpreted it that way, Dr. Feynman. I thought it would be all three, but the people were Mulloy and Reinartz who I felt would pass those on, and I presumed they passed them *all* on. I didn't see any reason why they wouldn't."

"Mr. McDonald," inserted Robert Hotz, former editor of *Aviation Week and Space Technology*, "you mentioned earlier that you thought this decision on launching should be an engineering decision and not a program management decision. How would you characterize the final decision to launch, from Thiokol? Was it engineering or was it management?"

"I guess I would have to characterize it as a management decision, the final decision."

18

Changing Hats and Minds

I was under the hot lights for two and a half hours before the Commission ran out of questions for me. A ten-minute recess occurred before swearing in the next witness, Jerry Mason. Now an observer, I was extremely interested in how Mason would handle his testimony, because he had planned it to come first and to be very structured with all his prepared viewgraphs.

The reality was that now that I had testified for two and a half hours, the Commission had far more questions for Mason than they would have had otherwise, making his canned talk even less viable. The Commission had asked me several questions about what had taken place during the caucus at Utah, and I had told them that since I had not been present, they needed to direct those questions to someone who was there.

That proved disastrous for Mason. He had to start his testimony in the middle of his planned presentation, and it didn't take long for Commission members to lose interest in his charts and start asking him some tough questions just as I had anticipated. Jerry got very flustered and really looked bad during his televised testimony.

At a critical moment in his testimony, Mason indicated that he conducted a poll and concluded that it was safe to launch.

"Could you tell us what the poll showed?" Chairman Rogers asked. "You say that you and Mr. Kilminster, Mr. Wiggins, and Mr. Lund were unanimous. How about the others?"

"We only polled the management people," Mason responded, "because we had already established that we were not going to be unanimous and we had already heard . . . "

"That wasn't the question," Rogers quickly interrupted, "The question was, what about the others? You testified before that you thought probably all of the engineers were against the launch, and now you say you took a poll and you only polled the managers. You didn't ask the engineers how they felt after their review of the data."

"In that discussion," Mason responded, "I felt that everyone had presented their opinion, and there were a number that I felt indicated, as we did, that with the consideration of the erosion margin and, well, the very factors that I just talked about, that they also considered that to be a reasonable . . . "

Rogers again interrupted, "Did they express a change of mind? Did they say, 'Well, I've changed my mind'?"

"Not per se, no. Excuse me, I just feel that it was the kind of discussion that we frequently have, in which all of the people express their opinions, and they make them clear."

General Kutyna started to bore in on Mason. "Mr. Mason, you have used the word 'uncertainty' now four or five times in the last five minutes and now you've just been unable to quantify this thing. That is the best thing you had going for you. I mean, every launch has a risk, but you take that risk because something must be achieved. What was driving things here? What was to be achieved that caused you to go? Why couldn't you wait a day?"

"Well, as far as waiting a day is concerned, we wouldn't have gained any more information."

"You would have gained temperature," Kutyna quickly noted.

Vice-Chairman Armstrong entered the discussion, also trying to understand the reason for Thiokol's actions: "Mr. McDonald has stated that he did feel some pressure in this meeting. Would you characterize your own feelings about pressure that may have influenced the decision process in any way?"

Obviously very uncomfortable with this question, Mason replied, "I've thought a lot about that, and there was some pressure, but I believe it was in the range of what we normally encounter whenever we're taking a position. NASA tests us on that position to explain it, justify it, that sort of thing. That's the way I perceive we're being tested, on how supportive or how well we can justify our position. So we responded in that fashion. Now, I can say, I get pressures in a lot of cases, from a lot of decisions, not just from NASA, but from many people I deal with, my boss and so forth. I think I am able to treat that properly and make a sound decision independent of that."

The puzzled faces of the commissioners made it clear they were in a state of disbelief. Jerry Mason, Senior Vice President and General Manager of Morton Thiokol's Wasatch Operations, was obviously trying to protect NASA by carefully avoiding anything that might accuse the agency in any way for exerting pressure on Morton Thiokol to reconsider its decision not to launch, despite the fact that this was the first time ever that NASA had challenged a recommendation from us stating why our hardware might not be safe to fly.

Alton Keel, the Commission's Executive Director, was particularly disturbed by Mason's comment that he didn't feel any pressure from NASA to change the recommendation. Yet he, Mason, had exerted a great deal of pressure on his own subordinate management. "What triggered the caucus that Thiokol asked for, according to the testimony, Mr. Mason, was the fact that NASA had expressed surprise and concern about the impact of such a recommendation for not launching below a 53° temperature for the O-ring, and particularly the impact on schedule. During the thirty-minute caucus, was there any part of your discussion that addressed that impact on schedule and what that meant to Thiokol? Any management concern on that?"

What Mason said next in response shocked me to the core, not only because it was not true but also because it was totally contrary to what Mason indicated that he would say when this subject came up on his chart the previous night before. Worse, he did not even use the chart indicating that the caucus had been called to evaluate the concerns from NASA's Mulloy and Hardy.

"Well, when you say what triggered the caucus was NASA's expression of concern," Mason said, "in fact, there's a bit of irony there, because from my viewpoint the primary trigger for the caucus was McDonald's comment about the fact that the secondary O-ring was in a preferred position, and that was a point that we had not actively considered in our previous discussion."

"Mr. Mason," Chairman Rogers interjected, "when you spoke to Lund and told him in effect to take off his engineering hat and put on his management hat, wasn't that pressure on your part to a subordinate that he should change his mind?"

"Well, I hope not, but I guess it could be interpreted that way."

There was no longer any doubt in my mind, based on what I had just heard Mason say, that our company had decided to do everything it could to protect NASA and totally discredit me if it could. Our corporate attorneys had declared earlier that our company's succumbing to pressure to change the launch decision could be considered negligence, but if our decision had been made on a disputed interpretation of the data, there would be no basis for litigation against us. I didn't really know if this was the course of action being taken, for just those legal reasons, whether it was being done to keep our company in good stead with NASA, or whether it was purely retaliation against me for bringing this whole issue to the attention of the Presidential Commission after NASA had been so successful in avoiding it.

Morton Thiokol's two most highly regarded seal experts, Roger Boisjoly and Arnie Thompson, testified next, basically reiterating everything they had told the Commission in the closed hearing of February 14. However, Boisjoly did address some of the issues contained in the written documents that he had turned over to the Commission after that hearing. Boisjoly was testifying about how he and Thompson had presented data opposing the launch to Thiokol management when Dr. Arthur Walker of the Commission, a professor of applied physics at Stanford University, asked him, "At this point did anyone else speak up in favor of the launch?"

"No, sir. No one said anything, in my recollection, nobody said a word. It was just being discussed among the management folks. After Arnie and I had our last say, Mason said we had to make a management decision. He turned to Bob Lund and asked him to 'take off his engineering hat and put on his management hat.' From that point on, management formulated the points to base their decision on. There was never one comment in favor of launching by any engineer or other nonmanagement person in the room, *before or after the caucus*. I was not even asked to participate in giving any input on the final-decision charts."

Walker then asked Boisjoly, "Do you know the source of the pressure on management that you alluded to?"

"Well, I can't speak for them, but I felt the tone of the meeting was exactly as I summed up, that we were being put in a position to prove that we should not launch rather than being put in the position to prove that we had enough data *to* launch. I felt that in a very real way."

"These were the comments from the NASA people at Marshall and at KSC?"

"Yes."

Before Boisjoly and Thompson left the witness stand, Sally Ride commented: "Just to be perfectly clear about this, is it fair to say that as engineers you didn't believe you had data or analysis to characterize the performance of the joint at a 30° temperature?"

Boisjoly responded, "That is correct."

"Is that true of you, too, Mr. Thompson?" Chairman Rogers asked.

"Yes sir."

Next up was Bob Lund and Joe Kilminster. As it turned out, the testimony from Lund, our Vice President of Engineering, was as confused as Mason's, and he, too, left the Commission members bewildered.

"How do you explain the fact, Mr. Lund," Rogers queried, "that you seemed to change your mind when you changed your hat?"

"We have got to go back farther in the conversations than that," Lund argued. "We've dealt with Marshall for a long time and have always been in the position of defending our position to make sure that we're ready to fly. I didn't realize until after that meeting—actually, for several days after that—that we had absolutely changed our position from what we had been before. But that evening, I had never had those kinds of things come from the people at Marshall that we had to prove to them that we were *not* ready. So we got ourselves into the thought process that we were trying to find some way to prove to them it wouldn't work, and we were unable to do that. We couldn't prove absolutely that the motor wouldn't work."

"In other words, you honestly believed that you had a duty to prove that it would *not* work?"

"Well, that is kind of the mode we got ourselves into that evening. It seems like we had always been in the opposite mode. I should have detected that, but I did not. The roles kind of switched. So after making, or listening, to the verbal presentation in the afternoon, they asked what Thiokol's position was. I looked around the room, and I was the senior person, and I said I don't want to fly. It looked to me like the story said that 53° was about it. Of course, we were then requested to go back and do something more, prepare detailed charts to show that in more detail. It was all in a real-time mode, and we were trying to absorb the data and put together the story. As a result of that telecom, I gave the charts that made the recommendation that we wanted to wait until the motor got to 53°. I didn't see anything different that I hadn't seen before. You've heard the story of what happened after that."

Neil Armstrong had a question about the 53°: "It has been often stated that the recommendation was at this point in time to stay within your experience base, but I find that to be a peculiar recommendation in the operation of any kind of system, because normally, you say, from our experience base, from our data points, from our analysis, and from our extrapolation, we would be willing at any time to go beyond our experience point out this far as a next step, and the only reason you say, 'I would stay within my experience base,' is that you had a problem at that point that said you dare not go any farther. So could you clarify why you said that?"

The first man on the Moon was identifying what was, indeed, the exact problem. We had seen evidence of the primary O-ring not sealing properly in the previous coldest launch (STS-51C), conducted one year earlier. Lund tried his absolute best to avoid this very logical conclusion. "It wasn't a question of that, Mr. Armstrong. It was a very definite question of conservatism. You know this mission has people aboard it, and we are very concerned about that. I want to make sure that, if there is any hint of a problem, we are not extending that. We had no data at that point that indicated we should go beyond that. Is that an answer to your question or do you want me to try again?"

Sally Ride reengaged: "Maybe a slightly different way of asking this is, normally, when you are trying to extrapolate beyond, maybe beyond, your flight experience, you rely on your qualification testing program; a system or subsystem is qualified to fly within a certain regime or a certain envelope. That would include environmental effects like temperature. Did you have any sort of qualification range on the temperature of the solid rocket motor?"

"Yes, we had development and qualification motors that were tested down into the 40s, as has been pointed out. The data, because of the horizontal assembly problems, what we were trying to do is put it back to a condition that would resemble what we would have in a vertical installation. So we had gone in and repaired the putty, because when you put those together horizontally you can't do it the same way. You just can't. So although the intent was to put them back to the vertical, there was some doubt that we could really do it adequately. So there was always the question, 'Was it perfect'? And the answer was, it was not perfect. My belief was that those 40° motors were probably valid and adequate, but there was doubt."

Once again, I couldn't believe what one of my corporate bosses was saying; more than that, I couldn't believe that Bob Lund himself really believed what he had just said. It was clear to everyone in the SRM program that static tests were totally invalid for qualifying the O-ring seals in the field-joints, because of the pretest packing of the putty in the observed blowholes. We may have thought for a while that the horizontal assembly created these blowholes on the static tests, but STS-2 showed severe erosion of the primary O-ring in the field-joint, the most severe prior to the *Challenger* accident. Eight other field-joints from returned flight motors also showed blowholes in the vacuum putty, so it was unfathomable that anyone could believe that the early static test program was valid for testing the temperature effects on the O-ring seals in the field-joints. During the January 27 telecom, a couple of people, including myself, had mentioned that the low temperature static test data was not valid for proving that the seals would work properly below the 53° launch recommendation.

Evidently, NASA and Thiokol management together had decided to try to convince the Commission that this data was legitimate in supporting the launch recommendation. They also wanted the Commission to believe that the SRMs were qualified to the CEI specification requirement of 40°, even though the motors were never actually qualified to this requirement.

"So going down below, say, 47° or 40°," Sally Ride wondered, "you could consider it, taking it below what the motor had been qualified to?"

"Correct," Lund stated. "One thing we need to make very clear. SRM-15 and SRM-22 were the ones that were the blowby motors. But keep in mind there were ten motors and thirty joints between those two temperatures that had no blowby at all. As NASA Marshall pointed out, as I think Mulloy pointed out, 'you know, the data it is just not conclusive at all,' and it wasn't because we had a low-temperature motor and a high-temperature motor, but we had ten motors in between that showed nothing."

Boeing Executive Vice President and Commission member Joseph Sutter asked Lund a very tough and insightful follow-up question: "Since your people were responsible for the design and were responsible for the testing and for the qualification, when it got to the point where there was a question of whether a launch was to be made, shouldn't you alone, or your organization alone, be the one that says yes or no? Why would there be a question coming from NASA because of not having data as well presented as they wanted? Why would they raise questions on your data, and why would you respond to that question? Why didn't you just tell them, 'It's our decision, and this is it,' and not respond to the pressure?"

"As a quarterback on Monday morning, that is probably what we should have done, but you know, you work with people and you develop some confidence, and I have some great confidence in the people at NASA we have worked with. We have worked with them for a long, long time."

"But what I think I've heard," Sutter continued, "is that your experts were developing a greater and greater concern, including writing rather powerfully stated memos, and it is hard to understand why they didn't get more attention."

Following a dramatic pause during which Lund had no reply to Sutter's incisive rhetoric, Arthur Walker from Stanford asked other questions about the qualification temperature of the O-rings that set up a big trapdoor for

Lund, and he fell through it: "McDonald testified that he had a discussion, I believe it was with Mulloy, about the meaning of the temperature range of 40° to 90°, whether it would apply just to the bulk temperature of the motor or whether it would apply to every part of the Shuttle. What is your understanding of that?"

"If you would have asked me a month ago, I would have told you the motor was 40° to 90° Fahrenheit."

"So there is no qualification in your mind on the O-ring temperature? For example, the O-ring temperature could have been quite cold, because the O-ring was certainly going to move in temperature with the metal case. The metal case could be quite a bit colder than the bulk temperature of the propellant if the weather had been cold."

"There were no full-scale motors fired below 40°" was Lund's answer.

"But what I'm asking is did you as the supplier of the system to the government have the specification on the O-ring temperature, or could the O-ring temperature have been anything?"

"I don't believe there was a specification on the O-ring temperature other than the material itself."

"And what was that specification?"

"Well, it is a specification that this material can withstand these kinds of environments."

"But it doesn't specify the operational aspects, does it?"

"To my knowledge, no."

"In fact, the milspec says specifically that the O-ring should be qualified for whatever uses they are put to. So did you take any steps to do that, that is, to qualify the range of temperatures over which these O-rings were to be used?"

"Yes. The development motors were fired from 40° to 84°, and qualification motors from 45° to 83°."

"So then 40° was the temperature limit for the O-ring?"

"In the full-scale program, that is correct."

"So when it was predicted that the temperature of the O-ring at launch of 51-L was going to be 29°, the O-ring was outside of the qualification temperatures by some 10°."

"That is correct."

"Then how could you make a recommendation to launch if you were 10° outside of your qualification?"

"Our original recommendation, of course, was not to launch."

"Well, I understand that, but your final recommendation was to launch."

"OK. What we need to do, then, is go through that rationale."

"So, Mr. Lund, you are going to answer my question then at the end of this discussion, hopefully?"

"If you want me to go through it now, I would be glad to do that."

Chairman Rogers had heard enough: "Well, I think we have heard what explanation you have given. I think the problem we're having is that this is not convincing. I mean, let me, if you don't mind. . . . I assume, Mr. Lund, you have great confidence in your engineers Boisjoly and Thompson and the others. They're probably as well qualified as anybody in the country in dealing with these problems of O-rings and seals and so forth, is that right?"

"Yes."

"And you had a long discussion in the telecom, and you decided, all of you, I understand, all of you decided that for safety reasons to oppose the launch. And thereafter, NASA, in one way or another, made it clear that they were displeased with that recommendation. I assume that you knew when you made your recommendation—that NASA would be displeased. Your recommendation in fact was going to determine whether that Shuttle launched or not, because NASA had indicated to you that they would not fly unless they had a written report from Thiokol saying you approve the launch."

"I didn't know NASA would accept that."

"You didn't know that?" Rogers asked, incredulously.

"No."

"Well, you must have known your recommendation was very important. You knew that if you voted against the launch it would not have been launched, didn't you?"

"Well, we had voted prior to it, and they didn't accept it, so I couldn't forecast what NASA would do."

"But you knew that that was the reason they asked you to reconsider. That is why you had a five-minute recess, didn't you?"

"That's a fair statement, yes."

"Now, knowing that, and knowing that the safety of the crew was involved, and knowing your own people, the engineers who you respected, were still against the launch, what was it that occurred in your mind that satisfied you to say, 'OK, let's take a chance'?"

"I didn't say, 'Let's take a chance,' because I felt there was some rationale that allowed us to go ahead."

"Well, maybe that wasn't fair. Then what was it that occurred in your mind that caused you to be willing to *change* your mind?"

"I guess one of the big things was that we really didn't know whether temperature was the driver or not. We couldn't really tell. We had hot motors that blew-by and cold motors that blew-by, and some very near either end of hot or cold that did not. The data was inconclusive. So I had trouble justifying in my own mind saying, 'By golly, temperature is a factor.'"

Commissioner Walker asked Chairman Rogers if he might interrupt for a moment. "Boisjoly has said," Walker reminded everyone, "that the thing which was compelling to him was that the blowby on the coldest motor was much more severe. He has emphasized how black the blowby was and how large an angle was over which it had occurred."

"But three or four degrees above that it was zero," Lund countered.

"So his discussion was not convincing to you?"

"It wasn't totally convincing because in two or three degrees it went from very, very bad to no problem at all, no blowby."

Richard Feynman couldn't restrain himself: "There were many seals, Mr. Lund, that didn't have any problem, as though it was a random effect. It depended on whether or not you got a blowhole or you didn't have a blowhole. If within a particular flight it happened that all six seals didn't get blowholes, that's no information. It didn't mean you were suddenly good because it worked and the next time when one went off it was suddenly bad. It seems to me that it has to be understood as probabilistic, a complicated situation. So you could never decide they were all going to break or they were all not going to break. When you look at it that way, it is a question of increasing and decreasing probabilities that we have to consider rather than did it work or didn't it work. And I would like you to explain it to me from that point of view."

Unwilling to talk probability theory with a Nobel laureate, Lund offered only, "Well, the question still is, is temperature the thing that caused the SRM-15 effect?"

"You have heard your engineers argue that there was an effect of temperature that looks like it made things worse," Feynman reminded him. "Is it not inconceivable that there's something else that sometimes produces blowby, that there's more than one effect and that temperature could still be an important effect and increase the probability in spite of the fact that at a high temperature it gets worse? Is there any evidence that temperature is not an important matter just because at some high temperature you have an accident?"

"Well, there is, as you point out, there are many variables in the thing, and it wasn't clear that temperature was the effect."

"But logically, Mr. Lund, and from the point of view of your engineers, they were explaining why temperature would have an effect, and when you don't have any other data, you have to use reason, and they were giving reasons."

"That's right, and that is what we did when we included in our rationale the fact that, sure enough, the temperature could be an effect. OK, it wasn't clear, but we said we will consider that to be so."

Robert Rummel, a member of the National Academy of Engineering and a former president of Trans World Airlines, expressed great difficulty with Lund's account. "In the usual practice, when there is any real doubt about flight safety," Rummel offered, "whether it is aircraft or whatever, you simply don't fly, and it seems to me that this was the reverse. I just have great difficulty understanding the question that has been asked before, that is, understanding any of your answer. I just haven't heard it as to why, if there is any doubt in your mind, why you went ahead, why you changed your mind. I just don't understand it, and would appreciate very much if you could explain that."

Finally, the Commission members gave up trying to make any sense of Bob Lund's rationale as to why he changed his mind or how he finally concluded it was safe to fly below the temperatures that he knew the motor and O-ring were qualified to. The only conclusion they could reasonably come to was that he changed his mind because his boss basically asked him to.

"Maybe we ought to go to Kilminster, if you don't mind," Chairman Rogers directed. "He signed the telefax. We have asked Lund a lot questions, and I think in fairness to Mr. Lund, Mr. Kilminster, could you attempt to explain to the Commission why, I guess, you changed your mind, too, didn't you?"

"Yes sir, based upon the discussion that we had and the rationale that was developed."

"How can you say you changed your mind," Rogers probed, "when you say temperature and data was not conclusive for predicting primary O-ring blowby? Did you have a feeling they had to prove that it *was*—that the burden of proof was on you to show that it *wasn't* safe?"

"No. We were asked to relook at the data, which we did."

"Could you tell me what data it was that you looked at that was different from the data you had looked at first?"

"There was one piece of data that we looked at which had not been discussed before and that was the erosion parameter, and the factor of three, but my evaluation was . . . "

"Was that the only new piece of data?"

"That was really the only new piece of data that we had not previously discussed on the telephone, but I think that the data that we did show, the fact that we had some subscale tests at 30° showing no blowby, was an indicator. As Lund pointed out, we had other flight motors at temperatures between the two in question, that is, between 75° and 53° O-ring temperatures that had shown no blowby, and we had static test motors at temperatures lower than 53° degrees that had no blowby. Now, I would like to discuss that just a moment . . . "

"Could you particularly, though, point out the *new* data, because the purpose for the recess, as we understand it, was to find out if there was any new data, or were you just asked to change your mind based upon the data you had?"

"I had one piece of data that was new in our discussion. That showed that if we did have blowby past the primary O-ring as it was being positioned to seal, and recognizing the fact that the cold temperatures could cause that timing function to extend, we had an opportunity perhaps of having some erosion occur. We looked at the erosion that had occurred on Shuttle 51-C and compared that to the data that we had developed both from cold hydraulic oil testing and from hot subscale testing, which indicated that that very flight, 51-C, had a safety factor of three over what would have to have happened in order to get to an area of questionable sealing capability. With that information in mind then, and also considering the fact that we had done some analytical work earlier to develop the limiting erosion parameters on O-rings, I was led to believe that we were in the condition of having a safe position for recommending a flight."

"How come that data wasn't available the first time around? How did it happen to show up at the recess?"

"It was just a matter of a discussion we had when we had the recess. When we said that, 'Well, if indeed the temperature is going to cause a longer time for the primary O-ring to function and there is the possibility of additional erosion, then where do we stand relative to 51-C, and how much more margin did that exhibit'?"

"So you had the data there before?"

"Yes."

"You had analyzed it. In other words, it was not new data. It was data that you hadn't properly analyzed, is *that* right?"

"It was data that we hadn't prepared or discussed on the earlier telecom or the earlier portion of the telecom."

"And that was through an oversight?"

"I don't know if it was oversight or just another piece of data that we were searching for in order to establish our position."

Always the intent listener, Sally Ride wanted more clarification: "You are saying, Mr. Kilminster, that you had a safety factor of three over the 51-C erosion problem. How were you attempting to analyze how the timing function would change and how much erosion that could possibly give you on 51-L?"

"Only on a subjective basis that if, as has been stated, the timing function under cold condition was to extend, and if what was observed under 51-C— not only the blowby but the erosion—were combined, then we developed the rationale about that safety factor of three."

"Mr. Kilminster," Rogers wanted to know, "did you have any feeling of pressure being put on you by NASA, or were you just calmly reassessing?"

"I felt the pressure that was put on us was to go back and look at the data, look at the detailed information that had been presented, to see if there was something that we were seeing that we were not presenting on the phone."

"You didn't feel they were trying to get you to change your mind?"

"I did not feel a significant amount of pressure in that regard."

Kilminster's testimony was not convincing to anyone, no more so than Bob Lund's.

Before the hearing ended, Chairman Rogers had one more question to ask. "Just one other question and I think I would like to adjourn for the day. Did either of you gentlemen, Mr. Kilminster or Mr. Lund, have any pressures, any outside pressures? Did anybody call you or anybody suggest that you should vote to launch? Was there anything of that kind that the Commission should know about?"

"No sir," both Kilminster and Lund responded. It was obvious to me that Rogers was trying to dispel the rumor that Senator Hollings was spreading about President Reagan pressuring NASA and Morton Thiokol to launch the day of his State of the Union address.

The hearing ended around 4:30. I was immediately surrounded by a flock of reporters asking for interviews. One man slipped a piece of paper in my hand asking me to appear that evening on the *McNeil/Lehrer Report*. I told everyone I didn't feel like talking to anyone and didn't really have anything more to say. I walked into the hallway and out the side door. I went directly back to the Morton Thiokol office in Crystal City, where the rest of our company people were discussing the day's events and speculating on what the NASA people might say the next day.

In the conference room where we had gone through our dress rehearsal the night before, I sat down next to Roger Boisjoly, who was sitting there alone. It had been a long, stressful day, and I knew he was as glad as I was that it was finally all in the open, if not over. The atmosphere was very tense, and none of the Morton Thiokol management in the office said a word to Roger or me.

Senior managers were holding a meeting in the manager's office. Jerry Mason finally came out of that meeting and announced that "all of the guys who are working on the failure team in Huntsville should go home on the company airplane tonight." I told him, "That may not be a good idea because rebuttal testimony from these folks may be needed." Joe Kilminster rebuked, "We all need to get back to work." "A day or two isn't going to make any difference," I dared to say. "We'll all be watching the hearings tomorrow anyway." CEO Ed Garrison finally came in and uttered the last words: "All of you plan on going home early tomorrow morning and get back to work. We need to rebuild this company, find out what really happened, and fix it."

I didn't give up. "Mr. Garrison, we need to stay here in Washington in case rebuttal testimony is needed."

Our corporate attorney, Lee Dribin, answered, "We can have rebuttal in writing based upon the transcripts of the proceedings."

"But I'd rather do it in real time, the way it was offered by the Commission."

Garrison would have none of it: "We *all* need to get back home and get our minds off of this and back to work understanding and solving the problems. We need to unite and get on with it."

Cal Wiggins then came in and announced, "Everybody is going back on the company plane in the morning." I asked, "Who all is going on the company plane?" "Everyone that came except Jerry Mason," Wiggins said. "He is staying for the hearings." I told him, "I thought Ed Garrison just said we were *all* to return home and get back to work, so Jerry should go, too. He shouldn't be standing around in these hearings." Wiggins gave me a dirty look and said, "We're going home. Jerry Mason is not part of our work team, and he can do whatever he pleases."

The meeting broke up, and people discussed where everyone was going to dinner. Roger Boisjoly, Arnie Thompson, Brian Russell, and I left the crowd and decided to go have dinner by ourselves at the hotel. Our company management were treating us as lepers, so we might as well act that way.

At dinner, the four of us all believed that our testimony had been very convincing to the Commission on why the original engineering recommendation not to launch had been perfectly logical. We had all testified that the only reason the caucus was held and the recommendation changed was because NASA management hadn't accepted our original recommendation not to fly below 53° and had pressured Morton Thiokol management to reconsider and change their minds. We all recognized that it was pressure from NASA management that caused the reevaluation, not any new data.

The three Thiokol vice presidents who had testified were riding in a much different boat. Supporting the final recommendation to proceed with the launch, they swore that they had not felt any pressure from NASA to reconsider the original recommendation; therefore, they had to find ways to justify why that had been a logical thing to do. Since it was not logical, and had totally ignored their engineering experts, they completely destroyed their credibility in the eyes of the Presidential Commission.

The next morning, the same quartet of lepers met in the hotel for breakfast prior to leaving for the airport. Picking up a copy of USA Today, I was shocked to see my picture on the corner of the front page with the caption "Dire Warning!" (Our Challenger testimony probably would have been the feature story on the front page of every newspaper in the country had it not occurred on the very same day that the Marcos regime fell in the Philippines and Corazon Aquino came to power.) The story on Challenger contained several excerpts from the prior day's testimony and ended with the statement that NASA would have a chance to respond today.

Climbing aboard the company jet, I noticed that neither my boss, Joe Kilminster, nor Jerry Mason was on the airplane.

Evidently, there was quite a bit of rebuttal for them to do.

Left: During the first public hearing conducted on February 11, 1986, Rogers Commission member Dr. Richard Feynman removed a piece of O-ring that he had squeezed in a C-clamp from a glass of ice water, thereby demonstrating to NASA's Larry Mulloy the loss of resiliency of a cold O-ring. (Photograph by Marilyn K. Yee/*New York Times*/Redux Pictures, reprinted with permission.)

Below: On February 25, 1986, I was sworn in as the first witness at the Presidential Commission's inaugural public televised hearing on the decision to launch *Challenger*. (Photograph © 1986, *Los Angeles Times*, reprinted with permission.)

PART IV

Obfuscation

Oh, what a tangled web we weave, when first we practice to deceive!
—Sir Walter Scott

The wise understand by themselves; fools follow the reports of others.
—Tibetan Proverb

A drowning man will clutch at a straw.
—Sir Thomas More

Never saw off the branch you are on, unless you are being hanged from it.
—Stanislaw Jerzy Lec

A lie told enough becomes the truth.
—Vladimir Ilyich Lenin

19

Rationalization and Innuendo

The morning following my testimony, my picture appeared on the front page of *USA Today*, the *Washington Post*, the *New York Times*, and *Florida Today*. It was interesting how different the editors had handled the headlines and captions. In the *Washington Post*, the main headline read "Thiokol engineers tell of being overruled"; the caption under the picture showing me at the witness table read "Thiokol engineer Allan McDonald: 'I felt I was pressured to approve launch,'" which didn't please me, because I thought I had made it clear that I *never did approve* the launch. On the front page of the *New York Times*, the story was "Rocket engineers tell of pressure for launching," with an accompanying picture of me testifying. Inside the front section of the paper a feature article told my story as "Tenacious Engineer— Allan J. McDonald." *USA Today* captioned its picture of me, "McDonald: Dire Warning." *Florida Today*'s coverage, "Engineers: We tried to halt lift-off," included a picture of me captioned, "Allan McDonald wanted bosses to make an engineers' decision, not a program management decision." The *Washington Post* also ran an article, "Testimony cheers Thiokol neighbors," datelined "Brigham City, Utah, February 25, 1986." The story read:

> For the first time in weeks, there were smiles today along with the pot roast at the Idle Isle, which bills itself as Utah's fourth oldest restaurant. "We were dead in the water till old Al got up there," said Ernie Apodaca, draped in an oversized blue Thiokol Space Shuttle team windbreaker. "We've been an easy target, and I'm glad some people got to set the record straight." The dramatic testimony of Allan McDonald before the Presidential Commission investigating the destruction of the Shuttle *Challenger* brought new cheer to this city. Morton Thiokol workers here made the solid rocket booster that appears to have played a role in the *Challenger* explosion. . . . To the people of this predominantly Mormon town of 17,000 on the flat edge of the Rockies, the aftermath of January 28 has been ugly at times. Last week, after Thiokol announced it would lay off 200 employees and put another 1400 on

4-day weeks, vandals painted the words, "Morton Thiokol Murderers" on bridges leading to the 20,000-acre plant west of town. Threatening phone calls to the company switchboard have prompted local police to patrol around the homes of Thiokol executives.

USA Today published a similar story, "Thiokol town anxious to get back to normal." The reporter had interviewed my son, Greg, in Utah, while he was working on his car. Greg was quoted as saying, "My father's anxious to get on with the Shuttle program, and everyone in the family is looking forward to quieter times."

Our plane didn't arrive back in Utah in time to see any of the NASA testimony on TV, but one of my friends in the photo lab at work videotaped the hearings and delivered a copy of the tape to my home that evening. The first thing I noticed was Jerry Mason and Joe Kilminster sitting in the front row *with the NASA management people.* Viewing the tape, I wished I'd been there to challenge not only some of the misleading comments but especially innuendoes and derogatory comments made about my testimony the day before. It was obvious that NASA management could not defend their own actions, so they tried to discredit mine. It appeared as if all of the NASA witnesses had been briefed by the same sort of defense attorneys as had briefed us at Morton Thiokol, only this time by U.S. government attorneys representing the space agency. The NASA guys also employed a viewgraph-presentation approach that Morton Thiokol management had pushed. No doubt both sets of lawyers had coordinated their clients' testimony to eliminate contradictions and controversy. It did not take me long to understand that Morton Thiokol senior management surely had sent me home so that I could not challenge NASA's testimony, and this had probably been coordinated with NASA as well.

The first NASA witness had been Larry Mulloy, my NASA counterpart, who was the SRB Manager at Marshall. Immediately, he attacked my previous day's testimony: "Mr. Chairman and members of the Commission, yesterday in the testimony that was given before this Commission, and, before that, in more dramatic statements attributed to Thiokol personnel by the media, a picture has been painted of the events of January 27th that I think at best may be misleading." Mulloy then attempted to explain why his actions in challenging Thiokol's data and engineering recommendation not to launch below 53° were perfectly rational. "Thiokol's data wasn't even transmitted to Kennedy Space Center in order for KSC to make the launch decision," Mulloy stated. "The data only came to KSC because some of the Marshall people were there."

Under the impression that KSC had the final authority for launch, Commissioner Joseph Sutter expressed a confusion that was obviously felt by his fellow panelists as well: "If the Thiokol people say, 'Don't fly!' does that stick? Can it be overruled? Who is the final judge to make a recommendation to whoever has to say 'Launch'? I'm confused and would like to get it clarified."

The good soldier, NASA's George Hardy, Deputy of Science and Engineering, tried to get back on point: "Mr. Chairman, I think I understand the question, and it is certainly a legitimate interest to have. If I might suggest that Mr. Mulloy and myself be permitted to proceed with our statements, I believe that some of the players and where they fit would be a little clearer to you."

Mulloy continued: "After we had looked at all of the data, the conclusions and recommendations charts that Mr. Lund had prepared came in from Utah along with the logic for his recommendation, which did not specifically address, 'Don't launch 51-L.' What it said was that, within our experience base, we should not operate any solid rocket motor at any temperature colder than we have previously operated one, which was 51-C."

"Didn't you take that to be a negative recommendation?" asked Chairman Rogers.

"Yes sir. That was an engineering conclusion, but I found the conclusion without basis, and I challenged its logic. Now, that has been interpreted by some people as applying pressure. I certainly don't consider it to be applying pressure. Any time one of my contractors or, for that matter, any of Mr. Hardy's people, comes to me with a recommendation or conclusion based on engineering data, I probe the basis for their conclusion to assure that is sound and logical. I found this particular recommendation not to launch an SRM at a temperature below 53° to be . . . it came somewhat as a surprise to me. The reason it did, after 51-C, which was when the observation was made about the blowby (which has been testified to), we dealt with the observations on 51-C very carefully in going into the next launch readiness review. And the conclusion was that 51-E was acceptable to launch with full expectation that we might again see exactly what we saw in 51-C. We tested that logic and rationale in great detail."

If I had been present to rebut Mulloy's testimony, I would have made clear that what Mulloy was purposefully not saying was that 51-C had been

246 PART IV. OBFUSCATION

preceded by the three coldest days in Florida history, a point emphatically made in the FRR for the subsequent launch of 51-E. The conclusion at that time was that we would never encounter that kind of weather condition again and certainly not in Florida in April when 51-E was scheduled for launch. Everyone knew that STS-51C, although not a desirable condition, probably represented the worst case. This logic was not the same as that for 51-L, where it was known that the temperature would be much colder than 51-C, and the conditions for sealing much worse. The risk we had accepted in the launch of 51-E was far lower than for STS-51C, not higher. We had interpolated data for 51-E because it was a Shuttle that would be launched at a temperature that was already in our experience base, while 51-L was clearly way outside of our experience.

With twisted logic, Mulloy was attempting to confuse the Commission. His defense that Morton Thiokol's recommendation was not logical demonstrated just how illogical his own rationale actually was. Forced to defend his position because he could not and would not admit that he had pressured Morton Thiokol into changing its recommendation not to launch, all the pot could do was call the kettle black.

Next Mulloy told his version of how the Morton Thiokol caucus came about: "Mr. Kilminster then requested an off-net caucus. It has been suggested, implied, or stated that we directed Thiokol to go reconsider these data. That is not true. Thiokol asked for a caucus so that they could consider the discussions that had ensued and the comments that Mr. Hardy and I and others had made. At that point Mr. McDonald interjected, making his first comment of the entire teleconference. Mr. McDonald stated that he thought what George Hardy said was very important—that the secondary O-ring would be in a position to seal during the time of blowby and would do so before significant joint rotation had occurred. Mr. McDonald asked Mr. Kilminster to be sure to consider Mr. Hardy's comment during the course of the discussions. Morton Thiokol then went into their caucus, asking for five minutes."

At home in front of the VCR, I almost came out of my seat. What Mulloy said just was not quite true, because it had not been my first comment. Earlier in the teleconference, I had discussed why the static test data at 36° was not a valid test for the O-rings. It was not a valid test for the O-rings, I had said, because the blowholes observed in the zinc chromate putty after assembly and leak-check had been filled manually prior to static test. However, Mulloy admitted in his statement that it wasn't a comment from me that had *triggered* the request for a caucus, as Jerry Mason had indicated in his testimony the day before. Mulloy (and later George Hardy) testified that I made my remarks about the action of the secondary O-ring *after* Morton Thiokol decided to caucus, not before.

Commissioner Arthur Walker asked Mulloy whether he had interpreted my comment as a statement in favor of proceeding to launch. Though very quick to respond, Mulloy immediately began to backpedal: "Yes, I certainly did, because Mr. McDonald was sitting close to me, and it clearly was a supportive comment. I would not say he was *recommending* launch. What he was saying was that what we were asking them to do was a very important consideration, and, as I remember, he said, 'I think that's very important.' I took it as a supportive comment to the rationale. He certainly didn't say, 'I think this will support a decision to launch.'"

Robert Hotz of the Commission wanted to know if Mulloy had, in fact, made the comment to Morton Thiokol during the teleconference, "Do you expect us to wait till April to launch?"

"It is a statement taken out of context, Mr. Hotz. The way I read the quote—and I have seen it many times, too many times—the quote reads, 'My God, Thiokol, when do you want me to launch, next April?' Mr. McDonald has testified to another quote that I allegedly said: 'You guys are generating a new launch commit criterion.' Now, both of those quotes, I think, kind of go together, and that is what I was saying. I don't know whether those comments occurred during the caucus or subsequent to it. I simply can't remember. But the total context in which I may have used those words was meant to say that there are currently no launch commit criteria for SRB joint temperature. 'What you're proposing to do,' I was saying, 'is to generate a new launch commit criteria on the eve of the launch, after we have successfully flown with the existing launch commit criteria twenty-four previous times. With this LCC—that is, "do not launch with the temperature lower than 53°"—we may not be able to launch until next April. We need to consider this carefully before we jump to any conclusions.'"

"But it was pretty clear to everyone that you and Mr. Hardy were very unhappy about the recommendation of the engineers," Chairman Rogers reiterated. "As we understand it, the recommendation of the engineers was 'Don't launch,' and you expressed your displeasure. And somewhere along the line, Morton Thiokol decided to have a five-minute recess, which seemed very odd to me from the first time I heard it. Why five minutes on a matter of such major importance? Why would anybody say, 'Let's have a five-minute recess'? I would have thought they would want a five-hour recess on a matter of such major importance."

"I hope I have not said that I was *upset* by a recommendation *not* to launch," Mulloy answered. "What I was challenging were the conclusions that had been drawn. The recommendation not to launch or to launch at that time wouldn't upset me one way or the other."

"But that was the whole purpose of the discussion, Mr. Mulloy, whether you would launch or not."

"Yes sir, the end result would be that. But the purpose of the discussion was to understand the data and the logic of conclusions being drawn from those data. That was the way I was working."

"Larry, I have a problem with that," spoke up General Don Kutyna. "You had a briefing in July that talked about O-ring resiliency and then a briefing in August at NASA headquarters that discussed the resiliency of those seals as a number-one concern. Now, how can you say that the seriousness of that concern wasn't transmitted to NASA? It's on the conclusions sheet of the meeting: 'Conclusions: primary concerns, resiliency.'"

"I've looked at that report, General Kutyna, for the substance behind that conclusion, and I can't find it."

Very quickly, it was apparent that Mulloy would have serious trouble, not only in his response to Morton Thiokol's recommendation, but as to why he hadn't raised concerns about O-ring resiliency any higher within the different levels of the NASA Mission Management Team. "So far, Mr. Mulloy," probed Commissioner Hotz, "testimony seems to indicate that the risk on the cold temperatures and the O-rings was not transmitted to the highest level of NASA. How do you explain that? Was all this information transmitted up to Level II or Level I, to NASA management?"

"All the information that was discussed on the night of January 27th was not transmitted beyond Dr. Lucas at the Marshall Space Flight Center."

Commissioner Acheson: "Did you have any feeling or apprehension that a delay of launch date for reasons related to the propulsion system would reflect on you or the Marshall organization?"

"No sir. My decision to proceed with the launch as recommended by the Thiokol official responsible for making such recommendations was based solely on the engineering data presented by Thiokol engineering and the Marshall engineering evaluation of those data. I can assure you, because I am absolutely certain, that no extraneous consideration, such as schedule, came into that decision process."

Mulloy didn't really answer David Acheson's question, nor did he reveal that he personally had been held responsible for a major delay on the previous Shuttle launch, which reflected very badly on him and the Marshall Space Flight Center. In my view, there was no question that this was the reason why no one at Marshall had informed anyone from Level II or Level I about this meeting, and why MSFC had pressured Morton Thiokol to reconsider its original launch delay recommendation.

Mulloy then volunteered that he had no knowledge of my being asked to sign the documentation ultimately giving Morton Thiokol's OK to go ahead with the launch. "That would have been unusual," Mulloy testified, "because Mr. Kilminster signs all flight readiness documentation."

Once again, I nearly came out of my seat; I had *not* testified that Mulloy asked me to sign that document. What I testified to was that I expected to be asked, in my position as the senior manager representing Morton Thiokol at KSC, to sign off on the launch. That was my reason for being at KSC. That's why I had said to Mulloy that I would not sign the document and that it would have to come from the plant.

"Now, after the teleconference was complete," Mulloy continued, "Mr. McDonald informed Mr. Reinartz [Marshall's Shuttle Project Manager] and me that if the Thiokol engineering concern for the effect of cold was not sufficient cause to recommend not launching, there were two other considerations: launchpad ice and recovery area weather. I stated that the launchpad ice issue had been considered by the Mission Management Team before deciding to proceed and that a further periodic monitoring of the icing condition was planned. I further stated that I been made aware previously of the recovery area weather and planned to place a call to Mr. [Arnold] Aldrich [Manager of the National Space Transportation System Program Office at Johnson Space Center] and advise him that the weather in the recovery area exceeded the launch commit criteria. The concern I had for that was not loss of the total booster but a loss of the main parachutes for the booster, which were separated at water impact, plus loss of the frustum of the boosters, which has the drogue parachute on it and which comes down separately, because with the fifty-knot winds we had out there, and with the kind of seas we had, by the time the recovery ships got back out there, there was little probability of being able to recover those. I informed Mr. Aldrich of that, and he decided to proceed with a launch using that information. I did *not* discuss with Mr. Aldrich the conversations that we had just completed with Morton Thiokol."

Once again, Mulloy was demonstrating just how illogical his thoughts about the launch-decision process actually were—and that he was still trying to defend. It was known by all well previous to the launch of *Challenger* that if the O-rings failed, the entire Shuttle system would explode, destroying the orbiter and everything in it, including seven human beings. But this catastrophic possibility didn't even warrant an "advisory call," whereas the possibility of losing the SRB parachutes after the launch did? It was easy for the MSFC people to discuss the benign parachute-recovery problem, but not the potentially grave O-ring problem, so they intentionally chose not to discuss the matter of the O-rings with Level II because they may have been overruled.

"Could you explain why you didn't talk to Mr. Aldrich about Thiokol's original recommendation not to launch?" pressed Chairman Rogers.

"At that time, and I still consider today, this was a Level III issue—Level III being an SRB element or external tank element or Space Shuttle main engine element or the orbiter itself. There had been no violation of launch commit criteria. There was no waiver required, in my judgment. We work many problems on the orbiter, SRBs, and external tank level that never get communicated to Mr. Aldrich or Mr. Moore. It was clearly a Level III issue, which had been resolved."

This was baloney. Both Mulloy and Hardy knew they were in clear violation of the LCC because the launch was going to fall outside of the qualified operating environment of the solid rocket motor. Launching outside the qualified operating environment of the SRM, which was 40° to 90°, as acknowledged on the teleconference by Stan Reinartz (Mulloy's boss and member of the MMT), required both Level II and Level I approval.

"Mr. Mulloy," Arthur Walker spoke up, "could I return for a moment to your conversation with McDonald. Yesterday he stated that he had a discussion with you about the meaning of the temperature on the launch commit criteria of 40° to 90°, whether it applied just to the solid bulk temperature or whether it applied to every element of the Shuttle system. Do you recall that conversation?"

"Yes, there was some discussion of that. I believe it occurred during the caucus, or it may have occurred after McDonald stated that 'we ought to at least get the joint temperature to 40°.' He indicated that at 40° he would feel more comfortable, because we had a spec that said we were good from 40° to 90°. I didn't find that argument to be very logical at all, though, because, based upon the data, the engineers were recommending that we not launch below 53°."

"Why is that not logical?" Chairman Rogers wondered. "Why didn't you require at least a 40°? You say you didn't think it was logical. It seems very logical to me."

"It wasn't logical, not on an engineering basis, sir. If one was concerned about the engineering data that said at temperatures below 53° we had an unsafe condition, there was certainly no logic for accepting that at 40°."

"Well, its more logical than 30°," added Rogers.

It was obvious from the expression on the faces of Chairman Rogers and Vice-Chairman Armstrong that they were in a state of disbelief at Mulloy's response.

"To what temperature are the solid rocket boosters qualified?" Sally Ride wanted to know.

"On the end-item specifications upon which the solid rocket motors are procured, there are two specs related to temperature. The first spec states that the motor must be capable of providing a given thrust/time trace, within limits, from zero to 200,000 feet at a temperature of 40° to 90°."

"That is basically on the propellant, right, Mr. Mulloy?"

"Yes, at a propellant mean bulk temperature of 40° to 90°. There is another requirement for the solid rocket motor that states that the SRM must be capable of meeting the natural environments specified in *JSC Document 07700*, Volume 10, Appendix 10.10. That document has a lot of environmental data in it. The one of interest here is what was specified as the ambient temperature for the launch site, which was picked up for the launch commit criteria that we discussed earlier. In Appendix 10.10, it states that the SRM must be capable of operating at an ambient temperature down to 31° while being exposed to a 5° sky. Now, you have to get into issues related to heat transfer to apply that to what the temperature would be on the SRM."

Apparently, Mulloy was not aware of the controversial 31° launch criteria when we discussed it the evening before the launch because it was never brought up; furthermore, he and his associates, by the end of that meeting, had supported a launch at 26°! There was no way for that decision to go ahead without a waiver approved by both Level II and Level I.

"So what you're saying," Sally Ride stated, "was that there was a spec that NASA imposed saying that the SRM should be qualified to launch at 31°. Was that taken into account in the qualification test program for the SRM?"

"Dr. Ride, we did not come into this discussion today with all those specifics. I've tried to go back and pick up most of the things that we discussed and what we were using in terms of qualification. May I have the last SRB chart on the screen, please? As you can see, there was then data presented on two cold gas tests at 30°—with the O-ring pressurized at the motor pressurization rate, at 30°—which indicated that an O-ring would operate before joint rotation at 30°."

This time I almost jumped through the TV screen right into the hearing room! These were not qualification tests by any stretch of the imagination! These tests would never have met the stringent requirements of NASA's own system for qualification status. Furthermore, the tests were conducted with a fixed joint that could *not* rotate during pressurization and therefore could not provide any effects of cold temperature on O-ring resiliency, which was the key issue.

It was a great relief to me that Sally Ride picked up on this deficiency. "Was that actually in a joint?" she asked Mulloy.

"No, it was not. It was a full-scale O-ring and full-scale groove in a sub-scale test device, with the pressure rise rate on that O-ring being zero to 900 psia in 600 milliseconds at a temperature of 30°."

"And, of course, before those tests, the putty was modified, right?" Walker surmised. "The putty was not just laid up and the seal made. The putty was smoothed out or some attempt was made to remove the volcanoes in the putty, I should think."

"It was the horizontal assembly of the test that caused that," Mulloy ventured, another misleading statement since examination of the vertical assemblies showed the same problem. We had already continued to see blowholes in the putty from returned flight segments, indicating the problem was not unique to horizontal static test motors.

Mulloy pressed on: "There's one other significant point on this chart that we *did* discuss that evening, for which we did not have the quantities on the 27th, and I mentioned this earlier. We have had 150 case-segment proof tests involving a large number of joints and simulation of a cold O-ring. . . . So there was this certification data, all of which we didn't discuss. The two cold-gas tests we had done, the segment proof tests we had done, and the development qualification motor test, we had done all of it as a basis for understanding what we could expect to happen at colder temperatures on the joints."

How uninformed and ignorant did Mulloy think the Commission was—or would remain? The segment proofs were pressurized with oil or water and took literally minutes to reach maximum pressure inside the SRB cases, compared to just a fraction of a second from actual motor ignition; therefore, these tests were basically useless for determining the effect of cold temperatures. What Mulloy also failed to state was that early attempts to hydro-burst the case resulted in so much water leaking past the O-ring seals that we couldn't get sufficient pressure to burst the case. An oversized O-ring had to be used to keep the joints from leaking. None of this data was applicable for qualifying the ability of the joints to operate satisfactorily at cold temperatures. If anything, the data suggested there could be a serious problem. Grasping for straws, Mulloy had the gall to say the data and rationale presented by our Morton Thiokol engineers didn't make any sense!

Though untrained in engineering, Chairman Rogers was a keenly intelligent man who could see that a smokescreen was being put up. "Mr. Mulloy, I want to let you finish, but let me tell you what troubles me very much. I see the charts and I've heard your presentations before, and I recognize your expertise and knowledge in this field. What is troubling, very seriously troubling, is why this is such a convincing matter to you. You are certain of these things, you are sure it's OK. How come then in a matter of such major importance involving lives of seven astronauts, you apparently were not able to convince any of the engineers at Thiokol, who were working on this on a daily basis, that you were right?"

"Sir, I was not aware they were not convinced. I had no knowledge of what went on in the thirty-minute caucus at Thiokol, and when asked—when Reinartz asked, 'Are there any other comments?' there were none."

"Would your opinion have changed, Mr. Mulloy, if you had known all of the engineers—or substantially all of the engineers at Thiokol—took another position and were opposed to launch?"

"Sir, I cannot speculate on what my decision would have been given certain other data."

"Well, why can't you right now, based upon the knowledge that you gleaned yesterday from the testimony, why can't you say whether you would have been influenced one way or the other by the fact that all of those engineers seemed to be opposed to launch?"

"I would like to answer your question, sir, except that that is so foreign to the way that NASA does business that I would have to think a long time about an answer to that question."

Unimpressed with this rationalization, Rogers tried to understand why the known problems with the joints had not been properly communicated within NASA. Mulloy's answer, I believe, was a bold-faced lie: "Sir, I would draw your attention to the August 19th detailed briefing that was given here in Washington last year. At that meeting everything related to the joint, to our experience with the joint, the testing and analysis that had been done, and the rationale for continuing to fly in the face of that evidence, was covered. Moreover, in July 1985, I ordered billets to allow the incorporation of a capture-feature on the steel cases, designed specifically to eliminate the joint rotation."

What a crock! Back at this time I had told my case project manager, Howard McIntosh, to put a hold on some of the existing forgings at our vendor at risk to Morton Thiokol, but NASA Marshall, in the fall of 1985, had disapproved our request. Mulloy's program office didn't approve this action until April 1986—two months *after* the *Challenger* accident.

"Mr. Mulloy, you make it sound as if this was just a routine discussion about data," Chairman Rogers challenged, "and that you always had a free exchange of information about data and so forth. But this matter was different, it seems to me. You had a clear recommendation of no-launch; the chart showed that Thiokol recommended against launch. Do you remember any other occasion when the contractor recommended against launch that you persuaded them they were wrong and you had them change their mind?"

"No sir."

"*That* is the part that is troublesome. That's what the Morton Thiokol engineers said yesterday, that they thought that you wanted them to change their mind and that is why they had to rely on management. They had their contract coming up for renewal, sometime pretty soon this year. So they were under a lot of commercial pressure to give you the answer you wanted. They construed what you and Hardy said to mean that you wanted them to change their minds. They didn't construe it to mean you had a fair discussion about the data. They construed it to mean, just as Lund said, that that's what he thought you wanted. And what you're saying is that wasn't what you wanted? You just wanted an intellectual discussion about the data?"

"No. 1, Mr. Chairman, I cannot conceive how Thiokol felt any pressure for the renewal of its contract, because it is our sole source for solid rocket motors at this time—and that contract was going to be renewed. There was no alternative, given the mission model. So that certainly wasn't a pressure factor for them."

"Well, in one of the memos I read yesterday, Mr. Mulloy, there was a reference to that fact—that they were concerned that NASA might be looking for another contractor."

General Kutyna sought to clarify: "Larry, I think what we're talking about is the dual source. You have responses from the contractors due on 14 March, I think, and there is some leeway as to how much they're going to buy from Thiokol versus the dual source. Isn't that true—that there is to be procurement of a minimum of six SRBs from the dual source?"

"In the solicitation of interest that is on the street, yes."

Sally Ride turned back to the data and what it was saying to the engineers: "The engineers' main problem during all this was that they felt they didn't have the data. They felt that their temperature data was inconclusive and were worried about it, but didn't have the data to quantify what problems the temperature could cause. Wasn't that a major basis for their recommendation not to launch in the first place, that they simply felt they just didn't have the proof that it was safe? Did you think *you* had the proof? Did you think that you had the database to show that it was safe at these temperatures?"

"No. All I did was recite the data we had available to us and ask that we consider that decision in the light of all of that data."

"Could you go to the telefax," Chairman Rogers instructed, "and see what in the telefax satisfied you? Because the telefax said the evidence was *not* conclusive. That is what Dr. Ride is pointing out. They said, 'It's not conclusive.' They weren't saying, 'We think it's safe.' They said, 'It is not conclusive.' And you say the data was conclusive to you?"

"The engineering assessment is what I relied upon in the telefax. Here it is [holding up a copy of the fax): 'Engineering assessment is that colder O-rings will have increased effective durometer,' meaning they will be harder. We had no argument with that."

"Cold will make it harder?"

"Harder O-rings will take longer to seat, Mr. Chairman. We had no argument with that. More gas might pass the primary O-ring before the primary seal seated relative to SRM-15. We had no argument with that. The demonstrated sealing threshold was three times greater than the 0.038 inches of erosion experienced on SRM-15. That was a fact. If the primary seal did not seat, the secondary seal would seat."

"That was pretty positive, wasn't it?"

"Yes sir."

"But the critical items list did not say that, did it?"

"No sir, nor should it have. The whole concept of redundancy is that if the primary system doesn't function the secondary will."

"But, Mr. Mulloy, the critical items list warns that the possibility is loss of sealing of the secondary O-ring. It says the opposite of what this telefax does."

"You see, that is a *waiver* to the requirement to have a redundant system."

"Could I just finish on this telefax, Mr. Mulloy, because it says, 'Temperature data not conclusive on predicting primary O-ring blowby.' That suggests to me that there was a possibility of *primary* O-ring blowby. Isn't that right? Do you accept that?"

"Yes sir."

"It goes on to say, 'Pressure will get to secondary seal before the metal parts rotate.' Doesn't this continue to indicate that they were putting some reliance on the secondary seal?"

"I don't read it that way. That is where we keep diverging, I think. What they were saying . . . "

"What does this say to you, then?"

"Well, it says to me that we have redundancy at ignition. We can expect blowby at any temperature—at *any* temperature."

Larry Mulloy was trying to hide behind NASA's complex system of rules and regulations to justify his actions, on the one hand, and excuse his actions, on the other, but finally he got trapped. The snare came from Arthur Walker:

"Mr. Mulloy, let's try an analogy and see if you think it's an apt one. Suppose you are a manager of a baseball team, and you have an ace relief pitcher. Suppose that relief pitcher has a sore arm, and you've put him on the disabled list. You know that if your starting pitcher gets into trouble in the fifth inning, you cannot bring in your ace reliever because he is not available to you by the rules. Wasn't that the situation with this waiver? You had a redundant seal, but according to the waiver the redundant seal was not effective; the waiver didn't give you the excuse, in any way, to use the redundant seal. Now, I understand that, in the back of your mind, you know, 'Well, that redundant seal is still there,' even though, as far as the paperwork is concerned, it is *not* still there; in the back of your mind, you know how it's going to function. But in fact, by a strict interpretation of that rule, you can't rely on the secondary seal. That's the way I understand it."

Though he must have known by now that he was fighting a losing battle, Mulloy persevered, trying once more (as George Hardy tried as well) to explain to increasingly agitated Commission members the convoluted logic as to why it was unacceptable to NASA to impose a new set of launch commit criteria on the eve of the launch and why he talked to Arnie Aldrich only about the potential loss of the SRB recovery system because the LCC required he make an advisory call to Level II on this condition but not on the possibility of losing the complete Shuttle! It was clear from the expression on the faces of the commissioners that this type of thinking was totally irrational: to justify why the O-ring concern was a Level III issue only, even though the decision to proceed with the launch was a violation of a Criticality 1 and required a waiver in the system concerning the secondary O-ring seal. But, still, Mulloy didn't think it was necessary to address this issue with Level II or Level I!

Sally Ride zeroed in on the definition of a Criticality 1. "The primary O-ring is what is defined as a Criticality 1 item. Now, all Criticality 1 items are reviewed and signed off all the way up the NASA chain, always up to Level I, and have to be signed off and understood at a very high level. It would concern me if I thought that, on the day before launch, or even a week before launch, that engineers were allowed to decide—even based on good engineering data—that well, 'It's OK to consider that a Criticality 1R, because we have added up the tolerances and done this sort of analysis and so we think we have a redundant seal during these 160 milliseconds.' They may be right, but they just haven't reached the visibility that the original waiver had. And that decision hasn't been signed off at the levels at which the original decision had been signed off. It would concern me to think that criticalities could be handled that way by our system."

Right on! Sally Ride was absolutely right. To ignore the existing waiver in the system to arrive at a flight rationale did require that a new waiver be processed. In *not* doing this for STS-51L, there could be no doubt: *MSFC was negligent.*

Kutyna followed up with a very telling comparison: "Larry, you said you made the decision at your level on this thing. If this were an airplane, an airliner, and I just had a two-hour argument with Boeing on whether the wing was going to fall off or not, I think I would tell the pilot, at least mention it. Why didn't you escalate a decision of this importance?"

"I did, sir."

"You did?"

"Yes sir."

"Tell me what levels above you."

"As I stated earlier, Reinartz, who is my manager, was at the meeting. And on the morning of the launch, about 5:00 a.m., in the operations support room where we all were, I informed Dr. Lucas on the content of the discussion."

"But Dr. Lucas is not in the launch decision chain."

"No sir. Reinartz is in the launch decision chain, though."

"Is he at the highest level in that chain?"

"No. Normally it would go from me to Reinartz to Aldrich to [Jess] Moore [NASA's Associate Administrator for Space Flight]."

Chairman Rogers wanted to go back to Sally Ride's question as to why reliance on the secondary O-ring was not processed to a higher level within NASA, which really was the heart of the matter. Ignoring the question before, George Hardy now sidestepped it: "As far as the levels of review, I really can't comment because that is not within my purview. I have made my comments primarily as things related to the engineering assessment of the performance of the seal, and particularly as it was related the night before the launch to the matter of temperature."

"The reason that I think her question and your answer is very helpful, Mr. Hardy, is because it was clear that that critical items list was based on the fact there was *no redundancy*—that you had to rely upon the primary seal," Rogers responded. "Then you had this discussion back and forth on the eve of launch during which the contractor writes 'temperature data and weather not conclusive on predicting primary O-ring blowby,' because the data was not conclusive. You may have primary O-ring blowby. So your contractor himself gives you a signal in this telefax that this may violate the critical items list and may be a catastrophe."

"I think we have a different interpretation of that, Mr. Chairman; it is quite obvious we have a different interpretation of that. Yes, on 51-L, we had to be prepared for primary O-ring blowby. But that was true also on every other flight we'd ever had."

Watching the videotape of the hearings at home, I wondered how the commissioners would have responded had they found out that this whole discussion about Criticality 1 versus Criticality 1R *never came up* in the teleconference the night before the launch. The only explanation for that omission was that MSFC intentionally ignored the issue because bringing it up would have reinforced a no-launch decision. At a minimum, MSFC would have had to process a real-time waiver through Level I before it could launch.

Neil Armstrong wanted to know, "Do you always have some responsible persons from the contractor sign off on the recommendations?"

"Yes sir," Mulloy answered, "that is the normal process in any flight readiness review or any issue related to flight readiness. We require it at every level."

"So the fact that Kilminster signed this thing was typical, that it would be signed by him and not the man on the scene, McDonald?"

"That is correct, Mr. Armstrong. We require the element project manager to certify flight readiness and not anyone else."

Balderdash! That was not true at all. This was the first and only time in my memory that the contractor had to sign a document to proceed with a launch after the L-1 flight readiness review. The man on the scene was always responsible for that; that was why he was there. (In subsequent testimony dealing with Rockwell's concern relative to launchpad icing for *Challenger*, it would be made clear that Rockwell's approval for the orbiter came in oral form only and required neither written approval by Rockwell's man on the scene nor from a higher-level program manager back at the Rockwell plant in Downey, California.)

Also, the teleconference was not a flight readiness review, as Alton Keel pointed out to Mulloy. "All FRRs might require a written statement, but this wasn't an FRR . . ."

"It was, in a sense, an FRR," was Mulloy's retort. "It was a review like an L-1 dealing with issues related to the flight."

In his obfuscation, what Mulloy failed to mention was that in all of those FRRs requiring an official sign-off from the contractor (including Level III FRRs, involving the Shuttle projects manager and the Marshall center director), the SRB project manager, namely Mulloy himself, also had to sign the readiness sheet, as he also had to do with all Level II and Level I FRRs. In this case, MSFC management signatures were noticeably absent from the January 27 teleconference, as would have been the case if it had been regarded as any sort of FRR! NASA only required a *Thiokol signature*.

Mulloy couldn't have it both ways. It was either an FRR where both NASA Marshall and the contractor signed off, or it was not an official FRR but rather a technical interchange meeting. If the latter, the man on the scene normally gave a verbal go-ahead without signing a document, if there was no change to the launch recommendation. If we at Thiokol had originally recommended launching rather than not launching, we would not have had to sign anything because such a discussion would not have affected the current plan by Level I to proceed with the launch. That's why Marshall management rationalized not saying anything to Level I about the teleconference.

We weren't requested to sign the first recommendation when it was presented, even though it recommended no-launch. We were only required to sign off on the prolaunch presentation because there had been a no-launch recommendation made previously, and NASA Marshall had to cover its backside.

In yet another effort to discredit my testimony, Hardy rejuvenated the idea that something I had said triggered the change in Utah from a no-launch to a go-for-launch recommendation: "Somewhere about this time, Kilminster in Utah stated that he wanted to go off the loop to caucus for about five minutes. McDonald, the senior Thiokol representative at KSC for the launch, suggested to Kilminster that he consider a point that I had made earlier, which was that the secondary O-ring was in a proper position to seal if blowby of the primary O-ring occurred. I clearly interpreted this as a somewhat positive statement of supporting rationale for launch. Any other interpretation of that statement by Mr. McDonald is a case of 'convenience of memory.' . . . To suggest that flight safety was regarded, or not properly regarded, in the discussions on the night of January 27th is, in my opinion, a great disservice to the many dedicated, committed professionals within NASA, many of whom have literally put their blood, sweat, and tears into the Shuttle program."

The tarring of my name and my testimony, which I don't believe George Hardy would have made if he hadn't known I was not in the audience, didn't go unnoticed by Chairman Rogers.

"Could I interrupt, Mr. Hardy, because you just said something that was disturbing. When you referred to McDonald, you referred to 'convenience of memory,' which suggested that you thought he was making something up for the record or some such. Was there anything that McDonald said that you thought was not true?"

"No sir, I did not mean to imply that."

"Then why did you use 'convenience of memory'? That is what it sounded like."

"That may not have been a good or proper term to use. . . . Many of us have gone back and asked ourselves questions and second-guessed ourselves. We have really probed to understand the issues that existed on the other side of 51-L. That is not easy. It is not easy just within your own heart and mind to try to separate yourself from what you know has happened. I don't mean by that the cause or failure. I'm talking about the tragic incident itself. I am not suggesting that anyone, in any testimony before this Commission, has knowingly in any way presented untruth. But I have talked to some of my colleagues, and I have found that they found it very difficult to remember precisely not only everything said and done but, even more than that, some of the motivations or some of the other thoughts that took place at that time."

"I respect your views, Mr. Hardy, and believe what you just said. But I also have the same feeling about Mr. McDonald, and I would like to have you withdraw those words, 'convenience of memory,' because I don't think that is what he did."

"I would be pleased to do that."

In testimony the following day, it would become known that inside Marshall Space Flight Center, there was one—but only one—engineer who also voiced an opposition to the *Challenger* launch. His name was Ben Powers, and he sat in the same room with George Hardy during the teleconference. Chairman Rogers wanted to know if he, Powers, had also been "appalled" by the Thiokol no-launch recommendation, as his bosses had been. "Sir," Powers would testify, "I fully *supported* the Thiokol engineering position and was in agreement with it."

Unfortunately, engineer Powers himself was a victim of Marshall's clenched-fist management system.

"And you made that known to Hardy?" Rogers asked Powers.

"No sir. I report to John McCarty. I also report to Jim Smith, our Chief Engineer. I report to my boss and to my associate project management in engineering. . . . I don't want to confuse this."

"Did you *report* that you agreed with the Thiokol engineers?" Rogers wanted to know.

"Yes sir."

"To whom did you report it?"

"I'm sorry."

"Who did you report that to?"

"John McCarty. He is my . . . well, he is not my immediate supervisor. He is my Deputy Lab Director, but he was the senior man in line at that time. I reported to him that I thought that the temperature would reduce the margin of safety for the joint performance."

"Were there others in that telecom that agreed with you, that you know of?"

"I can't identify anyone joining me in that position, sir. I cannot make that statement."

"Have you talked to them since, any of the people that were in that telecom, to find out how they stood on the issue?"

"Yes sir, I have."

"And what did you find out?"

"Some of the other engineering people have mentioned that they, too, were concerned, primarily with the temperature effect on the O-ring resilience, the spring-back ability of the O-ring."

"Was there anybody who agreed with Hardy or Mulloy, as far as you remember, on that telecom?"

"There was no dissent with Hardy, to my knowledge, other than the discussion that I had. I was the only dissenting engineer."

"The others remained quiet, I assume?"

"Yes sir."

It was a remarkable piece of testimony. Engineer Ben Powers had followed his chain of command and, in an atmosphere approaching secrecy, only voiced his objection to the launch to his deputy lab director, John McCarty. Powers had not voiced his objection to the entire group at Marshall Space Flight Center that evening. That also meant that no objection had been voiced by Morton Thiokol's senior representative at MSFC that evening, Boyd Brinton, our Chief Project Engineer for the solid rocket motor. Brinton had not taken any position relative to not launching below 53° or to the change in our recommendation to proceed with the launch. No doubt, Brinton, Powers, and Powers's colleagues at Marshall had all been intimidated by the strong objections of Mulloy and Hardy.

I learned later that another NASA Marshall engineer present at the teleconference, Wilbur Riehl, head of the Non-Metallic Materials Department in the Materials and Processes Laboratory at MSFC, told the Commission in a private interview that he, too, had been totally surprised at Marshall management's reaction to Thiokol's recommendation not to launch unless the temperature was at least 53°. During the caucus, Riehl passed a note to an associate sitting next to him.

The note read: "Did you ever expect to see MSFC want to fly when MTI-Wasatch didn't?"

20

31 Degrees?

At this point in NASA's first day of testimony, Chairman William Rogers thought it was time to talk about the weather: "Do you think that at some point the coldness of the weather makes a difference on the risk?"

George Hardy could not be specific: "Well, I'm sure there must be some point, because there is some point at which the structural integrity of the O-rings just wouldn't be maintained."

"At what point would that be?"

"I think that would be somewhere in the -40° to -50° F range."

Unbelievable! It was impossible to believe that NASA Marshall's Deputy Director of Science and Engineering could really think that we would have to be looking at a temperature as low as 40° to 50° below zero before we would have a concern with the O-rings sealing—and only then because of the structural integrity of the O-ring material!

"But it wasn't just a discussion about the data, Mr. Hardy. It was a discussion about the no-launch recommendation. This wasn't just an intellectual exercise. Your statement about being 'appalled' and Mulloy's about 'April' and so forth, all gave the Thiokol people the idea that you strongly disagreed with them on their no-launch recommendation, and that is why they wanted to recess for five minutes. It wasn't just a discussion among engineers about a theoretical matter; it was a very practical question: were you going to launch the next day or not? That is the problem I am having with it."

Clearly, Hardy wasn't being any more convincing than Mulloy as to why the original no-launch recommendation wasn't correct. NASA's "logic" for proceeding with the launch brought puzzled expressions to the faces of all the Commission members.

Before declaring a recess, Rogers wanted to ask one or two more questions on the subject of political pressure on NASA, rumors about which were spreading like wildfire through the press due to the trouble being stirred up by Senator Fritz Hollings. "Has there been any pressure, from any source, on either of you, Mr. Hardy or Mr. Mulloy? Did anybody urge you to get this launch off? Was there any intercession on the part of anybody to ask you to be sure that you worked hard to get the launch off or anything of that kind at all?"

"No sir," answered Mulloy. "None whatsoever, as far as I'm concerned," answered Hardy.

"No outside interference or attempt to put pressure on you at all?"

"No sir, quite the contrary," replied Mulloy.

Following the break, it was Stanley Reinartz's turn to testify. NASA's Shuttle Project Manager at MSFC immediately alienated the commissioners by trying to avoid the question why no one was supporting the launch at Thiokol until Marshall management revealed their displeasure over the no-launch recommendation. They were not too happy, either, with his doing his best to take a few jabs at me.

"How many of the Thiokol people on that telecom voted in favor of launch?" asked Chairman Rogers.

"As far as I know, sir, there wasn't any voting at that time."

"Well, maybe not a vote, but I mean the expressions that were made by the Thiokol people up until the time of the recess. The information we have is there was nobody on the Thiokol side that was urging a launch. Do you remember anybody who urged a launch on the part of Thiokol, up to that point?"

"They did not vote or provide . . . "

"I'm not talking about a vote," Rogers interrupted. "I'm talking about did anybody express a view, that we think you should launch the Shuttle—the *Challenger*—at that point?"

"They had provided a recommendation that it not be launched below 53°."

"Did anybody in the telecom say, 'We disagree; we think you should go ahead and launch'?"

"No sir."

"So, as far as you knew, based upon what was said, it seemed to be a unanimous view of the Thiokol people on the telecom that they were recommending no-launch?"

"At that point in time, yes sir. The senior Thiokol representative at KSC, Allan McDonald, did inject one significant comment, just *after* Kilminster had asked for a caucus. This comment by McDonald was perceived, I believe by all parties—and I believe as testified by Thiokol in the hearings yesterday [this was the comment by Jerry Mason that I had objected to the night be-fore the hearings]—as a supporting point for a positive launch recommen-dation, or at least a positive point. McDonald did not make a statement, 'I recommend launch.' He said that Kilminster should consider a point made by Hardy earlier, that the secondary O-ring was in the proper position to seal if blowby of the primary O-ring occurred. And I believe Hardy tried to elaborate on the importance of the secondary O-ring."

"Mr. Reinartz, during the caucus did you have a discussion with McDon-ald where he voiced his strong opposition to the launch?"

"McDonald, during the time period we were on that, discussed some of the same concerns that were then covered during the telecom."

"And didn't you draw the conclusion from that, Mr. Reinartz, that Mc-Donald was *opposed* to the launch?"

"At that point I did not draw that he was completely opposed to the launch, having made the statement that 'we should consider the secondary seal.'"

"Didn't he recommend to you at that time that the launch be delayed until late in the afternoon, until the temperatures reached 48° or 50°?"

"I'm not sure whether that was done during the caucus or whether that was in the statement that McDonald made after the teleconference was complete, when he said there were three things that should be considered for a launch delay."

"Are you really trying to convince this Commission that you didn't know that McDonald had serious questions about the launch and didn't really want the launch to occur, are you?"

"No sir. I'm trying to relay what came across that evening during the dis-cussion, Mr. Chairman. As has been testified, Marshall elements in Hunts-ville and those elements at Kennedy Space Center had no knowledge of the internal Thiokol discussions during the thirty-to-thirty-five-minute caucus that preceded their launch recommendation. At that point, McDonald, who had indicated some concerns, did not, as far as I know, take that opportu-nity to inject any of his thoughts or concerns via private telephone into the internal Thiokol discussions going on during that caucus. When I asked all parties collectively if there were any disagreements with the final Thiokol recommendation I received from Kilminster, there was no statement or comment from McDonald, at that time sitting with Mulloy and myself here at the Cape."

"So your testimony is that, after the decision was made, you were satisfied that McDonald had no question in his mind, that he went along with the decision to launch?"

"At that point, Mr. Chairman, when I asked very clearly and very deliberately on the telecom, while all parties were involved, if there were any disagreements among any of the parties, including McDonald, who was sitting right across from me, there was no comment, no objection, or anything, raised at that time."

"I accept that, Mr. Reinartz. But later on didn't you know he continued to object and that there was a lot of opposition on the part of the Thiokol people?"

"I did not know that a lot of the Thiokol people opposed it."

"Did you know that McDonald opposed it?"

"After the teleconference, after we had completed it and hung up, as Mulloy has testified, McDonald said, 'Well, if there are not sufficient reasons related to the concerns on O-ring temperature, then how about the ice situation and how about the recovery sea states?' Mulloy has testified as to comments relevant to the ice and recovery situation. We participated jointly in the telecom with the KSC personnel and with Aldrich about the possible loss of SRB parachutes and frustums resulting from the sea state."

"But McDonald testified yesterday under oath that he had a discussion with you in which he pointed out there were three reasons why he was opposed to launch—one was the O-rings and the other two were the things that you just mentioned. Didn't you know that?"

"Yes sir. The way that I recall McDonald stating it, he said, 'If that is not enough of a concern for you . . . if I were the Launch Director there would be three things . . . '"

"Mr. Reinartz, the reason I'm pressing you, as I think you know, it's very difficult for the Commission to understand how this serious matter was discussed by so many people just prior to the launch, when there still were serious questions, obviously, in the minds of a lot of people, how those concerns were not conveyed to the people who had to make the ultimate decision whether the Shuttle would be launched or not. That is what is very difficult for us to understand."

"I can appreciate the difficulty you are having, Mr. Chairman, with each one of us, collectively, trying to remember back to that point in time. But I think it is important to differentiate between a couple of items. We had the knowledge that evening, Mr. Chairman, of the data that Thiokol had presented in the charts and the discussion that was in that telecom. I did not know of any memos that had been circulated, that has now been said to have circulated to some higher levels within Thiokol. I had none of that knowledge. I had none of the knowledge of any dissension that was going on in the internal discussion at Thiokol. None of that was available. McDonald had the opportunity to inject anything to that Thiokol internal that he chose to. As far as I know, there was nothing constraining him. But the only thing that I had was the recommendation from the Thiokol Program Manager, Kilminster. I had Marshall's engineering support, and I had the SRB manager's decision that he wanted to proceed, with which I concurred. Those were the inputs that I had that evening. And then McDonald, he did not make any statement that said, 'Hey, I want to go take this to somebody else,' or 'I have a difficulty with this.' What he said was, 'If I was Launch Director, there are three things that I would be considering for tomorrow.'"

Too bad Stan Reinartz's memory wasn't better, because what I said, verbatim, was, "If I were the Launch Director, I would *cancel* the launch for three reasons," not "consider." Actually, I think his memory *was* better, but it did the NASA case no good for him to recall the actual content of what I had said.

Talk about "convenient memory."

Back in the early 1970s, NASA had introduced the concept of the "lead center" specifically for the Space Transportation System (STS). Managerial responsibility for the STS program was divided among the NASA centers, with Houston's Johnson Space Center put in charge of the orbiter and Huntsville's Marshall Space Flight Center responsible for the orbiter's three liquid-fuel main engines, the external tank, and the SRBs. Kennedy Space Center was to assemble the components, check them out, and conduct the launches. Besides giving it the job of handling the development of the orbiter, NASA also handed overall responsibility for overseeing all STS matters over to JSC, designating it the "lead center."

Although the concept of a lead center seemed sound at the time, in practice it led to a high level of intercenter tension and administrative difficulties, especially between Houston and Huntsville, between whom there had never been much love. For example, the Shuttle Project Manager in Huntsville (Stan Reinartz at the time of *Challenger*) really had two bosses: his own Level III Center Director (William Lucas) plus the Level II STS Manager in Houston (Arnold Aldrich).

In the *Challenger* hearings, General Kutyna now zeroed in on this troublesome arrangement. "Mr. Reinartz," Kutyna said, "you informed Dr. Lucas [about Morton Thiokol's initial recommendation not to launch]. Dr. Lucas is not in the reporting chain, is he?"

"No sir."

"If I could use an analogy: If you want to report a fire, you don't go to the mayor. In his position as Center Director, Dr. Lucas was out of the reporting chain, much like a mayor. If it was important enough to report to him, why didn't you go through the fire department and go up your decision chain?"

"General Kutyna, it is a normal course of our operating mode within the Center that I keep Dr. Lucas informed of my activities, be they this type of thing or another."

"But you did that, Mr. Reinartz, at 5:00 a.m. in the morning. That's kind of early. It would seem that's important. Why didn't you go up the chain?"

"Five in the morning is the time when we basically go in to the launch— so it was not waking him up to tell him that information. It was when we normally go in to launch in the morning. Based upon my assessment of the situation as dispositioned that evening, for better or worse, I did not perceive any clear requirement for interaction with Level II, as the concern had been worked and dispositioned with full agreement among all responsible parties as to that agreement."

"Mr. Reinartz," Commissioner Hotz interjected, "are you telling us that you, in fact, are the person who made the decision *not* to escalate this to a Level II item?"

"That is correct, sir."

"Do you think the system should be changed now?" Chairman Rogers asked. "Do you think that this Commission should make a recommendation to correct what appears to be inadequacy?"

In another telling example of avoiding admission of bad judgment and supporting the party line that the system was OK and not broken, Reinartz said he "would not give a very quick, off-the-cuff answer to that. I would like the opportunity to study that question some before I would recommend any major changes to the system."

"Did you appreciate Mr. Reinartz, that this was a Criticality 1 system that was being discussed?" Sally Ride wondered.

"Yes, Dr. Ride, I did."

I found this unbelievable! It was almost as if Reinartz was not aware that, by admitting this, he was basically conceding that he should not have accepted the recommendation to launch or he should have signed a new waiver against the Criticality 1 issue, because the flight rationale *depended* on the secondary seal to work like a Criticality 1R design!

In the next hour of the hearing, the Ice Team that had gone out to the pad the morning of the launch painted a very dismal picture. Only once before had the Ice Team ever gone out to the pad in the middle of the night prior to a launch—a year earlier, on STS-51C, NASA decided to delay the launch for twenty-four hours because of cold temperatures and icing concerns. This time the Ice Team conducted no less than three complete inspections—the first two lasted over an hour and involved every means possible to break up and remove the ice, and the third inspection was not finished until around 11 a.m. By then, it was approximately T-38 minutes.

Taking temperature readings on the SRBs, as was normally done, the Ice Team during its last inspection found that the difference in the temperature between the two boosters was approximately 14°, with the right SRB reading 9° and the left SRB reading 23°. Subsequent to the accident, however, NASA had found the temperatures to be higher—19° on the right and 33° on the left—and had "corrected" them. Curiously, the head of the Ice Team, Charles Stevenson, did not discuss those specific temperatures with any other members of the overall launch team. When asked why, Stevenson answered: "Well, for two reasons, I guess. Number one, the vehicle was operating within the red lines that we had. When we go out to make our inspection, we're required to report any anomalies. But, number one, having no anomalies, you report on any points of interest. Since the vehicle was operating within my guidelines, within its launch commit criteria, within the requirements, that was not a point to report."

To this day I cannot understand why the recorded 9° temperature on the right-hand SRB was not considered an anomaly—as was not the fact that the right-hand SRB was 14° colder than the left-hand SRB. Even more suspicious, why weren't these apparently incorrect low temperatures changed or corrected until *weeks after* the accident? Never before on any previous launch had the IR pyrometer data been corrected, including the launch attempt the previous day (January 27, 1986) or the cold launch of a year earlier, for STS-51C.

More astonishingly, it sounded like the head of NASA's Ice Team realized there was a specified temperature constraint for launching the Shuttle; Mulloy, Reinartz, and Aldrich had all stated in their earlier testimony that the Shuttle had a launch commit criteria specific to launchpad temperature, and that that temperature was 31°. Clearly, the temperatures measured by the Ice Team on various areas of the vehicle on the morning of the launch were well below the supposedly well-known minimum, so why didn't Stevenson report it?

My startling conclusion, based upon Charlie Stevenson's testimony, was that *nobody* on the day of the *Challenger* launch really *knew* that there was such a temperature constraint! One would certainly think that if anybody knew of such a constraint, it would be the head of the Ice Team! I came to believe that, *after the accident*, NASA searched all of its specifications in hopes to find *something* that could be at least interpreted as a launch commit criteria that was *below* Thiokol's recommendation of 53° SRB launch constraint. I also came to believe that the IR pyrometer data was corrected to make it *appear* that the launch actually *complied* with the alleged 31° requirement. I seriously doubt that Mulloy, Reinartz, and Aldrich were aware, either, on *Challenger*'s fateful day, of the 31° launch commit criteria.

Ironically, the Ice Team *did* report to the Launch Director that it was still finding ice, during the final inspection, on the left-hand SRB. Why would that be reported but not the extremely low temperature of 9° on the right-hand SRB? According to NASA's own official launch commit criteria, there *was* an LCC of 31° on the vehicle, but there was *no LCC* on *ice* on the SRB. Of course, some ice was always present on the external tank, and it was a judgment call on each launch to determine if the quantity of ice was OK from a weight performance or as a potential source of damaging debris. But ice on the bottom half of an SRB was *not* a potential debris problem, and ice on an SRB did not penalize performance anything like it did on an ET. So why report SRB ice, which wasn't so important, and not report extremely cold SRB temperature, which was?

A little puff of black smoke seen emanating from the right-hand SRB during ignition and liftoff (in precisely the area where the cold temperatures had been measured prior to launch) was the Commission's next focus. In response to continued questioning, members of NASA's Ice Team answered that the smoke indicated a leak—one "that may or may not be related to temperature"—most likely coming out the aft joint of the right-hand SRB. During the last ice inspection, the Ice Team "could see absolutely nothing" that indicated any kind of a leak of cryogenic vapors coming from the ET.

This testimony, from B. K. Davis of the Ice Team, interested me a lot. From it, I concluded that there had been no leak of cryogenic fluid that contributed to the problem. This eliminated one of the concerns I had raised back in my failure analysis of February 4, but left open the question as to why the temperature measurements made on the right-hand SRB near the ET were well below local ambient temperature. If there had been no leak, the extremely cold temperatures of the aft field-joint had to be a byproduct of cryogenic venting or local cooling of the ambient air near the bottom of the vehicle from cryogenic vapors.

Responding to a request from Vice-Chairman Armstrong for an estimate of the actual temperatures of the right-hand SRB's troublesome O-ring seal area at the time of launch, Davis gave an answer that absolutely reinforced my notion that the NASA launch support team (and its contractors) were *not* familiar with the 31° launch commit criteria at the time of *Challenger*'s launch: "Taking the 19° corrected measurement . . . and taking into account the change in ambient temperature," Davis calculated, "it comes out that the temperature should have been approximately 28° to 30° at the time of launch." Even this post-facto calculation by NASA showed that the temperatures of the O-ring seal areas of the SRB—when the original launch window opened at 9:38 a.m., two hours before the actual 11:38 a.m. launch—remained below the 31° launch commit criteria, even after the corrections were made to the IR pyrometer data weeks after the disaster. In truth, the actual temperatures at the time of launch were probably closer to 14° to 16°.

NASA's apparent insensitivity to weather concerns, piggybacking as it now did on top of clear evidence of lack of communication between the NASA centers, brought the commissioners' distress over what they were hearing from NASA to a boil. What Dr. Rocco Petrone, President of Rockwell Shuttle Operations (and former Director of NASA Marshall), recalled about the morning of *Challenger*'s launch hardly cooled things down.

Petrone himself was not at the Cape, having returned to his Downey, California, office on Monday, January 27, after the scrub that morning. In his stead, his two program managers, Bob Glaysher and Marty Cioffoletti, remained at KSC to watch over the orbiter. Following a top-level telephone discussion held at approximately 8:30 a.m. EST (5:30 a.m. PST) between the Rockwell men at the Cape and company management in Downey, Petrone concluded that it was not safe to launch. There was just too much chance that falling ice debris might do significant damage to the orbiter. "We had not launched in conditions of that nature," Petrone explained to the Commission, "and we just felt we had an unknown. . . . I said, let's make sure that NASA understands that Rockwell feels it is not safe to launch, and that was the end of my conversation. With that, I turned the matter over to my program managers and my site manager."

In answering follow-up questions, Rockwell's Bob Glaysher stated that his company had no data on which to base judgments about such a complicated ice profile that could affect the orbiter. "This is the first time this had occurred," Glaysher admitted. "It was not a design condition for the orbiter."

Curiouser and curiouser! For Rockwell, the possibility of ice formation from cold weather was "not a design condition for the orbiter," whereas the NASA folks had just said the launch commit criteria for the vehicle required a minimum temperature of 31°, one degree *below freezing*. I always thought "below freezing" was likely to mean ice, especially in a high humidity environment like Florida! We at Morton Thiokol had been raked over the coals by NASA for *not* testing the SRBs to these temperature extremes, but apparently the orbiter wasn't even analyzed or tested for that launch condition either! How come Rockwell and JSC had ignored these criteria for the orbiter? It made me all that more suspicious that 31° wasn't really a design constraint or a launch commit criteria, and not a necessary launch or operating condition, but rather just an ambient exposure condition for the vehicle. Sadly, maybe it all came down to how different people interpreted the 07700 JSC specifications.

"My initial quote to NASA," Glaysher recalled, "was that 'Rockwell cannot 100 percent assure that it is safe to fly,' which I quickly changed to 'Rockwell cannot assure that it is safe to fly.'"

At a 9:00 a.m. meeting with NASA's Arnie Aldrich, Rockwell's other program manager at the Cape, Marty Cioffoletti, told Aldrich that Rockwell's serious concerns about the ice dangers to the launch could not have been interpreted by NASA as anything other than a "no-launch" recommendation. Chairman Rogers wanted it to be even more emphatic: "But did you convey that to NASA in a way that they were able to understand, that you were not approving the launch from your standpoint?" Cioffoletti answered, "I felt that by telling them we did not have a sufficient database and could not, in particular, analyze the trajectory that the ice on the mobile launch platform would take at SRB ignition, I felt Mr. Aldrich understood that Rockwell was not giving a positive indication that we were for the launch." Glaysher seconded that assertion: "I stated more than once during the meeting Rockwell's position that we could not assure that it was safe to fly. It was my first statement at the meeting, and it was my last. I also reiterated the statement several times."

"Had Rockwell ever taken that position before on previous launches," Sally Ride inquired, "yet the launch still occurred?"

"No," answered Glaysher. "This was the first time where we had been in a position where we really had no database from which to make a judgment, and this was the first time that Rockwell had taken the position it was unsafe to fly . . ."

"But clearly, when they resumed the count," Neil Armstrong declared, "you knew that your recommendation essentially had been either considered and overruled, or dispositioned in some way?"

"That's right," Petrone admitted.

"Subsequent to that point in time, when the count resumed, did any of your people take any other opportunities for yourself to express your opinion again?" Armstrong asked.

"Mr. Armstrong," Petrone said, resolutely shaking his head. "I felt we had expressed our opinion to the proper level, on the proper occasion of the meeting that had been set up for it. I felt we had done all we could do."

21

Houston, You Have a Problem

Of the sixteen NASA people on the line during the teleconference the night before the launch, none of them came from Houston, the Cape, or NASA headquarters; they were all from Huntsville. Within an hour after the meeting, Larry Mulloy, the SRB project manager from Huntsville, telephoned Arnie Aldrich, NASA's overall Shuttle Program Manager from the "lead center" in Houston. Mulloy reported that bad weather at sea would likely delay recovery of the SRBs, but he made no mention whatsoever of possible problems with SRB joints due to the cold weather or, for that matter, about any of the discussions with Morton Thiokol. As Mulloy explained to the Commission, "At that time, and I still consider today, this was a Level III issue. . . . We work many problems . . . that never got communicated to Mr. Aldrich."

But Mulloy did mention Thiokol's concerns to Dr. Lucas, the Marshall Director, someone who was not even part of a Shuttle launch's direct chain of command.

At the end of the first day of public hearings for NASA, it was time for Mulloy's two bosses, Aldrich and Lucas, to answer questions about what they did and did not know about *Challenger*.

Chairman Rogers: "At that time, did they tell you, Mr. Aldrich, that there had been serious concerns expressed by Thiokol and Thiokol engineers, and that they had had a long teleconference on the subject, and that first Thiokol had recommended against launch and, secondly, management, in the person of Mr. Kilminster, had changed its mind and Thiokol decided to recommend launch? Did you know any of that sequence at all?"

"None of that was discussed," answered Aldrich, "and I did not know until *after* the 51-L launch that there had been such a meeting."

Mr. Hotz: "Did Mr. Stevenson at that time report to you anything about the extremely low temperatures they were recording on the solid rocket boosters?"

"No sir, he did not. The detail characteristics of the ice on the Fixed Service Structure were discussed in quite an amount of detail for all of the Mission Management Team that was assembled—not in a telecom but among the various engineering people who were there in Florida [with representatives from the Ice Team, including Mr. Stevenson, from JSC, from Marshall, from KSC, and from Rockwell]."

This comment made it clear in my view that had the Mission Management Team been sensitized to Thiokol's concerns about low temperatures, at least a few of the men at this meeting could have sounded an alarm about the low temperatures measured on the SRB. Two of the Huntsville managers who were there with Aldrich—Cecil Houston, the resident manager at the Cape for MSFC projects, and Jim Kingsbury, the MSFC head of engineering—certainly knew of Thiokol's concerns, but said nothing to the Mission Management Team, just like Marshall's Stan Reinartz had not done, who was a member of the MMT.

"Following these discussions," Aldrich explained, "I asked for a position regarding proceeding with the launch. Richard Colonna, JSC's Orbiter Project Manager, and Horace Lamberth, KSC's Director of Engineering, recommended that we proceed. At that time, Robert Glaysher, Rockwell's Orbiter Project Manager, advised me that he had been listening to the entire discussion and, while he did not disagree with either the JSC or KSC analysis, that Rockwell would not give an unqualified 'go-for-launch' as the ice on the launch complex was a condition that had not previously been experienced, and posed a small additional, but unquantifiable, risk. Glaysher did not ask or insist that we not launch, however. No other comments or recommendations were offered by the large group assembled with respect to concern for proceeding with the launch. At the conclusion of this review, I felt reasonably confident that the launch should proceed. However, I was concerned that, in the time between when the Ice Team had been out at the launchpad, starting at 6:30 a.m., and the time of launch, the conditions at the launchpad might have changed. Thus I asked for an additional ice inspection team visit to the pad, to be performed as close to launch as possible. . . . The report of the Ice Team following this final pad inspection indicated no significant changes."

Watching his testimony, I was struck by the fact that Aldrich, serving as NASA's Level II manager in flight readiness review, had relied on the senior management *contractor representative* at the launch site to provide a "go" or "no-go" launch recommendation, rather than the highest level program manager who was at the contractor's plant, which in this case happened to be Dr. Rocco Petrone, the former Director of the Marshall Space Flight Center, who had originally raised the ice debris concerns.

It is also interesting to note that Aldrich did not require the launch recommendation to be signed by anyone; he only conducted an oral poll with the contractor management people on site, which was the normal procedure in spite of what Larry Mulloy had said earlier in reference to Thiokol. This was the way the system was *supposed* to work and the reason the MSFC people chose to testify that they dealt only with my boss, Joe Kilminster, instead of me, a tactic to discredit my testimony and cover their own backsides. You can bet that if I had volunteered to sign the launch recommendation and then done so, they would have dealt only with me. Since I didn't sign it, and continued to try to persuade them to cancel the launch even after the "no-launch" recommendation was changed, they sought to convince the Presidential Commission that I was some kind of nonentity. Even though they admitted the Thiokol teleconference wasn't a flight readiness review meeting in a normal sense, they all stuck by their story that they dealt only with Joe Kilminster on all flight readiness recommendations.

Similarly, it was interesting that Aldrich knew that Rockwell had concerns about launching due to the ice debris issue, but he believed that the people on the scene were responsible for representing the contractor. Aldrich asked both managers from Rockwell what their position was because they were both there; he didn't just deal with a single individual as the MSFC people had stated. Aldrich also acknowledged that this was the first time that a contractor had raised such concerns at the last minute, but he did not request a launch recommendation from the contractor in writing. He obviously thought that he was making a decision to launch based upon his knowledge of the situation and the assessment of the ice debris risk made by his people at KSC. He may have had some concerns, but he believed that the risks were acceptable based upon what he had been told and did not need a written CYA ("Cover Your Ass") recommendation from the contractor in case something went wrong.

The MSFC people knew, but would not admit, that they had pressured Morton Thiokol management into changing their original recommendation not to launch; therefore, they had to have a signed written recommendation *just in case*. They obviously had some real concerns themselves or they wouldn't have required a signed recommendation to launch; after all, they had never required such a document from the contractor the day before the launch on any of the previous twenty-four Shuttle launches. The Marshall people also knew that this situation was so unusual and controversial that they did not want to raise the issue with the Level II or Level I Mission Management Team.

This decision not to inform the Mission Management Team was probably the worst decision among many bad decisions that were made concerning the *Challenger* launch, for there was absolutely no good reason not to share this information with the MMT. If they had done so, most likely it would have prevented the launch. Even if the Mission Management Team had accepted the final launch recommendation signed by Thiokol, it would have been sensitized to our concern about the effects of cold temperatures in the areas of the SRB field-joints and O-rings. In turn, this would have sensitized the Ice Team so that when it made its report at 9:00 a.m., it would have addressed the SRB temperatures, recorded to be as low as 7° to 9° on the right-hand SRB aft field-joint.

There was no doubt in my mind that everyone in the Mission Management Team, including Stan Reinartz, would have stopped the launch that morning if that information had been available, or that if they would have asked anyone from Thiokol, our engineers or our managers, if 7° to 9° was OK, they would have received a unanimous, resounding NO!

Unlike Arnie Aldrich, William Lucas, the Marshall Center Director, had been informed by Mulloy about the teleconference with Thiokol, but in his testimony to the Commission Lucas declared there was no way for him to have known about the Thiokol engineers' concerns about cold temperatures and O-rings, since Mulloy could not have been aware of them due to the nature of Thiokol's off-line caucus. Lucas did hear about them the next morning, when he went into the control room at the Cape and Stan Reinartz told him that "an issue had been resolved." "There were some people at Thiokol who had a concern about the weather," Reinartz had told him, "and that had been discussed very thoroughly by the Thiokol people and by the Marshall Space Flight Center people, and it had been concluded agreeably that there was no problem, that he had a recommendation by Thiokol to launch and our most knowledgeable people and engineering talent agreed with that. So, from my perspective, I didn't see that as an issue."

Agreed, except for the fact that Thiokol first recommended not launching and that I had continued to support that position with Reinartz after the launch recommendation was changed!

Chairman Rogers: "I gather you didn't tell Arnie or Jesse Moore [the NASA Associate Administrator for Space Flight] what Reinartz had told you?"

"No sir. That is not the reporting channel. Reinartz reports directly to Aldrich. In a sense, Reinartz informs me, as the institutional manager, of the progress he is making in implementing his program, but I have never on any occasion reported to Aldrich."

So large was Lucas's ego that he could not stop from exposing it, wanting to make it perfectly clear that he *never* reported to Aldrich, who was lower on the NASA totem pole than Lucas. What might also have been said is that Lucas used his superior position as an excuse not to say anything ever to Aldrich.

"And you had subsequent conversations with Moore and Aldrich prior to the flight and never mentioned what Reinartz had told you?" Rogers queried.

"I did not mention what Reinartz told me, because Reinartz had indicated to me there was not an issue, that we had a *unanimous* position between Thiokol and the Marshall Space Flight Center. There was no issue in his judgment—and not in mine as he explained it to me." "Unanimous" my eye! Reinartz at least knew that I had complained about the decision and recommended to delay the launch as acknowledged by Reinartz in his own sworn testimony.

Rogers: "Did you know they came and recommended against the launch is the question, Dr. Lucas?"

"I knew I was told on the morning of the launch that the initial position of *some members* of Thiokol—and I don't know who it was—had recommended that one not launch with the temperature less than 53°." *Some* members had recommended not launching! Initially that had been Thiokol's total position as presented by the Vice President of Engineering Bob Lund and supported by the Vice President of Space Booster Programs Joe Kilminster; it was our position in *writing*! Lucas was making it sound like our objection to launch was some casual comment made by some insignificant person in the back of the room.

"That didn't cause you enough concern, Dr. Lucas, to pass that information on to either Mr. Moore or Mr. Aldrich?"

"No sir, because I was shown a document signed by Kilminster that indicated that the temperature would not be significant, that it would be that much lower, as I recall it."

"Do you know how they happened to show this telefax to you?"

"When I asked what had been the resolution of the concern."

"So you knew then that the original recommendation was not to launch by Thiokol?"

"I knew that some members of Thiokol—I don't know who—had expressed reservation against launch."

"Didn't you ask who?"

"No sir, I did not, because I did not know the Thiokol people well enough to make a judgment. I do know Kilminster. Kilminster is the person who participates or has participated in all of our flight readiness reviews. He is a man I respect and believe to represent the position of Thiokol. I had no reason to question that."

Chairman Rogers didn't back off: "As we pointed out yesterday, an argument could be made because of this piece of paper that Reinartz lived up to the book, but there was no application . . . of common sense. I mean, this obviously was a very serious matter, and by insisting that this piece of paper was giving everybody the blessing of Thiokol, when the fact was that the top people who made the decision never knew about what happened in this long telecom and didn't know that the decision to launch—recommending a launch—was made really by just a couple of people, Kilminster and Mason, and maybe one other. The fourth gentleman, Lund, said he was in a position where he couldn't prove that it was not safe and therefore he put on his management hat and changed his mind. So this piece of paper, which really resulted in the launch, was made by just a couple of people, and apparently none of you gentleman knew about it."

"I did not know that it was made by a couple of people, Mr. Chairman. As I said, I recognized Kilminster as the senior Thiokol individual for the space booster programs."

"Did you know," Robert Hotz asked Lucas, "that Kilminster earlier had formally recommended against launch and then had reversed his position?"

"The testimony I heard was that, based upon the engineering evaluation, or some engineering evaluation, where the engineers say we don't propose to launch if the temperature is less than 53°, that Kilminster said, 'Well, I can't go against my engineering' or something to that effect, 'so let us now have a caucus and discuss this.' It is not unusual in our system for one or more engineers to raise a concern and then have those concerns discussed and thrashed out."

"Mr. Lucas, that's not the testimony, though," Chairman Rogers countered. "The testimony was they had a long teleconference, and Thiokol made a formal recommendation *against the launch*. It wasn't just a casual conversation; they had a chart there. The chart said 'we recommend against launch.' It wasn't what you said it was at all. It was something quite different. They made a formal presentation, 'no-launch,' and then they had a long off-the-record or off-the-telephone conference caucus, and it turned out that Mason, Wiggins, and Kilminster supported this document, and all the engineers were against it. And Lund said well, I'm chicken; I have to go along. I can't figure out a way to prove that it is not safe."

"Mr. Chairman, I heard that testimony this week. I did not have that testimony, though, at the time this happened."

"No, but you're describing the telecom as though it were just sort of one of those ordinary things, and I don't believe that is accurate."

"That was my perspective of it at the time. I would conclude also on the basis of what I have heard this week that it was not an ordinary situation."

"Well, in any event, at no time did you pass on the information you had, even though it was sketchy, to either Moore or Aldrich?"

"No sir, I did not. As I indicated, the channel—the project channel—is from Reinartz to Aldrich. On the basis of what he and Mulloy had told me, I didn't consider it an issue. I considered it in line with the kinds of decisions that had already been made relative to launch."

"You had occasion, though, to talk to both Aldrich and Moore before the launch. And whether it was in the line of authority or not, you had ample opportunity to pass on the information that there has been serious concern about the seal. Isn't that right?"

"Yes sir, I had the opportunity to talk to them."

After Lucas's testimony, the commissioners asked several members of the Mission Management Team who were present at the launch to testify as to whether they had any prior knowledge of concerns about the SRB O-ring seals. Arnie Aldrich stated that he had not known that STS-51C in January 1985, the coldest launch until *Challenger*, had experienced the most ever blowby and most erosion. Nor was that realized by Jess Moore, NASA's Associate Administrator for Space Flight at NASA headquarters: "I knew that 51-C was a cold launch because I remember we scrubbed a day on the 51-C flight because of excess ice on the external tank, and we were worried about the thermal protection system. . . . But I did not recall any correlation between temperature and the erosion experience that we had seen on 51-C."

Aldrich in his testimony persisted in his belief that 31° was a firm launch commit criteria for the solid rocket booster without his coming to terms with the fact that, if people had considered it so, then why hadn't it been necessary for the Ice Team to report temperatures measured below 31° on the SRB and orbiter? Instead, it was never mentioned that the temperatures were below 31° at the opening of the launch window and that the SRB seals were well below this. Aldrich was also completely wrong about what the document trail related to the SRBs would show. "My understanding," Aldrich testified, "is that the SRB project has certified to those conditions, and we can provide you a documentation trail that indicates that. In any event, in the launch environment, I certainly rely on the extra effort and time we have put into the launch criteria so that they tell us what constraints we have to honor and when things are acceptable." In fact, the only document trail that existed showed very clearly that the SRBs were not certified to operate at 31°, and that there was never any intent to do that even by analysis because it was not understood to be an operating requirement.

Gene Thomas, Director of Launch and Landing Operations at KSC, tried to support Aldrich's claim: "I might point out that for each mission we have an amendment to the LCC that picks up changes, some of them as late as the day before, and history has shown that Marshall is the most conservative, that they cover everything. Their history shows that they cover everything with an LCC when it is required."

But Thomas's point supported just the opposite view. "But I gather that is what Thiokol was saying," Chairman Rogers retorted. "We think now that we should not launch—or at least the engineers thought that. 'We don't think we should launch in this temperature'? Then, according to the testimony, Mulloy said, 'Well, you can't do that. You can't change the launch commit criteria on the *eve* of launch. This is Friday night, you can't do that.'"

"I'm not sure what he meant by that," Jess Moore answered. "You would have to ask him his intention there. But, as Gene [Thomas] said, there are launch commit criteria that *are* changed."

"Well, we heard Mulloy yesterday," Chairman Rogers responded, "and one might think he was saying the same thing Aldrich is saying here today. But as it turned out, Thiokol, however late in the game, then apparently said, 'We don't think the criteria is right. We think it would be dangerous to launch at this temperature.' That was their first recommendation."

"Mr. Chairman, you certainly can raise objections to launch commit criteria," Moore responded, making it clear that Mulloy's assertion was not correct. "There's nothing in the program that says if someone has a problem with the launch commit criteria, you can't raise objections. Until it is T minus 0, you can raise an objection to it. . . . All of our major contractors participate in our flight readiness reviews and in our L-1 day reviews. They're all polled individually. A senior member of that contractor team is polled individually to see if they are ready to launch." In other words, NASA normally would have required from Thiokol not Joe Kilminster's recommendation to proceed with launch but only that of the senior management representative who was present at the launch site—namely me, Al McDonald. And Jess Moore was saying he didn't require it in writing—and he was the chairman of the Mission Management Team.

Sally Ride wanted to know if Moore would have expected the problem to have been brought up to Level II. "Well, I will answer from my perspective," Moore offered. "Looking back on everything and the amount of discussion that went on, even though the people decided that it was judgmental and thought it had been put to bed, I would have thought it would have been brought to Level II, if you want my honest opinion."

Dr. Ride then refined her question: "Would you expect any new concern about a Criticality 1 item to be brought up to Level II?"

"A new concern on a Criticality 1 item, I think, should be brought up to Level II. At this point in time in the program I believe Level II is our repository for everything on the critical items list. So, yes, any changes and deviations and so forth to that, I certainly expected to be brought up to Level II."

"And would you expect to hear if a contractor originally gave a 'no-go' and then subsequently gave a 'go'? Would you expect to be informed of that?"

"That is hard to say. I would hope that I would have been informed of those kinds of things, but I am not sure about that when the people involved figured they had made a judgment on this and had satisfactorily resolved the whole thing. I have nothing in the program that says it is required."

"Why not?" Chairman Rogers asked. "Wouldn't that be the easiest thing in the world to put into the program? 'The contractor recommends no launch, please advise us.'"

"That would be very easy to do. But let me just add one comment. I have never felt that anyone in the program has been reluctant to speak up. I have operated under the ground rules and principles that people who have issues in the program or in the process of reviews could, in fact, speak up and raise those issues. I think we have all tried to be very open about it. In many, many of our reviews we have very lengthy discussions on various kinds of issues, involving not only the headquarters people but the field center people and the contractor people." In other words, it was clear to the Commission that Jess Moore, head of the Mission Management Team, would, indeed, have wanted, and expected, the Marshall representative on his team to have brought forward Morton Thiokol's concerns.

With Jess Moore's implicit criticism of the rationalizations provided by Mulloy, Reinartz, and Lucas, a door was open for additional negative comments about NASA's chain of command. In response to a question from Commissioner Hotz about the availability of information pertinent to the relationship of temperatures to the seals in the solid rocket booster, Gene Thomas answered, "If you're referring to the fact that the temperature according to the launch commit criteria should have been 53°, as has been testified, rather than 31°, yes, I expect that to be in the LCC. That is a controlling document that we use in most cases to make a decision for launch. There are some other judgments that we make based upon the clock, the hold times that we have remaining, the window for the day, other things like that that are not normally in the LCC. We get that information from the program elements late in the timeframe before launch. But most of the 'go/no-go' criteria, we expect to see in the LCC."

"So you're not really very happy about not having had this information before the launch?" asked Hotz.

"No sir. I can assure you that if we had had that information, we wouldn't have launched if it hadn't been 53°."

In a question directed to Richard Smith, Director of the Kennedy Space Center, Vice-Chairman Neil Armstrong hit upon another critical point very damaging to the whole NASA Shuttle management structure—demonstrating a total lack of common sense. "Mr. Smith," Armstrong asked, "has your launch facility been designed to handle freezing temperatures, and if not, why not?"

"Neil, early in the program it was recognized that we were not equipped, coming out of *Apollo*, to handle freezing conditions in the water systems on the pad."

"But why would that be?" Armstrong wondered. "Certainly any civil engineer knows that he has to handle the normal environments where he is building his building."

"Well, the answer to that would be, yes, except the history in the past ten or fifteen years has been that freezing pipes in Florida, in that area, has been a very rare thing. The decision was made to not implement insulation and so forth on the water systems but preclude rupture of lines; for the risk, the cost was not justified."

What Armstrong and others might also have been coming to understand was insulating pipes was also not justified because no one really expected to launch Space Shuttles in that sort of cold weather! The probability was too low, and it was easier just to postpone a launch for a day or two every few years!

Why this reality wasn't established as an LCC, knowing the limitations of the launch facility, is still a mystery to me. Why NASA would have insisted on having a launch commit criteria of from 31° to 99° for the Shuttle vehicle, but a facility for servicing the vehicle that simply could not operate within those temperature limits, just did not make sense. It also provides further supporting information that the 31° temperature at the launchpad was never really intended to be a launch condition but only a launch vehicle exposure requirement.

Finishing my review of the videotapes of the February 26 and 27 public hearings of the Presidential Commission, I was even more disturbed I had not been allowed to attend those sessions. Because of the heavy workload facing the Commission, it would take another month before I would even receive an official transcript of the hearings, let alone have an opportunity to respond to the testimony. What was even worse, I later found out that the Morton Thiokol legal department in Chicago prevented my responses, which were quite damaging to NASA and Morton Thiokol management, from ever reaching the Commission members.

22

Finger Pointing

None of the public scrutiny did anything to humble NASA Marshall management. Both at the hearings and in follow-up television and radio interviews, they seemed wholly unrepentant. The very day MSFC management returned to Huntsville from the hearings in Washington, they called a press conference to refute the Commission's conclusion that the decision-making process for launching *Challenger* had been "clearly flawed." "I'm not sure what Mr. Rogers means in terms of the decision process being flawed," a brazen Dr. William Lucas declared. "I do not believe that the decision process that I know and understood was flawed."

Larry Mulloy also would not give an inch: "I know that the decisions that I made based on the data that we had up to January 27th and through the launch countdown until the liftoff occurred on the 28th, all of the actions were proper. . . . The launch-decision process is sound." Asked by the press if he believed he had exercised good judgment on the evening before the launch, his answer was, "Oh, absolutely."

Stan Reinartz, Marshall's Shuttle Project Manager, sniped, "I can't agree [with the Commission]. . . . We thought that the problem was acceptable and did not endanger the safety of flying." Dr. Judson Lovingood, Reinartz's deputy, warned that before anyone thinks about making changes in the launch-decision process, "you've got to remember this is the same process that was used in twenty-four successful Shuttle launches, as well as earlier Moon landings, *Skylab* and other manned flight." While Reinartz agreed there was always room for improvement, he said the decision process was a pyramid, and "everything can't go to the top of any organization—in industry or government."

I couldn't believe that anyone who had heard the testimony of the past three days could possibly disagree with Chairman Rogers's assessment that the decision-making process was clearly flawed; I was astounded by Marshall management's posturing so publicly with its insistence that there was nothing inherently wrong with the NASA process, especially doing so while the Presidential Commission hearings were still in progress.

In his public statements, Reinartz did what he could to transfer more blame to Morton Thiokol: "Should we have known more from Thiokol, or should Thiokol have provided more information—that's another question. I find it strange that Thiokol was ready to launch on the morning of January 27th when the temperature was 40° and twelve hours later they were not ready to launch below 53°." His deputy, Lovingood, also vigorously disputed statements from Thiokol officials that NASA had reversed traditional safety considerations by demanding that the contractor prove it was unsafe to fly. "That one thing bothers me more than anything else. That's absolutely incorrect," Lovingood said, his face reddening and his voice rising.

George Hardy, Marshall's Deputy Director for Science and Engineering, chimed in: "There is an apparent difference in interpretation between Marshall officials on the one hand and the Commission and news media on the other hand with respect to Thiokol's initial recommendation that the Shuttle not be launched in temperatures below 53°. If one wants to be highly technical, Thiokol recommended *against* the launch early on the evening of January 27th, but it was very unusual for Marshall and Thiokol engineers to be discussing a problem simultaneously with Marshall and Thiokol managers." "Normally," Hardy explained, "the government and contractor engineers discuss things with each other, then each group of engineers talks to its own management privately, and finally the two managements speak with each other. Because that was not the procedure followed on January 27th explains why NASA had not viewed Thiokol's initial position as a formal recommendation against launch." What a phony explanation this was for not accepting Thiokol's no-launch recommendation, which was provided by two vice presidents from Thiokol to the appropriate management at MSFC.

Asked about the Commission's performance thus far, Hardy demurred. "Some of you may be able to judge their fairness," he told reporters. "I may have something to say about that when they finish."

As a result of appearing on the televised hearings, with excerpts of my testimony appearing on the nightly news of every TV station, my life suddenly changed. My nephew, Kevin McDonald, a U.S. Army soldier sitting in a bar in Seoul, Korea, on February 25, 1986, heard portions of my testimony over the local TV channel. He said to his drinking buddy, "That's my uncle!" "Sure it is," his buddy said.

I got calls for interviews from every TV network and national news magazine; I turned them all down. I received numerous letters of support from engineers, scientists, and managers all around the country complimenting me on my integrity and courage. My brother, John McDonald Jr., a professor at the University of Montana law school, told me that they were using the videotapes of my testimony before the Presidential Commission as an example of credible, believable, and knowledgeable testimony expected from an expert witness.

That was the good news. The bad news was that I also received some very negative letters. One nut from Prescott, Arizona, accused me of being a murderer; I turned this letter over to our security office, which notified the Federal Bureau of Investigation (FBI).

It was clear that the tragedy of *Challenger* was a very emotional event for many people—not just those involved in the launch. As with President Kennedy's assassination and the first Moon landing, nearly everyone could tell you where they were and what they were doing at the precise moment it had happened. For many of us, *Challenger* became the "C" word, still difficult to say over twenty years later at NASA or in the aerospace industry; at Thiokol, it has been totally avoided.

It was a very difficult time in my life. As depressed as I was, supporting letters from my alma maters of Montana State University and the University of Utah, along with letters from professional engineers and aerospace management, buoyed my spirits and hardened my resolve.

One of the remarkable telephone calls made to me came from someone who was in a management position at Rockwell. This person did not identify himself, but he told me that he found himself in a similar situation nineteen years earlier, at North American (later Rockwell) during the *Apollo* capsule fire that killed *Apollo* astronauts Gus Grissom, Ed White, and Roger Chaffee on January 27, 1967. North American had been responsible for the design and certification of the *Apollo* capsule. This fellow told me that some people in his company had been trying for some time to get NASA to change the 100 percent oxygen environment and some of the materials in the *Apollo* capsule because of the extreme fire hazard, but they had been unsuccessful in convincing NASA to do this. After the accident, the only investigation board that was formed was an internal NASA board, and, as a result, North American became "the fall guy" for the accident. To pacify and deflect blame from NASA, the company fired several high-level managers and made generous liability settlements with the families of the fallen astronauts. NASA came out clean. This fellow told me that he wished he had possessed the courage I had exhibited during the *Challenger* accident hearings. He wished he had told the *Apollo* investigation board all he knew about the circumstances that led up to that deadly fire. He had lost a lot of sleep over the past nineteen years, because he had remained silent. He maintained that he and many others inside his company came to believe that Rockwell was later awarded the lion's share of NASA's Space Shuttle program because of its loyalty in protecting NASA from any blame. (In the early 1970s, Rockwell was awarded the two largest contracts in the Space Shuttle program—for the orbiter and the Space Shuttle main engines.) Both of these programs were subjected to a very intense competitive procurement process, and the award of both contracts to the same company brought a howl of protests from the other bidders; NASA senior management dismissed all the protests. This fellow told me that even though it all worked out well for Rockwell in the long run, he personally felt bad about some of his colleagues losing their jobs and ruining their careers to protect NASA management.

It was also rather ironic that Dr. Rocco Petrone, Director of Launch Operations at KSC at the time of the *Apollo* accident, became a high-level executive at Rockwell in charge of the Shuttle program after Rockwell won the orbiter contract from JSC and the SSME contract from MSFC. Petrone was a member of the *Apollo* 204 Fire Accident Review Board and went on to become Center Director of MSFC before going to Rockwell.

The most troubling reaction to what I had done came from inside my own company, from some of my fellow employees at Morton Thiokol who found it an opportune time to distance themselves from me and from Roger Boisjoly in order to improve their own positions with senior Morton Thiokol management.

One of these people, Jack Kapp, whom I regarded as one of my friends, was the manager of our Applied Mechanics Department. I always knew he was very supportive of his boss, Bob Lund, Vice President of Engineering, who held a higher position in the Mormon Church than he did, but I never expected to hear such supporting words he would utter about the rest of senior management who had treated me so unfairly and were an embarrassment to the whole corporation. A meeting was held by Morton Thiokol management at the completion of the Presidential Commission's public hearings on the decision to launch *Challenger*.

All of Thiokol's senior management—Charles Locke, Ed Garrison, Jerry Mason, Cal Wiggins, Joe Kilminster, and Bob Lund—looked awfully bad during their testimony before the Presidential Commission or dealings with the press. Their actions before and after the accident had, in fact, been deplorable, and they only made things worse by the remarks they made to the news media. Nevertheless, their intentions in calling a meeting in late February 1986 were good: to try to unify the troops, put the accident behind us, and resolve us all to work for the common good, which was necessary for a successful redesign of the SRM and a return-to-flight for the Shuttle. I accepted this rationale for the pep rally of a meeting, thinking at the time that it was the only good decision that senior management had made for several weeks.

After the pep-rally speeches were given, the floor was open for questions and remarks. Several people had legitimate questions about the status of Morton Thiokol as a company and about the future of the Space Shuttle program. With these questions answered, the meeting was about to conclude when Jack Kapp stood up and said he would like to make a comment "for the group." Saying that he was speaking "for all the folks in this room who were particularly *proud*" of the way that Morton Thiokol senior management had handled "this very serious problem," he wanted them to know that "we feel that you have really done an outstanding job, considering the circumstances." Kapp wanted to thank them personally for their effort.

I almost threw up. Looking around the room, I saw many others with the same bewildered look on their faces. I felt personally insulted that Kapp said he was speaking for all of us in the room, when I didn't know of *anyone* who was proud of Thiokol management's actions before, during, or after the accident. Our management had many ardent supporters standing behind them because it was politic to do so, but there wasn't anyone who was actually "proud" of them.

As it turned out, it appeared that Jack Kapp's speech was effective, because Jack was the first one to be promoted in the new organization, chosen to be Director of Engineering Design for the redesign effort—even though some twelve years earlier he had been most responsible for the poor SRB field-joint design in the first place. More than anyone else, Jack had selected the original field-joint design patterned after the Titan SRM, and it had been his analysis that said the joint should close during pressurization, when in reality the joint opened. That mistake was known by MTI and NASA management and should have been corrected long before *any* Shuttle was ever flown.

Immediately upon my return home from the Presidential Commission's hearings, I and several of my program managers worked to organize our response to the crisis in the Shuttle program. On March 3, 1986, I wrote a memo for distribution assigning all of my managers to various tasks in support of the failure analysis, requests for support from the Presidential Commission, and eventual redesign and recertification of the SRB. The very next day, Cal Wiggins, my boss's boss, reorganized the entire program office, rescinded my memo, and reassigned me; I was now to be in charge of Improvement Program Plans. On March 6, I prepared a presentation for senior management on what the ground rules should be for the redesign activity. Wiggins took the presentation to Ed Garrison, but removed my conclusions and recommendations, based as they were upon my assessment of the cause of the *Challenger* accident.

On March 11, Ed Dorsey was appointed as the new Vice President and General Manager of Space Operations, reporting directly to President Garrison. Dorsey had been in charge of the original development, demonstration, test, and evaluation for the Space Shuttle SRM and, prior to his retirement from Morton Thiokol in the fall of 1983, was the Vice President of Space Booster Programs. Ed had been replaced by my boss, Joe Kilminster.

Ed Dorsey was a real southern gentleman, a good engineer, an excellent program manager, and a man of high integrity. However, I was very surprised they brought him out of retirement for this job, because he was in charge of the program that accepted the original field-joint design that had failed. Even though the design had been developed in the mid-1970s with the best analytical tools available in the industry at the time, the fact remained that Thiokol's original analysis indicated that the joint would close during pressurization when, in fact, early hydro-tests indicated that it really opened. This was the most fatal flaw in the SRB design, and it came when Dorsey had been in charge of the Shuttle program for Morton Thiokol. I feared that some of the vultures in Congress or the press would bring up this issue and criticize Morton Thiokol even more severely for making this move. I really expected to read in the newspaper, "Morton Thiokol receives sharp criticism from Congressional committee for naming the same person responsible for the defective *Challenger* field-joint design to head up the redesign activity."

As far as mending some of the wounds in the company, Ed Dorsey was the best possible choice. He was held in high respect by all Morton Thiokol employees as well as by NASA management at all levels, and fortunately his approval of the original Shuttle design was never revealed to the public during these difficult times.

Coming back on board, Dorsey chose not to rescind my March 3 memo or change my position in the company. Still, I was coming to realize that my bosses were doing what they could to push me aside. A few weeks earlier, my boss, Joe Kilminster, who replaced me on the SRM Failure Analysis Team in residence at MSFC, had named my good friend Lee Bailey, an engineer who used to work for me, as the acting Vice President of Space Booster Programs. Soon Bailey released a memo concerning an upcoming meeting with some of the Presidential Commission members in Utah with a list of presenters and planned attendees from Morton Thiokol; my name was on that list but with a question mark next to it. Bailey had made the notation on his draft memo, but his secretary forgot to remove it when it was reproduced for distribution, so I received it that way. It reinforced my suspicion that management did not trust me and that my involvement in the ongoing *Challenger* accident investigations was to be as limited as possible and carefully controlled and monitored. I believe that my good friend Bailey really thought that I should be involved, but he also knew that his management thought differently.

The meeting with the visiting Presidential Commission members—Joseph Sutter, Arthur Walker, and Robert Rummel—was to occur on Monday and Tuesday, March 17 and 18. The Commission asked for the meeting to better understand the original SRM development, qualification, and certification program, and interpretation of the contractor-end-item specifications that had been provided to Morton Thiokol by NASA. It was considered such an important meeting that a preliminary meeting was requested by NASA to review the material with J. R. Thompson, the new person heading up the *Challenger* Failure Analysis Task Force for NASA. After hearing so much evasive and contradictory testimony from various NASA managers involved in the *Challenger* launch-decision that was made against some strong objections by both Thiokol and Rockwell, Chairman Rogers had all those people removed from the NASA Failure Analysis Team. Thompson, from Princeton University, was selected to replace Jess Moore as the head of the NASA Failure Analysis Team.

This preliminary meeting was to take place on Saturday, March 15. NASA/MSFC management was also asked to participate so as to provide their interpretation of the specifications, and they agreed, not wanting there to be any surprises in front of the Commission members the following Monday.

Their intent, it was clear to me, was to establish "the party line" for us to follow, so there would be no controversy or major disagreements between NASA and Morton Thiokol in front of anyone from the Commission.

I attended the preliminary Saturday meeting even though I wasn't formally invited. While there, I questioned NASA Marshall's interpretation of the SRM specification and qualification requirements. Even though I was not involved in the original SRM development program, I was the only one to challenge NASA's position on this issue. Bob Lund, Vice President of Engineering, was involved in the original design, but he didn't make any comments. As I had been going through the certification process for the new FWC-SRM just prior to the *Challenger* accident, I believed I was more familiar with this issue than any of those who were on the original development, demonstration, test, and evaluation program some ten years earlier. Even though Ed Dorsey, our new General Manager, did not participate in the meeting, he must have agreed with me, because in the aftermath of the meeting, he specifically requested that I attend the follow-on meetings with the Commission members.

The meeting on Monday, March 17, dealt with the specifications and qualification of the SRMs. Larry Mulloy presented the NASA position. I sharply disagreed with Mulloy's interpretation of the CEI specification and the qualification limits and told him that was not consistent with what we had been doing recently for the FWC-SRM in the qualification for that system relative to environments, which were supposed to be identical to the steel-case motor. Mulloy was obviously very upset with me for challenging him in front of the Commission panel, but he did not waver on the issue and exhibited his typical arrogant attitude. The meeting ran rather late because of all of the questions from the Commission members, and it was decided to reconvene early the next morning in the main conference room of our Management Information Center, which was just across the hall from my office.

The next day, about fifteen minutes before the meeting was scheduled to start, Mulloy came storming into my office. He slammed the door behind him so hard that it rattled the walls. Larry walked right up to the front of my desk and said, "What in the f— are you trying to do, McDonald, by giving information directly to the Presidential Commission before going through your own management and NASA? Just what in the f— is your motivation for doing this?"

Indeed, I had challenged him the day before in front of the Commis-

sion members, even though the preliminary Saturday meeting had been designed to avoid that. Without going through the established "channels" set
up by our legal department to log all information provided to the Commission prior to transmittal, I had also given the Commission members a copy
of my February 4 failure analysis scenario. Still, I was absolutely shocked by
Mulloy's abusive and threatening demeanor.

"I sure did give them a copy of my assessment of why the accident occurred, Larry, but I didn't think I needed a note from my mother to give
them anything I wanted to."

"I just want to know what in the hell was your motivation for doing this?"
Mulloy growled.

"I don't have any."

Strangely, Mulloy's mood then seemed to change. He said he respected
my technical abilities and hoped there would be no hard feelings. He shook
my hand and walked out the door.

Totally befuddled, I walked out of my office to go to the meeting that was
scheduled to start in a few minutes across the hall. My secretary, Larene
Thompson, asked, "What was that all about? I thought your door was going
to fall off its hinges he slammed it shut so hard!" Joe Kilminster's new secretary, Diane Palmer, who had been working alone in the next-door office,
was also there to inquire about the noise. "Why, is that any way of conducting a meeting?" she asked.

I went into the meeting and sat down. To show Mulloy that I could not be
intimidated, I persisted in my questioning of his interpretation of the qualification of the SRM. I knew that had been the very reason why he came into
my office the way he did that morning. He wanted to intimidate me, so that
I would not continue to embarrass or challenge him in front of the visiting
Presidential Commission members. He should have known me well enough
to know better, but again his arrogance got in his way of good judgment.

At the lunch break, I went into our executive dining room with astronaut
Bob Stewart, who said he wanted to talk with me. Actually, I wasn't invited
to be there, as Cal Wiggins had intentionally left my name off the luncheon
list for both days of the Commission visit. Wiggins saw me walking in, and
if looks could kill, I would've been a dead man.

The next day, Wednesday, March 19, I was interviewed by an investigator
for the Presidential Commission, Emily Trapnell. Regularly employed as an
investigator for the Federal Aviation Administration on airplane crashes,
Trapnell had been assigned temporarily to the Presidential Commission to

help investigate the *Challenger* accident. She was very professional and a very nice woman. She conducted a taped interview with me lasting several hours, asking me many questions concerning my earlier testimony. Nothing really new was discussed.

The following week, Ed Dorsey created a new organization, reassigning me as Director of Special Projects to support the STS-51L investigation. I was also to assist Joe Pelham in developing a Joint Environment Simulator (JES) for determining the response characteristics of the SRM field-joints, using full-scale hardware. Even though Dorsey had indicated that I would support the STS-51L failure investigation, he did not include me as an integral part of the Morton Thiokol/NASA STS-51L Failure Analysis Team. Still, I was glad to work on the recovery activity and to be helping Pelham on the JES. Pelham was Chief Scientist for our Utah operations, and I considered him my mentor during my long career at Thiokol. I learned more about solid rocket motors from him than anybody else and held him in very high esteem. He was one of the best solid rocket engineers in the world and a man of high integrity. The Joint Environment Simulator itself, which later proved to be a very valuable tool, was Joe's idea.

A couple weeks later, Larry Mulloy drafted a memo outlining what he thought should be the ground rules for the redesign approach. I strongly disagreed with some of his ideas, so I wrote a memo that commented on Mulloy's SRM recovery planning schedule and challenged several of his ground rules for the redesign.

When I attended my first meeting at Wasatch with MSFC personnel on the redesign, I was very uncomfortable. I felt like many of the NASA personnel were blaming me for getting them into so much trouble and casting the management of the Marshall Space Flight Center in such a bad light. Even though several Marshall people personally called to tell me that I had done the right thing and they were behind me all the way, many others sympathized with MSFC management and treated me almost like a traitor. It was part of the "good-ol'-boy network" that I had exposed, and many of them stayed totally loyal to their longtime associates.

On March 7, 1986, nearly six weeks after the accident, a group of NASA Marshall officials, led by Dr. Wayne Littles, MSFC Associate Director for Engineering, provided a status report on the findings of the MSFC Failure Analysis Team to the Rogers Commission. I watched the hearings on C-SPAN and, once again, could not believe what I was hearing.

Dr. Littles did everything he could to rationalize why temperature was not a contributing factor in the joint failure and to implicate inadequate O-ring inspections, joint mating and assembly, and leak-check port plug installation as the primary suspects in causing the failure. "We have been looking very carefully at both the left- and right-hand boosters for 51-L," he told the Commission. "There are three joints on each vehicle, of course, and we have looked very carefully at similarities and differences between those joints—because we had *one* anomaly on *one* joint; we had none with the other five, as far as we know. If you look at the temperatures on all those joints, you find there really is not a lot of difference in the type of ranges that we're talking about. . . . You see roughly the same kind of temperatures, and *so the temperatures were roughly the same for all of the joints.*

"There are two things that we have discovered that are unique relative to the aft field-joint on the right. One of them is that we did have to use the rounding fixture to mate these segments. I have a note here that says, in fact, that *this was the maximum reshaping to date [suggesting] potential damage to the primary and/or the secondary O-ring at assembly.* In a close-up view of this area [from a joint close-out photograph made just prior to mating the aft segment], a feature of interest is the apparent increase in the distance between the edge of the O-ring and the top of the O-ring groove. From that, it looks as though there may be an anomaly in the O-ring."

It certainly appeared to me from Dr. Littles's testimony that Marshall management was conspiring to lay the blame for the accident on SRB assembly at Kennedy Space Center and on inspections by Thiokol, rather than on anything they may have done wrong. That would allow them to extricate themselves from this mess. But the alternative explanation being cooked up by Littles was not very convincing, particularly his second suggestion that the real culprit behind the accident, if it was not a defective O-ring, due to inadequate inspection, may have been a faulty leak-check port installation.

"The [second] scenario deals with a leak in the leak-check port. In our mind at this point, it is still a viable scenario, the hypothesis being that you get a leak through the leak port at an early time, like half a second [after ignition], to generate the puff of smoke, and then you continue leaking through that leak-check port until you damage the secondary O-ring or you erode the port enough to blow the port out and then start growing the leak there."

"But I don't know *anyone* who thinks smoke came out of a leak-check port," exclaimed General Don Kutyna. "Who do you know who thought smoke came out of a leak-check port?"

"Well, I did, for one," Littles replied. "I'm not an expert, and that is why I asked my experts to go off and put together a story and come convince me, because when I look at the black-and-white photographs, I, with an untrained eye, admittedly, can see what I think is white smoke emanating from near that leak-check port."

To add further confusion to the failure analysis, several of the charts presented by the MSFC official also implicated the external tank, that structural failures in the ET might be the real cause of the failure. This persistent effort by Marshall to deflect attention away from the effects of cold temperature on O-rings began to enrage Chairman Rogers: "Each time representatives from Marshall have testified, they have pointed to the external tank as the number one suspect. I noticed Dr. Lucas also said that in his press conference the other day. And yet, ostensibly, it seems as if the joint seems to be the number one suspect. And I don't quite understand it."

In fact, it would not be until the March 21 hearing (nearly two months after the accident), when MSFC was reviewing the dynamic O-ring test results using full-size cross-section field-joints at Thiokol, that anyone from Marshall management finally admitted that the SRB field-joint was the *total problem* in the *Challenger* accident.

Not only did NASA Marshall officials keep pointing the finger at the Rogers Commission in disagreement with their conclusion that the launch-decision process was flawed, but they also tried their best to place more of the blame on Morton Thiokol for the accident. But NASA Marshall wasn't the only one pointing fingers at Morton Thiokol's defective field-joint design; so, too, did our corporate competitors.

Glancing through the new issue of *Missilani*, the company newspaper of Hercules Inc., in March 1986, I saw a picture of an American flag flying over the Hercules facility in Bacchus, Utah. The caption said it was "the *Challenger* Memorial Flag," the same flag that had flown over the nation's capital at half-staff (from January 29 to February 3, 1986), at the request of Utah Senator Jake Garn.

To arrange to have that flag fly over our competitor's home office was an insensitive and irresponsible action by the senior senator from our state. It was a personal insult to all employees of Morton Thiokol. Its message was clear: Senator Garn wanted to associate himself with one of his biggest political supporters back home, but at the same time not have anything to do with us at Morton Thiokol, because we had been involved in the *Challenger* accident. Flying that flag over Bacchus threw salt into the wounds of a lot of good people in northern Utah who were already hurting under all of the bad press concerning the tragic accident. These were the same people who had produced the rockets that had successfully launched Garn into orbit on the Space Shuttle *Discovery* just a year earlier.

I was already upset with Garn for a comment that he had made on television following a Rogers Commission hearing in February. He told the media that if Thiokol's Allan McDonald or Roger Boisjoly had been so concerned about the weather conditions for launching the Shuttle and weren't satisfied with NASA's decisions, then why didn't we pick up the telephone and call him! The implication was that Boisjoly and I were in some way negligent in discharging our duties, when nothing could have been farther from the truth. Roger Boisjoly was particularly hurt by Garn's comment. I told Roger that Garn's statements were asinine and that what Garn was saying was just political rhetoric, and he should ignore it.

To better understand *Challenger*'s SRB joint failure and to assist in the planned redesign, the Rogers Commission had requested that an independent assessment team be established to review the test program and test articles that Morton Thiokol had been using. Organized by General Kutyna, this team consisted of air force people from the Space Systems Division, air force consultants from the Aerospace Corporation in Los Angeles, and experts from the Air Force Rocket Propulsion Laboratory at Edwards Air Force Base. Basically, they were the same people who had been involved in the very successful Titan III program, initially launched in the mid-1960s, in which there had not been a single SRB flight failure in over seventy launches. During Kutyna's visit to Thiokol with his team on March 19, I told the general that I was surprised that the SRBs for the Titan did not show O-ring distress, because the Shuttle joint design was not just very similar to it, it was actually patterned after the Titan.

"According to our most recent findings, General, we would expect that the Titan SRB joint design should also have an O-ring erosion problem. In a meeting with the Aerospace Corporation back in June 1985, we requested some information on the Titan III O-ring experience, but we've received no response. We were told that the Titan didn't have the O-ring problems observed in the Shuttle program, because they had a compression fit of the insulation at the joint." Kutyna answered that he would make sure we received all of the air force's data on the Titan.

Three weeks after Kutyna's visit, on April 12, 1986, we received from NASA a copy of an engineering change proposal recommending a change in the Titan IV's SRB field-joint to incorporate a second O-ring groove in the clevis leg; the letter had been sent from the Chemical Systems Division of United Technologies Corporation to the Air Force Space Systems Division. The justification for the change was "to provide additional seal capability in the event the first seal develops a heat path due to hot gas erosion. Seal erosion has been observed in previous Titan static tests and similar Space Shuttle static test and flight units." Attached to the ECP were three charts relating some of the history pertaining to Titan III and Titan 34-D O-ring erosion.

What this data showed was that the *frequency* of O-ring erosion and O-ring charring in the Titan SRB field-joints was *three times worse* than with the Shuttle's SRB. The date on the ECP was October 18, 1985! Yet it was never reported or revealed to NASA.

This was unfortunate, but I do not believe that this was intentional. I believe that the Titan O-ring erosion data, some of which was over twenty years old, was totally forgotten about because it wasn't considered a major concern at the time it was first discovered in the early 1960s. What one has to remember is that the criterion for success of a solid rocket firing in the 1960s was very simple—the motor had to be still in the test stand in one piece. There was so much pressure at the time to maximize performance— that is, range or payload for an ICBM or payload-to-orbit for a space launch vehicle—that any extra margin was considered detrimental to the design because it increased inert weight that penalized the performance of the vehicle. I remember going out to inspect early Minuteman Stage I motors after static test, and the best ones were those that showed several areas of "hot spots" because it was considered a more optimum design for meeting performance objectives. I worked on the Minuteman internal and external insulation design, and I never recall being criticized for not having enough margin in the insulation design; I only remember being criticized for having too much virgin material left, implying it was overdesigned. I believe this was also the case with the Titan SRMs designed in nearly the same time frame, and as with the Minuteman, these early observations were quickly forgotten and not passed on to others who came into the program some ten to twenty years later because these rockets were very successful and had demonstrated a high reliability.

I had raised the same concern with the O-rings in the four hinged nozzles on the Stage I Minuteman after the *Challenger* accident. According to our post-*Challenger* O-ring erosion analysis, we would predict that the O-rings should periodically experience erosion and charring in the grease-filled split lines of the hinged nozzles. All of my old Minuteman compatriots claimed we never observed such a problem, but when we went out and examined a recently fired "aging and surveillance" motor, we observed some erosion and charring of the O-rings in the nozzles. This was never recorded as an anomaly or problem because it was considered perfectly acceptable.

I believe the same thing happened with the Titan SRM field-joint O-rings because the bottom line is that they had worked just fine. In fact, the ECP request to incorporate a second O-ring in the Titan SRMs was rejected by the air force. It was rejected based on the adage that "better is sometimes the enemy of good enough," and "if it ain't broke don't fix it." This, by the way, is a very basic philosophy of the Russian space program that has served them well over half a century of manned spaceflight. The Titan SRMs had never experienced an O-ring failure, and therefore the ECP was rejected, which I believe was the correct decision.

The other major factor that resulted in accepting these anomalies of O-rings eroding and pocketing erosion of nozzles in the Shuttle SRMs was a by-product of the solid rocket motor industry not understanding what "man-rating" was really all about. Prior to the Shuttle, the only "man-rated" rockets ever flown by the United States or former Soviet Union were all regeneratively cooled liquid engines, which looked pristine in the test stand after firing. Solid rockets are basically a pile of charred rubble in a hot metal shell after firing. The whole motor contains charred rubber and phenolic insulating materials. Since the O-rings are also rubber, it was deemed acceptable to experience some erosion or charring as long as some virgin material was left to maintain a seal. This was the philosophy that was also accepted in the Titan III program, which was initially intended to be the first "man-rated" launch vehicle using solid rocket boosters. This air force program, called Dyna-Soar, was canceled in the early 1960s and replaced by the Manned Orbiting Laboratory, which was canceled in the late 1960s. The SRM technology that was to be used in these early "man-rated" launch vehicles was simply adopted by Thiokol and the solid rocket industry for NASA's Shuttle a decade later.

An April 6, 1986, article in the *Orlando Sentinel*, "Shuttle Booster Design Couldn't Pass Titan Test," reported on this background history, but not without passing along a number of significant misstatements, errors, and blatant falsehoods whose intention was damning to Morton Thiokol:

The Shuttle booster that failed and triggered *Challenger*'s destruction has design features that were rejected as unreliable more than 20 years ago by the maker of a similar booster used on the unmanned Titan rocket. Joints in the Shuttle's twin segmented boosters were modeled after those in the Titan boosters, designed and built by United Technologies Corp. in Sunnyvale, Calif. However, NASA and Shuttle booster maker Morton Thiokol modified the joint design to make testing and manufacturing easier, to save money and to reduce weight so heavier payloads could be carried aboard the Shuttle. The Shuttle booster also was not as thoroughly tested as the Titan booster, particularly in cold weather. "It was, and is, my belief," said one of the Titan booster designers, "that the Titan booster is more reliable—and in this case, substantially better." A United Technologies engineer said his company experimented with reducing the thickness of steel and insulation materials to reduce weight, but rejected the idea when a joint leaked during a test firing. Engineers from United Technologies also rejected use of a backup O-ring, because they said their experience showed that if the first O-ring failed, the second ring would be unable to plug the leak. Unlike Morton Thiokol's SRB, the Titan was test fired in a vertical position: the booster turned upside down so exhaust gases would shoot into the air. "Testing horizontally is fundamentally wrong," said Walt Lowrie, a former Mission Success Director of the Titan for Martin Marietta Corp., the contractor in charge of the launch system. "You induce loads that you don't see in a real launch." United Technology has a site in Sunnyvale where Titan boosters were sometimes chilled to 30° before being fired. United Technologies, which bid unsuccessfully for the Shuttle booster contract, uses a more expensive technique. Insulation pads in the segments are jammed against each other when the segments are joined, creating a tight barrier between the hot gases and O-rings. Putty is applied to the pads before the segments are connected, serving as a glue to even out any minute irregularities in the pad surfaces. "My own view is that a compression fit is a must," said one of the Titan booster designers. "I would never approve a design that didn't have it."

It is questionable whether United Technologies ever really experimented with a thinner steel case for the Titan SRB—one more like the Shuttle SRB—and that that joint had leaked during a test firing. I don't believe, either, that the Titan boosters "were sometimes chilled to 30°" before being fired, because neither Chemical Systems Division nor the air force had facilities capable of doing this.

Nowhere in the article was it mentioned that O-ring erosion problems on the Titan SRMs had, in fact, been observed, nor that United Technologies, less than six months earlier, in October 1985, had submitted an ECP to the air force indicating the use of two O-rings for redundancy, which it now claimed it had always known was a "bad design." As for criticizing Morton Thiokol's horizontal testing of the Shuttle SRMs compared to their vertical testing ("nozzle-up") for the Titan SRMs, later analysis (done during the redesign of the Shuttle SRMs) would come to verify that horizontal testing was more severe and more representative of the Shuttle flight environment and that the most benign way to test was nozzle-up, as with the Titan SRMs.

Talk about kicking a man when he's down! At the time, I was very disappointed in my colleagues from Chemical Systems Division (and others in the solid rocket community) so willing to take such cheap shots at Morton Thiokol, when their own closet was so full of skeletons on similar issues, but I later came to realize that these really weren't my colleagues speaking. It was a few opportunistic people who wanted to see their names in print; they certainly did not represent the people I knew at CSD.

But as the old proverb goes, "If you live in a glass house, don't throw stones." Just twelve days after the article premiered in the Orlando newspaper, a Titan 34-D, upon launch from Vandenberg Air Force Base in California, exploded at 800 feet—and did so because of a failure of a Titan SRB. Ironically, the failure was also a field-joint failure—only this time it was an insulation failure in the region of the field-joint rather than an O-ring failure. The failure was a result of the insulation design (after touting the insulation design for the *Orlando Sentinel*), the manufacturing approach, and very poor quality control and inspection. The insulation was unbonded from the case wall, and the compression fit of the insulation at the field-joints that

CSD claimed was so much better than the Shuttle's was a major contributor to the failure. (It provided an added load on the insulation, which aggravated the insulation unbond problem that caused the failure.) What was more amazing was that the quality-control procedures for the Titan's SRB did not require inspections of the insulation bonds. No X-rays were ever taken, and the insulation was secondarily bonded to the case rather than vulcanized like the Shuttle's.

As it turned out, this particular Titan SRM was over five years old and had been sitting around in uncontrolled storage for that long, and no one had properly inspected the motor before assembly and launch. Other Titan segments examined after this failure revealed massive insulation-to-case unbonds. NASA and Morton Thiokol certainly deserved criticism for not fixing the Shuttle motors when serious O-ring erosion problems were first observed, but there were plenty of things to criticize the air force and its contractors for as well. The air force never X-rayed a single Titan SRM in the entire history of the program. Had the air force done so, it would have clearly revealed this problem. All of the Shuttle segments were X-rayed; in fact, I was told that the Shuttle SRMs were the largest users of Kodak film in the world. The air force quickly installed an X-ray facility at the launch sites to prevent this problem in the future.

We all live and learn by our mistakes.

23

The Apocalypse Letter

One day in April 1986, my secretary buzzed, "Mr. McDonald, a Mr. Wiley Bunn from Marshall Space Flight Center is on the line. He says he's the Director of the Reliability and Quality Assurance Office at Marshall, and he wants to know if he could have a minute to talk to you."

I knew who Wiley Bunn was, so I took the call.

"Mr. McDonald," Bunn said, "I don't mean to scare you. But I've heard some talk around here that leads me to think that Marshall management may be plotting against you. If I were you, I'd watch your backside."

Shortly after that disturbing phone call, I received a letter in a plain brown envelope; its return address was simply "Huntsville, Alabama." The mysterious letter, which was several pages long, was signed by an anonymous author who referred to himself only as "Apocalypse." I learned later that this same letter was also sent to the members of the Presidential Commission and that underground copies of the letter surfaced in the hands of many people who wanted to know the true story behind *Challenger*.

The "Apocalypse Letter," as it came to be called, was extremely critical of Dr. William Lucas, MFSC Director. For me, it confirmed some of my own suspicions about what was taking place at the highest levels of the Marshall Space Flight Center and explained a lot about the misleading and evasive testimony of MSFC management before the Presidential Commission.

Undated, the letter read:

Dear Sir,

I am deeply troubled by the *Challenger* tragedy and the events immediately preceding and following the loss. I am a senior manager at the Marshall Space Flight Center and answered as a member of the MSFC flight readiness review board for all 25 shuttle launches. I have watched in despair as William R. Lucas, MSFC Center Director, has lied time and time again in national news conferences. I have watched in disbelief as he has *engineered a cover-up of the MSFC involvement*. When I add this to the deep concerns that I and many other senior managers secretly shared preceding the launch relating to the management philosophy associated with determining flight readiness, my conscience tells me to speak out.

The Marshall Space Flight Center is run by one man, William R. Lucas. His style of management can be best described as feudalistic. In his ten-year tenure as Center Director, he established a personal empire built on the "good ol' boy" principle. The only criterion for career advancement is total loyalty to this man. The loyalty to country, NASA, the space program, means nothing. Many a highly skilled manager, scientist, or engineer has been buried in the organization because they underestimated the man's psychopathic reaction to dissent. In my own case, I have played the game. I have learned not to debate with him. Not to argue, but to compliment and flatter him. We have learned that we must tell him he is right even though he is wrong in order to survive. Lucas is a classic godfather mentality who corrupts the merit promotion system at MSFC to hold total power over the careers of 3,000 people. In addition to violating the merit promotion system, an investigation would reveal a great many other abuses of power such as using the government aircraft at his disposal for personal trips. Combine this institutional corruption and dishonesty with the usual faults of the government bureaucracy and one has a problem situation of immense proportions. Morale at the Marshall center even in the best of times is very low. Those of us who know him are not really surprised that he would resort to a *cover-up* in order to preserve his power.

There is no major decision which is made at MSFC which Lucas does not approve personally. Everyone within the agency looks to this man concerning the readiness of the shuttle main engines, the external tank, and the solid rocket boosters because the systems are so complex. No one outside the MSFC and its contractors have an in-depth understanding of the systems. This man selected the contractors, selected the government managers (and in some cases the contractor managers), approved countless system and subsystem designs, and many other aspects of this system. To this day, he is still the final approval authority for the continuing stream of design changes to the SRB. No one man in NASA has been more involved in the development, testing, and certification of the SRBs than this one man. No one individual within NASA has done more to place his personal stamp on a piece of hardware than this one man. No one man in NASA has done more to so completely suppress dissent to further his own power than this man. Unfortunately, no one man can hope to make these types of multi-discipline decisions successfully. This man was the individual who okayed the *Challenger* SRBs for launch over the objections of Thiokol engineers who thought they were too cold. This man, more than any other individual in NASA, is singly responsible for this terrible loss and the deaths of seven good men and women.

Lucas could[n't] care less. It is common knowledge among the senior management at MSFC that his primary concern is simply to cover his tail. He engineered a two-phase cover-up at the MSFC. All data and people were locked into the Huntsville Operations and Support Center. Lucas personally ordered his senior management that no one at the center was to talk to the press without going through him first. The cause of the tragedy was generally known within a few hours on the same day it happened. Yet, under Phase I of the cover-up, information was to be withheld as long as possible then fed to the press piecemeal. It was reasoned that the longer the information could be covered-up the better, as the course of world events would eventually tend to dilute the initial shock and public reaction. Some MSFC officials privately indicated that they were praying for another hijacking or something similar "to take the heat off." Once data could no longer be held back, Phase II would be to present as much highly technical data as possible letting the situation in the general public's mind be diluted by various confusing theories which were sure to result. Stories were to be planted which would serve to shift blame away from MSFC to Thiokol and the contractors doing the processing at the Cape.

It has been apparent for some time that the flight readiness review process developed by Lucas and other senior NASA managers simply was not doing the job. It was not determining flight readiness. Rather, it established a political situation within NASA in which no center could come to the review and say that it was not ready. To do so would invite the question "If you are not ready, then why are you not doing your job? It is your job to be ready." At each flight readiness review every Center and every contractor are asked to vote on readiness. It is a "no win" situation. For someone to get up and say that they are not ready is an indictment that they're not doing their job. As a consequence the center gets up and basically "snows" headquarters with highly technical rationale that no one but the immediate experts involved can competently judge. Lucas has made it known that under no circumstances is the Marshall Center to be the cause for delaying a launch. Hence, MSFC hardware is now flying with nozzles which have eroded to within five seconds of burnthrough in one case. No one to this date understands what caused the near catastrophe. MSFC's SSMEs are flying with potentially cracked turbine blades due to the same schedule and political pressures. If one of these blades were to come loose, it would be catastrophic.

There are dozens of such problems which have been uncovered on the SRBs and the main engines and which none of the experts involved truly understand. Many of these have been glossed over simply because we were able to come up with a theoretical explanation which no one could disprove. We are flying shuttles based on the flawed management philosophy that if no one can prove the hardware will fail then we launch.

And, we vote. It is understood that Lucas expects his board to vote for launch. In the twenty-five Center level flight readiness reviews that I have participated, never has there been a single negative vote from a Marshall board member or a Marshall contractor that we are not ready. This amounts to hundreds of yes's and not a single no. This speaks for itself. This process is an exercise in self-disillusionment and illustrates the tremendous gap between MSFC and NASA management decision making and what happens in the real world. The contractors understand that Lucas can get them fired; hence they tell NASA what they know NASA and MSFC want to hear: "We are ready."

And, how many times has it been now that we have not been ready? We have had a main engine shut down milliseconds before the SRBs fired (incidentally, one of these occurrences was never fully understood). We have had an abort to orbit; we have had numerous delays and numerous problems. And, now we have had a catastrophe.

Perhaps, the most significant tragedy of this Orwellian nightmare that many of us find ourselves trapped within is what happened on the first launch this year. After many delays we finally achieved a successful launch. However, but for the grace of God, we would have destroyed that shuttle and crew. To make a long story short, we experienced an anomaly and aborted launch for that day. In recycling the count we discovered an undetected failure in which a sensor had broken off and prevented a valve from closing. Had we not experienced the other unrelated anomaly, we would have gone ahead with an undetected problem and blown up an engine which would have substantially blown up the external tank destroying the vehicle and its crew. At the *Challenger* flight readiness review, this problem was discussed and then dismissed. The Commission must look at this event as well.

Something must be done here at the management level. Lucas should retire gracefully and a new manager brought in from the outside to reorganize the center. I know of no other way to deal with the good ol' boy Mafia that has evolved over the time Lucas has ruled. Their ineptness and incompetence have contributed to this tragedy.

I am planning to retire soon and for obvious reasons do not plan to disclose my identity at this time. However, I have enclosed a copy of a presentation which Thiokol presented to Lucas in July of 1985 clearly outlining the problems with the seal. Nobody really understands the problem and hence no one could disprove the theory. Lucas proceeded at risk rather than have the MSFC delay the program, a classic management error in shortsightedness.

God help us all,
"Apocalypse"

I strongly suspected that Wiley Bunn was "Apocalypse." Besides his phone call, not long after I got my copy of the Apocalypse memo, I also received a letter written in a similar style from Bunn admiring my courage and honesty and sympathizing with how he imagined I was probably being treated at Morton Thiokol:

April 29, 1986
Dear Al,

It is my contention that one of the worst things a man can do is to fail to express his admiration to those who exhibit characteristics which exemplify traits worthy of emulating.

Such is my feeling for you and for the manner in which you have conducted yourself the past several months. I have some idea of the fashion in which you are being treated. My conjecture is that you are classed somewhere between a leper and a child of questionable lineage at a family reunion. I don't mean to make light of what I suspect is happening, but after some thought those are the words that come to mind.

If those conclusions are in error, then please treat these words as an expression of concern for your well being. However, if my jumping to conclusions is somewhat accurate I offer these words from Kipling:

> If you can dream—and not make dreams your master;
> If you can think—and not make thoughts your aim;
> If you can meet with Triumph and Disaster
> And treat those two imposters just the same;
> If you can bear to hear the truth you've spoken
> Twisted by knaves to make a trap for fools;
> Or watch the things you gave your life to, broken,
> And stoop and build 'em up with worn-out tools;
> Yours is the Earth and everything that's in it,
> And—which is more—you'll be a Man.

It seems to me by Kipling's standards you qualify as a man. A man of character, a man who speaks his convictions, a man who I wish I knew better, so that I could name him as a friend.

Sincerely,
Wiley

One of the labels that the press and media gave me in their reporting of the events surrounding *Challenger* was "whistle-blower"—a title I had always absolutely despised because of its negative connotations. Most people believe that a whistle-blower is some kind of professional snitch or troublemaker whose credibility is always suspect, whose motives are unclear, and whose allegations are generally hearsay. I was a very active participant in what happened the night before the *Challenger* launch, and my sworn testimony was concerned with my own actions and direct observations.

It was interesting to me also that the news media always referred to me as an "engineer" opposed to the launch when, in reality, I was the lone representative of "management" either at Thiokol or NASA who opposed the launch. I was the Director of the Space Shuttle Solid Rocket Motor Project and the senior management representative of Morton Thiokol at KSC at the time of the launch.

Even though I was no longer an official member of the Morton Thiokol/NASA Failure Analysis Team, I believed it was my duty and obligation to provide my own conclusions about the cause for *Challenger's* failure. On

April 23, 1986, I prepared a draft memorandum to my new boss, Ed Dorsey, Vice President and General Manager of Space Operations at Morton Thiokol. Some of my conclusions were based upon the camera coverage of the launch and the data obtained in conjunction with the launch as well as other data obtained from testing conducted after the accident. I was particularly concerned over NASA's aversion to admitting that the low temperature was the primary contributor to the O-ring leaking. In an attempt to make it appear as if the joint assembly damage or contamination was the primary contributor to the joint failure, the NASA people were avoiding the influence of temperature altogether. My memo pointed out the shortcomings of the NASA Failure Analysis Team's conclusions based upon what I had heard. It also assessed the open questions that I thought still remained. My memo concluded by asserting that the field-joint design was marginal even under warm temperatures and needed to be redesigned under any circumstances.

What I didn't realize at the time was that the memo was going to get me into more hot water with my management, because the memo didn't totally agree with the NASA/SRM Failure Analysis Team Report. This report supposedly reflected Thiokol's opinion as well as NASA's because MTI personnel, including my prior boss Joe Kilminster, were members of the joint SRM Failure Analysis Team.

What's more, the very day I sent the memo to Ed Dorsey, he and Kilminster, along with one of the corporate attorneys, Lee Dribin, were in Huntsville reviewing a draft of the official NASA Failure Analysis Team Report, then in its final stages for presentation to the Presidential Commission. What Ed and Joe had to say about that report pretty much ignored what I'd said in my memo.

The following week I attended the annual meeting of the American Institute of Aeronautics and Astronautics in Washington, D.C., finishing my two-year term as chair of the AIAA's Solid Rocket Technical Committee. While there, I made a point of attending a session on space transportation chaired by the Undersecretary of the Air Force, Pete Aldridge. I wanted to hear what Secretary Aldridge had to say about the sad state of the U.S. launch vehicle industry in the wake of *Challenger* and the air force's Titan 34-D failure at Vandenberg. As the father of USAF's complementary expendable launch vehicle program, Aldridge occupied a lofty position in the arena of space transportation. He had also trained as an astronaut, one in line to fly the first Shuttle launch from Vandenberg using the FWC-SRM, originally planned for this spring of 1986. Aldridge had taken many bows for initiating the CELV program prior to the Shuttle accident.

Under the Reagan administration, the Space Shuttle was to carry all DOD, civil, and commercial payloads to orbit (referred to as the "Shuttle Only" policy), with DOD payloads receiving priority. Aldridge had initiated the CELV program with the air force to provide an alternate launch vehicle for DOD payloads in the event of a Shuttle failure or grounding.

Following *Challenger*, Secretary Aldridge had strutted with his thumbs under his suspenders saying, "I told you so!"—that is, until the Titan 34-D failed, since his new CELV was really just a stretched, but higher-risk, version of the Titan 34-D. (The CELV later became the Titan IV using a stretched tank for the liquid core and seven-segment SRBs—rather than the five and a half—on the Titan 34-D.) Like a lot of other people, I was interested in what Secretary Aldridge was going to say about the U.S. space transportation situation, but the man didn't show up. At the last minute, in his place, he sent General Don Kutyna from the Air Force Space Systems Division, who was a member of the Presidential Commission. General Kutyna didn't have much to say other than the generic statement that the recent failures had put our country into a terrible situation that we must recover from as quickly as possible.

General Kutyna noticed me sitting in the front row. When the session ended, he came over. "You don't have time to be sitting in these sessions, Al," he said. "You need to be spending all of your time figuring out why the Shuttle failed the way it did." I didn't disagree.

"Al, do you agree with the conclusions in the NASA Failure Analysis Team's report?"

"I haven't had the opportunity to see the report, General Kutyna, but I've drawn some conclusions of my own based upon the data that I've seen. Just last week I sent a memo to my boss, Ed Dorsey, giving my thoughts on the accident, but in doing so, I told him I had not been privy to the recent work of the NASA Failure Analysis Team. Still, based upon what I had seen relative to NASA's presentations to the Presidential Commission just a few weeks ago, I couldn't agree with their conclusions." General Kutyna then asked me if he could have a copy of my memo, and I gave him one.

"Al, what NASA presented to us a few weeks ago was, indeed, its final findings. Those of us on the Presidential Commission are certainly under the impression that these findings are based upon conclusions from a team jointly composed by NASA and Morton Thiokol personnel, and that that team included both Roger Boisjoly and yourself."

"Well, that's wrong! Neither Roger nor I have been directly involved in the failure analysis activities. Not for some time. We were both removed from that team by Morton Thiokol management."

"When did this happen, Al?"

"It happened right after our public testimony to the Presidential Commission on February 25th."

"That really disturbs me, Al. The Presidential Commission really wants to hear what you and Boisjoly have to say about the NASA Failure Analysis Team Report before we draw our own conclusions. Furthermore, we want to hear from Thiokol management exactly why you and Roger were *removed* from the Failure Analysis Team. I'll make sure this gets addressed in a closed-door executive hearing of the Presidential Commission when we reconvene at the end of the week."

As much as I appreciated General Kutyna's concerns for Roger Boisjoly and myself and his confidence in our technical abilities to understand the root causes of the *Challenger* accident, I knew that this chance meeting with him would get me into more trouble with my management. Once again, I had provided information directly to a member of the Presidential Commission without going through the established Morton Thiokol and NASA management channels.

Not to my surprise, the May 2 hearing of the Presidential Commission turned into a very ugly affair for me. Larry Mulloy tried to blame me for closing out the O-ring erosion problem so that it wasn't visible in the flight readiness reviews for *Challenger*, even though he personally, without anyone's knowledge, had waived the issue for every flight for the past six months. In response to a question from the Commission, I had to tell them about Mulloy's attempt to intimidate me after my testimony before the Commission.

The Commission severely admonished Morton Thiokol management for removing Roger Boisjoly and me from the SRM Failure Analysis Team, seeing it for what it was: retribution for our testimony before the Commission. The commissioners also sternly told Thiokol management that they intended to have Boisjoly and me return to Washington the following week to review and comment on the NASA Failure Analysis Team Report.

As a result of this Commission hearing, stories highly negative of Morton Thiokol appeared in the *Washington Post*, *New York Times*, and *Wall Street Journal*.

Roger and I already felt like lepers, but when we returned to Utah following the May 2 session, our colleagues treated us as if we had just been arrested for child sexual abuse. Few wanted to talk to us, fewer to be seen with us. Everyone knew we were on management's blacklist. No one wanted to risk sharing our company.

24

"Trouble with Your Logic"

In the first week of May, the Presidential Commission wound down its investigation. The panel decided to have one more closed hearing before starting to write its report to President Reagan. The NASA Failure Analysis Team had delivered its final report to the Commission, a report that concluded that the accident had, indeed, been caused by hot gas leaking from the aft field-joint of the right-hand solid rocket booster.

The Commission was very interested in all of the historical documentation generated by NASA and Morton Thiokol over the years concerning the known problems with the O-ring seals in the nozzle and field-joints of the solid rocket boosters. Of particular interest was how MSFC and MTI had dealt with these problems and why they weren't as well known to the highest levels of management as they should have been, especially Level I staff in Washington, D.C., and Level II people in Houston. And how did the problems remain totally unknown to the astronaut corps?

The closed executive hearing started in the morning with testimony from key Level III people from Marshall Space Flight Center. Thiokol personnel were explicitly prevented from being in attendance at this hearing. Learning the contents of their testimony, we found a lot of surprises. Certainly unbeknownst to me, Larry Mulloy had imposed a launch constraint on the last half a dozen flights of the Shuttle because of the O-ring erosion and blowby problem noted in the nozzle of STS-51B when it was removed in Utah in June 1985. Without anyone knowing, Mulloy had also personally removed the launch constraints for each Shuttle flight since June 1985. Even more surprising was that the launch constraint didn't even appear for *Challenger*.

Most upsetting for me was Mulloy's assertion that this launch constraint had been inadvertently removed because of a request from me!

In response to a line of questioning from Chairman Rogers as to how the launch constraint was erroneously closed out prior to STS-51L, Mulloy answered that it happened because of a letter from me, dated December 10, 1985. When this letter came to Mulloy's attention, "My reaction was, 'we are not going to drop this from the problem assessment system because the problem is not resolved and it has to be dealt with on a flight-by-flight basis.'" Unfortunately, "the people who run the problem assessment system erroneously entered a closure for the problem on the basis of this submittal from Thiokol. Having done that then for the STS-51L review, it then did not come up in the flight readiness review as an open launch constraint."

"Who made the error? Do you know?" Rogers asked.

"The people who do the problem assessment system," Mulloy stated. His sidekick, Larry Wear, Manager of the SRM Project Office at Marshall, added, "At the incremental flight readiness reviews, there is a heads-up given to the quality representatives [one from Thiokol and one from NASA's Shuttle Project Office] for what problems the system has open, and they cross-check to make sure that we address that problem in the readiness review. On this particular occasion, there was no heads-up because their PAS [problem assessment system] considered that action closed. That is unfortunate."

"Assuming that you were advised, as you were by Thiokol, that they opposed the launch on the 27th because of weather," Rogers queried, "would you have reacted differently if this had been closed?"

"No sir," Mulloy replied. "Frankly, I was not aware that this erroneous entry had entered in the PAS, because it did not come up."

"To you, what does a constraint mean, then, Mr. Mulloy?"

"A launch constraint means that we have to address the observations, see if we have seen anything on the previous flight that changes our previous rationale, and address that at the flight readiness review."

"When you say 'address' it, I always get confused by the word. Do you mean think about it? Is that what you mean?"

"No sir. I mean present the data as to whether or not what we have seen in our most recent observation, which may not be the last flight—it may be the flight before that—is within our experience base and whether or not the previous analysis and tests—that previously concluded that was an acceptable situation—is [sic] still valid, based upon later observations."

Mulloy's explanation left the Commission more confused than ever. What Mulloy and the Marshall people were trying to do was defend the thoroughness of their problem assessment system and flight readiness review process when, in fact, all the information they were analyzing indicated that the system did not work! Just in terms of the teleconference the night before the launch, errors had been made, information had not been communicated to all parties, and the basis for developing a database for the PAS was totally ignored.

But the brazen demagoguery of the Marshall men could not be mistaken. "A lot of people must read these constraints," said Commissioner Joseph Sutter suspiciously, "and a lot of people could read this one as saying, 'Hey, that's signed off; don't worry about it anymore.' Who reads this? It's in the books."

"Let me explain how this occurs," answered Larry Wear, with more than a touch of a patronizing attitude. "There is what is called a Problem Review Board meeting that is held within each project. There's one for each of the projects at Marshall where they go over these items. The reason you see Jim Thomas in this document, is he's chairman of that board. That is discussed at that time and that is where these are recognized as being within our database or perhaps not within our database. It is covered in the flight readiness review, and that is the process."

"So at the flight readiness review, the people read this and understand it?" asked Sutter.

"Jim Thomas or I," said Wear, "in doing this activity, know if it is in the flight readiness review or if it's not, as the case may be. Plus the other participants there know that. Therefore it is listed as a result of being presented and discussed in the flight readiness review, and it would be within our database or not."

Wear was being disingenuous, to put it mildly. All of these PAS items were *not* reviewed with the FRR boards on each flight, and he knew it. They were reviewed by Morton Thiokol and by the MSFC/SRM Project Office, and *only* those determined to be worthy of consideration by the MSFC/SRM Project Office were carried forward for review with any FRR board.

As usual, Chairman Rogers had the smarts, disposition, and wherewithal to cut right to the chase. "What do you mean by 'database,' Mr. Wear, because Thiokol keeps saying that experience was not within your 'database' on flight 51-L, and you all say that it was within your 'database'? Which is right?"

"The Thiokol that I addressed, sir, says it is within their database," Wear ventured.

"I think the Chairman's point, though," clarified Commission Executive Director Alton Keel, "is with respect to the night of the telecom, Thiokol was arguing just what you are arguing now with respect to erosion. They were arguing that 'we want to stay within our database and we want to go with an O-ring temperature not any lower than 53°,' just like you're arguing now that 'it was OK to fly because we were within our database,' implying you wouldn't fly outside of your database."

"So what do you mean by 'database'?" Rogers repeated.

"The database to me, sir," Wear offered, "is the previous test and flight experience that we have, as supported."

"You've never had experience in this cold weather, Mr. Wear," Rogers responded, "so when you're talking about your 'database' on flight 51-L, you didn't have any! It seems to me it is used as sort of a slick way of just getting over the problem, saying it's 'within our database.' But you hadn't had any experience of that kind before and so you can't say it was within your 'database.' The engineers at Thiokol were saying, 'it is not; we have never had that experience. We warn you: Don't do it. We don't know what's going to happen.' And we keep hearing from NASA, 'it was within our database.'"

"Well, going to that particular evening, which I think what you are referring to, Mr. Chairman, is the January 27th evening, and that particular evening, in my mind process, we had faced a cold launch the year before in which we had some erosion."

"The worst experience you had," added Rogers.

"And they, the Thiokol that you're referring to, they had said that that condition—on that particular occasion the year before—had, I forget the exact words but in effect, aggravated the situation, but that it was acceptable and would perform. That was the conclusion."

"But you agree, though, Mr. Wear, it was not within your database, don't you?"

"That experience of the prior year was within my database—yes, sir."

"Could I say on that one, though, you had your worst result on that one?"

"Yes."

"So you can say, well, 'we almost had an accident, but we didn't quite. Therefore, it is within our database.' Then you get to another date when it is colder, and you still argue it is within your database. Thiokol said it is not in your database because you had never tried it in this cold of weather. Now, what I'm asking is how do you explain that controversy? How do you explain that conflict?"

"If I may continue," Wear pleaded, "their engineering organization and my engineering organization had agreed the prior year that that experience was—could be—acceptable on the next launch: that if the same condition occurred, which was the finding made at that time, if that same condition existed it could be accepted. And that was the same engineering organization that was talking on January 27th, so its conclusions and report to us on the previous year was [*sic*] that at those conditions and what we observed on that particular launch, that if that condition—meaning that type of that entire condition, that weather condition, whatever—occurred again, that that was acceptable."

"But you didn't have those conditions," charged Commissioner Robert Hotz.

"Let me finish," Rogers himself pressed on. "That was 'within your database,' you argue. But what I'm saying, you've got a new condition now which was not 'within your database.' Now, how do you keep relying on it to be 'within your database'? The Thiokol people said it's not. The engineers said it's not. We've never tried it at this cold a temperature. How can you keep saying that it is 'within your database'? It wasn't. It *exceeded* your database. You never had that kind of experience before."

Wear stuck with his party line: "Well, on that particular evening, I heard what the Thiokol engineering people stated. I also know that they caucused and when they came back they stated that the conditions could be accepted. So I have to conclude some engineering people must have changed their mind."

"I'm not taking issue with that," Rogers admonished. "I'm taking issue with the slickness of the words 'within our database,' as if that excuses everything. What I'm pointing out is I don't think you ever had a database of this kind in flight 51-L."

"That's true," Wear finally conceded.

"Well, that's really all I was pointing out."

Reading through this testimony, I was really impressed by how Chairman Rogers pulled the rug out from under Larry Wear. It had been like pulling teeth. NASA was clearly trying to confuse the issue of its purported careful handling of databases for justifying continued flights as best it could. In the case of the approaching *Challenger* launch, NASA had totally ignored the database.

How could launch constraints be implemented, then waived, and then removed *without anyone knowing about it*? The Commission still wanted that mystery explained.

"So as far as you're concerned," Alton Keel asked, "you were still operating as if this remained a launch constraint?"

Larry Mulloy said they were.

"But as a matter of practicality," Keel continued, "even though this was a launch constraint it was being waived for *each* launch?"

"That is correct, on the basis of the presentation at the flight readiness review."

"So you in effect waived this for 51-L, Mr. Mulloy?"

"That is correct."

"But it doesn't show up on your summary?"

"No, because the man assumed when the closure came in from Thiokol that this was going to close the problem, and that required Project concurrence—Mr. Wear's concurrence—which Mr. Wear and I had discussed it, and that was not going to happen."

"As far as you were concerned, though, you still considered it a constraint in spite of this document?"

"Yes sir."

The inconsistencies in the NASA logic were staggering. Both Mulloy and Wear were trying to defend how good their PAS system was, yet neither of them knew that an error had been made that closed out the O-ring erosion problem! At the launch of *Challenger*, they were still assuming it was open! It demonstrated that no one, including Mulloy or Wear, paid any attention to the PAS system. Furthermore, Mulloy was waiving this problem every time, anyway, so it *didn't make any difference* if it were closed or open. The only reason they were trying to defend the importance of the PAS system was because I had requested that the seal erosion problem be closed on that report. Mulloy and Wear wanted to use this as an excuse to blame me and not MSFC for lack of visibility for this important problem prior to launch of the *Challenger*.

This transferring of blame was entirely consistent with the contents of the "Apocalypse Letter," discussed earlier.

Larry Mulloy tried hard to control the damage: "I would like to go back to the point Mr. Wear made, because I think it is important in addressing your concern; it's a very important point that Mr. Wear made that you do not see an accepted closure on that last entry. That is an entry that the guy who makes entries into the PAS system made, and you do not see a project signature concurring in that. So, in essence, it *is* still open. It is open until Larry Wear concurs that it is closed."

What a cop-out! The fact that Jim Thomas from Larry Wear's SRM Program Office had signed it was normally considered sufficient, because as Wear stated, Thomas was the chairman of the Problem Review Board. But the real reason neither Wear nor Mulloy signed it was because they didn't consider it that important, and they were too busy trying to get the prior launch up and off after it got delayed because of Mulloy's problem on the SRB.

Commissioner Sutter wanted to know, "Do the people who read these look for all the signatures?"

"No sir," answered Mulloy, preparing again to pass on the blame. "I agree it is unfortunate that that error was made by the gentleman who made the entries into the PAS system on the basis of a submittal from the contractor that this was no longer a problem of significance to carry in the problem assessment system." Rather than me or anyone else at Thiokol removing the concern from our list, the truth was we considered it so significant that we had a full-time task force working the problem and planned to address the progress on it in each FRR, rather than in the PAS system that nobody read—or let the item be waived by Mulloy at his wishes.

Marshall's misdirection continued. Asked about a statement that Jim Thomas had made in an interview with the Commission's staff about Marshall's decision to close out all issues that were over six months old, Larry Wear indicated that the intent was not to stop discussion of any problems prematurely but to "get off our duff and work and reach solutions for these problems, not continue to drag them out for six months. . . . That was the thrust of my direction to Thiokol." In actuality, Wear's intent was to get these problems off the record in order to imply *progress* and reduce the size of the PAS report.

Asked whether Thiokol had or had not ultimately recommended taking the SRB joint problem off the problem report list, Wear declared: "That's the nature of McDonald's letter. . . . As I understand the thrust of the letter—the way I read it now and the way I read it then—was that 'we are discussing these problems in the flight readiness review; let's not also put them someplace else' so that he, McDonald, in effect has to report them in two places. As I recall his letter, he's saying 'we're doing the tracking job in the flight readiness cycle. Let's take them out of this other tracking system.' He's saying let's do it once."

"What was Marshall's response to that?" Commissioner Keel wanted to know.

"My response to that was, no," said Wear.

"Mine, too," chimed in Mulloy.

"We wanted to keep it in the system," Wear explained, "because it is a formal check and balance, to be sure something didn't slip through a crack someplace. That's the way I've always looked at it, to create a double check by the Quality organization to see to it that something wasn't overlooked by the Project. Perhaps the Project might not even be aware that it could be overlooked, that it had to be faced."

Wear's answer might have momentarily satisfied a few Commission members, but most knew it was a farce. Neither Wear nor Mulloy had ever intended *not* to accept my recommendation to drop these off the PAS report and report progress by the O-ring task force at each FRR. I never received a negative response from them verbally or in writing to that effect. If the *Challenger* accident hadn't happened, closing out this item would surely have been accepted because they were getting pushed by their management to reduce the size of the PAS report.

"Nobody answered this letter of Mr. McDonald's?" Richard Feynman asked.

"No," Wear admitted, without saying that he had had a month and a half to respond.

"That's exactly the point," heralded Sally Ride, "because you got the system that records open problems, and you have to have some way of distinguishing 'unimportant problems' from 'important problems' from 'very important problems,' and it seems to me the one that says 'launch constraint' next to it must be the 'very most important problem.'"

"That's right, Dr. Ride," Wear offered, "and that is why it has to be cleared by this PAS system before we can proceed."

"What I'm trying to understand, Mr. Wear, is how many problems are in the launch constraint category in your system?" Ride asked.

"I can't give you a precise answer."

"I would think that if you have waivers and there weren't many, you would remember them all," suggested a surprised Chairman Rogers.

"Have you ever refused to waive a launch constraint because you thought the problem was so serious?" wondered Commissioner Arthur Walker.

"No," Wear conceded, revealing just how worthless the PAS system had become. It was a paper-pushing exercise only. Everyone with any authority whatsoever had been totally ignoring the PAS report because it was so thick and overwhelming. NASA Marshall management had even subcontracted the effort to Rockwell so that NASA would not have to waste its own manpower on this activity.

Chairman Rogers and everybody else associated with the Presidential Commission had trouble following the lame attempt to explain how all this was supposed to work. "There is a launch constraint put on by somebody, some decision," said Chairman Rogers, mystified.

"By me, in this case," Mulloy indicated.

"By you, OK. Now, who has the authority to waive it?"

"I do," Mulloy answered.

"You put it on and you take it off?"

"Yes sir."

"Now, when the manufacturer then said, 'we recommend *don't launch*' to begin with, did that cause you any concern, particularly in view of the fact that there was a launch constraint on it?"

"Yes," answered Wear. "When the Thiokol engineering people expressed their concern, yes, that caused me some concern."

"And they suggested a slight delay until the weather was better."

"That is what the Thiokol engineering people stated that night."

"And you were willing to go ahead even in the face of that recommendation, even though the weather was not good, and even though you had no database that would say it was safe to do it under those conditions, you still were ready to go ahead and launch?"

Wear fidgeted in his seat and offered another untruth that suited the MSFC post-*Challenger* party line: "Yes. Let me explain one thing, though. In my dealings with Thiokol, I deal principally with Mr. Kilminster. He is not just a mimic from them, and therefore I have to depend on him to present what Thiokol concludes." In truth, Larry Wear usually dealt with me.

"If you read the documents submitted," Chairman Rogers charged in summing up this line of questioning, "it seems to me everything was almost *covered-up*, ever so slightly noted, when it seemed to be such a serious problem. The papers reflect that a lot of you thought it was serious, and yet it doesn't seem that serious when you read this documentation. In other words, others that we have questioned said they didn't realize it was serious and apparently didn't realize it *was* a launch constraint. Well, anyway, that is not really in the form of a question."

After reviewing all of the documents that were available to NASA prior to the launch of *Challenger*, the Presidential Commission had become even more confused as to how NASA could have become so insensitive to the potential impact of cold temperatures on the ability of the already troubled O-rings to properly seal. Larry Mulloy had tried to defend the bad decision to launch because the O-ring erosion data that had been presented by Morton

Thiokol didn't indicate that lower temperatures were any worse on O-ring erosion, but that had never been the issue: low temperatures dramatically increased the potential for *blowby* of the O-ring seals, *not erosion*, and when blowby happens, you run a high risk of no seal at all.

Larry Mulloy continued to do his best *not* to make this crucial engineering clarification: "Going back to the extensive look we gave to STS-51C after experiencing the coldest temperature in Florida history back in January 1985, and that the erosion on it was not outside of what we'd experienced at the warm temperature of 80°, and also the conclusion that that type of erosion—which we all understood—could be anticipated in the future, and because we knew we could get paths through the putty and hot gas impingement through those paths, causing O-ring erosion, and our conclusion that STS-51E had been able to fly under those circumstances. . . . Then you get to the night of the discussion of January 27th, where the engineers were essentially citing the same data relative to the effect of resiliency, and that their concern was increased blowby of the primary O-ring seal, which I took to be what engineers always do—i.e., realize what risk you're taking, because we could get increased blowby, thus higher erosion . . . "

"You talk about erosion," said Commissioner Keel, stopping Mulloy's tangled soliloquy, "but the footnote in the conclusion of your documentation is a reference to *blowby*. This was the worst case of *blowby*."

"That is correct."

"But the Thiokol engineers didn't say that the worst condition would be acceptable. They didn't say that if the weather was even worse it would be acceptable. When the weather got worse, they said, 'we recommend against launch.' So I have trouble with your logic."

"I understand, sir."

25

The Monkey Changes Backs

That afternoon the Presidential Commission called in the Morton Thiokol people to testify in the closed executive hearing. Unlike the morning session where the Morton Thiokol people were excluded, the NASA people who had just testified were allowed to be in attendance. Initially, the Commission was interested in finding out if Morton Thiokol knew about the O-ring launch constraint that Larry Mulloy had imposed and personally waived on each flight and why Thiokol had written a letter asking to remove the O-ring erosion problem from the problem review board list.

It was my turn to explain.

"I wrote the letter, and I would like everyone to read the letter to see what it says. It says: 'The subject critical problems are ongoing problems which will not be resolved for some time.' So, right away, I don't think that tells anybody that we're going to forget about this and take it off of anything. It also says that 'we request that subject critical problems be closed and removed from the next PRB [Problem Review Board] agenda list'—the reason being that we spend more time each month going through reading all of that same thing that we've been reading every month for two years, because somebody colors in a square down at Marshall, 'please keep track of this each month.' We said 'we've got a full-time task force working this problem. We have weekly meetings. We need more people solving the problem and less people keeping track and status-ing it, and there's no sense for us to continue doing that when we got a very heavy activity in doing that.'"

"We then got a letter a couple of weeks later from Larry Wear," I continued. "In fact, I was called *before* that letter, saying 'we're going to get it,' that Marshall was very upset with us because we've got problems that have been on there for several years and we haven't gotten them off of this list, because it keeps getting thicker and thicker. 'If you ever want to get something where nobody will read it, you get it so thick that they finally pay no attention to it' and that was exactly the thing that the Problem Review Board was doing. It was getting so thick that it had problems on there we knew we weren't going to solve for some time.

"So, we said, 'OK, if you want to get them off the list, then just take them off and we'll handle them through this other mechanism that we're addressing with everybody that *really knows* about this problem.'

"I was unaware that a waiver was ever written for every flight . . ."

"So, your idea, Mr. McDonald," said Chairman Rogers, "really was to cut down the paperwork?"

"Yes. We were spending a lot of time in going through this matter and others where we had nothing to add, nor did anybody else."

"Mr. McDonald, your letter was a request," said Dr. Feynman. "Did you ever get an answer that permitted you to do this?"

"No, I didn't get any answer that permitted me to do that, nor did I know that anything was done about it. The letter just said, we've got a *lot* of things to do. This is one thing we don't need to do."

Extending the line of questioning, Chairman Rogers then wanted to know the current assignments for Roger Boisjoly and myself inside our corporation. It was very uncomfortable for Roger and me to say, but we responded as truthfully as we could, which really upset Larry Mulloy and my boss, Joe Kilminster.

"Mr. McDonald, are you working on the new SRB design?" Rogers asked.

"Part-time, when I can; I've had some ideas that I've turned in on how to fix it."

"Has your assignment changed since the accident?"

"Yes."

"What are you doing now?"

"I used to be Director of the SRM Project. Now I have a title called Director of Special Projects. The people that all worked for me now work for somebody else. I am involved in reviewing some of the failure data that is provided to me, and in coming up with some ideas on how to fix it and defining a test vehicle that will give us meaningful information."

"Were you given any reason for the change in assignment?"

"Well, that is my second change in assignment since the hearings started. The first change came after my February 14th testimony in Florida. I was pulled out of my position and given the assignment of scheduling."

"Was any reason given?"

"No reason other than that was what I was going to do. My people were put aside and assigned under somebody else. And I wasn't to be involved."

"Who notified you to that effect?"

"Mr. Cal Wiggins. He was General Manager of the Space Division at that time."

"Were your people given a different assignment, too?"

"They were put under another individual—in fact, one who used to work for me."

"What was your second assignment?"

"When Mr. Dorsey came in and took over as the new General Manager of the Space Division, and Wiggins was made his assistant, he gave me the assignment as Director of Special Projects reporting to him rather than to Wiggins."

"What do you do in that capacity?"

"Dorsey told me I could work on reviewing some of the information that has been provided on the failure analysis generated at Thiokol and Marshall, and help Joe Pelham in coming up with a test article for the recovery program—a full-scale-type test article for the seals—and to feel free to make what recommendations that I might want to make relative to improvements in the program."

"Mr. Kilminster, did you concur in the decision to change McDonald's assignment?"

"My assignment was also changed, Mr. Chairman," Kilminster replied. "I've been located at Marshall, working on the investigation team. That was done while I was down at Huntsville, working there."

"Do you have any reason, Mr. McDonald, to think that you were given another assignment because of the testimony you gave before the Commission?"

"Yes, I do. I feel that I was set aside so that I would not have contact with the people from NASA again, because management felt that I either couldn't work with them or that it would be a situation that wouldn't be good for either party. So I was taken out of the failure analysis work that I was doing at Huntsville prior to that assignment."

"So, you were in fact punished for being right?"

"I feel I was."

"Mr. Boisjoly, your assignment hasn't changed at all?"

"In one respect, yes, and in one respect, no. I have been designated as 'Seal Coordinator.' I have been preparing a lot of information for input. But I, too, have been put on the sideline in that loop with relationship to the customer."

"Do you feel that may be in retaliation for your testimony?"

"I think that is a possibility, a distinct possibility."

"Does anybody from Thiokol want to comment on what McDonald or Boisjoly has said?," Rogers asked.

Joe Kilminster spoke up: "Mr. Chairman, I would like to comment that since the new General Manager has come on board, Mr. Dorsey, there has been a basic organizational concept change, a structure change, in that we have Engineering now, which in the past had been in a support organization—supporting all the organizations in the plant, including SRM. Now we have identified specific individuals from that core organization, and they now report directly into the Space Shuttle SRM program."

"You know what I'm driving at, Mr. Kilminster. If it appears that you're punishing the two people—or at least two of the people—who were right about the decision and objected to the launch, which ultimately resulted in criticism of Thiokol, and then they're demoted or they feel that they are being retaliated against, that is a very serious matter. It would seem to me, just speaking for myself, they should be *promoted*, not demoted or pushed aside. Do you want to comment on that?"

"There was certainly no demotion involved that I know of."

"Well, you heard what McDonald said."

I thought I needed to clarify: "I was not demoted. They just took my people away and gave me a more menial job, as far as I was concerned."

"All right, Mr. McDonald, I withdraw the use of the word 'demoted.' But it sounds as if you were demoted."

"I felt like it."

"Mr. McDonald," said Chairman Rogers, moving on to a different subject, "as I remember your testimony in public session, it was to the effect that you refused to sign the telefax, or refused to go along, and normally when you were at Kennedy you were the one that would have signed. Is that correct?"

"I said I hadn't had such an experience and wasn't aware of one where anybody was ever asked to sign anything, but I felt that it was my responsibility if it ever came up, because I was the senior official for Thiokol."

"Kilminster said that he normally was the one to sign, which seemed to conflict with your testimony. Is that what you testified, Mr. Kilminster?"

"I think what I said, Mr. Chairman, was that that whole business was unusual in the sense that we were talking about something being signed the night before the launch. However, under other circumstances, the piece of paper signed to identify flight readiness, as far as it concerns Morton Thiokol that is a document I normally sign."

"Even when McDonald is there?"

"Yes."

"Is that correct, Mr. McDonald?"

"That is true for every formal flight readiness that we have prior to the launch. The reason I was there and Joe wasn't there is we *alternate* doing that. In case something comes up after the last formal review—the L-1 review—then it isn't signed by *anybody*. If anything comes up after that, then that's why I'm there. And the other contractors have people like me there, who can resolve those issues and are responsible for going ahead with a launch. As I said in my earlier testimony, Jess Moore takes an oral poll subsequent to the L-1 review, because in many cases there are items brought up at the L-1 that finally get resolved, and then Moore takes a poll to ask each of the contractors if they are ready to fly, because now everything is supposed to be in and OK. I'm the one that has to answer, 'Yes, Thiokol is ready to fly.' I have never been in the situation where we ever *had to sign anything*; I don't think there was ever any situation where we had to sign anything after an L-1. Because of this telecom, I expected that I would have to do that, because I felt I had the same responsibility at the launch site as Kilminster had when he was there. Otherwise, I shouldn't even take the time in going."

"You told them that you would not sign, Mr. McDonald?"

"I told Mulloy I wouldn't sign and that it would have to come from the plant. Now, Mulloy didn't *ask* me to. I just told him I wouldn't do it because I felt I was going to have to do it. That was my responsibility. That's why I was there."

"Did you know that, Mr. Kilminster?"

"No, I did not."

"Why did you think that NASA required your signature in this case? You said you had never done it before, Mr. Kilminster."

"Because we had not had a later than L-1-identified problem."

"Why would it have to be in writing?"

"The only thing I can say is I was not surprised when Mulloy asked for that, but again the whole thing was unusual."

"Didn't it occur to you that they might want to put the monkey on your back in writing?"

"I wonder," said Commissioner Robert Hotz, "if we could poll the

Thiokol delegation as to whether they were or were not aware that a launch constraint was being waived as a formal waiver for each of the flights after 51-B."

"I was not aware" was the reply of every one of us: Ebeling, Kilminster, Russell, Boisjoly, and myself.

The Presidential Commission then inquired as to our understanding of the behavior of the SRB field-joints, particularly as it related to the presentation I had made at NASA headquarters in August 1985. I stated to the Commission: "The meeting came about as a result of the problem with the nozzle O-ring eroding through, and that is what drove the meeting. Headquarters wanted to hear about it. We had lost a primary seal and eroded some of the secondary in a nozzle joint. We sat down with our engineering people and put that presentation together, which said, we ought to address the whole seal issue, not just that one failure, because we all felt that if that ever happened in the field-joint we were in bad trouble, because the nozzle has a much better secondary seal than the field-joint does. We decided to put it together as a total pressure seal presentation, and to highlight the field-joint, even though it was a nozzle joint that had caused the problem that drove the restriction to launch—the one I had been unaware that we constrained."

"This meeting at NASA headquarters was called because of *Thiokol's* concern that the joint was really in trouble?" asked Commissioner Sutter.

"No," I answered. "We had another meeting scheduled at headquarters at the time for reviewing the mixer fire at our plant earlier in the year. I believe Mike Weeks called Joe or I and said, while you're here, you ought to come and address a couple of other issues that have happened recently that we're very interested in. . . . They were aware that we had violated the primary seal in the nozzle and wanted to hear about that and what our rationale was to continue. I called Mulloy and told him we had been requested for this presentation and he said, 'you're going to have to review that with us in Huntsville before you can go to Washington,' which was, in fact, the normal sequence of things. So we were prepared to go down and review it with the Marshall people. Eventually we held that on a telecom and faxed down all the charts we planned to take to Washington and reviewed it with the Marshall people before going to Washington. Mulloy met us here in Washington."

Regrettably, what I failed to mention in my testimony was that Mulloy had requested three changes be made to our recommendations for that presentation. The first was to add a schedule to the planned hot-fire and cold-gas subscale tests, and one other was to tone down the first statement in the recommendations. That statement read: "The lack of a good secondary seal in the field-joint is most critical and needs to be fixed by reducing and/or eliminating joint rotation." At Mulloy's request, we changed that to say: "The lack of a good secondary seal in the field-joint is most critical and ways to reduce joint rotation should be incorporated as soon as possible to reduce criticality." The last was to eliminate a statement that said "*data obtained on resiliency of the O-rings indicate that lower temperatures aggravated this problem,*" but we could leave in the statement about our concern about joint deflection and secondary O-ring resiliency.

As the questioning proceeded, Chairman Rogers grew more and more agitated by Morton Thiokol's lack of openness with the Presidential Commission and its treatment of Roger Boisjoly and me for testifying honestly.

"I want to make a comment not just to Kilminster but to the company as a whole. I am very upset about the testimony that McDonald has given. It's a very serious matter. In this kind of an accident where people come before a Commission and tell the truth and then they are treated—as McDonald believes he has been treated, in some way as punishment or retaliation for his testimony—it is extremely serious. The whole idea of the program is to have openness and to have an honest exchange of views. And in this case, McDonald, Boisjoly, Thompson, and others, were *right*. If their warnings had been heeded that day and the flight had been delayed, there's no telling what would have happened. We might never have had the accident. And to have something happen to McDonald that seems to be in the nature of punishment is shocking, and I just hope you convey that, Mr. Kilminster, to management. I don't know how the other commissioners feel, but that is how I feel. I would think you would want him *in all of your discussions*; Boisjoly and he shouldn't be treated this way. He should be treated the other way that he was right and you were wrong, and others who changed their decisions were wrong, and they were right. To have something that seems to me to be in the nature of punishment is very, very distressing, and I just wanted you to know that."

When Chairman Rogers's strong statement to MTI management about the demotion of Roger Boisjoly and me were released to the press, it resulted in several scathing newspaper articles and cartoons concerning our apparent nondemotion.

"Mr. McDonald, I want to ask you one more question," the chairman added. "When you said before you had answered my question, that there was reference to how your customer would feel about you, what was the basis for that?"

"Well, prior to the testimony I gave in Florida, as I indicated, I was spending full time at Huntsville. Subsequent to that time I was not allowed to go back to Huntsville."

"Was any reason given?"

"The only reason given was they didn't think it would be in the best interests of either party that I do that."

"Who told you that?"

"I talked with Ed Dorsey for a while on that, and he felt I would be better off working with problems in the plant."

"Did he indicate anyone from NASA had said that to him?"

"No, he did not. I do know that, even after the recovery team started and I had submitted some ideas on how to fix the problem. Before we went to our first formal meeting at Marshall, one of the fixes we proposed was one that I had developed. I wasn't asked to even go and support *that*."

"Any reason given?"

"No. In fact, I didn't even get a copy of the presentation. I had to go borrow somebody else's. I also did not get copies of some of the material that was being generated in the failure analysis. I had found out in the hallway about that work being done, so I went and found some of the data. But I got the distinct impression that it was not being sent to me on purpose, because some people knew that I did not agree with all of the conclusions that were being drawn on some of the data."

"Mr. McDonald, subsequent to the accident, in your testimony before this Commission, did you receive any personal comment from people at NASA about your testimony?" This was a question from Commissioner Hotz.

"I'm not sure of the exact meeting—I believe it was when Mr. Sutter's team came out to Utah to review the development, qualification, and certification of the SRMs. It was a two-day meeting. The second day I was there, Mulloy came into my office and slammed the door, and as far as I was concerned, was very intimidating to me. He was obviously very disturbed and wanted to know what my motivation was—and I won't use his exact words—for doing what I was doing, and I asked him, 'What's your problem? Do you mean what my testimony was?' And he said, 'No, as I understand it you're giving information to the Commission without going through your own management and without going through NASA. What's your motivation for doing that?' I told him to calm down, that I didn't think I had to get a note from my mother or anyone to give anybody information—and I felt it was *appropriate* to give them information. I asked him what information it was specifically that bothered him, and he said it was the information I had passed on to Commissioner Covert that I had generated when I was in Huntsville on February 4th, in which I was critical of what was being done in terms of the failure analysis on *Challenger*. In doing that, I guess I had not gone through the proper channels or something. Mulloy was very upset about that and was very intimidating. But I ignored him. On the other hand, he said, 'You know, I've never been against you and I have a high regard for your capability.' But I could see no reason for him saying what he did."

"Did you get the feeling that there might be some feeling on the part of the Huntsville people that they wanted to control this flow of information to the Commission?" Hotz asked.

"I got the feeling that that was happening from things I was reading, the data I was looking at, and the conclusions being drawn—and how they were being drawn—instead of focusing on the real potential causes, to me they were making it fuzzy."

"Mr. Mulloy, do you have any comment to make on McDonald's statement?" asked Chairman Rogers.

"Yes sir, I would be glad to. As often happens, when we have a meeting, we don't all remember it being the same. It must've been a different meeting.

"I came to McDonald's door and asked him if he had a moment, and he said he did, and then I closed the door. I didn't realize I slammed it. I was not upset. I started by shaking Al's hand and said, 'Before I say anything, I want you to know that I don't have any personal feelings one way or the other about what has occurred subsequent to launch. However, I have a curiosity, and my curiosity is, why you have taken the approach you have in circumventing your own management and the customer in voicing concerns about the launch of 51-L that you never voiced to your management or to me?

"His response was that he was very upset about the way the investigation had been going at Marshall when he was down there, because when he went to the HOSC after the accident, he had found NASA organizing a great number of teams and laying out a broad spectrum of areas to look at, including the external tank, the SSME, and the SRB, and was forming teams to look in a broad way across the total system, to try and determine what caused the accident.

"Al mentioned to me that he had obtained the information about the temperature readings on the right-hand SRB from the IR gun at 9°, and that he was trying to introduce that into the failure analysis team to which he had been assigned, which was Schwinghammer's SRM team, and he had become very frustrated, because the team seemed to be more interested in getting a structure to look across the total spectrum at the possible failure causes as opposed to picking up on the most obvious thing, which was the low temperature on the right-hand booster.

"That is the way I recall the discussion."

"May I ask," said General Kutyna, "who directed you to look across the whole spectrum of the accident structure? Where did you get that guidance?"

"I did not do that," answered Mulloy. "I was not participating."

"Mr. Mulloy, our panel gave you the guidance to look across the whole spectrum of the accident structure, to look at every facet of it rather than hone in on something anyone might think was the conclusion."

"I think that was the proper approach."

"But the one-on-one discussion that McDonald has referred to was initiated by you?" posed Chairman Rogers.

"Yes sir, it was."

"And what was the purpose of it again? Don't go through the whole thing—just what was the purpose? Why did you do it?"

"I have known Al for some time and had worked with him for some time. Why I did it was based upon his testimony that he had objected to the launch and had continued to protest the launch after the discussion on the night of January 27th, and that was not a fact."

"In other words, you were challenging his veracity at this point?"

"I wondered why he said that, when he had not passed that on to his own management, nor had he passed it on to me, nor had he at any time during the launch process when he was on the console during that morning. At no time did he make any comment or continue to object to the launch. As a matter of fact, as he himself has testified, he left his console and left the loop during the launch process."

I was shocked by Mulloy's statements here, because most were blatantly—knowingly—false. His own sworn testimony and that of Stan Reinartz acknowledged that I had disagreed with the qualification temperature limits of the SRM and that I had requested that NASA delay the launch. Their sworn testimony further verified that I had recommended not launching for three reasons, not just one, and that they couldn't accept the final, changed Morton Thiokol recommendation to launch because it was outside the qualification temperature limits of the SRM and therefore violated the launch commit criteria.

Furthermore, I did not leave the console or the loop during the launch process; I was at an adjacent console with Carver Kennedy.

The only thing Mulloy said that was true was that I did not, on the night of January 27, 1986, convey the same thoughts to Morton Thiokol management.

I think the real reason why Mulloy was so upset was because he knew the buck had stopped with him to proceed with the launch, not with Joe Kilminster, our signer of the launch recommendation; Mulloy knew it because of what I had said to him shortly after the launch recommendation was changed. Had Kilminster and Thiokol management known that I had several good reasons for not recommending a launch, they may still not have changed their minds. But Mulloy did know them, because I had voiced the objections to him directly. True, I didn't do that with my management or with the Level I NASA Mission Management Team on the net the next morning; however, I did voice the same objections to Mulloy's boss, Stan Reinartz, who was a member of the Mission Management Team. If I would have voiced the same objections to my management and to the MMT, then Mulloy and Reinartz would actually have been off the hook, because everyone else then would either have been supporting or preventing the launch on the same set of available information. Had Mulloy or Reinartz carried this same information to the Mission Management Team on the morning of January 28—like they should have done, and their job required—the same result would have been accomplished. In his frustration and demonstrated lack of good judgment, Mulloy tried to lay the blame on me.

"So you had no particular motive to start this discussion other than to satisfy your own curiosity?" Rogers asked.

"Yes sir. As to why Al felt there was a concern now that he did not consider worthy of passing on to his management or to the customer at the time."

As the tense back-and-forth testimony between Mulloy and myself came to an end, Alton Keel made an important point for the record: "Mr. Chairman, we did give directions with respect to the broad investigation, but we never gave a direction with respect to anyone who had information that they thought should be passed to the Commission, but they could pass it directly to the Commission. There was never any direction it should go through management, or through NASA, or through anyone."

Chairman Rogers pressed that point a bit, asking me, "At that time of your meeting with Mulloy, he had been advised by this Commission not to have anything to do with the investigation, and not to take part in the investigation. And your testimony is that that he advised you about how you should convey information to the Commission?"

"He asked me why I was doing it, what I was doing, and why I was taking the liberty of giving information directly to the Commission without going through my own management and NASA first. That is exactly what he asked me—and it wasn't in a nice voice."

"Do you feel now that you have had the opportunity to present as much information as you want to the Commission? We want to be sure that anybody that has any comments or information, that they have direct access to us. Do you feel you have had?"

"Well, in violation of Mulloy's concern again, I gave some other information directly to the Commission without going through the proper channels. I sent a copy to Dr. Keel and also to General Kutyna about a memo I had drafted to my boss about a week ago. And just last night I informed our company's Mr. Garrison about my conclusions on the failure analysis based on the data that I have had access to, recognizing that I haven't seen everything, as to what may have happened and what seems very conclusively to have happened. We need to make sure that we understand those things that are well substantiated versus those that may be speculative."

"Will you feel free, Mr. McDonald, to give the Commission any information you want to give us? And if you consider what Mulloy told you as a 'direction,' forget it, because *we* are running this investigation and you have access to us any time you want. That applies to all of you. . . . Any instructions from Mulloy about how the investigation is going to be conducted, or how information is going to be conveyed, is directly opposite to what we have told everyone."

Mulloy tried to defend himself: "From my vantage point, Mr. Chairman, what I was talking about was not any information related to the investigation. The discussion I was having pertained to why McDonald didn't express any of his concerns initially. In the closed-door hearing at KSC in the initial discussion, he stated he had continued to protest against the launch after the decision was made, and my question was why, if he had those concerns, did he not express them to his management on the night of the 27th or on the morning of the 28th? What I just heard him say was that he interpreted something different than that."

"It wasn't an interpretation," retorted Rogers. "He said you questioned him about why he was giving information directly to the Commission and not going to NASA and his own management. Did you say that to him, Mr. Mulloy?"

"Yes sir. What I said was related to the events of January 27th. My question was, if he had all those concerns, why he did not relay them to his management on the night of the 27th and up to launch time on the 28th."

"So there was a misunderstanding," said Rogers, admitting the possibility. "You were talking about why he didn't do something on the 27th, and he construed it to mean something else."

"And waiting until the Commission hearings on that Friday to express those concerns," Mulloy repeated, "when they had not been expressed to his management in time to do anything about the launch of 51-L."

Even if this had been the case, which it wasn't at all, Mulloy was still implying that I should never have told the Commission anything that NASA or Thiokol's management hadn't already been aware of. He was basically saying that I should have withheld this information from the Commission members just like NASA was doing! Mulloy himself had plenty of opportunity to tell the Commission about our conversation before I ever did, but he and NASA deliberately chose not to do so, just like they chose not to tell the Commission about the original recommendation of Morton Thiokol not to launch below 53°.

Commissioner Acheson could tell this was becoming a rather heated discussion and decided that it was time to change the subject. He said, "I don't want to stop this episode, but I do have two questions for Kilminster. One goes to the design of the joint. A lot of the material we have received, one reads that the designers, presumably both the corporate designers and the NASA supervisors, believe that the joint was designed to compress and seal in the gas tight under combustion pressure. And it turned out very quickly in the joint history that it did the opposite. It opened up. I just don't understand why the program then decided to go into a lot of little fixes to see if you could compensate for a fundamental error, and maybe you could explain that to us."

Joe responded, "Again, I think we have to refer back to the experience base that we used to design that configuration in the joint, and that was the Titan joint."

Commissioner Walker quickly interjected, "You are telling me this accident was the fault of the air force and the Titan program?"

"No sir, I am just answering the question."

Ironically, it was at least partially the fault of the air force and the Titan program. Most likely, no one in the Titan program remembered the Titan SRM O-ring erosion data when the Shuttle program started a decade later, but it should have been retrieved from the archives when Morton Thiokol specifically requested it in June 1985.

The Commission then started to ask me some questions about my concerns over the conclusions being drawn by the NASA Failure Analysis Team.

Richard Feynman asked the first question: "Mr. McDonald, Marshall has been making some tests, and so have you been making some tests of the seal and so on in various kinds of jigs and small-scale and that sort of stuff. Do you have something to tell us which you presumably disagree in some way with the conclusions or the answers that we have been getting, because we got most of our information directly from them and not directly from you, and I wanted to correct that, if possible, if you had something to suggest about the way we interpret the results, and have you seen the kind of results that they have had?"

"Yes, I have seen, I believe, most of them. I don't know if I've seen them all. I don't know what their final recommendations—I guess my comments would be based on the last presentation I'm aware of that was made on the 10th of April. I think it was made by Marshall to the task force, and the conclusions from that presentation were that if you look at the conclusions, they don't even mention temperature in there at all, but even where they do mention temperature as having an effect, that it by itself couldn't explain the problem. It had to be in conjunction with other things."

"Do you mean like the seal fitting into the groove?"

"That is correct, and it is obviously, I think, biased towards potential assembly problems, either through the assembly itself or combination of things that one can't specifically prove other than that there is no indication in any of the records or any of our prior history that this was outside of that, and if you look at their own chart on the dynamic O-ring tests, you can go across that chart and see at 25° degrees where they never had a single success. It was 100 percent failure in both the primary and secondary O-rings. If you go across the same chart and looked at 55° and up, there was not a single failure. There were 100 percent successes, and it was like seventeen tests out of seventeen at higher temperatures and ten out of ten at the lower temperatures."

General Kutyna then piped in, "But, Al, if I can interpret, the chart was presented as a compilation of your data and their data, and what is the ordinate on that chart? It's temperature, right? And the bottom line of that chart is, boy, when it gets cold, you start failing. Is that not true?"

"Well, that is absolutely true. That is exactly the point I am making. And how you can look at that chart and then not conclude—I mean, you are making my point. I look at that chart and I don't know how anybody cannot conclude that that wasn't the major driver, if not the whole thing."

The general responded, "I think we are saying the same thing."

Vice-Chairman Armstrong ended this hearing making sure that I knew that may be a conclusion of NASA's but not necessarily the Commission's. "Well, Mr. McDonald, it is this Commission that is going to make the conclusions."

At another commission hearing, Chief of the Astronaut Office Captain John Young made some very perceptive comments about the range safety system. John thought that the range safety system should come off, and I agreed with him. Who was going to be the one who pushes the button to blow up the entire Shuttle and crew? *Challenger* was a good example that the system is no good. It wasn't activated until over half a minute after the accident, and it shouldn't have been activated then. The SRBs had done a complete 360° flip in the air after the explosion, righted themselves, and continued on their upward trajectory before the range safety officer hit the destruct while the motors were just about to burn out. If the destruct system had not been activated, there was a good possibility that the failed right-hand solid rocket booster would have come down intact possibly under its main parachute. This would have enhanced the findings for the failure analysis.

John told the Commission, "I recommended thrust termination back in the early days in order to avoid the range safety system problems that you have when you separate—when you activate the range safety package either on purpose or inadvertently. I find the range safety package, if we have to carry one, should be one that doesn't tear up the whole piece of machinery, including the crew. I will tell you, we fought this long and hard to even have a range safety package on the vehicle in the early 1970s, and we were never successful to get it removed. In fact, we had sort of an unwritten agreement that when we took the ejection seats out, the range safety package would come off, and it just never did."

The day after our May 2 session with the Presidential Commission, I met with Tom Davidson back in Utah. "Al," Davidson said, "Ed Garrison doesn't think you have any corporate loyalty. Ed said that Cal Wiggins hates you for what you have done, as does Jerry Mason, and he doesn't know about Bob Lund." On the other hand, a number of our people, including Joe Pelham, Phil Dykstra, and John Thirkill, really admired me for telling it like it is.

This was more good news than bad. Even though the most powerful men in the corporation disliked me for what I had done, the people for whom I had the most respect supported me. Tom Davidson himself was also one of my good supporters; he offered me a job as Director of Engineering working for him for the next sixty days, until the Presidential Commission's report was completed and the congressional hearings over. Tom said he really didn't know what Morton Thiokol was going to do when this whole thing was over.

I thanked Tom for his support, but told him that I didn't want to change jobs at this time, because I enjoyed working for Joe Pelham on the Joint Environment Simulator. I also told Tom, "I don't agree with Ed Garrison's assessment that I have no corporate loyalty because of what I have done. I did every one of these things out of corporate loyalty, because they were the right thing for the corporation to do and in the best interests of the corporation, its shareholders, and employees. I truly believe that if the corporation would have taken the same position as I did on all these issues, we would all be better off, employee morale would be higher, and it would be costing the corporation much less money. Our stock would be more attractive for investors, and we would be facing far less litigation from the *Challenger* accident in our misguided effort to protect NASA."

Puzzled looks troubled the faces of Rogers Commission members upon hearing NASA's Larry Mulloy trying to explain why it was illogical to restrict launch to 40° F, which I had recommended, but somehow that it was perfectly rational in his view to launch at 30° F. (Photograph by Diana Walker, 1986, *Time*, reprinted with permission.)

seals. McDonald
liction, about 22°,
 serious," called
president for engi-
urge a full-scale
f the seals could
 temperature. It
eering decision."
"not a program-

dson Lovingood,
huttle projects of-
 a teleconference
managers at the
 Utah to discuss
fore it began, Lo-
sion that Thiokol
o seek a flight de-
Stanley Reinartz,
r at Marshall who
o tell Arnold Al-
le manager at the
in Houston, who
 a flight delay was
ided to wait until
nding of the situa-
Aldrich.
parations for the
vere rushed. Lund
ed on any correla-
re and the amount
in the O rings on
y worried in par-
ssion 51-C in Jan-
 seal temperature
igh the air had
time of launch).
rs were recovered
Boisjoly described
 coal" was found
in one booster, in-
d blown past the
sion had also been
warmer tempera-
posed to overnight
d more extensive
knew that lower
the resiliency and
At 29°, the antici-
e O rings the next
t getting the rings
nts might be like
 into a crack," al-
"a sponge."
tant teleconference
rence Mulloy, chief
 at Marshall, had
McDonald at the
rk. Lovingood and
lle. In Utah, Lund
ninster, vice presi-
ms; Jerald Mason,
d Calvin Wiggins,
 projects. A dozen
tah were also par-
sented six charts
ed to the others and
perature was a fac-
nance. Lund, the
cer, said flatly that
eached at least 53°,

led the NASA chal-
. Hardy said that
the reasoning be-

TO LAUNCH OR NOT TO LAUNCH

NASA TOP MANAGEMENT

Jesse Moore
Associate Administra-
tor for Space Flight
*Had the final decision to launch
but did not know of any no-go
recommendations*

Arnold Aldrich
Space Shuttle
Manager at Johnson
Space Center
*Knew only of Rockwell's
reservations*

MORTON THIOKOL

Jerald Mason 📞
Senior Vice President
*Asked for a management, and not
an engineering, decision*

Joseph Kilminster 📞
Vice President for
Booster Programs
*Signed memo telling NASA to go
ahead*

Robert Lund 📞
Vice President for
Engineering
*Initially strongly opposed launch
but was persuaded to vote for it*

Allan McDonald 📞
Director of Solid
Rocket Motor Projects
*Thiokol engineer at Kennedy who
opposed launching*

Roger Boisjoly 📞
Head of Seals Task
Force
*Expressed deep concern about
launching in low temperature*

Arnold Thompson 📞
Engineer in Utah
*Joined Boisjoly in opposition to
launching*

Brian Russell 📞
Engineer in Utah
Opposed the launch

TIME Chart by Joe Lertola/Photographs by Diana Walker

MARSHALL SPACE FLIGHT CENTER

William Lucas
Director, Marshall
Space Flight Center
*Was aware of problems but was
outside launch chain of command*

Stanley Reinartz 📞
Manager, Shuttle
Projects Office
*Did not inform Aldrich or Moore
of Thiokol's concern about cold*

Lawrence Mulloy 📞
Chief of Solid Rocket
Booster Program
*Challenged Thiokol's engineers to
prove their case*

George Hardy 📞
Deputy Director of
Science & Engineering
*Was "appalled" at Thiokol's rea-
soning for not launching*

ROCKWELL INTERNATIONAL

Rocco Petrone
President, Space
Transportation
Systems Division
*Objected to launch because ice
might damage the orbiter*

Robert Glaysher
Vice President
*Told NASA that Rockwell "could
not assure it was safe to fly"*

📞 Participated in teleconference
on evening before launch

In this collage, just as it was on the night before and the morning of the launch
of *Challenger*, those who were against the decision were on the bottom, and those
in agreement to launch were on the top. (Photographs by Diana Walker and chart
by Joe Lertola, both from 1986, *Time,* reprinted with permission.)

PART V

Commissioners and Congressmen

Honest criticism is hard to take; particularly from a relative, a friend, an acquaintance, or a stranger.

—*Franklin Jones*

We have met the enemy and he is us.

—*Walt Kelly*

The first principle is that you must not fool yourself—and you are the easiest person to fool.

—*Richard Feynman*

Congress is so strange. A man gets up to speak and says nothing. Nobody listens, and then everybody disagrees.

—*Boris Marshalov*

26

The Green Ball Theory

The second week in May, Roger Boisjoly and I returned to Washington. The Rogers Commission had asked us to review the SRM Failure Analysis Team Report, document our comments, and be prepared to present our findings to a closed executive meeting of the Commission consisting only of Chairman Rogers, Vice-Chairman Neil Armstrong, General Don Kutyna, Dr. Eugene Covert, and Dr. Alton Keel, Commission Executive Director. In addition to Boisjoly and me, other Morton Thiokol personnel in attendance were to be Ed Garrison, Ed Dorsey, senior engineer Arnold Thompson, and Lee Dribin, our corporate counsel. Roger and I flew in the evening of May 7 and early the next morning headed over to the Commission's headquarters, located in a building across the street from NASA headquarters. We were taken to a room containing several volumes of the various reports submitted by NASA in conjunction with the *Challenger* accident and told we could examine any or all of these reports, but that the Commission particularly wanted our comments on the SRM Failure Analysis Team Report. We rolled up our sleeves and began the arduous task of digesting all this material.

It was around noon when I asked Boisjoly if he was ready to go eat some lunch; we certainly needed a break. Walking out the door, Richard Feynman asked if he could join us. We said, sure, and proceeded to a nearby café.

Sitting down, Commissioner Feynman commented that he held us both in high regard and that we were kind of like him in a way, a sort of maverick willing to "say it like it is." Such a statement coming from a former Nobel Prize winner in physics, who was held in very high esteem by his students and peers, was a high compliment, indeed. Feynman then turned to me, "Al, I've been examining this whole issue concerning the decision to launch the *Challenger* for some time now and there's one thing that really puzzles me, and that is, I don't really understand why Morton Thiokol management changed their minds after they made the initial recommendation not to launch. Could you enlighten me a little bit on that?"

"Sure, it's very simple. Thiokol management changed their mind because of the green ball theory."

"What in the hell is the green ball theory?"

"You mean to tell me that you've won a Nobel Prize in physics and you don't even know about the green ball theory?! Well, before I explain it, let me just tell you about a few things that were going on at the time. About a year ago, Thiokol was in the process of submitting a sole-source proposal to NASA for a Buy III production contract for the next sixty-six flight sets of Space Shuttle solid rocket motors. Under pressure from our competitors, NASA decided to evaluate the cost-effectiveness of introducing a second source for SRM production to meet NASA's projected twenty-four shuttle launches per year by the end of 1988. NASA requested that other interested members of the solid rocket industry submit a bid to produce the SRMs assuming that Thiokol would provide the reusable metal hardware for the case, igniters, and nozzles. We were then asked to submit our Buy III proposal in a work structure format that would allow NASA to compare our projected cost directly to those being proposed by our competitors. Since this was going to be a cost decision only, we then decided to submit an option for our Buy III proposal on a fixed-price contract to close out the competition. We felt comfortable that we could offer NASA a fixed-price contract that was lower in price than the cost-plus-incentive-fee contracts requested from the potential second-source suppliers, because we knew exactly what it cost to produce these motors.

"We submitted our Buy III proposal to NASA in the late spring of 1985. It was nearly the end of the year before NASA completed their evaluation of the second-source proposals. NASA finally announced that, based on that evaluation, it did not appear there was any cost-benefit of establishing a second source for the SRM. We were ecstatic over these conclusions and expected we'd now be able to complete the Buy III proposal negotiations and receive a sole-source contract for the next production buy of Shuttle motors. However, by the time of the *Challenger* launch, we still had not negotiated the Buy III production contract. In fact, on January 21, 1986, just one week before *Challenger*, NASA publicly announced that, even though it was not economically beneficial to establish a second source for the SRM, they were going to proceed by requesting bids for a second source based upon the national security concern of only having one source for these large solid rocket motors."

Not a patient man, Feynman wanted to know, "So what about the green ball theory?"

"I'm getting to it. Remember these two things: we still did not have a signed contract for the next buy of SRMs, and our number one customer just announced that it was going to proceed with a solicitation for second-sourcing the SRMs. So, what do you have when you have a green ball in your left hand and a green ball in your right hand?"

"I don't know. What do you have?" Feynman replied.

"Complete and absolute control of the 'Jolly Green Giant,' and that's why Morton Thiokol's management changed their minds."

Feynman burst out laughing: "Now I understand the green ball theory."

We were back from lunch only a few minutes when General Kutyna came into the reading room where I was and asked, "What in the hell is this green ball theory that you just told Dr. Feynman?" I told him the story; he laughed and said he had to agree.

Roger and I completed our review of the NASA Failure Analysis Team Report that day. The documentation was so overwhelming that I only reviewed the executive summaries for each of the failure team reports involving the Shuttle's systems and launch operations along with some of the supporting reports, such as the Ice Team's. Roger concentrated on the details related to the SRM only.

The next morning, May 9, the Presidential Commission met in executive session to hear our comments. As it turned out, the commissioners also wanted to find out more from Thiokol senior management why Roger and I had not been more involved in the SRM failure analysis activities.

General Kutyna set the tone: "As we started this accident investigation, we wanted to look at the cause of the failures on the Space Shuttle. My past experience on both the Titan and the Inertial Upper Stage was such that, as we went through these investigations, it would be a very close marriage between the contractor and the government. I assumed that was going to happen in this investigation. We set up the team with Marshall; I saw the contractor representatives there; we had government representatives and independent testers all in the room and a lot of people in back of us. I assumed, as I was given data, that that data had been agreed to below our level and had been freely discussed below our level, and that there was a consensus. As we neared the end of the investigation, however, I got feedback from various sources that maybe that wasn't true. There were still some voices out there that weren't being heard. That is when I went to Mr. McDonald and Mr. Boisjoly and got the feeling they were not fully up to speed as to where this investigation was going. They had not seen all the data and had not had all their inputs heard. Chairman Rogers suggested that we hold a hearing in Washington. So that is kind of where we are."

Chairman Rogers also wanted to make a point: "Obviously, this is somewhat different than a typical accident investigation, because this is a Presidential Commission. When Don Kutyna speaks about the way we're proceeding, it should be clear that the Commission as a whole hasn't come to any conclusions as yet. We're in the process of drawing those conclusions from the data we have, but we're not bound by any conclusion of anybody else. We're free to make our own conclusions, and that is what we plan to do. As Don says, the purpose of this meeting is to be sure that Morton Thiokol has an opportunity to tell us anything that they should tell us, and we want to be sure that we have gotten the information from *all* the appropriate sources. This may be technical, but I would like to listen to some of it."

My presentation to the Commission was pretty straightforward. I told the panel that the volumes of reports that Boisjoly and I had read through yesterday were substantially the same as the presentation we reviewed earlier. Again, however, I emphatically disagreed with their main conclusions relative to what caused the loss of the seal; after all, the mismatch of the assembly of the right-hand aft center to the aft SRB segment was far from the worst that we had seen. (The NASA Failure Analysis Team had reported to the Commission that the joint that failed had the worst mismatch of any joint that had flown; I found eight other joints that had flown successfully that were worse, and I gave this data to the Commission.) A number of technicalities related to the fitting of the tang and the clevis leg were being grossly overemphasized, I thought, whereas the matter of temperature, though it was addressed in the document, never surfaced as the prime suspect. In one place, NASA's Failure Analysis Team even made the outright statement that temperature wasn't the prime factor, when it was very evident to me, Roger Boisjoly, and others that temperature was the primary, if not the only, factor causing the accident.

I also reported that I was a little disturbed by a few things I saw, and a few things I did not see, in the Ice Team document. In the data, I couldn't find a temperature difference between the right and left boosters of a Shuttle that prior to *Challenger* had been more than 4°, which I thought was significant. I didn't think the report looked closely enough at the possible effects of local cold pockets on the SRB joints, and I questioned the accuracy of the different instrumentation used at KSC for measuring the temperatures on the vehicle, launch platform, and water troughs, and the basis for making corrections to those readings. My point was that the temperature at the field-joint might have been colder than measured and that we really needed to be thinking carefully about that when we redesign that joint.

Another item of major importance that I thought needed to be examined further was the aerodynamic loading on the Shuttle at the time the final problem really occurred, which was in the fifty-four-to-seventy-three-second time period. I expressed concern about how the loads had been taken individually, how somebody should have compared them and determined the combined load, and how structural implications from those loads could very well have been a major contributor to the final failure of our seal.

I also took issue with a statement that said that, prior to the *Challenger* failure, we at Morton Thiokol did not realize that the putty could seal, which was not right. That was exactly the reason why we kept increasing the stabilization pressure on our leak-check, because we had been concerned that the putty sealing might be keeping us from checking the O-ring. Somebody could have left the O-ring out, and we might have concluded it was there because the putty was sealing. So that was an absolutely untrue statement.

Another problem I had with the NASA's Failure Analysis Team Report was that it put far too much emphasis on the assembly of the SRB segments at Kennedy. We looked at all of them under a microscope in the de-stacking of STS-61G, and they all exhibited some blowholes at all of the joints through the putty, which was not surprising. Other than that, all the joints looked just fine and perfect, indicating there was no problem with the joints or the O-rings in any of those. And that was a sister stack to 51-L. At least the left side was very close to the same mismatch, and we used the same rounding tool and the same type of assembly, in nearly the same time period. In my view, this was another major piece of information from the Failure Analysis Team that I thought should take a large amount of the concern off of the assembly. One could never say there wasn't something unique about *Challenger*'s assembly that nobody knew about when it was put together, some contamination. But as far as the data we had, there was no reason to believe it was different than anything done previously.

General Kutyna agreed but wanted everyone to keep thinking about it as a potential problem: "We've talked about the mating and the potential for damage, and the bottom line is I don't believe it caused this accident, but I sure don't want the next one to happen because of it, so I really want to be cautious on it because there are potentials."

After I completed my report, Chairman Rogers asked whether Roger Boisjoly and I had been involved with the people at NASA Marshall making the analysis of the accident. Boisjoly answered that he had been in Huntsville from January 31 to February 26, but after that had zero contact with what was going on. I indicated that I was in Alabama until February 14 and then had no further contact.

"Wouldn't they have been normally involved?" Rogers asked, turning to his panelists.

"I would have thought so," replied General Kutyna.

"Suppose I say to you, Mr. McDonald, that the cause of the *Challenger* accident was the failure of the pressure seal on the aft field-joint of the right solid rocket booster. It was due to a combination of factors as follows. Up to that point, would you agree?"

"I would agree, Mr. Chairman."

"And what would be 'as follows,' in your mind?"

"Well, the first one would be the cold temperature. The second one would be due to the high aerodynamic loading on a damaged O-ring from the initial leak. The initial leak was due to cold temperatures, and the failure of that damaged O-ring was from high aerodynamic loading, which was also unique to 51-L. And the third one would be the possibility of either damage or contamination of the seal, but that is unsupported. Finally, there is always that potential for an assembly problem that isn't recorded."

"Apparently in this case there is no evidence of it."

"There is no evidence."

"What about the possibility of ice in the joint?"

"I was going to mention: I think there is a very high probability of ice in the joint also affecting that seal, especially the secondary seal capability."

Now that Thiokol's testimony on the conclusions of the NASA Failure Analysis Team was basically over, General Kutyna stated, "This is a super session. I am really glad it took place. I am sorry that it was not a process that was going on continually from day one in the accident investigation, but I am sure there are reasons for that. But I would like to make a couple of general comments. I think the greatest thing that I have gotten from this morning's session is that with very few exceptions, I think you will find that we all agree that everything you are talking about has been considered. If you could read our findings right now, I think you would be very happy that many of the things that have been brought up are reflected almost verbatim, almost as if you helped us write them in these particular findings. That is number one. I think you will find very little differences, interpretations, emphasis, yes, will differ. But the bottom lines, I think, are going to be very much the same."

I was pleased that Chairman Rogers then asked for a written copy of my comments so they could be included in the Commission's report.

Commissioner Covert chimed in and said, "I would like to make one more comment, General Kutyna, and I would like the record just to show, Mr. Dorsey, that the support that I have personally received from your engineers has just been outstanding. Whatever I have asked has come and has come promptly, and has come complete, and I think it is a credit to your organization, and it has been a privilege to work with your people."

Following the testimony on the failure analysis, Ed Garrison thought he needed to defend Thiokol again against the charge that it had demoted Roger Boisjoly and me. In recent days, after reporters got wind of what had been said in the Presidential Commission's closed executive hearings of May 2, the media's already stinging indictments of Thiokol (and NASA) had grown scathing. The headline "Burying the Truth at Morton Thiokol" ran on the front page of the *New York Times* on May 13. The story read:

The maker of the Space Shuttle's flawed booster rockets, Morton Thiokol, has a crippling management problem: On the eve of the disastrous *Challenger* launch last January, two company engineers argued urgently against proceeding because the unusual cold would have degraded the seals even more. Morton Thiokol's management overruled them, and the *Challenger* was lost with all of its crew. When the engineers, Allan McDonald and Roger Boisjoly, appeared in February before the Presidential Commission investigating the *Challenger* disaster, their articulate and thoughtful testimony was a credit to the company. . . .

[But] the company's management couldn't see when it was ahead. It understood only that its bad decision had been exposed by its own employees. They had, instead of lying or dissembling before the Commission, told the truth, an offense Morton Thiokol was not prepared to forgive. The management couldn't even wait until the Presidential Commission had disbanded. It promptly punished the two engineers by transferring them to jobs they considered menial. Fortunately, the

Commission formed to root out the truth about the Shuttle disaster is still in business and able to raise a shout of protest at Morton Thiokol management. Chairman Rogers said he wanted to tell the company as a whole of his distress that the engineers had been punished for testifying. That's distress the American public is wholly entitled to share.

"Truth in Space" was the ironic headline that ran that same day in the *Washington Post*. The night before the launch of *Challenger*, the story said,

Engineers at Morton Thiokol begged officials not to go ahead. They feared precisely what appears to have occurred: that cold weather at the launch site would stiffen troublesome O-rings used as seals in the spacecraft's boosters; that the stiff rings might not seal, letting hot gases break through the boosters and walls—and a leak of that kind could cause a catastrophe. But NASA officials insisted on launching anyway, and Morton Thiokol management finally took NASA's side. Thiokol's chief engineer was told to take off his engineer hat and put on his management hat. He did, the launch was approved, the Shuttle blew up, and seven people died. That is hardly a record of which the company can be proud. You can bet it won't make the annual report. It suggests that the desire to curry favor with the big bucks' customer overcame all other elements of judgment, and the worst kind of organizational imperatives inside both NASA and Morton Thiokol were allowed to crush the truth. Bad enough, but now Morton Thiokol has one-upped itself. It appears to have put on the shelf two of the engineers who resisted that night and three weeks later went public to disclose what happened and make clear that the O-rings have long been recognized as a weak spot in the boosters and a source of potential disaster. The two were Allan McDonald, who was Director of Thiokol's Solid Rocket Motor Project, and Roger Boisjoly, who had been put in charge of a task force on the O-ring problem. In closed testimony on May 2nd, the transcript of which was released last weekend, Mr. McDonald told the Presidential Commission investigating the *Challenger* accident that he has been given a new title, the Director of Special Projects, and that the people that all worked for him now work for somebody else. He was asked by Chairman Rogers if he thought he'd been reassigned because of the testimony he gave before the Commission. "Yes, I do," he replied. Company officials quickly said there had been no effort to punish engineers for testifying. There had been a general reorganization after the accident, they said, a lot people had their responsibilities change, but, "We haven't demoted anyone and, well, everybody can't be in charge." Chairman Rogers found this unconvincing, and rightfully so. He blistered Thiokol management, calling its behavior shocking. In the same hearing, just as correctly, he admonished NASA for having pretty well glossed over and almost covered up the problem with the O-rings at various points along the way.

The editors of the *Washington Post* believed there was an important object lesson in this sad affair for America's children:

> We tell our kids in this society not to lie. We encourage them to believe that if they tell the truth they'll be supported. We owe them the example of practicing what we preach. The most important thing at issue in this case is no longer how well the space program has been managed or even whether *Challenger* should have gone up that day. It is the maintenance of an insistence upon integrity. Go get 'em, Mr. Rogers!

Knight Ridder stated:

> It was ironic timing that Ronald Reagan handed out the Presidential Medal of Freedom, the highest civilian honor, to six Americans, including a football coach, an actress and his good pal, publisher Walter Annenberg. "You're a group of happy rebels," Reagan told the honorees. He quoted George Orwell, "Freedom is a right to say no." Empty words—unless Reagan tells the space bureaucrats to lay off Allan McDonald, a real rebel who said no. Pick up the phone, Mr. President, or one more whistle-blower is ticketed for the garbage dump.

Not all of the bad press was targeted at Morton Thiokol executives, however; some of the press was aimed at the NASA hierarchy, as indicated in this *Ogden Standard Examiner* article:

> Earlier this month, it was disclosed that last August 9th an engineer for Morton Thiokol, the manufacturer of the booster rockets, wrote a letter to officials at the Marshall Space Flight Center in Huntsville, Alabama, calling their attention to the results of a test showing that safety seals in the booster rockets could fail at temperatures of 50 degrees or lower. The rupture of those seals on the right booster rocket is believed to be the cause of the disaster that killed the Astronauts. The

engineer's letter was not turned over to the investigating Commission until investigators independently learned of its existence, presumably because some NASA employee confidentially tipped off the Commission. The letter contradicted testimony to the Commission in February by Lawrence B. Mulloy, the head of the booster rocket program at the Marshall Center, that no one at NASA was convinced of a strong link between low temperature and the seal's reliability. "This is not a piece of paper that they were in a big hurry for us to see," an investigator said. It's now well known that two senior engineers for Morton Thiokol protested heatedly against the launching on the previous day. When reports of their protests began to leak into the press, Senator Ernest Hollings (D-South Carolina) at a hearing of the Senate Subcommittee on Space, Aeronautics, and Related Sciences on February 18, asked William Graham, then the acting head of NASA, about these reports. Hollings asked, "With respect to the Morton Thiokol opposition to launch—is there or is there not any evidence that you know of to support that report?" Graham replied: "No, Senator, the evidence is in the other direction." Hollings and Senator Albert Gore (D. Tenn.) subsequently denounced Graham for giving *false and misleading testimony* and demanded his resignation.

In response to all the negative press about Morton Thiokol senior management, Ed Garrison wrote a letter that was to appear in the *New York Times* and *Washington Post*. In it, Garrison stated that I was working full time on matters related to the failure investigation, which simply was not true, even though I had testified four different times in Washington, twice in Utah, and once in Florida, and had been responding to calls for information from the commissioners and or their staff. When the Rogers Commission finished its report, Garrison indicated, I would remain one of his key senior managers responsible for the redesign of the SRM joint.

At the Commission hearing on May 9, Garrison endeavored to clear Thiokol from the charge that it had retaliated against Boisjoly and me by demoting us and leaving us out of the failure investigation. "I'm afraid we've left the wrong impression," ventured Garrison. "We have made no attempt to freeze anyone out of the data. We haven't demoted anyone. We've changed a lot of duties, but we haven't demoted anyone. . . . I feel comfortable that we haven't violated any sacred thing, excluding or setting aside people so they can't be involved. We did ask that any written work that was submitted be logged and run through management. But we have made no restrictions. I think we played it completely wide open. We're in trouble and need all the help we can get. We certainly need the support of all our people."

General Kutyna didn't let Garrison's defense simply pass. "You know, Mr Garrison, in the Department of Defense, conflict of interest is a very heinous crime, so we stay away from any conflict of interest to the utmost. They'll get you even on the *appearance* of conflict of interest; you don't even have to have it, just the appearance. It is the same in this case. There is certainly some impression that these gentlemen have been punished or lowered in their responsibilities as a result of the proceedings since the accident."

"Well, everybody can't be in charge," Garrison retorted, an impolitic statement to make to the Commission right after he had done a good job of presenting his position and reasons for reorganizing. The insensitivity of it, because it reflected his real feelings, clearly irritated many of the Commission members.

"But I caution you against the appearance of that, even," Kutyna repeated.

"We are very sensitive to that, General, and that is the reason we have a legal counsel and a corporate counsel monitoring this thing very closely. We are being very careful in that area." You bet they were.

On the way out of that final meeting with the Rogers Commission, Don Kutyna pulled me aside and told me he hoped that everything would work out OK for me. Although he was glad it was all over but the paperwork, he was going to miss many of his fellow panelists as well as Roger Boisjoly and myself. "Stay in contact, Al, and when you're in the Los Angeles area, stop by and see me. My wife has watched all of the Commission hearings on TV and she'd really like to meet you. I don't want this to go to your head, but she admires what you've done and thinks you're the cutest thing on television." "No wonder," I jested, "look who she's comparing me to!" The general gave it right back to me: "You must mean Roger Boisjoly!" I gave him one of my gold-embossed Space Shuttle business cards to give to his wife. After all the negative comments I had endured from Morton Thiokol and NASA over the past few months, these remarks from General Kutyna really made me feel good.

I returned home that Friday evening only to leave again the following Monday for Elkton, Maryland, where one of our other rocket divisions was located. Tom Davidson, our Corporate Technical Director, had asked me some time back to participate in a technical proposal review at Elkton related to using some dual-chamber controllable SRM technology that I had worked on earlier in my career in support of a kinetic energy weapons system being studied as part of the Strategic Defense Initiative, or "Star Wars" program. I was supposed to be in Elkton for a week but had to return home early because I was notified that I was needed for a teleconference with the Presidential Commission. There was also a meeting in Ogden that afternoon with our senior corporate management. Called by Bob Ebeling to try and iron out the bad feelings that had been created between Thiokol senior management and those of us who had opposed the launch and thereby reunite the company for all the hard work we were going to have to do to crawl out from under this mess, it was important for me to be there, especially since Charlie Locke, Ed Garrison, and Ed Dorsey would be attending.

At the meeting, our company leaders were fairly congenial, but the atmosphere was still very tense. The only thing that broke the ice was when Charlie Locke leaned back on his chair and fell over backward. Obviously quite embarrassed, he composed himself before he started to speak. He told us that all would be forgiven and that our reward would be based upon how well we did from here, not what we had done in the past. I stood up and said directly to Locke and Garrison that openness was good for the company, that they shouldn't try to control people, and it was not a good idea to follow the golden goose over the cliff.

What I did not know at the time was that Locke had just the day before said in an interview with the *Wall Street Journal* that Roger Boisjoly and I were just screwing off, running around the country and not doing any productive work. When the article with those quotes ("Thiokol's Chief Denies It Demoted Aids, But Hints Displeasure Over Controversy") came out the next day, it was clear that the company was speaking out of both sides of its mouth. "They travel all over the country at our expense to appear before Commissions or just make idle trips to talk to somebody in Washington," Locke told the reporter. "We've been very open and above board with that. But once this Commission issues its report and this thing is closed, it's going to be a different situation, because people are paid to do productive work for our company and not wander around the country gossiping with people."

Locke's comments were totally contradictory to what Garrison was telling everyone. Though Garrison's statements seemed conciliatory, I knew that the statements that really counted were Locke's. He was the real top dog in Morton Thiokol. Garrison followed Charlie around like a lost puppy, and

After the *Challenger* accident, Thiokol corporate management from Chicago visited the Utah plant to get a better understanding of how the O-rings were made. (Cartoon by Cal Grondahl, 1986, *Standard Examiner*, reprinted with permission.)

I knew my life was going to be more miserable because Ed would quickly mimic Charlie's attitude, which he did.

If I had known about those comments at the time, I wouldn't have even participated in the "goodwill" meeting. However, naive and hoping for the best, I went away from that meeting feeling more comfortable that at least I wasn't going to lose my job.

27

Commencement

Shortly after the final session with the Presidential Commission, several congressmen requested that the NASA Inspector General's (IG) Office conduct an investigation to determine if Roger Boisjoly and I had been demoted for our testimony before the Commission, and whether NASA and its contractor, Morton Thiokol, had done anything illegal or inappropriate during the conduct of the *Challenger* investigation. Supporting that call for a comprehensive probe was a letter to NASA Administrator James Fletcher from twenty-eight U.S. senators. MTI spokesmen once again denied that anyone had been reassigned in retaliation for their damaging testimony, issuing statements from corporate headquarters that many job changes had occurred at Thiokol following the *Challenger* accident, but only in a sincere effort to switch from supplying NASA with booster rockets to redesigning the rocket joints. At a news conference, Senator Albert Gore Jr. (D-Tenn.), a member of the Appropriations Subcommittee, called Thiokol's explanation "gobbledygook" and expressed "outrage" at the company's behavior. The future vice president and presidential candidate said he was considering introducing legislation that would make it a felony to retaliate against witnesses testifying before a presidential commission, as such a law already protected congressional witnesses.

Within days, representatives from NASA's IG Office in San Francisco phoned, wanting to visit with me at my home. My company advised me that I should tell the IG, if I preferred, that the interviews would be best held at my workplace. The corporate legal staff also told me that I was entitled to counsel wherever or whenever I met with the IG, and that MTI wanted to provide me with an attorney. Because of my great distrust of the company at the time, I declined its offer for legal counsel and chose to talk to the IG representatives at my home in Pleasant View, Utah, one evening after work.

For the meeting with the Inspector General's people, I prepared a very detailed chronology, highlighting events and memos that supported the position that I had been effectively removed from my position and placed into one of inferior stature and responsibility. I also gave the IG a copy of the memo I had sent to the Rogers Commission on April 1 relative to statements made by NASA and Thiokol management during the Commission's public hearings from February 25 to 27. I had given this memo, with some minor corrections to my own testimony of February 25, to our corporate legal counsel, Lee Hales, for him to transmit to the Presidential Commission, as this was the process the Thiokol legal department had set up to assure they were aware of all information transmitted to the Commission. This was the only material I provided to the Presidential Commission through the "proper" channels established by Morton Thiokol and NASA; all other information I provided directly to the Commission members. My memo contained a considerable amount of information that in a court of law could be very damaging to both Thiokol and NASA management. The IG staff also visited me at the plant the next day, where they also interviewed several other people. After going through all the questioning from the IG Office, I felt like I was back at ground zero with the company, viewed once again as some sort of snitch.

Shortly after the IG visit, and just as the Presidential Commission's report was to be published and presented to President Reagan in a Rose Garden ceremony at the White House, Thiokol senior management publicly announced that they had chosen me to head up the redesign of the solid rocket motors for the Space Shuttle. My new title was to be Director of the Solid Rocket Motor Verification Task Force. My job was to define and direct a program that would "successfully redesign, requalify, and recertify" the solid rocket motors for use on the Space Shuttle. Not a small task at all, but I welcomed the challenge, because I could now devote all my energies toward something positive rather than reflecting on all the negative things associated with the *Challenger* accident and its aftermath. I felt a great deal of apprehension about being on center stage—and in a media fishbowl—but I knew that came with the assignment.

I was even more nervous about going back to Huntsville for the first time since all of my controversial testimony. Since that time, however, most all of the key players both at NASA and Thiokol who had been associated with the decision to launch *Challenger*—especially those involved in the infamous teleconference the night before the launch—had been removed from the Space Shuttle program. These people had resigned, been reassigned, or were in the process of one or the other. On the Thiokol side, Jerry Mason, Senior Vice President and General Manager of Wasatch Operations, the man who had made the final decision at Thiokol to proceed with the launch, had been forced to retire—but not without a very lucrative supplemental retirement package. Moving over to become the Assistant General Manager of MTI's Strategic Division was Cal Wiggins, former Vice President and General Manager of the Space Division, a man who, while supporting his bosses, for some strange reason had escaped all of the hearings and testimony, thereby earning him the nickname "Mr. Teflon." Wiggins was eventually sent to the Redstone Arsenal to serve as Vice President and General Manager of MTI's Huntsville Division. Bob Lund, the Vice President of Engineering who had changed his mind when he changed his hat, was transferred to a corporate office position. Lund was eventually promoted to Vice President of Advanced Technology (later called "Science and Engineering") at a salary and bonus level that was significantly higher than what he had been making at the time of *Challenger*. My previous boss, Joe Kilminster, Vice President of Space Booster Programs, was transferred to Thiokol's Automotive Products Division in Ogden. Joe became Director of Engineering and Quality Assurance for the blossoming car air-bag market, which was to have a much better future than solid rockets.

On the NASA side of the ledger, at headquarters in Washington, D.C., Dr. James Fletcher came back for another stint as NASA Administrator, replacing Dr. William Graham. Serving as his new Associate Administrator for Space Flight was former astronaut Admiral Richard Truly, who replaced Jess Moore. Michael Weeks, the Associate Administrator under Moore, took a reassignment in a non-Shuttle program, while both Moore and Graham retired. The Center Director at Kennedy Space Center, Richard Smith, also retired, and was replaced by USAF General Forrest McCartney, and the man who had sat as Launch Director for the *Challenger* launch, Gene Thomas, stepped aside to be replaced by Bob Sieck.

No significant personnel changes were made in Houston, with Arnold Aldrich remaining as Manager of the National Space Transportation Systems Program Office. At NASA Marshall, however, wholesale changes took place. Center Director Bill Lucas was asked to retire; he was replaced by J. R. Thompson, a previous MSFC employee who had left for a post at Princeton University and had headed up the overall NASA Failure Analysis Team. Stan Reinartz resigned as Manager of the Shuttle Projects Office; replacing him was Bob Marshall. Larry Mulloy resigned as Manager of the Solid Rocket Booster Project and was replaced by Gerald Smith. The Deputy Manager of the Shuttle Projects Office, Dr. Judson Lovingood, retired and went to work for Thiokol in its Huntsville Division. Royce Mitchell took Larry Wear's job as Manager of the SRM Project Office, with Wear transferring to another MSFC program. Jim Kingsbury, Director of Science and Engineering, retired and was replaced by Dr. Wayne Littles, and Kingsbury's Deputy Director, George Hardy, was replaced by Jim Odom. John Thomas, former Manager of the Spacelab Program Office and a member of the NASA Failure Analysis Team, was anointed as kind of a task force director for the SRM redesign, reporting directly to the new Center Director, J. R. Thompson. Both Thomas and Mitchell were temporarily relocated to Utah to oversee the redesign activity at Morton Thiokol. In the days just before the Rogers Commission released its formal report, the nation's newspapers picked up on all of these changes in personnel.

For me, all of these changes collectively meant good news/bad news. The good news was that I would be dealing with a whole new set of faces during the redesign, and I wouldn't have to confront those whose careers had been damaged by my testimony. The bad news was that the "good-ol'-boy" network was still in place, and a lot of people who were close friends, allies, and sympathizers of those who had been forced out at Thiokol and NASA could still be blaming me for ruining these people's careers. I didn't know exactly who all these people were, but I had a pretty good idea about most of them. I also knew that I had to prove myself anew with NASA and Thiokol management because of the reputation I had gotten as a whistle-blower. No one likes a whistle-blower because he or she may jeopardize people's jobs. Some people also believed that the only reason I did what I did was for the publicity, which was far from the truth. I had always avoided the press as best as I could, sometimes at great personal sacrifice, by staying with my friends and not even going home at night. But there were many out there who thought differently, because after each time I had testified, my name flashed across the national media.

I was extremely nervous about attending the first technical interchange meeting on the redesign, held at NASA Marshall from June 9 to 11, 1986. The meeting was set up to go over the ground rules and establish the contractor-end-item specifications for the redesigned solid rocket motor. The day the meeting started was the day that the Rogers Commission was to deliver its formal report to President Reagan, and anticipation was high. I was very concerned that the nationally televised event would provoke a lot of controversial questions from the media that could involve me, but fortunately that did not happen. I was also concerned that the Commission's report could very well disagree with some of the conclusions I had presented to the panel. I had told the Commission what I believed was the exact cause of the accident, why I thought the decision to launch was not logical, and the reasons why I believed that NASA management had pressured Thiokol management into changing their minds after initially recommending not launching. It was a very stressful day to say the least.

As the technical interchange meeting began in one of MSFC's large conference rooms, I felt like everyone was staring at me. I remained very uncomfortable until the meeting got down to the business of going through the old contractor-end-item specifications for the SRM and suggesting what changes needed to be made to improve the SRM's overall reliability. When the meeting ended, I immediately rushed to my motel room and turned on the local news to see what had happened at President Reagan's press conference with the Rogers Commission.

On the screen, I saw President Reagan shake hands with several members of the Presidential Commission, congratulating them for a job well done. The newscaster reported that the Commission had concluded that the *Challenger* accident had been caused by the failure of an O-ring in a joint of the right solid rocket booster, which had been precipitated by the cold temperatures the morning of the launch. The process leading to the decision to launch had been "flawed," with Morton Thiokol management buckling to pressure from their biggest customer, NASA.

I was greatly relieved that my name was not mentioned. Wanting to read the entire Commission report as soon as possible, I called our Washington office and was told that the first volume of the report was on its way to my office in Utah. Four more volumes were to come, but they wouldn't be available for a few weeks. I asked for the volume to be sent in the overnight mail to me at Marshall so I could read it the next morning during my lunch break. Instead of eating my meal, I went over to our MSFC office and read as much as I could, starting with the summary section and the section on the decision to launch. I saw that the volume contained several statements from my sworn testimony, and I was particularly pleased that the Commission not only agreed with my conclusions as to the cause of the accident but also with my view that the decision process had been flawed, with pressure from NASA being the principal source.

The Rogers Commission report was not without controversy, however. Prior to the release of the report, the media announced that the report may be issued without the signature of one of its members, Dr. Richard Feynman, who refused to sign because he thought the report was too kind to NASA management. In his opinion, NASA management had exhibited very poor technical judgment and had appeared to be more interested in public relations to gain public support and government funding at the expense of honest risk assessment and flight safety. In a reliability assessment of the Shuttle that he had personally conducted with various NASA personnel, he discovered that some NASA officials were short in technical ability as well as honesty and integrity. Feynman thought that the Commission's report, as written, did not deal directly with these concerns.

In fact, the report ended with some concluding remarks *supporting* NASA, not admonishing them as Feynman thought was more appropriate. The Commission urged that NASA continue to receive the support of the administration and the nation. As a national resource, the agency played a critical role in space exploration and development while providing a symbol of national pride and technological leadership. The Commission applauded NASA's spectacular achievements of the past and anticipated impressive achievements to come. The nation's most important objective was to fix the problem with the Space Shuttle as soon as possible so that the Shuttle could return to flight. That could only be successfully accomplished with full support from the administration and public.

While Feynman agreed with the need to correct the problems and return the Shuttle to safe flight, he didn't feel comfortable praising NASA based upon his personal experience with many of the NASA officials. He finally agreed to sign the report if he could include his own personal observations on the "Reliability of the Shuttle" as an appendix, an idea to which the Commission's leadership agreed with only mild reluctance.

A few excerpts from Dr. Feynman's personal observations on the "Reliability of the Shuttle" follow:

> It appears that there are enormous differences of opinion as to the probability of a failure with loss of vehicle and of human life. The estimates range from roughly 1 in 100 to 1 in 100,000. The higher figures come from working engineers, and the very low figures from management. What are the causes and consequences of this lack of agreement? Since one part in 100,000 would imply that one could put a Shuttle up each day for 300 years expecting to lose only one, we could more properly ask, "What is the cause of management's fantastic faith in the machinery?" We have also found that certification criteria used

in flight readiness reviews often develop a gradually decreasing strictness. The argument that the same risk was flown before without failure is often accepted as an argument for the safety of accepting it again. Because of this, obvious weaknesses are accepted again and again, sometimes without a sufficiently serious attempt to remedy them, or to delay a flight because of their continued presence.

An estimate of the reliability of solid fuel rockets was made by the range safety officer, by studying the experience of all previous rocket flights. Out of a total of nearly 2,900 flights, 121 failed (1 in 25). This includes, however, what may be called earlier errors, rockets flown for the first few times in which design errors are discovered and fixed. A more reasonable figure for the mature rockets might be one in 50. With special care in the selection of the parts and in inspection, a figure of below one in 100 might be achieved, but one in 1000 is probably not attainable with today's technology. (Since there are two rockets on the Shuttle, these rocket failure rates must be doubled to get Shuttle failure rates from Solid Rocket Booster failure.) NASA officials argue that the figure is much lower. They point out that these figures are for unmanned rockets, but since the Shuttle is a manned vehicle, "the probability of mission success is necessarily very close to 1.0."

The phenomenon of accepting for flight, seals that had shown erosion and blowby in previous flights, is very clear. But erosion and blowby are not what the design expected. The fact that this danger did not lead to catastrophe before is no guarantee that it will not the next time, unless it is completely understood. When playing Russian roulette, the fact that the first shot got off safely is little comfort for the next.

Let us make recommendations to ensure that NASA officials deal in the world of reality in understanding technological weaknesses and imperfections well enough to be actively trying to eliminate them. Only realistic flight schedules should be proposed, schedules that have a reasonable chance of being met. If in this way the government would not support them, then so be it. NASA owes it to the citizens from whom it asks support to be frank, honest, and informative, so that these citizens can make the wisest decisions for the use of their limited resources. For a successful technology, reality must take precedence over public relations, for nature cannot be fooled.

Myself, I was extremely glad to see Feynman's comments added to the Commission's Report.

Richard Feynman was a unique individual, and I felt very fortunate to have had the opportunity to make his acquaintance (he died in 1988)—even though the circumstances bringing us together were not pleasant. One of my prized possessions is his book *Surely You're Joking, Mr. Feynman! Adventures of a Curious Character* (1985), which he autographed for me at the conclusion of the Presidential Commission hearings. Three years later, I was gifted a copy of his next book, *What Do You Care What Other People Think?* (1988). The book was given to me by the College of Engineering at Carnegie Mellon University in Pittsburgh on the occasion of a speech I gave there concerning progress in the redesign of the Space Shuttle solid rocket motors. Feynman credits my testimony for enabling the Presidential Commission to find the real cause of the *Challenger* accident:

Then something happened that was completely unexpected. An engineer from the Thiokol Company, a Mr. MacDonald, wanted to tell us something. He had come to our meeting on his own, uninvited. Mr. MacDonald reported that the Thiokol engineers had come to the conclusions that low temperatures had something to do with the seals problem, and they were very, very worried about it. On the night before the launch, during the flight readiness review, they told NASA the Shuttle shouldn't fly if the temperature was below 53°—the previous lowest temperature—and on that morning it was 29. Mr. MacDonald said NASA was "appalled" by that statement. The man in charge of the meeting, a Mr. Mulloy, argued that the evidence was "incomplete"— some flights with the erosion and blowby had occurred at higher than 53°—so Thiokol should reconsider its opposition to flying. Thiokol reversed itself, but MacDonald refused to go along, saying, "If something goes wrong with this flight, I wouldn't want to stand up in front of a Board-of-Inquiry and say that I went ahead and told them to go ahead and fly this thing outside what it was qualified to." That was so astonishing that Mr. Rogers had to ask, "Did I understand you correctly, that you said . . . ," and he repeated the story. And MacDonald said, "Yes sir." The whole Commission was shocked, because this was the first time any of us had heard this story; not only was there a failure in the seals, but there may have been a failure in management, too. Mr.

Rogers decided that we should look carefully into Mr. MacDonald's story and get more details before we made it public.

On the Friday following the release of the Presidential Commission report, I returned to my alma mater at Montana State University, where I was to receive an Honorary Doctor of Engineering Degree. I always enjoyed returning to Montana, because that was where my roots were. I agreed to a one-on-one interview with Charles Kuralt of NBC; the interview was conducted in the backyard of my sister and brother-in-law Juanita and Jim Vollmer's home in Bozeman, Montana, prior to commencement exercises. He asked me several questions about my involvement in the circumstances surrounding the decision to launch the *Challenger*, and he was interested in my response to the conclusions from the Presidential Commission's report relative to the cause of the accident, and the flawed decision to launch the *Challenger* in such cold weather. He also informed me that there was considerable public support for me from around the nation. The recognition I was about to receive by being awarded an Honorary Doctor's Degree for my actions was very appropriate. NBC also videotaped portions of the "Honors and Awards" dinner and reception the night before the commencement, as well as portions of the commencement when I was receiving my Honorary Degree and addressing the graduating class.

That beautiful spring day was the first totally pleasant experience I had had for nearly five months. I felt the world had just been lifted off my shoulders now that the hearings were over and the report released to the public. The inscription on my honorary doctorate read: "In recognition of your creative and original work on sophisticated rocket systems, for your contributions to the advancement of technology and ability to transform ideas into reality, and especially for your demonstrated courage and integrity for always standing on principles you know to be right and just."

My alma mater asked me to give the commencement address to the College of Engineering and, prior to that, to provide the "Charge" to the entire graduating class of 1986. Unfamiliar with what a "Charge" was, I called the commencement committee to find out. They told me that it was a short (not to exceed three minutes) motivational speech to all the graduates before they went to formal graduating ceremonies for their particular college. Some of the charges given in the past had been very serious, some rather humorous—the choice was up to me. I decided to go with the humor.

Degree candidates, parents, relatives and friends, distinguished guests, members of the faculty: it is my privilege and pleasure to present the charge to the graduating class of 1986. . . . As you leave this University, your life will be a cycle of bull markets and bear markets. You will find that there is a lot more bull out there than there is bear. Much to your surprise, you will find that the whole world out there lives by the Golden Rule, which is, "Whoever provides the gold also makes all the rules." It doesn't matter whether it's industry, government, or education. I know that most of you have a goal to make a million dollars a year, but I can tell you, a million dollars a year doesn't guarantee happiness. I know a guy who makes $5 million per year and he is no happier than the guy who makes only $4 million per year. A million dollars per year just doesn't seem to make that much difference.

In all seriousness, you have spent four hard years here, maybe more, getting to this point in your life. It has cost you thousands of dollars, thousands of hours of study and many nights of lost sleep studying for exams. You and your parents have made a heavy investment in your future—the best investment one could ever make in their entire life. The return that you receive on that investment is all up to you.

My charge to you is a simple one, and that is to put forth that same dedication and perseverance to your chosen profession that has gotten you to this graduation. Continue to be yourself, be honest with your colleagues, faithful to your family, and help those that need help. Never forget those that have helped you along the way. If you do this, the investment in your education will have been a good one. You could do no greater honor for yourself, your family, your friends, and for this University than to respond to this charge.

I was so thrilled to give the commencement address that I committed the entire thing to memory. For redundancy, something we engineers always wanted, I had a typed copy of the address on the podium, but I never looked at it once. I began the speech by saying I was maybe "*the most successful failure in history*":

I received my notoriety by trying to delay the launch of STS-51L, the Space Shuttle *Challenger*, on January 28, 1986, because of my concerns with the sealing of the O-rings in the joints of the solid rocket boosters at cold temperatures, *but I failed*. I then came forth to the Presidential Commission and told my story of the events that transpired the night before the launch that eventually led to the decision to proceed on with the launch on that cold chilly morning of 28 January against my objections and those of many good engineers. I was sitting at the console in the Launch Control Center at the Kennedy Space Center at 16:39:13.3 Greenwich Mean Time when this nightmare unfolded right in front of my eyes. My worst fears had been realized; the *Challenger* had exploded. Because of my continued objections to the launch and my honesty with the Presidential Commission, I became an overnight hero, but I do not consider myself a hero. There were no heroes associated with the launch of *Challenger*; the only real heroes were the seven astronauts who went up in flames that day. I grieve for the families of those seven astronauts. It will be a day that I will never forget as long as I live, because in the final analysis it was *my* rocket motor that failed. But I have rededicated myself to fix those joints in the solid rocket boosters, so that they can never fail again.

I never had an opportunity to meet any of the *Challenger* astronauts, but I have had the opportunity to meet several others. Without exception, they are all anxious to get America back in space safely and finish the work of the *Challenger* crew, because they know that is what the crew would want them to do.

The Space Shuttle will fly again, and will be safer because of the sacrifice of Jarvis, McAuliffe, McNair, Onizuka, Resnik, Scobee, and Smith. The space program, NASA, Morton Thiokol, Congress, and the American public have learned a powerful lesson from this tragic experience. That lesson is that one cannot legislate, dictate, or veto the laws of physics, and politics has no place in technical decisions. The Space Shuttle will never fly without risk, but it should never fly with risks that we do not have to take.

We are living in an age of rapidly changing technology—an era where science and engineering will affect people's lives more than ever before. Just looking back over recent events at Bhopal, India, where thousands of people were killed by methyl isocyanate leaking from a nearby pesticide plant, the loss of the *Challenger* crew in the most complex machine ever built by man, and the loss of lives due to radiation poisoning from the recent fire at the nuclear power plant at Chernobyl in the Soviet Union has made it very visible to all the world the importance of technical understanding and sound engineering judgment. This country and the world badly needs good engineers and scientists to continue our technological progress into the twenty-first century while understanding and minimizing the dangers that are ever so present. New technology is exciting and can become man's greatest blessing, but if improperly designed or used, it can also bring great tragedy.

Since I first received my bachelor's degree from Montana State University some twenty-seven years ago, there have been more advancements in technology than there were since the evolution of mankind, and I expect that changes in the next twenty-seven years will far exceed those of the past. Since I've been out of school, we've landed men on the Moon and returned them home safely, replaced human hearts with mechanical devices, and developed computers to solve problems in seconds that used to take years or could not be solved at all. By the turn of the century, computers will be as common in the home as TVs are today, robotics will move from industry to the common household, new sources of energy will be developed to begin replacing fossil fuels, artificial organs and limbs will be on the verge of becoming commonplace, and we will be preparing for man's first trip to Mars. You are the engineers that will make all this happen.

Many people perceive a college degree as some kind of certificate of knowledge, but that is a misconception. Knowledge without application is utterly useless. Computers store more knowledge than any of us can comprehend, but we do not award degrees to computers. Yes, there will be many scientific truths and engineering principles that one should never forget, but your degree really means that you have demonstrated that you are capable of learning and applying the knowledge that you have learned. What you learn outside the classroom will far exceed what you have learned here the past four years, but what you have learned here will serve as the foundation of your future professional growth.

Hopefully, you have learned more than just facts and figures. You have learned how to identify a problem, the steps necessary to solve the problem, and how to implement your game plan to arrive at the correct solution.

Never compromise your principles, use good engineering judgment, learn as much as you possibly can about your product or your work, be professional, and work hard. This is the true path to success. There is no substitute. There has been no time in history where there has been such great need for good engineers. The future is indeed yours. I congratulate you all and wish you all good luck.

Following commencement exercises, I agreed to participate in a general press conference. This was the first press conference I ever participated in, and it was a real eye-opener for me. It was clear that the media were trying their best to accomplish two objectives: one, open old wounds by trying to get me to condemn Morton Thiokol management for how they had treated me, and, two, get me to disagree with some of the conclusions from the Presidential Commission's report. As for the second agenda, I told them I had not read all of the report, but that I had agreed with what I'd read. As for the first, I emphasized that Morton Thiokol management had appointed me to head up the task force for redesigning the SRM, which must show their confidence in me. Pressing me hard, I did say that I couldn't agree with the recent statement from NASA Administrator Dr. James Fletcher indicating that the Shuttle would be back flying by the next summer, 1987. Candidly, I told them I didn't agree with that, because I thought it would be at least two years before the Shuttle flew again.

This disagreement was the main item highlighted in all the press releases. Needless to say, this got me into hot water again with Thiokol senior management, because I had publicly disagreed with the head of NASA.

Graduation weekend in Bozeman was as enjoyable as it was short-lived. The next weekend I was back aboard an airliner heading to Washington, D.C., where I was to testify before the House Committee on Science and Technology about the Commission's findings. I had lulled myself into thinking that the inquisition was over and I could devote all my time and energies to something positive, but my ecstatic vision was fleeting. The congressional hearings proved to be worse than those of the Presidential Commission, because congressmen weren't as interested in what caused the *Challenger* to explode as they were for having a TV appearance that would help in their reelection.

Long live "Truth, Justice, and the American Way"!

28

Commodore Locke Assailed
in Washington

Going into the congressional hearings, Morton Thiokol senior management was very nervous. Ranking members of the U.S. Senate's Subcommittee on Space, Aeronautics, and Related Sciences were making comments for the press about bringing charges of criminal negligence against participants in the ill-fated *Challenger* launch decision. Singled out by some of the politicians for improper behavior and "willful gross misconduct" were Larry Mulloy and William Lucas, but other individuals were sure to be named. "No remorse, no misgivings, no understanding of individual responsibility," Fritz Hollings said about Bill Lucas. As for MTI, the South Carolina senator named no names yet, but definitely was out for blood. "I'd like to give a good government award to the Morton Thiokol engineers and a kick in the 'you-know-what' to management there."

The report of the Presidential Commission on the *Challenger* accident had just been released, and now, the afternoon of June 10, 1986, the two extremely distinguished and popular leaders of that august body, Chairman William Rogers and Vice-Chairman Neil Armstrong were in front of a highly suspicious and agenda-laden congressional panel, chaired by a Democrat, to answer a series of leading questions as to whether NASA personnel, particularly those at MSFC, had cooperated fully with the Presidential Commission and whether, based on their knowledge, any litigation against some individuals involved in the decision to launch *Challenger* was appropriate.

Any hope that the Marshall guys might get off easy was crushed when Democratic Congressman Tim Valentine of North Carolina immediately fired an opening salvo. "Do you feel, Mr. Rogers, that you received complete cooperation from all of the people at NASA with whom you came into contact, you and other members of the committee, whom you interrogated?"

Rogers broke his answer into two parts. Most people from NASA, particularly Admiral Richard Truly, who was going to continue on with NASA as its new Associate Administrator for Space Flight, had cooperated with the Commission "fully in every respect." The exceptions were the people at NASA Marshall, who did more than just withhold certain salient facts, especially about launch constraints, *almost going so far as covering them up.* As for Morton Thiokol, Rogers said, "They were very cooperative," which I thought was a little too generous given that my bosses had not provided the written comments I had made on the February 25–27 public hearings, but unfortunately Chairman Rogers did not know that at the time—it wasn't revealed until the IG completed its investigation a month later.

In all of his public statements following the conclusion to the Presidential Commission, Rogers made it clear that he did not believe that any criminal negligence had occurred. "I don't believe there was any venality here. I don't believe there was gross negligence. I think there was a misunderstanding about what each person was responsible for. In the case of NASA, it's such a big organization, with so many people, that it has all the evils of bureaucracy. Responsibility is pretty diffuse. The Marshall people thought they had the right to make final decisions because they were Level III, while those at Level II did not think that was the case. . . . That was the mistake of the system. But in terms of criminal negligence, you could certainly say that the people at Marshall *thought* they had the right to close out this problem. They *thought* they had the right to decide it, and they decided it. And there were a lot people involved in that decision. It wasn't just Larry Mulloy. There were a lot of other people, too. So I think it would be very difficult to prove the willfulness or even the gross negligence that is required for a criminal prosecution. Furthermore, I don't think it would be in the national interest."

Although Rogers generally spoke highly of Morton Thiokol's cooperation, some members of Congress did not feel good about allowing the company to continue providing the Shuttle program with its SRBs. "It's only in the aftermath of the accident that we're getting the sense that Thiokol is devoting total commitment to a redesign and only now projecting the 'can-do' attitude about fixing the SRBs," charged California's Democratic Congressman Norman Mineta. "Is it possible that we have a moral responsibility to look elsewhere for the redesign and the resupply of the SRBs, given what was Thiokol's interest in making money over safety, and given its somewhat unrepentant attitude and also what it's done to its two dissenting engineers who testified before the Rogers Commission? I'm just wondering whether we should be looking elsewhere for the redesign supply of SRBs."

Dr. James Fletcher, returning to the job as NASA Administrator, made the point that the "entire nation" would be involved in the SRB redesign in that various NASA facilities as well as a specially constituted task force from the National Research Council (NRC) would be making contributions and overseeing the effort.

With the hearings being televised, everyone knew the atmosphere would be very politically charged. Already there were rumblings in Congress about "throwing the bums out," meaning to get someone other than Morton Thiokol building solid rocket motors for the Space Shuttle. MTI leadership fully understood that the company needed to project its best image in these hearings, so they scheduled an all-day practice session with a Washington, D.C., public relations firm specializing in preparing for congressional hearings and other legal proceedings.

It was a very expensive operation, these mock hearings, with detailed questioning of the planned participants deliberately conducted, videotaped, and reviewed with the experts. Everything about it was done to invoke the same emotional responses and duplicate the same near-circus atmosphere that was bound to occur at the congressional hearing, where some people might be sleeping and others surely talking to their neighbors at the table.

The simulations were tough, with questions being best guesses as to what would be asked by a hostile congressman who had already made up his mind that Thiokol was at fault for the Shuttle disaster. We wanted our people to feel more at ease, so that everyone who testified would come across more believable and appear less confrontational. A week before the practice session, each of us who was to testify prepared written answers to the questions we anticipated—the goal being to provide the same verbal response as written response. Clarity and conciseness were desired, and the manner in which the responses were made was considered very important. Witnesses sat together at a table just as they testify before Congress. Anticipating that some of our testimonies would overlap, we practiced reinforcing each other's testimony where we could to present as complete a picture as possible.

I had absolutely no problem with the manner in which the experts set up our mock hearings, but that wasn't the case with Joe Kilminster, Ed Garrison, or Charlie Locke. Our teachers excused me early while they went through some additional coaching with this trio; in fact, they took Locke aside to provide some individual coaching in a private room.

I don't know what the bill was for all of this training, but I'm sure it was very expensive.

The hearing started at 9:30 the morning of June 17, 1986, in Room 2318 of the Rayburn House Office Building. Presiding over the House Committee on Science and Technology was Robert A. Roe, a Democratic congressman from New Jersey. After opening comments by Congressman Roe and Congressman Manuel Lujan (R-N.Mex.), the committee's ranking minority member, in which they both made clear their committee wanted more information about "what went wrong in the chain of communications between the contractor of the solid rocket booster and NASA," Charlie Locke read his prepared statement on behalf of Thiokol:

> Thank you, Mr. Chairman and good morning. I am Charles S. Locke, Chairman of the Board and Chief Executive Officer of Morton Thiokol, Inc., and seated with me at the table here are Ed Garrison, President of our Aerospace Group; Joe Kilminster, Division Vice President; and Al McDonald, Director of our Solid Rocket Motor Verification Task Force. Also with us, and seated here in the front row, are Carver Kennedy, a Division Vice President, Space Booster Programs; Roger Boisjoly, Staff Engineer; and Arnie Thompson, Supervisor of Structures Design. We have two prepared statements to make and then would be happy to answer your questions and those of the other committee members.
>
> We at Morton Thiokol share the anguish this country feels as a result of the *Challenger* tragedy. Indeed, the accident and loss of the crew have been particularly painful for each of us since, *in the final analysis, it was our solid rocket motor that failed.*

I had intentionally not sent a copy of my commencement address at Montana State University to corporate headquarters, so I am not sure that Locke borrowed that last statement from my speech; but it was almost verbatim. It would not be long before Locke would deeply regret his use of that phrase, "it was our solid rocket motor that failed," whether he came up with it on his own or not.

He continued:

> Nothing we can say or do will bring back those extraordinary people whose lives were lost, but I pledge that Morton Thiokol will do everything in its power to be sure that such a tragedy does not happen again.
>
> We congratulate the Presidential Commission on an excellent job in reviewing the Shuttle accident and establishing the framework for

a safer space program in the future. We are in full agreement with the Commission's recommendations.

I should also say that we take pride in the contributions of our employees who testified—Joe Kilminster, Al McDonald, and Roger Boisjoly, Arnie Thompson—as well as many others. The Commission's report is evidence of the candor of these men and their engineering knowledge was of great value. This policy of openness will not change as the space program regroups and moves forward.

With the benefit of hindsight, it is clear that some decisions made the evening of January 27th were wrong—that mistakes were made. Our space program experts, confronted with reports that the weather would be substantially colder than for any previous launch, reviewed the available data and initially concluded that a launch should not occur at an O-ring temperature lower than 53° Fahrenheit, the lowest previous launch temperature. But we all know that NASA questioned Morton Thiokol's decision. Our engineers could not prove that it was unsafe to fly at less than 53°. Thus, after reviewing the data further and evaluating the concerns of a number of engineers, our managers, each of whom has a technical background, came to the judgment that it was safe for our booster motors to fly. I might add that, had we known how very cold the right aft joint of the motor really was—it may have been, in our opinion, as low as 16°—we believe our judgment surely would have been different.

I do want to comment on some of the personnel changes that followed the accident. Besides bringing in new management, we did substantially reorganize the responsibilities and jobs of many others in the Division. In the course of these changes, we came to believe that Al McDonald, who had spoken candidly, but harshly, about NASA in the investigation, should operate in an environment where he could continue to do important work, but in which he would be less likely to interact directly with the agency. We could not afford the possibility of friction, which would be counterproductive to the important work ahead. Similar concerns existed concerning Roger Boisjoly, one of our seal experts. In retrospect, we must criticize ourselves for not being sufficiently sensitive to how these actions would be perceived.

I should also say that I am sorry about some remarks I made, which were reported in the press. These remarks grew out of my frustration over the misperception of the actions we took with the respect to these two gentlemen. I hope subsequent events have demonstrated that we had no intention of punishing anyone. Such action would be totally contradictory to what our company has ever done or stood for. The task force which Al McDonald heads will lead the redesign effort. It has already begun to coordinate with NASA and the National Research Council oversight committee. Solving the problems and redesign will be a complicated process.

Mr. Chairman, the Presidential Commission concluded its report by observing that its findings and recommendations are intended to contribute to the future NASA successes that the nation both expects and requires as the 21st century approaches. We embrace the report with that goal in mind and pledge that we will do our part to support NASA's efforts.

Finished, Locke turned the table over to Joe Kilminster, who read his own prepared statement briefly summarizing the history of the SRM field-joint design and the prelaunch situation as Thiokol experienced it. Wrapping up, Kilminster admitted, "Obviously, we were wrong. We did not have the safety margin necessary to cover some things we were not aware of—temperature of the joint lower than 29°, perhaps as low as 16°; potential for ice in the joint; putty behavior at cold temperature; and the effects of violent wind shear conditions. In hindsight, we all wish we could reverse the judgment we made. The decision we made that night has been constantly on my mind since the morning of January 28th."

Congressman Roe then wanted to hear from me and from Roger Boisjoly. As it was never my style to read a prepared statement, I just responded to questions from committee members. Essentially, I repeated what I had previously told the Rogers Commission. My one important amplification came in response to Roe asking, "When the decision came back to you from the caucus that lasted approximately half an hour, were you satisfied with that decision or did you still doubt in your mind whether that was the wise thing to do?"

"I was not satisfied with that decision. I was a bit taken aback and surprised because the rationale presented did not indicate to me that we had run the calculations to convince me that we had a good number again. In fact, we didn't come back with a number, which bothered me a little bit. It just said that we would recommend proceeding with the launch. In reviewing the chart that Kilminster eventually had to sign, there were about nine items on that chart, and five of the nine were reasons not to launch. They were the concerns raised earlier in the telecom and only four of those were items that you could say, or may say, 'it is all right.' There were still more unknowns than knowns. That is when I raised the issue with both Mulloy and Reinartz—my concern that I didn't think they could accept that recommendation."

It soon became clear that what some of the congressmen most wanted to do was put our senior management on the spot, so they quickly turned their attention away from Roger and me back again to Ed Garrison and Charlie Locke. Democrat James H. Scheuer from New York was especially eager to get at CEO Locke.

"Mr. Locke, you are quoted as saying to a newspaper man recently—as a matter of fact, on the first page of your statement today you talk about the pain and anguish of this tragic accident. Is this what you were referring to when you mentioned to a newspaper correspondent recently, and I quote, "This Shuttle thing will cost us ten cents a share this year."

"Yes sir, it was, and I would like to clarify the circumstances under which those remarks were made. This was . . . "

"Make it very brief, because we only have five minutes and I have some other questions to get on with."

"This was an article by a *Wall Street Journal* reporter who asked me to give his paper a financial analysis of where the company stood as a result of the Shuttle incident, as well as all other factors of the company. So I was simply responding directly to his questions."

"From the national point of view, Mr. Locke, would you agree that it cost every shareholder in the American company—every man, woman and child in the United States—not ten cents a share but perhaps $20 or $25 a share? And that is not counting . . . "

"Well, sir," a red-faced Locke tried to stop the congressman.

" . . . not counting the incalculable loss of time that we cannot put a monetary value on: the trauma to the American people, the incalculable loss in lives, and the seven lives that were lost. Would you say that is a true financial loss of the accident, $25 per shareholder in the American enterprise?"

"Sir, I don't believe you can put a financial value on this tragedy at all."

"You certainly can't, and I would say that your statement that this Shuttle thing cost you ten cents a share has to go down in the annals of most regrettable history. In 1882 American railroad tycoon William H. Vanderbilt, in an answer to another newspaper reporter's question, said, 'The public be damned!' Now, for over a century that remark has stood unchallenged and unparalleled for its gross insensitivity, for its banality and tastelessness, but I believe you have finally outdone it. You have finally moved Mr. Vanderbilt over in that profane hall of corporate leadership. You have done it. Let me ask you another question, on the first page of your statement you said it was your solid rocket motor that failed. That was your flat statement. In your contract with NASA—the Morton Thiokol contract with NASA—doesn't it provide that, in the event of a failure of the solid rocket motors to perform in compliance with the specification requirements of the contract, there will be a fee reduction of $10 million and the loss of the flight success incentive fee. Would you say you're [sic] failure, as you describe it—the solid rocket motor that failed—should trigger that $10 million fee and the loss of your flight success incentive fee?"

"Sir, that is a contractual matter that I will just have to defer to other people to conclude."

"Well, what was the clear meaning of your words this morning? This morning you said it was your rocket motor that failed. But let's leave all of the Philadelphia lawyers out of this. You said your solid rocket motor failed. Okay? In this contract it is perfectly clearly stated that, in the event of a failure of the solid rocket, that a $10 million penalty would be triggered plus the loss of your flight success incentive fee. Is that a fair reading of the contract and a fair reading of your words this morning?"

"The contract is a very complex document."

"It is not complex. It is very straightforward, Mr. Locke."

"I don't have any other comments."

"Mr. Locke, you have said on page two of your testimony that this policy of openness in which you respect the candor of these men and their engineering knowledge will not change as the space program regroups and moves forward. Yet you told this same reporter that—once again, I quote—'Once this Commission issues its report and this thing is closed, it is going to be a different situation because people are paid to do productive work for our company and not wander around the country gossiping with people.' You were critical, Mr. Locke, of engineers who travel all over the country at your expense to appear before Commissions or take what you called 'idle trips' to talk to somebody in Washington. Now, here you have taken this trip; I hope you don't think it is an idle trip to come to Washington to talk to us, because a number of your staff have come, too. Your words are open, but I detect from your comments to the *Wall Street Journal* that you don't consider Commission hearings or congressional hearings to be a very constructive part of the legislative process. You don't look on them favorably. Now, which is it?"

No amount of practicing with experts in a private room could have prepared Locke for this.

"Well, sir, the interview with that particular reporter was a very long interview and they selected only certain parts of my comments to report. They did not report the entire . . . "

"They never do, but we don't write the stories, do we, Mr. Locke? We find that out up here. Apparently you were not misquoted when you said they travel all over the country at our expense to appear before Commissions or just to take idle trips to talk to somebody in Washington. You must not consider your visit here today to be legitimate corporate activity. Do you really believe that? Are you here today just to talk to somebody? Is this idle trip just to talk to somebody in Washington?"

"Those remarks were made in connection with the conclusion of the Presidential Commission's report and the conclusions of all of its investigations. All I was simply trying to say was that, after all of the investigations are over, we have a very big and very complex job to do and we've got to get on with it."

"Does that mean you will cooperate with whatever continuing, ongoing oversight there will be and that you won't cast a damper or cloud or bring any pressure on outstanding fine Americans like Boisjoly and McDonald in their efforts to help us understand what happened and prevent something like this from happening again?"

"Sir, I think our record is very exemplary. In fact, Mr. Rogers himself said before your very committee that he had gotten complete cooperation from our corporation during his investigation. He was very, very complimentary of us, and I am very glad that he did say that to this Commission."

"I think we have pursued this line far enough," declared Chairman Roe, cutting off the hostile exchange.

Next the heat was applied to Joe Kilminster by Missouri's Democratic Congressman Harold L. Volkmer.

"Mr. Kilminster, did you see in the documentation provided to us the activity report from Boisjoly dated October 10, 1985? In it he wrote that he

needed to add several people to the O-ring task force—a specific manufacturing engineer, a quality engineer, a safety engineer, and four to six technicians—in order to do an important series of tests on a noninterfering basis
with the rest of MTI's operation. This request was deemed not necessary
when Joe—I believe you are Joe . . ."

"Yes sir."

"Joe decided that the *nursing* of the task approach was adequate."

"We will supply you, Congressman, with a summary of the activity that
was conducted over that time period."

"I don't want a summary; I want it detailed."

"We will provide it."

"I just don't believe you. I don't believe that you devoted substantial time
and energy to that effort."

Steaming under the collar, Kilminster decided to pass off all the tough
questions to me. "I would like to defer to McDonald. Al, would you comment on that?" What could I do but try to answer.

"There were some conflicts in getting some things done in the plant, and
I shared Boisjoly's frustrations. I think we all did. What one has to remember, when you have problems, you normally can solve an engineering type
problem by going into the laboratory and solving it, because the hardware is
small, you can go test. But with the Shuttle, the hardware is so large. We have
to use actual flight hardware to run some of the tests, and we have to be very
careful that we don't do anything to that hardware that makes it unusable
for the flight. Boisjoly was getting very frustrated that we were handling the
engineering assessment effort just as if we were getting ready for a flight. In
many cases we had no option but to do that, because the hardware could go
back into flight."

Throughout the morning, the House Committee on Science and Technology kept the heat on Thiokol management, which in turn kept trying to
deflect the answers to many of the contentious questions to me. When it was
his turn, Congressman F. James Sensenbrenner of Wisconsin zeroed in on
how Thiokol management had allowed the ultimate criterion of launch to
change from proving that it was safe to fly to proving that it was not: "Mr.
Locke has said in his statement that Thiokol's engineers could not prove
it was unsafe to fly at less than 53°, saying they couldn't prove it was unsafe to fly at less than 53° is not the same as saying that it was safe to fly at
less than 53°. We then get testimony that there has never been any testing

done at temperatures as cold as what existed at the Cape at the time of the launch, and that there wasn't an engineering computation. It seems to me, Mr. Kilminster, that you opened a heck of a big crack for the *Challenger* to fall through, and it did. Now how did that happen?"

"I believe that we did point out all of the data that was available to us. We had shown that we had blowby at 75° on one launch and 53° on another. We had prior to that conducted static tests down to 47° with no blowby."

"My time is up, Mr. Kilminster, but I observe that after listening to your testimony, even with 20/20 hindsight, you would do it all over again the same way. Is it your testimony here this morning that the impositions placed on you by NASA at the time of the fatal launch were no different than previous launches—that your obligation was to show it wasn't unsafe to fly?

"That is what I believe."

It wasn't what I believed, which brought the focus back on me. "My understanding, Congressman, was exactly what Bob Crippen has said to the Presidential Commission, that you always have to prove it is safe to fly. My job was to argue why I felt our rocket motor was safe to fly but all of a sudden we reversed roles; I couldn't understand that. It was a total role reversal of all I had ever seen in the two years I had been in the program."

Congressman Sherwood L. Boehlert (R-N.Y.) expressed his confusion with many of the Shuttle's technical aspects but even greater difficulty understanding the decisions of what he called the "technocrats" leading up to launch: "The Presidential Commission has concluded, has it not, that Thiokol management reversed its position and recommended the launch of 51-L at the urging of Marshall and contrary to the view of its engineers in order to accommodate its major customer. Mr. McDonald, you were at the Cape. I get the impression that you felt the pressure building up down there. Would you expand on that a little bit?"

"Yes, I definitely felt the pressure, and it was because this role had been reversed. I had stood up probably in front of more readiness reviews in the past two years than anyone, and I always had to justify why my hardware was ready to fly. I personally signed off on all the defects in the hardware, all the critical defects outside of our experience. I had to explain why it was OK to fly with those, i.e., what data and tests did I have to support that. My feeling was these previous matters were more minor than what we were facing that night. So when I heard comments like they were 'appalled' at our recommendation and 'when are we going to fly this thing—in April' . . . "

"Who made those comments?" asked Congressman Boehlert.

"Hardy from Marshall made the comment he was appalled at the recommendation and Mulloy about not being able to fly until April."

"So you had the distinct impression the pressure was being applied from NASA to Thiokol?"

"Yes, I did."

"Did you feel any pressure as that caucus was under way from NASA—you're internally talking to the boys trying to make a prudent decision—did you feel under great pressure? In your mind did you think that NASA wanted you to proceed with launch come hell or high water?"

"No," Kilminster added. "My perception was that they wanted us to look at the technical data and come to a technical recommendation."

"Then you had a different perception from that of McDonald, because I think he had the distinct impression from his vantage point that NASA wanted to go—no questions asked."

"Yes," I said.

"My opinion is different from McDonald in that regard," Kilminster offered.

"One last question to follow-up, Mr. McDonald. You were there and you can tell us; a lot is lost when a transcript is printed in black and white. Allegedly Mulloy said—and I will read the words exactly without any particular inflexion, 'My God, Thiokol, when do you want me to launch, next April?' Would you say that was said in heat—'My God, Thiokol, when do you want me to launch, next April?!' Or was it, 'Come on, guys, when do you want me to launch, next April'?"

"It was the louder of those two," I replied.

"So you had a real gut feeling, Mr. McDonald, that this flight should not launch?"

"Well, I was more concerned about what we didn't know than what we knew."

"You didn't answer my question. With all the data that was available to you, did you have a gut feeling that this flight shouldn't launch?"

"I had a gut feeling that it wasn't prudent to launch and that we could wait a day or two and get rid of any concerns."

"You stated to the Rogers Commission, 'I am the one that has to answer, yes, Thiokol is ready to fly. I am the guy who is going to have to get up and say, yes, we are ready to fly.' Mr. McDonald, if you felt strongly that you shouldn't launch and you had this responsibility, why didn't you go make a statement that, 'If you launch, I cannot accept responsibility,' or make a statement like that? If you knew Mulloy wasn't going to do anything about it, essentially, why didn't you go to Mr. Moore?"

"I thought they had addressed that with Arnie Aldrich on the teleconference. I was *sure* they had. I was amazed that they never went further, because I knew that if I was going to have any influence on whether to launch or not, after management had come back and recommended to proceed, I was talking to the right gentlemen to influence that. One, Reinartz, was a member of the Mission Management Team, so I had no reason to believe they hadn't discussed it thoroughly with Aldrich and Moore. I was shocked to find out they hadn't."

The morning session with the House Committee on Science and Technology finally over, I left the Rayburn House Office Building with the small Morton Thiokol entourage heading for lunch. On the street, two big black limousines awaited us, arranged for by Charles Locke. Opening the heavy car door for our ostentatious CEO was Hugh C. Marx, our Vice President for Human Resources at our Chicago headquarters. "Commodore Locke, your limo is ready," Marx said to him, feigning chivalry.

I couldn't believe Marx said this. Congressman James Scheuer had just totally embarrassed Locke in the hearings by comparing him to Commodore Vanderbilt. I figured the Vice President for Human Resources must have dozed through that part of the testimony, must be ready to retire, or knew something about Charlie Locke's sense of humor that was never apparent to me.

29

The Buck Stops Here?

On the afternoon of June 17, 1986, after our testimony before the committee, it was NASA's turn. Chairman Robert Roe forced Larry Mulloy to admit that NASA simply did not have enough test data to make the magnitude of decision that it made, not when the temperature of STS-51L was 29°, 24° lower than the previous coldest launch. Congressman Ronald C. Packard (R-Calif.) then wanted to hear more about what I had conveyed to Mulloy the night before the launch.

"Well, sir," Mulloy answered, "the three major concerns that were communicated to me on January 27, was, first, the concerns that engineers at Thiokol had expressed about the ability of the O-ring to seal at the predicted launch temperatures. As I've said, McDonald didn't enter into that discussion during the two-hour teleconference. Subsequent to that, he said, while he would agree there was some question about the validity of the recommendation of 53°, he didn't understand how we could operate out of what he thought the motor specification was. It was 40° to 90°. The second concern expressed was for the ice that was on the pad, what McDonald understood about those conditions out there. And the third was the fact that retrieval ships for the solid rocket booster were in an absolute survival mode and coming back toward the shore at very low speed and that they would not be in a position to recover the solid rocket boosters."

"And which of those three concerns did you convey to those above you?" asked Congressman Packard.

"Well, Reinartz, who is my superior, was in the meeting and he also heard those three concerns. . . . We had placed a telephone call to Aldrich that night because, in the launch criteria, there is a statement relative to the recovery area that if there's the possibility that the boosters cannot be recovered, we need to make that as an advisory call. It does not say we *cannot* launch under those circumstances. I took that to be my responsibility, to advise Aldrich that we might not be in a position, possibly, to recover the solid rocket boosters, because they would be some forty miles from the impact area."

"Was the concern for the O-rings discussed with Aldrich?"

"No sir, it was not."

"Why not?"

"Well, sir, I testified to this in the Commission. It is in the record. I will repeat it here. The O-ring and other special elements of a Level III system are considered in the management system to be a delegation to the Level III Project Manager, for him to make dispositions on any problems that arose on those. Our judgment was that there wasn't any data presented that would change the rationale previously established for flying with the evidence of blowby. Since that data was inconclusive, plus the fact that we had redundancy at the time when blowby would occur, which was less than 170 milliseconds, there was no reason to change our rationale. Therefore, there was not any requirement to have the rationale approved by Level II or Level I. That was the judgment."

"Did you share any of this information with Mr. Moore at any time?"

"No, I did not."

One of the most fiery exchanges came when members of the House committee attempted to get NASA to agree with the Presidential Commission's conclusion that Morton Thiokol had reversed its decision from "don't launch" to "proceed with launch" at the urging of NASA, and the NASA folks tried their best not to admit to it.

Republican Congressman Robert S. Walker of Pennsylvania asked William Graham: "One of the conclusions of the Rogers Commission says that Thiokol management reversed its position and recommended the launch of 51-L at the urging of Marshall. Dr. Graham, do you agree with that Rogers Commission contention?"

"Mr. Walker, my information on that is derivative. It is basically information I have been told by others who were participating in that, so I am going to ask . . . "

"No, Mr. Graham, I want an answer from each of the people at the table as to whether or not they agree with that finding of the Rogers Commission."

"I have no independent information to either confirm or deny that particular statement, so I would only be giving information based on the secondary sources that I have. I've not yet found those to be of sufficient specificity that I can give a completely conclusive answer."

"But it is the conclusion of the Rogers Commission that it took place that way. I'm asking whether or not, based upon everything you know, whether NASA has also arrived at that conclusion."

"I have not independently arrived at that conclusion, but I accept that conclusion of the Rogers Commission."

Congressman Walker then asked Admiral Richard Truly to respond.

"That is the same way I was going to answer it, Mr. Chairman. I have accepted the conclusion of the Rogers Commission. The Commission took all the testimony in closed and open public testimony, came to that conclusion, and I accept it."

It was now the turn of the MSFC Center Director, William Lucas, who alone among the others remained stiffly diffident: "Based upon my own knowledge, I have no knowledge that NASA did influence Thiokol. I was not in the meeting. I have talked to all of my people who were, and they do not believe that they influenced Thiokol or insisted that Thiokol change their position on the matter."

"So you disagree with the Rogers Commission, Dr. Lucas?"

"No sir. I don't have all the information from the Rogers Commission and would not be in a position to disagree with them."

"If you don't have it, who does?"

"I believe there are still a few volumes that have not yet been released."

"So you are saying that you do accept that conclusion by the Rogers . . ."

"I accept the conclusion."

Mulloy was next on the spot.

"Yes sir, Mr. Chairman, I think that is a conclusion that the Rogers Commission drew from all of the testimony. I have not seen all of the testimony. As Dr. Lucas said, there are four volumes yet to be released, and I don't know the total basis by which they reached their judgment. I was not aware that I was trying on the night of the 27th to influence Thiokol to reverse their position. However, Thiokol has testified . . . some individuals felt, that is what they perceived, and I think that is the basis of the Commission's judgment, and I have no argument with the Commission's judgment."

"McDonald said something today, earlier, Mr. Mulloy. He said that he told someone at NASA, if you go ahead with this decision, I wouldn't want to face a Board of Inquiry if something happened to the Shuttle. Who did he say that to?"

"Sir, he said that to Reinartz and myself after the decision was passed down from Thiokol. He made it in the context that if he was the Launch Director, because of the situation with the retrieval ships, he would not launch this vehicle—although he did agree that there was some question about the recommendations for it not to fly below 53°. He went on to say, 'If I was the Launch Director in making this decision.' . . . No, I don't believe that is what he said. Let me restate. He said, 'I would not want to appear before a Board of Inquiry,' and he explained why, because I would be flying the vehicle outside of the propellant mean bulk temperature specification limits."

Mulloy could no longer even get my statement correct, even though he had now heard it many times. He couldn't remember it even though he had confirmed what I had actually said in his earlier sworn testimony. Larry was trying to convince the House committee that I had objected to launching because of the PMBT, readings that truly were within specifications. *But that is not what I objected to at all, and Mulloy knew it.* I had argued with him that the *other components* like the O-rings would be much lower than the PMBT and therefore we would be launching outside the qualified limits—not of the propellant but of the other SRB components!

Congressman Volkmer continued the line of questioning: "At the time of the teleconference when Thiokol made its first presentation with their criteria for not going ahead with a launch, and you and your people there had said OK, 'we're going to scrub, we'll look at it again tomorrow,' do you think Thiokol would have turned around and said, "No, we're going to go ahead and have a little conference of our own, a caucus of our own, to change our mind?!"

"No sir, I do not believe they would have done that."

"I don't either, and that is what disturbs me, because in answer to the questions from the gentleman from Pennsylvania [Congressman Walker], you all said that you really didn't yourself know whether Marshall turned Thiokol around. It is obvious to me that the attitude, and what proceeded with Thiokol, and what occurred during that teleconference, maybe it wasn't said directly, but there's no question in my mind that it was there. I think Marshall has to accept that responsibility. If you gentlemen aren't going to accept that responsibility, that gives me great concern, great concern for the future for what is going to occur."

I was not sure he would ever admit it, but at this point in the hearing Larry Mulloy, having avoided acknowledging it for several months through all the hours and hours of questioning from the Rogers Commission, Congress, and news media, finally conceded, however grudgingly, the basic technical truth of the *Challenger* accident. His concession came in response to Congressman Harold Volkmer's question: "Mr. Mulloy, do you now feel that lower temperature has an impact on the operation and the sealing effect of O-rings, lower temperature?"

"Yes sir, I agree with the Commission's conclusions in that regard relative to the failure investigation. They list a number of causes of failure of the joint, and temperature is listed as a contributing factor, and I believe that is correct."

"What brought you to that conclusion?"

"The failure analysis has shown there was a lot about that joint that we didn't understand. We didn't understand the effect of overcompression on the O-rings. What we worried about prior to 51-L was being assured that we did have compression on the O-rings. What we found as a result of the failure investigation was that overcompression was also a contributing cause to the accident, in combination with temperature and other factors. We know a lot more about that joint today than we did before 51-L."

Incredibly, Mulloy was still trying to find some other reason that was more important than temperature for causing the accident, by implying that overcompression was a main contributing cause to the accident, which was not true at all. Overcompression of O-rings had nothing to do with it. In fact, none of the joints of STS-51L showed any evidence whatsoever of overcompression; they were basically the same as all the previous twenty-four successful flights.

The next morning, June 18, 1986, when the hearings resumed, Morton Thiokol was back on the hot seat. Congressman Jim Sensenbrenner of Wisconsin wanted to know if NASA officials at the Cape had talked to me on the morning of the launch, and if so, what they thought about my opinions. Joe Kilminster answered that he hadn't inquired about that, because when Stan Reinartz, who was at KSC for the launch (along with myself and Larry Mulloy), asked during the teleconference if anyone had any further comments to make relative to the recommendation to go-for-launch, "none were forthcoming."

Sensenbrenner probed deeper: "Yesterday, Mulloy testified that McDonald did not argue against launch during the telecom and that he sat and listened to the Thiokol people in Wasatch present its arguments. Mr. Kilminster, you had no conversations with McDonald, who was your representative at the Cape, during the telecom?"

"No sir, not specifically, just on the telecom as he was one of the people listening to what was being presented."

"Mr. McDonald, was your role in the telecom to be seen and not heard or something else?"

"No. As I indicated, I had requested the telecom because I felt it was an engineering decision that had to be made, I requested engineering to analyze the situation, come back and recommend a temperature we were comfortable to launch with. That was done by the engineers. The recommendation was done by the Vice President of Engineering. I had no reason to enter that telecom, because I agreed with everything that was said."

"Well, now that we have all of this testimony on the table, I am trying to find out how all of a sudden, the four senior managers in Utah pulled the rabbit out of the hat and recommended the launch, when the evidence pointed in exactly the opposite direction. One final question: Once a decision like this was made by what I will refer to as the 'Gang of Four' in Wasatch, was that decision reviewed by anybody further up in the Morton Thiokol chain of command, either Mr. Garrison or someone else?"

"Not to my knowledge," Kilminster answered. "That would have been left up to my superiors to carry that forward."

"Well, Mr. Chairman," Sensenbrenner concluded, "I would just like to make the observation, Harry Truman had a sign on his desk in the Oval Office saying 'The buck stops here.' Obviously, no one at Morton Thiokol had that sign. It seems to me that, when we are dealing with legislation to try to prevent this tragedy from happening again, we're going to have to be insistent there be a well-defined chain of command, because it is clear that no well-defined chain of command existed either at Morton Thiokol or within NASA. It was an accumulation of those factors that resulted in the tragedy."

Part of the morning for me was spent explaining Thiokol's interpretation of the design specification requirements for the SRB, a subject that Congressman and Shuttle veteran Bill Nelson correctly thought to be a pivotal one.

Every element of the Shuttle went through a preliminary design review, a critical design review, and a design certification review; in all of them, I explained, NASA was the principal "participator" along with the element subcontractor, but NASA would also bring in outside people to make sure that the element was qualified to the specifications it was supposed to be.

The final review prior to the very first Shuttle launch had occurred in April 1980 and produced the "Space Shuttle Verification Certification Propulsion Committee" headed up by General Thomas W. Morgan and Walt Williams, a longtime flight test engineer and NASA veteran who was serving as the Chief Engineer at NASA headquarters at the time. This panel, in arriving at its findings, had compared everything that had been done to get the Shuttle in its entirety ready for qualification. One of the major concerns expressed back in April 1980 was that there had been no tests at the temperature extremes of 40° to 90°. Specifically about the solid rocket motor, the committee's report said, "The SRM verification program does not include any full-scale firings or any instrumented storage tests at environmental extremes. These extremes would include short-term horizontal storage conditions at Utah and in shipment, long-term storage at 32° Fahrenheit to 100° Fahrenheit or firing at 40° Fahrenheit and 90° Fahrenheit."

This was not a document that *modified* the Shuttle's design requirements; it was a document that was written based on the data that was available in 1980 to determine whether Morton Thiokol and the Shuttle's other contractors were meeting the design specifications as NASA had interpreted them. For the members of Congress who still might not have understood my meaning, I emphasized that this meant, in other words, that everything that MTI had done in terms of our design approach on the SRM had to be preapproved by NASA.

I also updated Congress on our progress to date on the SRM redesign. I explained that, as head of the redesign task force, we were in the process of establishing criteria for various concepts whose purpose was to make sure that we had full redundancy for the full duration of the flight, no matter what type of seal we ended up going with. As best as we could, we were also going to eliminate any kind of rotation that could ever result from our joint design. As to seal types, we were looking at an improved O-ring type of redesign as well as some that did not involve O-rings.

Unfortunately for Joe Kilminster, Joe's inability to nuance his statements properly brought down the wrath of Congressman Michael A. Andrews (D-Tex.). When Joe declared, in response to a question from Andrews, that there had been no "failure of common sense" in what caused the *Challenger* accident, only "the result of different people looking at the same data" drawing "somewhat different conclusions," Andrews would have none of it: "Frankly I am just appalled that we have you sitting here today testifying before this committee that Thiokol management in that teleconference used good common sense, good judgment. . . . As we as a committee try to go about designing better public policy, we need some harder answers than that. We've got to avoid the kind of bad discretion and bad judgment that Thiokol used in that very, very critical meeting."

New Jersey's Robert D. Torricelli (who later in his career as an incumbent U.S. senator would face serious ethics troubles stemming from illegal 1996 campaign donations and questionable gifts) chastened my former boss with the prospects of lawsuits: "You find yourself in an interesting position, Mr. Kilminster, as one who invariably in the future could be involved in a civil litigation or worse. As I understand your testimony this morning before this committee, and I assume in the legal process that may follow, you are going to suggest to a court that you were engaged in a conversation and a decision to launch a $4 billion vehicle with seven lives aboard, and although one of your engineers was on the site and on the phone with you, you never really questioned his contrary judgment. You did so without real working knowledge of the conditions at the site, and without ever having adequately tested the equipment you sold to NASA. . . . I have seen the makings of negligence cases before, but rarely one that was so strong. You are going to find yourself in a unique position, if you and your corporation are going to claim that negligence has not been committed in this case."

Torricelli then asked Ed Garrison if it was the policy of his company to sell equipment to the U.S. government that had been certified to have met standards set forth in the bidding process, but which in reality had never adequately been tested to the limits of their specifications. "Could that be said to be the case with any other equipment that you have sold to the Government, Mr. Garrison, or shall we assume that this is a unique exception and that everything else you sold to the Government has, in fact, been tested to the limits of its specifications?"

"Mr. Congressman, I explained earlier that I do not believe we did that," Garrison pleaded. "I will repeat again, we interpreted the specifications. We submitted our plans, which were agreed to and approved. I feel we complied with the contract. You have to understand—and there may be some misconception in this room—that we take a contract and just go away and do our thing and deliver a product. That is not true, because everything we do is under scrutiny and approval of NASA."

Torricelli conceded that "whatever failures took place in the corporation," they were "done under the nose of NASA" and, "in fact, with their approval." But it was difficult for any layperson familiar with the evidence to come to the conclusion that the SRM had been tested to the limits of the specifications. "We believe we did that," Garrison countered.

As the exchange continued, anyone with an ounce of common sense had to feel exasperated by the disconnect between Congressman Torricelli's questions and Garrison's answers; it was almost like two ships passing in the night. Myself, I couldn't say which of the two passages bothered me more.

"Let me suggest to you then, you today, Mr. Garrison, sitting there.... Would you suggest that NASA fly these motors at 31°?"

"Absolutely not."

"But you tested them to 31°?"

"I think the record shows what the qualification program was."

"Again, I don't come to this as a scientist. I come to this as a layman."

"Our specification said the operating limit of the rocket motor was 40° to 90°, sir."

"And that is, you felt the limit of your obligations to NASA, that it be operational at 40°?"

"No. I was just commenting on what the specification requirements were. I thought you were referring to that."

"I see. But your view is that the limits of your responsibility to NASA, that it be operational at 40°?"

"That was the requirement of our contract."

"Would you fly the Shuttle today at 40°?"

"I would not fly at all until we've gone through a redesign of the joint."

"But yet you are telling me that you adequately, as required under the contract, tested at 40°. You have done no test since. You had those tests, but you now conclude that that was not the proper conclusion?"

"My statement was that I feel very strongly that we met all the contractual requirements."

"I think the point is sufficiently made."

Torricelli wouldn't let go. He told Garrison that according to Secretary of Defense Caspar Weinberger, the rockets and all of the related equipment that were vital to America's missile defense had been tested down to -60°. The congressman wanted to know why the standards for Shuttle technology would be so different.

Garrison had no idea, so I intervened: "I think I can address that, since I was the chairman of the Solid Rocket Technical Committee for the American Institute of Aeronautics and Astronautics. First of all, Mr. Weinberger doesn't have anything in his arsenal that has joints in it. The only other system that does is the Titan, which serves as a space launch system, but it is not a tactical weapon. So it is very unique, this . . . "

"So in military systems there are no joints?" Torricelli asked, incredulously.

"There is no *field-joint* in any military system. There are joints, but not field-joints."

"Because they separate at the joint?"

"No, no. We have joints in every rocket motor we build. We have to attach things like igniters and nozzles. They are different types of joints, and they do not deflect like the Shuttle's because they are not as big and not as long. The Shuttle's joint is unique to this type of an application, because there is a tremendous structure involved."

"Thank you, Mr. McDonald. That is helpful."

Members of the House committee wanted to know where the reference to a 40° threshold for the SRM had come from. It appeared to them to be a number that NASA had agreed to but that somehow floated in out of the thin air. I explained that was not the case. The number came from the specifications for the propellant mean bulk temperature, which was to operate between 40° and 90°. It had been in our specs from the start.

"So how had that specification seemingly been dropped to 31°?"

"That specification," I explained, and it was a distinction hard perhaps for the layperson to understand, "was a Johnson Space Center specification that said that all elements of the Shuttle 'must withstand 31° to 99°.' That JSC spec did *not* say all the elements of the Shuttle had to *operate* under those temperatures, only *withstand* them. Different elements of the Shuttle encountered certain environments it would have to withstand and other environments in which it would have to operate fully. They had to withstand storage over a wider temperature range than they operated over. They had to withstand sitting on the launchpad for some undetermined period under some extreme environments. They had to withstand influences from other parts of the vehicle. This was where the confusion came in earlier: people confused the environment in which the SRM actually operated and what it only had to withstand. But our technical documents, in my opinion, made the distinction very clear. They said 'withstand' and that was probably where the misinterpretation came from. There was no design specification saying the SRM should operate at 31°, which meant, with an ambient temperature of 36° at the Cape at the time of launch, we were 4° colder than the lowest temperature for which the SRM was qualified to *operate*.

"That was my argument with Mulloy and Reinartz after the teleconference, why I didn't understand why they could accept the recommendation, because it was my interpretation that the recommendation was outside of what everybody thought the motor was qualified to. Mulloy's insistence that the data was ambiguous and that the SRM could be exposed to much colder temperatures than 40° as long as the propellant mean bulk temperature dropped no lower than that, I thought that was asinine. It was asinine because the propellant in the motor was so large and acted as such a massive insulator that it could be exposed to -100° for several hours and only change the propellant mean bulk temperature by a few degrees! I knew the spec didn't really mean for us to do that. That was a comment I had made to him that night."

In our specs, we had interpreted the PMBT limits to be the specific operating temperature limits. We had received a withstanding temperature of 31° and a propellant mean bulk temperature of 40°, but nothing more clear. We didn't think we needed it. We thought we understood what the specs meant. Some people at NASA thought they meant something different. That's why I told the congressmen that it was a lousy set of specs.

Congressman Bill Nelson expressed concern over his own launch, which had occurred just before *Challenger* and had been delayed by several scrubs due to cold weather. "Mr. Chairman," Nelson said, "I could not help but reminisce as we've heard this whole conversation. . . . We have heard these temperatures, 40° to 90°, and we've heard many times a reference to 53° and about McDonald's concern about not launching below 53°, because that was the coldest temperature at any previous launch and he knew there had been severe O-ring degradation on that launch. So my question is, which I couldn't help but reminisce, my STS-61C mission that finally launched on January 12, 1986, we were scrubbed four times and, during several of those scrubs, the temperatures were less than 53°. My question is, did any of these same concerns with the temperature come up in discussions during the final checks before those attempted launches? I know the air temperature for the first scrub of 61-C, on December 19, 1985, was in the low forties."

"I am not aware that they had, Congressman," I answered. "I don't know. I wasn't at that launch, but I don't recall that it came up."

I don't want to make it sound as if the congressmen never had me on their hook. Even though I had explained myself fully to the Rogers Commission—or at least I thought I had—Congressman Don Ritter (R-Pa.) repeated the charge that perhaps I had been part of the problem that caused the *Challenger* accident: "Mr. McDonald, didn't you, during the course of the caucus, mention to Kilminister that he consider a point that the second-ary O-ring was in the proper position to seal if blowby of the primary O-ring occurred? Didn't that act in support of what they were saying? Weren't you wearing two hats at that time?"

"No, though they seem to have interpreted it that way. It was a point I made at the end of the teleconference when the decision was made to cau-cus to reconsider. At that point I mentioned that 'if we are going to make a recommendation that is anything other than 53°, then I think it is important to look at the secondary seal as well, because even though cold temperature is bad for both seals, it was not as bad for the secondary.'"

"Weren't you saying, as Hardy testified to the Presidential Commission, that there was pretty good redundancy? In that sense hadn't you supported the idea that the secondary O-ring would seal? Isn't that correct?"

"No, I didn't support the idea. What I said was that it was something important to look at, if we were going to recommend launching at some temperature lower than 53°. I said that at the time, and I still feel it is true, because it was very clear that the recommendation that we'd made for 53° was totally within our experience base from flights and that if we were go-ing to go outside that experience base then we needed to understand, to run some calculations as to how the sealing was going to be affected at lower temperatures. We had to consider both seals, because if we concluded that the cold temperatures were not going to affect anything in the first 170 mil-liseconds, so that we didn't get out of that regime that Boisjoly was talking about, it was probably all right to go colder. But if it did affect that timing in those first 170 milliseconds, and went beyond that, then we couldn't count on the secondary seal. We could count on the first 170 milliseconds."

"Mr. McDonald," asked Congressman Ritter, "with your high-level posi-tion, though, weren't you in the management loop—and not just in a techni-cal sense. When Reinartz said, 'Are there any further comments?' how come you didn't jump up and down and say no?"

"Well, I had made some notes of what Joe Kilminster had explained over the network as to what our rationale was to launch; I had jotted those down because the five-minute caucus—what was supposed to be a five-minute caucus—turned into a half-hour. I had no reason to believe that that half-hour wasn't taken up by running calculations to determine whether we could indeed launch at a colder temperature outside our experience. Only when things got to the bottom line, with no particular temperature recommended, and that we were to proceed to launch, that I realized what was happening. It took me by surprise. As I have testified, I was convinced that it was the pressure from NASA that caused the whole caucus in the first place. I knew that, and I knew that Mulloy had challenged our recommendations. Sure, he always challenged everything, but his challenges had always been on the other side, where, when we tried to say it was OK to do something, he always said, 'I don't think you can convince me.' Now all of a sudden, we had this list of three reasons why Thiokol said we ought to go ahead and launch, when in fact there were a lot more reasons not to. Mulloy didn't challenge those three reasons like he should have, I thought. At that point I knew that if I was going to have any impact on this thing, I was going to have to deal with Mulloy and Reinartz, because I felt that my management, if I talked to them, would say, 'Well, have you talked to Mulloy and Reinartz about that?' Furthermore, the last bit of telephone conversation was very brief. Joe Kilminster read through the three reasons. When he was told to put it in writing and sign it, I commented to Mulloy, because I thought I would have to sign it as our man at the Cape, that I would not sign it; it would have to come from the plant. I just couldn't live with that recommendation and immediately told them so. The rationale that had been presented did not convince me that it was all right to proceed. I didn't understand how NASA could accept that recommendation."

"So you did protest to Reinartz?"

"Oh, I did, very, very vigorously, but he and Mulloy were not then on the network. That came after everybody went off of the network. I protested for some time to both Reinartz and Mulloy."

At the end of that day, the congressional hearings for Morton Thiokol testimony were effectively over. The House committee did conduct hearings later the next month (July) with other members of NASA management. These final hearings resulted in some contradictory testimony from Jess Moore, NASA's Associate Administrator for Space Flight, and his subordinate, Mike Weeks, concerning the August 19, 1985, briefing I had given at NASA headquarters on the SRM's O-ring erosion problems. In a briefing document that Weeks had given to Moore on my presentation, a sentence read, "Efforts need to continue at an accelerated pace to eliminate SRM seal erosion." In that same document, Weeks's notes indicated that this needed be done "before another launch."

Moore tried to dodge the issue: "On August 19, I had planned to attend that briefing, but the morning that briefing was given I, in fact, was working with our Center Directors on an engine problem. I did not attend that briefing."

"That is not an issue, Mr. Moore," declared Congressman Scheuer.

"I did not get the report from Mr. Weeks. I did not read their report that was given to NASA headquarters on August 19. So if that was said, that was not true."

"Let me read you Weeks's testimony, given before this committee in this very room: 'I did brief Mr. Moore the evening of August 19. I told him about the briefing and showed him the briefing documents. As we left that evening, I told him I was still not quite satisfied and I wanted to call somebody that I had great trust in.'"

"If that was the situation, Congressman, that certainly is not commensurate with my recollection of the situation. Weeks did say he had attended the briefing, that he had reviewed the situation with Morton Thiokol and the Marshall people. But his judgment at that time was that it was OK to proceed in terms of our course of action. That is the extent of my knowledge of the August 19 briefing. The first time I saw the briefing documents was actually after the *Challenger* accident on January 28."

"Well, now, the Rogers Commission has concluded in its report—and I'm quoting—'The O-ring erosion history presented to Level I at NASA headquarters in August 1985 was sufficiently detailed to require corrective action prior to the next flight.' Do you agree with the Rogers conclusion that headquarters should have corrected the problem prior to the next flight?"

"In hindsight and looking back at the August 19 briefing and knowing what we know today, I would say, yes, we should have corrected all the O-ring problems before the next flight. Yes sir, I would agree with you."

"Did Weeks tell you that Thiokol was calling for accelerated pace to eliminate SRM seal erosion?"

"To my recollection, no."

"There seems to be a very serious conflict in recollection."

"I understand."

As this final testimony was being given, I was in an airplane returning home from Washington. While in the air, I wrote a series of thank-you notes to various members of the Presidential Commission who had supported me. The next day I mailed them to Chairman William Rogers, Vice-Chairman Neil Armstrong, Dr. Sally Ride, and General Don Kutyna.

A short time later I received two autographed photos from Sally Ride and Neil Armstrong. Sally thanked me for the help I had provided the Commission, and Neil sent me a picture he had taken of Buzz Aldrin standing next to the flag on the Moon. Previously, at the end of the Presidential Commission's hearings, I had asked Neil to autograph one of my business cards for each of my children, and he had graciously done so. Later I found out that Neil had decided not to sign autographs for anyone anymore. It had become too much of a burden, and especially bothersome when he found so many of his autographs showing up on the Internet for sale.

It was nice to have the autographs of these great Americans. But I would have given them up in an instant in exchange for not having had to have gone through, for the country not having had to have gone through, the calamity and trauma that was the *Challenger* accident.

The release of the Presidential Commission's report to the public closely followed by the televised congressional hearings led to a plethora of newspaper articles and cartoons depicting NASA's decision-making and launch communications problems.

At its annual company awards dinner, Morton Thiokol senior management recognized Roger Boisjoly and me for blowing the whistle on the O-ring problems while it continued to defend the position that we were never demoted because we had not incurred any reduction in pay or grade level. (Cartoon by Bob Englehart, 1986, *Hartford Courant*, reprinted with permission.)

"He Wasn't Demoted — Only Reassigned"

Morton Thiokol senior management explained that I was not demoted, just reassigned to a new position created for me at the plant. (Cartoon by Blannesy, 1986, *Baltimore Sun*, reprinted with permission of the Copyright Clearance Center.)

THE O-RING

The public's image of who was to blame for the *Challenger* accident was based upon testimony before the Presidential Commission that was often highly contradictory. (Copyright 1986, *Miami Herald*, reprinted with permission.)

Before a televised hearing of the House Committee on Science and Technology on the Rogers Commission report on June 17, 1986, Morton Thiokol personnel swore an oath to tell nothing but the truth: *left to right*, Arnold Thompson, Joe Kilminster, Ed Garrison, Charles Locke, and Allan McDonald. (Photograph by Bernie Boston, 1986, *Los Angeles Times*, reprinted with permission.)

"Ten seconds to launch... are the seals holding?"

Humorists criticized NASA Marshall management for preventing the O-ring concerns of Morton Thiokol engineers from being heard by the NASA Mission Management Team, the group responsible for actually launching *Challenger*. (Cartoon by Bruce Beattie, 1986, Copley News Service, reprinted with permission.)

"Now run along and let management make the final launch decision, er, based, of course, on the reservations you boys in engineering have expressed."

One laughed but wanted to cry at how NASA and Morton Thiokol senior management overrode the engineering concerns that could have prevented the *Challenger* disaster from ever taking place. (Cartoon by Pat Bagley, 1986, *Salt Lake Tribune*, reprinted with permission.)

GREAT MOMENTS IN COMMUNICATION.

It was clear even to cartoonists that communication problems and turf battles led to the ill-fated *Challenger* launch, with NASA Marshall management making a conscious decision not to share Morton Thiokol's concerns about cold O-rings with any of the other members of the NASA Mission Management Team. (Cartoon by Dana Summers, 1986, *Orlando Sentinel*, reprinted with permission.)

PART VI

No Consensus

You can't build a reputation on what you are going to do.

—*Henry Ford*

Excellence can be obtained if you care more than others think is wise, risk more than others think is safe, dream more than others think is practical, and expect more than others think is possible.

—*Author Unknown*

The three great essentials to achieve anything worthwhile are, first, hard work; second, stick-to-itiveness; third, common sense.

—*Thomas Edison*

30

The Bell Tolls for Thee

With the congressional hearings on the *Challenger* accident over, the search for the guilty had finally ended—at least I thought that at the time. But the impact of all the hearings was starting to take its toll. Roger Boisjoly had taken comments from some of the members of the House Subcommittee on Space and Aeronautics very seriously—especially those that questioned why he and I hadn't escalated our concerns about the *Challenger* launch to higher levels. In his profound grief and heightened sense of guilt, Roger started to reflect on, yes, why hadn't he called up the Launch Director at KSC or, for that matter, the president of the United States, to tell him why we should not launch the Shuttle, because in retrospect, he thought, what was there to prevent him from doing just that? Roger sensed that if he had done that, he probably could have stopped the launch and prevented this national tragedy.

Boisjoly's guilt trip was so intense that it started to rub off on me. I certainly could have done the same thing; in fact, it would have been much easier for me, because I was sitting at a console with my headset on in one of the firing rooms of the Launch Control Center and was in direct voice communication with the Launch Director. I hadn't uttered a word on the network during those many minutes and hours of the terminal countdown. I had truly believed that Stan Reinartz had told the other members of the Mission Management Team about our concerns of cold O-rings. I had gotten more than a little upset when Utah Senator Jake Garn made a statement to the press that if I had been so concerned about the launch, I should have called him. It bothered my oldest brother, Arthur, an ex-Marine pilot in World War II and Korea, even more. Arthur called me from his home in Idaho Falls and told me he wanted to call Garn and give him a piece of his mind. My brother took it as an insult to me that I hadn't done enough to stop the launch.

I talked my brother out of making the call.

I did feel some real guilt, but I knew I had to get a good grip on myself if

I was going to be successful in my newly appointed job of leading the redesign effort on the solid rocket motors. I decided to immerse myself totally in my work, so that I would not have time to reflect on the past and what might have been. My new assignment was going to be so overwhelmingly busy that I wouldn't have time to reflect on anything other than getting the job done. Boisjoly was an important member of the redesign team, and I told him to take the same approach: "There's nothing we can do to change what has already happened, Roger, but by working together we can prevent such an accident from ever happening again."

But Roger was an emotional mess; he was really hurting. He couldn't sleep at night and thought that the management and many of his coworkers were very upset with him for being a whistle-blower. "I can relate to that," I told him, "but the company must still have some faith in us or it wouldn't have appointed both of us to the Redesign Task Force." He said he really needed to take some time off, because he was physically and mentally exhausted. "Fine, Roger, you should do just that."

What I didn't know at the time was that Roger was also suffering from ill treatment he was receiving from some of the people in Willard, Utah, the small Mormon community where he lived between Brigham City and Ogden. When my wife, Linda, and I visited Roger and his wife, Roberta, at their home, I was surprised to hear about the treatment he was receiving from some of his fellow members in the local Mormon ward. I had received great moral support from my Mormon neighbors and members of St. James Catholic Church in Pleasant View, but that was clearly not the case with Roger. Roger told us that many of the members of his own church had not only shunned him and his family, but someone had actually placed a dead rabbit or dead cat or something like that in his mailbox. He even claimed that he had several close encounters with cars forcing him off the street as he was taking one of his frequent walks around the neighborhood. What was even more surprising is that these were the same people who had elected Roger to be the mayor of this small community not long before. All this severely added to the mental anguish and guilt that Roger was feeling at work at the time and was the reason he requested some time off to try and recover. His doctor had told him that he was starting to show some signs of post-traumatic stress disorder.

Roger got his time off and then chose not to come back to work. After two weeks, he requested extended medical leave, and it was granted.

I often reflected on why Roger had a much more difficult time than I did coming to grips with the accident. I concluded it was because he had been put into a much more difficult position on the night before the launch than I was. Roger had been very instrumental in assembling and interpreting the engineering data used to support the original no-launch recommendation. He had continued to support that position during the infamous "five-minute" caucus with Thiokol senior management. Roger knew very well that there was no new engineering data to be developed during the caucus—and certainly none justifying changing the recommendation. He had tried his best to convince management that it was too risky to approve launching the Shuttle below 53°, based upon his personal observations of the STS-51C hardware a year earlier. His arguments not only had been totally ignored, but he could tell that the senior management was very irritated with his remarks, reflecting what they believed was his uncompromising position. Roger had known at the teleconference that his job could be at stake when Jerry Mason asked Bob Lund, Roger's ultimate boss, to take off his engineering hat and put on his management hat. Lund had supported MTI Engineering in its original recommendation, but after Mason's remark he changed his mind. At that decisive moment, Roger had sat down and said no more. For the rest of his life, it was a moment that he would want to do over.

Suffering perhaps from some element of post-traumatic stress disorder, after all the hearings were over, Roger reflected and reflected and reflected on what he had done and hadn't done. He knew that the launch decision had been changed because of pressure from NASA management during the telecom. He knew that Jerry Mason had decided to accept the risk because no one could prove that it wouldn't work and because it was obvious that NASA didn't like the no-launch recommendation. In his agony, Roger thought he should have escalated his concerns after that, because he knew the recommendation to launch was too risky and that there was no new engineering data warranting a change. He further knew that the recommendation changed from an engineering recommendation to a purely management decision.

Roger piled more and more guilt on himself because he believed he had failed not once, but twice. He had failed to convince management that the original no-launch recommendation was correct, and he had failed by not informing someone higher in the launch decision chain about this whole situation.

I could sympathize with Roger on his feeling of failure because I, too, had failed to convince Mulloy and Reinartz to delay the launch until warmer

weather. Unlike Roger, however, I had not been aware that the launch recommendation had been changed by senior management in Utah. I had
heard Bob Lund present the original no-launch recommendation and my
boss, Joe Kilminster, support it. Since Kilminster presented the final launch
recommendation after the caucus, I had reason—as did the NASA management who participated in the telecom—to presume that Joe was presenting
the final engineering recommendation. In my mind, the reason the caucus took so long was because the Morton Thiokol engineering people had
either conducted some additional analyses or found some additional data
to convince themselves that their original recommendation was indeed in
error. I didn't know that our senior management had turned this decision
around; in fact, I didn't know senior management was even present at the
caucus, because they never identified themselves on the network, and no
one acknowledged they were present.

I also felt absolutely sure at the time that both Mulloy and Reinartz had
escalated this whole issue to Dr. William Lucas at Marshall and to the key
decision makers on the NASA Mission Management Team; after all, Reinartz himself was a member of the MMT and was sitting at a console right
next to all of the key people for at least three hours before the launch. Mulloy
was in the same No. 1 firing room with these people, while I was located in
the adjacent No. 2, a separate area provided for supporting engineers and
contractor management where we could monitor all the data on a computer
to support the launch team. I had assumed that all these people had been
aware of the telecom and the change in Thiokol's recommendation and that,
after reviewing all the data, the MMT had decided that it was an acceptable
risk to proceed with the launch as scheduled.

These assumptions on my part, which later turned out to be totally erroneous, had made it much easier for me to rationalize why it was not necessary for me to escalate my concerns any farther up the launch chain.

Having heard Mulloy and Reinartz talking via telecom to Arnie Aldrich the evening before the launch after the launch recommendation was
changed, I had been confident that the three of them had discussed my
concerns—another erroneous assumption. I further thought that the only
reason that the MSFC management had insisted on the launch recommendation being signed by someone from Thiokol was because they felt very
nervous about the situation and wanted to make sure they were covering
their asses. Because they hadn't asked for a signature on the original recommendation not to launch, I was sure they had shared their uneasy position

with as many people as they could so that if anything did happen, there would be plenty of people around to share the blame.

As it turned out, I was wrong again, because I was naive about the poor relationship and lack of communication between MSFC and JSC and between MSFC and NASA headquarters. I hadn't realized that the turf battles between these various NASA groups were so severe.

In spite of all my rationalizations, I often pondered what I would have done if I had known that both Mulloy and Reinartz intentionally kept the other members of the MMT in the dark about this whole matter. I felt very confident that I would have called Jess Moore and Arnie Aldrich and told them about the situation. Anyone who knows me knows that I am not a bit bashful about stepping into the ring to do battle with anyone, especially when I feel I'm right; I'll always be armed to the teeth to support my position.

As bad as I felt about not doing more to prevent the launch, I knew that Roger Boisjoly had been in a much tougher position, which caused him to feel much more guilt than I felt.

It wasn't long after Boisjoly's departure from Morton Thiokol that another MTI employee who had been against the launch began to suffer from severe depression. Bob Ebeling, Program Manager for Ignition Systems and Final Assembly, had worked for me and was present in Utah during the January 27 teleconference; in fact, it was Bob who initially contacted me at KSC about the problem. He, too, never changed his mind against the original no-launch recommendation. Bob's depression got so bad that he suffered a number of nasty side effects. He saw a psychiatrist and requested a medical leave of absence. He, too, never returned to work, taking early retirement.

Somehow I managed to keep going. I did it by not looking back and totally immersing myself in the task of redesigning the SRMs.

As I had anticipated, the task was enormous. I spent very long hours and weekends at the plant. I got a lot of good help from the outside. The Presidential Commission had recommended that a committee from the National Research Council (NRC) of the National Academy of Engineering and National Academy of Sciences provide independent oversight to Morton Thiokol and to NASA during the redesign. Also, NASA Marshall formed its own oversight committee. As Director of the SRM Verification Task Force, I was not only the leader of the redesign team, but I was also the chief spokesman for Morton Thiokol with the press and the primary interface with both of these oversight committees.

The two oversight groups were made up of very knowledgeable and impressive individuals. On the NRC committee were Chairman Dr. H. Guyford Stever, former White House Science and Technology Advisor and Director of the National Science Foundation; Vice-Chairman Dr. James W. Mar, the Jerome C. Hunsaker Professor of Aerospace Education, Department of Aeronautics and Astronautics at MIT; Lawrence Adams, retired President of Martin Marietta; Dr. David Altman, retired Vice President of the Chemical Systems Division of United Technologies Corporation; Robert C. Anderson, retired Vice President of Thompson Ramo Woolridge (TRW) Inc.; Dr. Jack L. Blumenthal, Chief Engineer of TRW's Space and Technology Group; Dr. Robert C. Forney, Executive Vice President and member of the board of directors of the DuPont Company; Dr. Alan N. Gent, professor of Polymer Physics at the University of Akron; Dean K. Hanink, retired Manager of Engineering Operations for Detroit Diesel Allison Division of General Motors; Edward W. Price, Regents Professor of Aerospace Engineering at the Georgia Institute of Technology; Robert W. Watt, retired physicist from Stanford Linear Accelerator Group; and Dr. Myron F. Uman, member of the professional staff of the National Research Council.

Heading the NASA Marshall Oversight Committee was Al Norton of Martin Marietta. Some of his fellow panel members included Sam Tennent, of the Aerospace Corporation; the already legendary Max Faget, of Space Industries Inc.; A. O. Neal, of McDonnell Douglas; Dominick Sanchini, of Rocketdyne; John Young, NASA *Gemini*, *Apollo*, and Shuttle astronaut; Mike Card, of NASA Langley Research Center; Henry Pohl, of NASA Johnson Space Center; Dave Winterhalter, of NASA headquarters; Horace Lamberth, of NASA Kennedy Space Center; Charles H. Feltz, a former Chief Engineer for North American called out of retirement; and Len Harris, also from NASA headquarters.

The credentials of the members of both panels were so impeccable, in fact, that some members of the redesign team were a little intimidated. I wasn't intimidated, but I did have some serious reservations as to how helpful these impressive people could actually be. It didn't take long for me to realize that they could be—and were—very helpful. Not only did they ask many good questions, they also became very important in supporting many of the critical—and, in some cases, very controversial—decisions that had to be made. The NRC would interface with the press on these issues, really helping Morton Thiokol and NASA, both of which had lost a considerable amount of credibility with the media.

As helpful as the two oversight committees were, I wouldn't be telling the truth if I didn't say that these two groups hardly made my life easier. The reviews that were held with these bodies took a lot of preparation and time that I did not have, but I considered it a necessary evil. I started out conducting monthly reviews with both the NRC panel and NASA oversight committee. With the NRC, we sometimes met even more frequently than that. Some of the reviews were held at our plant in Utah, but most were held in Washington, D.C., with the NRC, or in Huntsville, with NASA. These frequent trips, along with the numerous technical interchange meetings and design reviews at MSFC and visits to our major vendors, kept me away from home again for much of the time.

At our first meeting with the NRC in Utah in July 1986, I was asked to present what I thought had caused the *Challenger* accident and to express any disagreements I had with the conclusions of the Rogers Commission. I presented a more polished version of the presentation I had given to the NASA Failure Analysis Team just one week after the accident. I told the NRC group that my only disagreement with the Presidential Commission involved some needed corrections to the IR temperature data that had been measured on the morning of the launch. I repeated that I believed these measured temperatures were real. Something we didn't understand had caused this controversial data, and we needed to investigate what it was. I'm not a strong believer in coincidences, and it was more than a coincidence that the only SRM joint of six that failed was the one that was recorded to be much colder than all the rest.

I had one other disagreement with the Commission report, but I chose not to discuss it with the NRC because it was really not relevant to the cause of the accident or the redesign. It centered on a graph in the report showing the number of incidents of O-ring distress in the field-joints as a function of temperature. This graph, which would appear in several technical magazine articles and books in the years after the *Challenger* accident, claimed that if Thiokol engineers had recognized the correlation and presented it in a similar manner, that NASA might have accepted the original no-launch recommendation. I knew the correlation was scientifically unsound. It was unsound because it included data from both erosion and blowby of O-rings and motors *that had no blowholes in the putty.*

I knew there was no physics or thermo-chemical-based relationship between O-ring erosion and temperature, but rather a strong relationship between O-ring blowby and temperature. O-ring erosion resulted from a hot

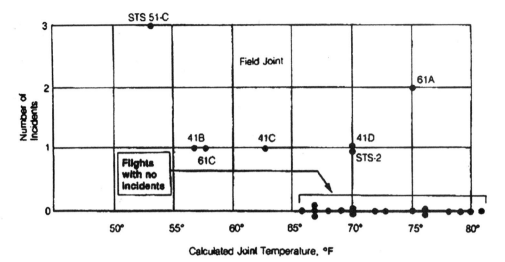

Misleading graph showing the correlation of the number of field-joint O-ring incidents on Shuttle flights versus temperature. (*Report of the Presidential Commission on the Space Shuttle Challenger Accident*, Vol. 1, June 6, 1986.)

gas jet impinging directly on the O-ring. This hot gas jet occurred as a result of a blowhole or tunnel in the zinc chromate vacuum putty ahead of the primary O-ring that was formed during the joint assembly process. The hot gas flowed through the blowhole or tunnel in the putty during ignition of the rocket motor and impinged on the primary O-ring in a process that was totally independent of temperature. O-ring blowby was a function of the ability of the O-ring to move across the O-ring groove, extrude into the seal gap, and maintain contact with both sides of the metal parts in the joint as the motor was pressurized. The ability of the O-ring to properly move across the O-ring groove was dependent on the viscosity of the O-ring lubricant. The ability of the O-ring to extrude into the seal gap depended on the hardness of the O-ring material, and the ability of the O-ring to maintain contact with both sides of the joint during pressurization was dependent on the resilience of the O-ring material. The viscosity of the O-ring lubricant (grease) aided in movement of the O-ring across the groove, and this got worse with lower temperatures. The O-ring hardness increased with lower temperatures, making O-ring extrusion in the seal gap more difficult. Most important, the O-ring resilience or spring-back capability became very poor at cold temperatures, making it more difficult to maintain seal contact in the joint as the metal parts started to separate during initial pressurization at ignition.

The data used in the Commission's graph also included all of the joints that showed no evidence of any blowholes in the joint putty, which was the preponderance of the data, as it involved seventeen of the twenty-four Shuttle flights. I was very surprised that this graph appeared in the Commission's report because Richard Feynman had correctly identified the inappropriateness of using this data for drawing *any* conclusions. The physicist had severely chastised Bob Lund, Vice President of Engineering at Thiokol, for using this exact same data as the basis for his rationalizing why O-ring temperature was *inconclusive* for predicting O-ring distress. Feynman had stated to Lund, "*It depends on whether or not you get a blowhole or you don't get a blowhole. So if within a particular flight it happens that all six seals didn't get a blowhole, that's no information.*" Therefore, the flights with no incidents of blowholes should not have been included in the graph. Their presence totally invalidated the correlation with temperature. The graph should have included only the blowby motors; there were only two (STS-51C and 61-A) of these, one at the coldest temperature (53°) and one at one of the warmer temperatures (75°). Ironically, this point would have supported the NASA position for claiming that the data was inconclusive relative to temperature affecting blowby.

The following month, August 1986, the NRC panel held its meeting at Kennedy Space Center in order to examine the recovered hardware from *Challenger* and become more familiar with the SRM hardware and booster assembly operations. The NRC was amazed that this was the first time that I had personally examined the recovered SRM hardware from the aft field-joint of the right-hand solid rocket booster, the one that had failed on *Challenger*. They did not know that I had been removed from the Failure Analysis teams before the hardware had been recovered and that I had been prevented from going to either Marshall or KSC.

I examined these rusty old pieces of metal very meticulously and was firmly convinced that they, indeed, verified the failure scenario I had originally postulated. It was very clear from the failed aft field-joint that was recovered that my focus on the inability of the O-rings to seal the joint at cold temperatures was correct.

We also had an opportunity to inspect the portions of the orbiter that had been recovered. It gave me goose bumps to see what was left of the vehicle. It had been reported that most, if not all, of the crew had still been alive until water impact; this was based on the fact that they had apparently used supplemental oxygen, as they had been launched in a shirt-sleeve environment in a pressurized crew cabin. Both the commander and pilot required activation of their emergency oxygen from one of the crew members behind them, and it had been reported that both were found with their emergency oxygen activated. It was also reported that an audiotape had been recovered from the crew cabin that had not been released to the public at the request of the astronauts' families. The recovery of the Shuttle debris from the ocean floor was one of the largest salvage operations in naval history.

All of the recovered hardware was later deposited into one of the abandoned air force Minuteman missile underground launch silos at Cape Canaveral, with the silo then sealed with a concrete cover. To me, this seemed a very appropriate resting place. My very first job with Thiokol had involved the U.S. Air Force Minuteman missile. I had been responsible for the external insulation design for the first stage of this ICBM. I had been one of the flight test group leaders for the first twenty-one flights of the Minuteman from 1961 to 1963—all but the first three of which were launched from the Minuteman silos at the Cape.

In August 1986, the NASA Inspector General's Office released a report that stated it did not find any evidence that either Roger Boisjoly or I had been demoted by our company even though both of us had been immediately reassigned after our testimony before the Presidential Commission. Neither of us had suffered a reduction in pay grade or salary. The IG accepted Thiokol management's response that it had only moved us out of our old jobs, for our own good, to avoid further conflict with NASA management. The report did scold Thiokol for one irregularity, however, and that was the failure of its corporate offices in Chicago to send on my memo containing my clarification of statements that various officials had made during the public hearings of the Presidential Commission on February 25–27. NASA's Inspector General confirmed that the Presidential Commission never received my memo, eventually finding it sitting in a box in a corner of a closet at corporate headquarters. Yet the IG accepted Thiokol's excuse for not sending this to the Commission. The excuse was that corporate management thought they had included my memo along with other correspondence sent to the Commission in the same time frame.

Personally, I do not believe that this was an *unintentional* error, because my document contained a considerable amount of rebuttal testimony to statements made by several NASA and Morton Thiokol management officials who had testified under oath immediately after my testimony on the morning of February 25. It was more than coincidental that I was sent home by my management and not allowed to attend these hearings even though I requested to do so. I had been told I could respond to the written record when it became available. I did just that a month later, and somehow it just happened to get lost in a box sitting in the corner of the closet of the corporate office's legal department in Chicago, of all places? I had taken a lot of flack from the company for providing several documents directly to Commission members without first going through our established legal channels. It was more than ironic that the *only* document I submitted through the established channels never reached the Presidential Commission.

Reading the IG report, I immediately called Dr. Alton Keel, the Executive Director of the Presidential Commission, and asked him if the Commission had ever received my document. Keel told me that he recently received a copy from the NASA Inspector General's Office. He told me that several of the commissioners had found many of the discrepancies I had referred to, and that those were reflected in the conclusions and recommendations of the Commission's final report, even though they had actually never seen my document.

Although somewhat relieved by this response, I also knew that if Thiokol ever got dragged into court over the *Challenger* accident, that my document, along with my presentation to NASA headquarters on the O-ring seal problems on August 19, 1985, would most likely be a focal point for any litigation.

Even though the IG report was very disappointing to me, I was so totally immersed in my job by now and feeling pretty good about myself since the company had selected me to lead the very important redesign. I knew how important this was to the economic well-being of the company, as the Space Shuttle program represented over one-third of its business base. I knew that it was critically important to be successful in the redesign in order to restore the Shuttle to safe flight as soon as possible, especially so since Thiokol had agreed to accept this challenge at no fee. It was unprecedented in the industry to accept $400 to $500 million of work at no profit. The company had already agreed to allow NASA to withhold $10 million of profit from the work conducted prior to *Challenger* in payment of the failure, without having to admit failure by the company, which would have resulted in this payment anyway in accordance with our contract. It was done to avoid further litigation on the contract, litigation that would have cost even more money and delayed the redesign for months, if not years.

My feeling good about being selected to head the redesign was short-lived, however. In August 1986 I received a telephone call from Congressman Edward Markey, a Democrat from Massachusetts who chaired the House Subcommittee on Energy Conservation and Power, part of the Committee on Energy and Commerce. Congressman Markey asked me if I was truly leading the redesign activity at Thiokol, or was that just a facade being promoted by the company while it actually kept me working on more menial and unimportant tasks. Taken aback, I asked the congressman why he was asking the question—did he know something I didn't know? He told me that he had followed the hearings and that he and several of his colleagues in the House were very disturbed over the treatment that Roger Boisjoly and I had received from our company for telling the truth. Markey wanted to make sure the bad treatment had stopped. I told him I was, indeed, in charge of the redesign, that I had not been demoted, and that Boisjoly was also part of the redesign team. "I'm glad to hear that," he said. "Maybe the legislation I introduced back on May 14 had been effective."

Congressman Markey also asked if I had received a copy of the letter he had sent to Charles Locke, Thiokol's CEO, back in May. I told him I had not. "Then I'll mail you a copy along with a copy of the House Joint Resolution 634, which I introduced into the House for the express purpose of protecting Mr. Boisjoly and yourself from further recriminations." Receiving Markey's letter to Locke, I read:

May 19, 1986
Dear Mr. Locke:

As you may know, on May 14 I introduced a joint resolution with Representative [Douglas] Bereuter (D-Neb.) that I would disqualify your company from future contracts with the National Aeronautics and Space Administration until such time as the Comptroller General affirms either that none of your employees were reassigned because of their testimony before the Presidential Commission investigating the accident of the shuttle *Challenger*, or that those employees have been restored to their former positions of responsibility. A copy of this joint resolution is enclosed for your reference.

In addition, consistent with the intent of this legislation, I am calling upon NASA to cancel a $200 million contract recently awarded to your company and to cease awarding such contracts until the matter of the reassignments is satisfactorily resolved.

These actions reflect my concern about how the treatment of certain Morton Thiokol employees, particularly Mr. McDonald and Mr. Boisjoly, will adversely affect the future of the Space Shuttle program. Any failure to maintain an atmosphere of support for objective scientific inquiry and honest criticism will decrease safety at a time when improved safety is desperately needed.

In Sunday's *Washington Post*, Mr. Ed Garrison, President of your Aerospace Group, suggests that the reassignments were made to provide "a fresh, objective review of the situation." Mr. Garrison also suggests that the reassignments were necessary to reduce friction with certain NASA employees.

I believe it would be helpful to a speedy resolution of this issue for us to meet at the first convenient opportunity. For example, to the extent that these reassignments were considered "necessary" because of friction with NASA, it may be necessary to consider that issue as part of restoring your own employees to their proper position. Please contact Mr. Richard Lehfeldt of my subcommittee staff at the above address and phone number at your earliest convenience.

I look forward to hearing from you.

Sincerely,
Edward J. Markey
Member of Congress

As for H.J. Resolution 634 (May 14, 1986), I knew nothing about it until it came in the mail along with Markey's letter. The resolution stated:

Debarring Morton Thiokol Inc. from contracting and subcontracting with NASA until a determination is made by the Controller General with respect to actions which were allegedly taken by such corporation against its employees because they gave certain information to the Presidential Commission on the Space Shuttle *Challenger* Accident.

Whereas the space program and the safety of all participants in the space program are national priorities;

Whereas the space program has been seriously hampered by major safety issues raised by the recent accident of the space shuttle *Challenger*;

Whereas a Presidential Commission was appointed to investigate the causes of this tragedy;

Whereas the success of the Commission depends on the freedom of individuals to communicate the whole truth, without fear of reprisal by employers who supply parts, service, or technical assistance in connection with the space shuttle program or who participated in the decision to launch the space shuttle *Challenger*;

Whereas the freedom of the witnesses depends on their expectation that revealing facts to the Commission, whether in support of, or damaging to, their employer, will not result in their punishment by demotion, transfer, or loss of job responsibility;

Whereas the punishment of individuals who tell the truth and cooperate with the Commission compromised the future safety of the shuttle program and threatens our capacity to identify and correct potentially fatal flaws in the shuttle program;

Whereas it is critical that those industries performing government contracts protect and encourage such cooperation among their employees;

Whereas published transcripts of proceedings before the Presidential Commission have revealed that Morton Thiokol Inc. has transferred certain of its employees to less responsible or desirable job situations after their cooperation with the Commission;

Whereas it appears that these transfers took place as punishment for the cooperation of these employees with the Presidential Commission in its efforts to identify and correct safety problems with the space shuttle; and

Whereas these actions call into question the ethics and integrity of Morton Thiokol Inc. and send a message to its employees that they should not cooperate with efforts to ensure the safety of the space program: Now, therefore, be it

Resolved by the Senate and House of Representatives of the United States of America in Congress assembled,

That Morton Thiokol Inc. is hereby excluded and disqualified from contracting and subcontracting with the National Aeronautics and Space Administration, and the Administrator of such Administration may not solicit an offer from, award a contract to, extend an existing contract with, or approve the award of a subcontract to Morton Thiokol Inc. unless and until the Comptroller General of the United States affirmatively finds that either—

(1) no employees of Morton Thiokol have been transferred, reassigned, demoted, or discharged, or have had their job responsibilities or staffs reduced, because they gave information to the Presidential Commis-

sion on the Space Shuttle *Challenger* Accident in connection with such Commission's investigation of the accident which occurred on January 28, 1986; or

(2) any employees of Morton Thiokol who were in fact so transferred, reassigned, demoted, or discharged, or whose job responsibilities or staffs were in fact so reduced, have been fully restored to the positions, responsibilities, and staffs which they held or administered immediately prior to such accident.

A few days later I received a copy of a letter addressed to NASA Administrator James Fletcher. Sent to me by Senator Thomas Eagleton (D-Mo.) and signed by twenty-seven other senators, the letter, dated May 13, read:

Dear Dr. Fletcher:

We are greatly disturbed by testimony of May 2 to the Rogers Commission that strongly suggests that Morton Thiokol, Inc., and at least one NASA official, have attempted to control the flow of information to the Commission through acts of intimidation and punishment.

In evaluating that testimony, it appears that a Morton Thiokol employee, Allan J. McDonald, has been punished by his company for telling the Commission last February that he argued against launching the *Challenger*. Although company officials deny that McDonald was punished, they do admit that after his appearance last February he was stripped of his former position as Director of the Solid Rocket Booster Program and "transferred" to "special projects."

McDonald also testified that after his February appearance he was confronted by Lawrence B. Mulloy, then head of NASA's shuttle booster rocket program. Mulloy reportedly questioned McDonald's motivation for testifying and suggested that he clear future statements with Morton Thiokol and NASA.

Another Thiokol engineer, Roger M. Boisjoly, indicated on May 2 that Thiokol has taken similar action against him for his testimony earlier this year regarding his warnings about shuttle launchings.

We agree with Chairman Rogers, who charged that Thiokol's actions against these employees seem to be "in the nature of punishment." Indeed, they appear to have been punished for telling the truth and doing what is right.

We cannot hope to have a complete understanding of the shuttle disaster if those with pertinent information can only expect retribution if they come forward with it. We urge you to fully investigate the charges of intimidation and retaliation by Morton Thiokol and the reports of intimidation by NASA management. Should these charges be true, we believe you should reevaluate your agency's relationship with Morton Thiokol. At this fragile moment in its existence, NASA must demand openness and honesty, not duplicity and vindictiveness, from itself and its contractors.

We would like to know your plans for investigating these reports, and what follow-up actions you may pursue, should they prove accurate.

Signing the letter to Fletcher along with Eagleton were some of the Senate's most esteemed members, including Dale Bumpers (D-Ark.), Lawton Chiles (D-Fla.), Alan Cranston (D-Calif.), Christopher Dodd (D-Conn.), Albert Gore Jr. (D-Tenn.), Tom Harkin (D-Iowa), George Mitchell (D-Maine), Sam Nunn (D-Ga.), and Jay Rockefeller (D-W.Va.).

Knowing these letters existed, dated as they were from May 13 to May 19, 1986, which was before I was selected by the company to be the Director of the SRM Verification Task Force, I really began to question in my own mind whether I had been given my important new assignment because my bosses thought I was the most qualified to lead the effort or because they had been responding to these letters. I was naive and believed, because I wanted to believe, that I really earned this difficult job, but in my heart I now knew that MTI senior management only gave it to me because of these threatening letters. The letter from Markey, in association with H.J. Resolution 634, had to be very intimidating to Thiokol senior management in that it threatened to prevent the company from receiving any NASA contracts now or in the future if either Boisjoly or myself was demoted or suffered any loss of salary or position as a result of our testimony before the Presidential Commission.

It was a good thing I didn't know about any of these documents earlier, or I very likely would have declined the job. Certainly I would have strongly suspected that it was being offered to me only to appease powerful members of the U.S. Congress.

By midsummer, however, I had already gotten too involved in the redesign effort to turn back. My immediate boss, Carver Kennedy, my friend and former Vice President of Space Services at KSC, assured me that, as far as he was concerned, I was the best person to take on this almost impos-

THE BELL TOLLS FOR THEE 419

sible task. Carver himself had volunteered to help in the recovery effort, knowing full well that there was going to be little meaningful activity at the Cape for the next couple of years until we were able to start launching the Shuttle again, so agreed to replace my previous boss, Joe Kilminster, as Vice President of Space Booster Programs. Carver accepted this as a temporary assignment until we could ship the first flight set of the redesigned hardware to the Cape, at which time he planned to return to KSC to return to his old position as the "Czar of the VAB."

Carver Kennedy was an excellent man for the job. He was a good technical manager, had considerable experience in program management, and was well liked by all the employees and well respected by NASA management and others in the industry. Carver was a good organizer, very logical, and not afraid to make the tough decisions that had to be made to accomplish the enormous task of returning the Shuttle to safe flight. Carver seldom filed anything of importance, and he worked from an unbelievably messy office. He'd stack papers in piles on his desk, table, chair, on top of the bookcases, even on the floor. He received a lot of kidding about his "piling system," but he could always retrieve what he needed in a minute's notice. He knew exactly which pile everything was in and how deep he had to go for any particular item.

Carver was probably the best manager I ever worked for. He allowed his subordinates to make the decisions, but if they were unsure of themselves, he wasn't afraid to do it himself. He had always supported me and could not understand why Thiokol management had treated me the way they had. He was a good buffer between me and corporate management. I enjoyed working for him so much that I soon found myself adopting his "piling system" so that my office began to look like his.

Kennedy and I made a good team, but the job that we had taken on seemed so overwhelming—especially with the horde of NASA people from MSFC, some from other NASA centers, the oversight committees, all of the air force people, as well as several others from academia and industry that NASA invited in to help us. I once mentioned to Carver that we were going to have to learn quickly how to play a "zone defense," because there was no way we could play "man-to-man" with the size of the community we were dealing with, most of whom were already ten plays behind in the playbook; we were going to have to bring them all up to speed. Some of our meetings would not turn out to be pleasant, as there were many controversial issues to deal with. We had to be ready to answer a lot of criticism.

Even before the Presidential Commission's report had been completed, some preliminary redesign activity had gone on; once the report was officially released, the workload increased dramatically.

One of the first things I did was request a briefing from the air force on the conclusions from its Titan 34D launch vehicle accident of April 1986, caused by failure of the solid rocket motor. It was important that the results of this failure analysis be communicated to the Shuttle SRM redesign team because the Titan SRB was very similar in design to the Shuttle motors.

In its typical fashion, the air force had been very closemouthed about its failure investigation, wrapping it in a cloak of "military secrets" so it wouldn't have to address it with the media. However, the air force panel, chaired by Colonel Nate Lindsay, had come to some preliminary conclusions on what had caused that accident, and at the request of NASA, agreed to give us a classified briefing on their findings at our plant in Utah. Colonel Lindsay reported that the accident was the result of an "unbond" in the primary case insulation in the joint area, a defect that allowed hot combustion gas containing aluminum oxide to contact the D6AC steel-case wall. Lindsay went on to say that examination of the failed case hardware that fell back on the launchpad, along with some tests that were run at United Technologies Laboratories in Connecticut, indicated that the failure had been due to the formation of some "eutectic" between the aluminum oxide and the D6AC steel that essentially had dissolved or eaten away the case in that area in a short period of time.

I had been in the solid rocket business for over twenty-five years and could hardly believe this failure scenario. I was even more surprised that the folks from the Aerospace Corporation, Air Force Rocket Propulsion Laboratory, and Chemical Systems Division of United Technologies, who made the rocket motors, apparently bought into this failure mode hook, line, and sinker. I commented to Colonel Lindsay that we didn't believe his failure scenario, because we had been using the same rocket propellants, insulation materials, and case materials ever since the start of the Minuteman program in the late 1950s and had never observed such a phenomenon. I told him that I suspected that the failure had been initiated by a large insulation-to-case "unbond" in the joint area, which in turn led to a case burnthrough because of backside heating of the propellant (or perhaps a failure of the inhibitor or stress relief filler material in that area), resulting in some type of hot gas flow. The manufacturing quality of the insulation, inhibitor, and propellant in the Titan 34D was very poor compared to the quality of manufacturing processes used on the Shuttle. It was routine to have numerous major discrepancies in these areas on the Titan that required repairs, while that was totally unheard of in the Shuttle SRM program.

The colonel didn't agree with us, not even after we told him that we had experienced several instances where we had had hot gas contact with the bare metal on the top of the inner clevis leg because of putty blowholes, but that we had never seen any heat-effect, much less metallurgical changes in the steel in contact with hot aluminum oxide. In fact, aluminum oxide had been plasma-sprayed on steel nozzle parts over twenty years ago in an attempt to make a steel nozzle that would resist erosion to solid propellant combustion gases.

Though the meeting ended with this disagreement on the Titan 34D failure mode, we received a ton of action items from NASA to assess the possible impact of this failure scenario on the Shuttle SRM redesign. Fortunately, it wasn't too much longer after this meeting before the air force woke up and discarded this ridiculous failure mode, concluding that its Titan accident had, indeed, been caused by hot gas flow behind a poorly bonded piece of insulation that may have been further propagated by their interference fit insulation design in the field-joint.

31

Redesign and Resurrection

Redesign activity was now moving very rapidly. My first priority was to duplicate the *Challenger* joint failure to remove all doubt. Joe Pelham, Chief Scientist for Thiokol's Wasatch Operations, came up with an excellent idea to assemble two field-joints in a short stack of full-scale hardware. Dubbed the Joint Environment Simulator (JES), the test article contained two half-length case segments separated by an ET attach section, the forward dome, and a small nozzle that was attached to the aft dome. The cylinders were loaded with an inert propellant to control the initial volume. Sufficient live propellant was bonded into a cartridge in the head-end and in the field-joint interfaces to duplicate exactly the pressure rise rate and hot gas generation distribution that would actually occur at ignition on a launchpad of a full-scale SRM. The motor burned for less than a second. The whole system was triggered via a standard SRM igniter that was attached to the forward dome. Designed and fabricated in record time, the JES-1A and its associated vertical test stand were scheduled for their first checkout in August 1986.

The test went perfectly: the measured ignition transient was just as predicted and nearly identical to a flight system.

The successful JES-1A test paved the way for refurbishment of our test hardware. Our second test, JES-1B, took place in October 1986, with the hardware configured to duplicate the *Challenger* failure. The test motor had two identical field-joints with prefabricated blowholes through the zinc chromate vacuum putty, which assured identical hot gas flow to the O-ring seals in both joints. We kept the upper joint at local ambient conditions, around 60°, while preconditioning the lower joint to 20°, plus or minus 5°. At ignition, our films clearly showed a large puff of dark black smoke coming out of the cold joint with nothing coming out of the joint at the warmer temperature. In fact, the smoke cloud, when superimposed on the *Challenger* launch at ignition, was nearly identical in color, size, and shape. I felt very good about the test because it verified the *Challenger* failure scenario I had always believed in and demonstrated that the JES hardware would be a valuable tool for understanding the joint behavior and verifying improvements in the redesign program.

Verifying the failure mode of *Challenger* was particularly rewarding because considerable doubt had been raised by numerous sidewalk rocket scientists who had written to NASA or Morton Thiokol, and in some cases to various members of Congress and even the president of the United States, claiming that the failure of the O-rings in the SRB had not really been the problem. Some of these folks had examined videotapes of the flight, while others had read the Presidential Commission's report. The failure mechanisms put forward by these pundits ranged from sabotage by an expert marksman shooting at the Shuttle at liftoff, to structural failure of the solid rocket motor case at liftoff, to a structural failure of the struts attaching the SRB to the external tank, to cracks in the solid propellant grain, to failure of the internal insulation in the SRM joint area, to leaks in the ET causing overheating of the SRB joint from a hydrogen fire, to exhaust plume interactions between the SRB and the SSMEs causing reverse flow along the side of the aft segment of the SRB, to failure in the SSMEs, and my personal favorite, a hard left turn of the huge mobile transporter that moved the Shuttle on the mobile launch platform to the launchpad. This last theory had been fully documented and analyzed by Ali Abu Taha, who submitted it to NASA in a lengthy report. Abu Taha claimed that since this had been the first launch off Pad 39B, the mobile transporter carrying the Shuttle stack had to make a left turn to reach this pad, whereas all previous launches, from Pad 39A, had been in direct line to the VAB, requiring no turns during transportation. Abu Taha claimed that the jerky movement of the transporter caused the Shuttle stack to wobble, causing added loads on the joints; then, when the *Challenger* stack turned left to go to Pad 39B, the whole stack leaned over, causing excessive loads on the aft field-joint of the right-hand solid rocket booster. This added load, which NASA had not accounted for in its loads analysis, was sufficient to crack the steel joint in the SRB, which then failed on ignition.

Unfortunately for Abu Taha's analysis, the Shuttle transporter did not move out to the pad at anything like breakneck speed: when this huge iron monster made its turn to go to Pad 39B, its speed down the track was less than one mile per hour. Abu Taha was so convinced that he had diagnosed the problem that NASA invited him to come down to KSC to witness the slow crawl of the transporter for what was to be the next planned Shuttle flight, STS-61G, which had already been stacked in the VAB at the time of the *Challenger* launch. NASA even instrumented the stack to determine the deflections and loads during transportation to the pad.

Needless to say, these loads were minuscule, and Abu Taha, after witnessing this event, withdrew his failure scenario.

But Abu Taha did not give up, publishing another new failure scenario and releasing it to the press. This theory was even harder to disprove, a sure-fire way of persuading the ill-informed press. Still convinced that a structural failure had occurred due to excessive loads introduced into the joint area by the struts during SSME start-up and bend-over of the Shuttle stack prior to SRB ignition, Abu Taha conducted an analysis that purported to show that the actual loads induced by this action were much higher than the design loads calculated by NASA, and that caused the SRM case to fail structurally in the area of the joint. Furthermore, Abu Taha said he had personally examined some of the films from the launch and claimed that he could see pieces coming off the Shuttle that disturbed the plume; he could see what he thought was hot gas leaking from the aft joint area, as evidenced by disturbances in the outer boundaries of the exhaust plume of the right-hand SRB for the full duration of the flight right up until the Shuttle exploded. Abu Taha claimed that the ET attach struts pulled a piece of the case loose in the area of the joint and that this piece had never been found because the salvage ships and submarines only picked up debris in the area of impact rather than closer in to shore, where the failed case section would most likely have been.

Ali Abu Taha's interpretation of a continuous leak was not supported by the flight films or the telemetry data. The joint did leak from 0.638 to 2.5 seconds during liftoff because of an O-ring leak, not because the case was broken by an excessive prelaunch load. There was no evidence of a leak again until T+59 seconds, and within 1 second a fully developed plume was seen coming from the joint area of the right-hand SRB. Within 14 seconds, the motor pressure dropped 20 psia and the ET exploded. There was no way the SRM could have been leaking near 6,000° gas from a broken case for the previous 59 seconds as Abu Taha claimed without *any* drop in pressure, without developing *any* observable plume, or without causing a major structural failure to the SRM or burning through the ET during that time frame. Contrary to Abu Taha's claims, the launch and flight films, the SRM chamber pressure data, the recovered hardware, the postaccident subscale and full-scale testing with joints at low temperatures and with flawed O-rings all supported the conclusions of the Rogers Commission.

Unfortunately, Abu Taha's second failure scenario turned out to be much tougher for us to put to bed; for the past twenty-two years after the accident, it has reappeared in one form or another in several magazine articles, often with some not-so-well-known expert endorsing it or attesting to Ali Abu Taha's academic credentials. In fact, George Washington University in Washington, D.C., offered a course titled "Failure Analysis of Complex Systems" taught by Abu Taha, which used his analysis of the *Challenger* failure as a case study.

The failure theories, outlandish and otherwise, abounded. Mostly what we got were letters from what I call the "know-how-to-fix-the-problems-crowd," for the most part well-meaning people who just wanted to help. But we also got letters from people with something to sell. Whatever the sources, every suggestion NASA or Morton Thiokol got, they transmitted to me for a formal response back to the originator.

I didn't have the time for this, so I delegated the responsibility to some of the fellows working for me. Our legal department also kept track of these ideas to make sure, if we used any of them, that we didn't violate anyone's originality or patent rights that could result in litigation.

One of my favorite suggestions sent in recommended that we encase both of the solid rocket boosters in firebrick, so that when a leak in the field-joint occurred it would not impinge on the external tank and cause a failure in the Shuttle system. Most of the ideas were more practical and involved ways to eliminate joint rotation or provide a better seal at the joint that was not so temperature sensitive. Most of these suggestions would have resulted in serious weight penalties, costs, or an excessive amount of time to test, verify, and implement.

I looked at all of these suggestions as truly a good news/bad news situation. The good news was that it indicated a genuine interest in the American people to support the space program and a desire to return our astronauts to space as soon as possible. The bad news was that someone from our already overworked organization had to respond to each of these suggestions. We were all badly overworked, spending countless hours of overtime during the past six months with no end to the work in sight. We were understaffed especially in engineering. Few people in our entire engineering organization had ever designed anything, and most of those in the engineering organization in Space Operations had really been in a production support role for most, if not all, of their careers.

Most of our engineering design talent had been transferred to our Strategic Operations to support the Navy Trident II (D-5) program, which was just completing development, and the Air Force Small Inter-Continental Ballistic Missile program, which was in the development phase. As a result, Space Operations had to recruit nearly 300 engineers as soon as possible to support the Shuttle redesign; most of these were brand-new college graduates. Training all the new hires during the height of the redesign placed an additional burden on our senior engineering staff.

In order to give immediate support to the program, especially in test planning, systems safety, and supporting engineering analysis, Thiokol hired a number of experienced "job shoppers." NASA was not concerned about how much money we were spending; its criticism was that we didn't have enough people on the redesign and were not spending money fast enough. For each day the Shuttle was out of flight status, it was costing NASA over $10 million—or $4 billion a year.

Through the late summer and early fall of 1986, there seemed to be a never-ending series of meetings at the plant and at MSFC for reviewing the design requirements, preliminary designs, planned test program, test articles, and, last but not least, schedules and projected costs. I was very much involved in the preliminary design and establishing the program that defined the test articles, test conditions (including how many tests should be run), what the pass/fail criteria should be, and what instrumentation should be installed to understand the joint behavior and verify the improved design. Carver Kennedy took on the formidable task of program schedules, vendor coordination, contracting, and cost estimating, and some semblance of cost control for the redesign program. He pretty much left the technical decisions to me while he worked the "programmatics."

As it was going to take at least a year to include a capture-feature on the field-joint to reduce the joint rotation, NASA and the NRC decided to redesign all of the components in the SRM that had shown any evidence from postflight inspection of not meeting the safety factors. It would have taken nearly two years to make all these changes in the case metal parts if we had not started this activity without NASA's concurrence six months before the *Challenger* accident.

This decision to improve all aspects of the SRM design expanded the scope of the redesign program far beyond that of just fixing the field-joint that had failed on *Challenger*. Many elements of the nozzle's metal parts, sealing areas, and ablative sections were redesigned, as were the igniter steel chambers and insulation, with insulation thicknesses increased throughout the SRM. Minor changes were made to the propellant grain design, and several manufacturing processes and adhesives were changed to improve bonding throughout the motor. All in all, there was hardly anything in the redesigned solid rocket motor (RSRM) exactly the same as *Challenger*'s. Safety margins were improved throughout. In addition to the redesign of the SRM, NASA required a safety reassessment of the entire Shuttle system, which resulted in many redesigned areas in other elements of the Shuttle. Of particular importance to us was a change in the shape of the SRB external tank attach ring, from a 270° ring to a full 360° ring, made by United Space Boosters Inc.; USBI also changed the SRB's aluminum aft skirt. The concern for safety brought a number of much-needed improvements in other areas of the Shuttle as well.

As important as it was to have such a strong concern for safety, the degree to which both NASA and Congress overreacted made it almost impossible to lay out a test program for the SRMs that was comprehensive enough to be endorsed by the emerging legions of the new safety community. NASA decided to terminate work on the FWC-SRM and prohibit the newly developed cryogenic Centaur G Prime upper stage from being carried in the cargo bay of the Shuttle because of safety concerns with flammable cryogens. A new foam-core systems tunnel for the SRM was also abandoned. This tunnel would have carried all of the electrical cables for the SRM as well as the destruct system. This tunnel would have been much lighter and reduced assembly time at Kennedy Space Center, and it could have been used on either the FWC or steel-case SRM. It had been already installed on several steel SRM case segments at KSC prior to *Challenger* and was ready for its first flight. These components were all removed and thrown away. The air force reaction was similar. Stunned by the one-two punch of the *Challenger* and Titan 34D failures, it decided to mothball its Shuttle launchpad at Vandenberg after spending $3 billion building it.

The redesign program was controversial enough within NASA and Morton Thiokol that it didn't need to generate any more from the outside. The approach selected for the redesigned field-joint was far from unanimous and had already created some conflict in the technical community and management ranks within NASA and Thiokol. Not wanting to miss anything and cover itself from additional public criticism, NASA had invited all of Thiokol's competitors in the solid rocket industry, as well as several technical organizations that supported NASA contracts, to review all our redesign activities. That proved especially troublesome given the strong difference of technical opinion that existed even within Thiokol engineering itself.

Many of our engineers at Thiokol recommended that the new field-joint design be vented to allow hot gas to pressurize the primary O-ring uniformly at ignition. This, they believed, would allow the O-ring, when pressure-actuated, to extrude properly into the O-ring groove, thereby preventing jets of hot gas from forming and impinging on the O-ring. That was the way most O-rings had been designed to operate in all previous solid rocket motors. All of the other members of the solid rocket community who had been invited to participate in the design review recommended this same approach.

But Roger Boisjoly and I were against it. We preferred to isolate all the hot combustion gases from the O-rings totally. Carver Kennedy agreed with us and supported our design approach even when several folks in Thiokol engineering wrote memos strongly supporting the vented design approach, memos that were endorsed in writing by most of design and project engineering management. Basically, those who disagreed with us were people who were concerned that we could not isolate the primary O-ring from hot gases 100 percent of the time; hot gas impingement would eventually occur.

In more normal circumstances, Roger, Carver, and I would also have supported a vented approach, but under the "new" contractor-end-item specification supplied to us by NASA, *any* heat effect or blowby of the primary O-ring was to be considered a violation of the SRM specification and would most probably cause shutdown of the Shuttle flight program. I told Carver, NASA, and Thiokol engineering that the only mathematical way to guarantee that hot gas would *never* heat-effect or blow-by a primary O-ring was to *never* let hot gas *reach* a primary O-ring. If we selected a vented design where hot gas loaded with aluminum oxide particles was guaranteed every time to pressurize all the primary O-rings, sooner or later there would be some significant heat-effects or blowby of an O-ring. We couldn't take such a chance. We needed to develop a field-joint insulation design that absolutely would not allow hot gas to reach the primary O-ring area.

Our idea was to add a third O-ring into our new capture-feature joint hardware that would stop hot gas from ever reaching the primary O-ring behind it. We designed the insulation in the joint with what was called a "J-joint." With a contact adhesive applied to the mating insulation parts, the

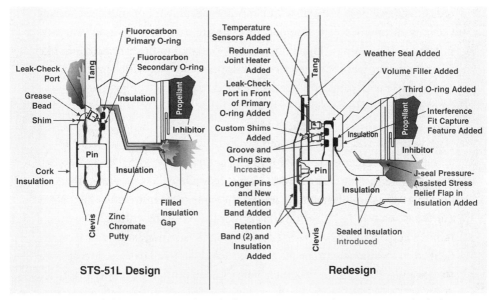

STS-51L Design

- Fluorocarbon Primary O-ring
- Fluorocarbon Secondary O-ring
- Leak-Check Port
- Tang
- Grease Bead
- Insulation
- Shim
- Propellant
- Inhibitor
- Pin
- Cork Insulation
- Insulation
- Clevis
- Zinc Chromate Putty
- Filled Insulation Gap

Redesign

- Temperature Sensors Added
- Redundant Joint Heater Added
- Weather Seal Added
- Volume Filler Added
- Leak-Check Port in Front of Primary O-ring Added
- Third O-ring Added
- Tang
- Custom Shims Added
- Insulation
- Interference Fit Capture Feature Added
- Propellant
- Groove and O-ring Size Increased
- Inhibitor
- Pin
- Longer Pins and New Retention Band Added
- Insulation
- J-seal Pressure-Assisted Stress Relief Flap in Insulation Added
- Retention Band (2) and Insulation Added
- Clevis
- Sealed Insulation Introduced

Redesigned field joint compared to *Challenger*'s. (Courtesy of NASA/Morton Thiokol.)

J-joint insulation would bond during the stacking process. During ignition, the open J-slot (formed in the insulation in the shape of a "J") would be the first to pressurize with hot gas, further increasing the compression load on the insulation, thereby preventing hot gas from entering the O-ring area. By introducing a third O-ring into the field-joint design, we provided two seals ahead of the primary O-ring, tightening with pressure from the motor and also providing a "true" seal redundancy. In addition, we added electrical heaters to keep the O-rings at 90° under all weather conditions. This "belt and suspenders approach" increased the reliability of the field-joint design by several orders of magnitude over the *Challenger* design.

The new field-joint design retained the primary and secondary O-rings in the inner clevis leg like the old *Challenger* design. The new capture-feature on the tang side of the joint restricted the opening of the joint to only about 15 percent of that experienced by the *Challenger* design. This design also provided a capture-feature O-ring with a zero gap that tried to get even tighter with pressure from the motor during ignition, thus being totally insensitive to temperature and requiring zero resiliency in the capture-feature O-ring. Any load that reduced compression on the capture-feature O-ring seal increased the compression on the primary and secondary O-ring seals on the other side of the clevis leg, and vice versa. We considered it a "truly" redundant seal design because our capture-feature and primary/secondary O-rings automatically responded in opposite directions, eliminating any possibility that a single load could unseat the total joint sealing area as like what happened with *Challenger*. Because the *Challenger* design had adjacent bore-seal O-rings, any load that tried to prevent the primary O-ring from sealing would do the same thing to the secondary O-ring, not allowing "true" redundancy to be maintained.

A very intensive O-ring testing program was conducted in order to establish manufacturing specifications and assure that the final design met all the CEI specifications. The Rogers Commission had recommended that any new joint seal should maintain at least the same minimum safety factor under the worst temperature conditions that the basic structure had to maintain: a safety factor of 1.4. NASA decided to increase that factor to 2.0 on the field-joint design, which meant that the joint had to be designed to operate successfully at the coldest launch temperature with a joint opening twice as large as expected from the design at the highest operating temperature when the pressure was a maximum. The O-rings, therefore, in dynamic testing, had to demonstrate sufficient resiliency to accommodate a gap opening over twice as large as expected.

Carver Kennedy and I felt very good about this new joint design in spite of the fact there were still numerous people in the technical community who thought it was a mistake to try and isolate the O-rings from hot gas. These folks thought that the rubber insulation J-joint design could not be manufactured to always maintain a compression fit and that contamination during assembly, or lack of adhesion during segment mating of the insulation parts, would likely result in periodic hot gas jets streaming through the insulation.

This became a very controversial issue with the National Research Council as well, but the NRC, after many days of discussion and debate, supported our approach. This wouldn't have been possible if key NASA managers from Marshall hadn't also supported our position. In order to return the Shuttle to flight as soon as possible, NASA had decided to relocate the key management responsible for the SRM redesign in Utah rather than Huntsville, thereby bypassing all the red tape involved with coordinating with all the different organizations involved at MSFC. In place of Larry Wear, Royce Mitchell was appointed Program Manager for the RSRM program; Mitchell reported to the new SRB Project Manager, Gerald Smith, who replaced

Larry Mulloy. Smith himself reported to the Shuttle Project Manager, Bob Marshall, who replaced Stan Reinartz in this position.

In addition to these changes, Marshall's new Center Director, J. R. Thompson, gave John Thomas the job of overall technical manager for the RSRM. Thomas, who also relocated to Utah, reported directly to Thompson and had the authority to approve all the key design and testing decisions. Thompson's replacing William Lucas as Center Director was a most welcome change. With a very open style of management, in stark contrast to Lucas's dictatorial style, Thompson encouraged active participation from all of his subordinates. Even though there were numerous people within the MSFC engineering organization who also supported a vented joint design, Thomas, Mitchell, and Smith preferred isolating the hot gas from the joint seals.

With so much controversy over the design approach, it was decided also to pursue a backup vented joint design as well as an alternate sealed joint design using a metal-to-metal seal in the joint area. The metal-to-metal seal was achieved with a large metallic ring that sealed on the inner surface of the capture-feature on the tang and on the inner surface of the clevis leg on the other side of the joint during segment mating operation. Designed by Veto Gray Corporation of Houston, Texas, this metal-to-metal seal approach was based upon a long history of success in sealing large-diameter pipes used in the oil-drilling business. It was a very novel sealing approach and was the one most favored by Roger Boisjoly before he left the company.

The vented joint design was pursued as a potential backup. This design had a very complex labyrinth in the insulation, with a porous carbon fiber rope at the bottom of the insulation to allow the gases to pressurize the O-rings, but would cool the gases and prevent aluminum oxide slag from depositing on the O-ring surfaces.

All of these approaches were analyzed and tested in full-scale-size components in short-duration test motors. The metal ring did encounter some assembly and sealing problems. By the time the vented design was tested, the selected J-joint design had been successfully tested in full-scale, full-duration test motors.

I was totally overwhelmed with my work and had little time to think about or do anything else. My wife was not working at the time and was able to stay at home with our four-year-old daughter, who was in preschool for a few hours each day. Linda noticed she was not being invited to many of the

Morton Thiokol upper management ladies' luncheons that were periodically held, and that she and I were frequently excluded from company parties that we had been invited to previously. Being a social pariah was much more difficult for her to accept than it was for me, because I was totally wrapped up in my job. She and I discussed the matter frequently and both came to the same conclusion that the company was only keeping me on because of the congressional letters—and because they knew that I would do a good job in the redesign effort. It was politically correct for them to keep me in a high-profile job on the redesign, but they were using me to their benefit. They would allow me to work long hours of noncompensated overtime because of my dedication to the redesign, but after the Shuttle started flying again, they most likely would have little use for me. They would subtly make it more uncomfortable for me at work, hoping I would quit.

I decided to test the waters and see if a better job might exist for me somewhere else. Some people had looked at my role in the *Challenger* investigation quite positively, so I thought another company would be interested in having me. But I was wrong. I quickly found that what I had done was far more of a deficit than an asset. I contacted several companies in the aerospace business and people in the government whom I knew, and with only one exception, they all gave me the cold shoulder.

I wasn't so naive as to be shocked by this, but I was really surprised and depressed when I couldn't even get any support or positive recommendations when I requested help from some of the members of the Presidential Commission and of the National Research Council. It was quite apparent that key aerospace executives were skeptical of my actions, because I had showed too much honesty and therefore couldn't be trusted, because I couldn't be controlled. I just might be considered a liability with many in NASA's "good-ol'-boy" network who were close friends with some of the people who had screwed-up during the *Challenger* accident and had been forced out of their jobs.

I did receive a very positive response from Aerojet Solid Propulsion Company in Sacramento, California. They invited me for an interview on a Saturday and suggested I bring Linda with me to look over the area. While I was interviewing at the Aerojet plant, a few of the wives of Aerojet managers, including several who had previously worked at Thiokol, took Linda to lunch and toured some choice residential areas. At the plant, I spent most of my time with Chuck Levinsky, who had just been named Vice President of Space Booster Programs after having been in charge of the engineering organization. A newcomer to Space Booster Programs, Aerojet had just decided to make it their number-one priority for new business. NASA had awarded several Phase A study contracts for an advanced solid rocket motor (ASRM) for the Space Shuttle, and Aerojet, along with everyone else in the solid rocket industry, including Morton Thiokol, had received one of these contracts. Also, the air force was developing a seven-segment solid rocket motor at the Chemical Systems Division of United Technologies to upgrade the capability of its Titan 34D launch vehicle, creating a new vehicle called the Titan IV. This program had formerly been called the complementary expendable launch vehicle. After the *Challenger* accident, the air force had decided to transfer most of its Shuttle payloads to the Titan IV. To deliver Shuttle-class payloads to orbit, the air force was also considering further up-grading the Titan IV with a new solid rocket motor. Aerojet thought that I could be a real asset in helping it to get into this very lucrative space booster market.

Since Levinsky's duties as Director of Aerojet's ASRM study program were consuming all his time, he wanted me to come to Aerojet and posture them to compete on the new air force space booster program. I would be used in a consulting capacity on the ASRM and also be Director of Air Force Space Booster Programs. Offered the position, I told Levinsky that I would need to have the same executive benefits package I had with Morton Thiokol, plus a salary increase of at least 10 percent to offset the higher cost of living in Sac-ramento compared to Ogden. I told him that my annual management bonus was typically worth 30 to 40 percent of my annual salary, depending on the profitability of the company and stock options. I also benefited from a salary continuation plan that provided my full salary to my wife in the event of my death until my youngest child (Meghan) reached the age of twenty-four, got married, or finished college, whichever came first. At that point in time, my wife would receive half of my full pension. Levinsky thought he could swing the salary and bonus, but that stock options in Aerojet were not offered to anyone below the vice president level. Also, he did not believe that Aerojet offered the key executive salary continuation plan to anyone as far as he knew unless it was at the corporate executive level.

With my being forty-nine years old and my youngest child being only four years old, it was critically important for my family to have that salary continuation plan. Levinsky said he would see what he could do, but he wasn't optimistic, because the position he was offering me was at the direc-tor level that reported to him, not at a vice president level. As far as I was concerned, I was more than qualified for his vice president position and suggested he should consider that as an option. He smiled and said, "Well, I will see what I can do."

Both Linda and I were impressed with all the people we met at Aerojet; they were very friendly, open, and down to earth. This contrasted sharply with Morton Thiokol, whose senior management were very stuffy and closed-mouthed about the company, and very social-status conscious. The next day on the flight home, she and I talked about leaving Utah and all our friends. We both thought it would be a good thing to move on.

A couple weeks later I received the formal offer from Levinsky in the mail. He offered me the job of Director of Air Force Space Booster Programs reporting to him. Aerojet was able to meet my salary and bonus requirements, but could not make my appointment at a high enough level for either the stock options or salary continuation benefit. Linda and I agonized for several days before concluding that I would be giving up too many benefits to take the job. Anyway, the air force programs might not even materialize, and with Levinsky already running the ASRM contract for NASA, I could be putting myself in a prime position for any reduction in force that might occur, because I would be Aerojet's highest paid new kid on the block.

I decided not to refuse the offer, but make a counteroffer that would enhance my position and provide some compensation for the added risk and for the benefits I would be losing. I mailed the counteroffer back to Levinsky, and within two weeks he responded. Although Aerojet was still quite interested in my employment, it was not in a position to improve on the earlier offer. Again, Linda and I agonized over the situation, finally deciding it would be best to stay at Morton Thiokol. If the redesign turned out to be very successful, maybe the company would change their attitude toward me.

No sooner had I made this decision when the news broke that the U.S. Justice Department was being asked to open a new probe into the Space Shuttle. Former NASA "whistle-blower" Richard Cook—the NASA budget analyst who wrote the memos released to the *New York Times* that resulted in the first open hearing of the Presidential Commission back on February 11, 1986—had requested Attorney General Edwin Meese to open a new investigation into the *Challenger* accident and to appoint a special prosecutor to run it. Cook believed that the probe conducted by the Rogers Commission had failed to study adequately the possibility that NASA had felt White House pressure to launch *Challenger* because President Reagan was to deliver his State of the Union address that night and was going to mention having a teacher in space. This was the same allegation that Senator Fritz Hollings had made earlier. Cook admitted in a news conference that he did not have any direct evidence of White House pressure, but he believed NASA officials might have perceived there was such pressure.

Not hearing anything further about this news story, I presumed that the request was dismissed by the Attorney General and the Justice Department.

The following month, November 1986, I had to give a detailed presentation on the cause of the *Challenger* accident and the status of the redesign program to George Rodney, the newly appointed Associate Director of Safety, Reliability, Maintainability, and Quality Assurance at NASA headquarters. I then toured our major steel-case component manufacturers with members of the NRC. These were the most critical components in the redesign and also the longest lead-time items to procure. The tour included Latrobe Steel just outside of Pittsburgh, the company that fabricated the D6AC steel billets, and Ladish Company in Milwaukee, which did the forging, forming, and rough machining of the steel-case components. Both vendors did an excellent job in the recovery program by delivering high-quality parts on a much-accelerated schedule.

Later that month, Congressman Bill Nelson (D-Fla.) and several of his colleagues and staff from the House Subcommittee on Space and Aeronautics visited Morton Thiokol's operations in Utah. Congressman Nelson was interested in seeing the SRM manufacturing processes and requested a detailed briefing on the redesign status and planned test program. After the briefing I gave on the redesign program, I joined this group on the bus for a plant tour. I sat down next to a very lovely woman on the bus and introduced myself. She told me that she already knew who I was. This didn't surprise me because of all the times that I appeared on television, but she then said she knew who I was from her husband, who had attended a conference in Israel over a year ago with me. I was very puzzled over that, so I leaned over and looked at her visitor's badge to see her last name. I was very surprised to see the name Susan Hauck on her badge, and I immediately said, "You aren't the wife of Astronaut Rick Hauck are you?" She replied that indeed she was. She was working on the staff of Congressman Nelson.

At the end of that week I traveled to Washington, D.C., to meet with the NRC, where I summarized the results of the preliminary design review that had been presented to the NASA/Preliminary Design Review Board at Marshall Space Flight Center in October. Tests of numerous joint assemblies had been conducted with full-scale hardware, and several of these assemblies had been pressurized to better understand joint assembly stresses and deflections. All of these tests had involved numerous strain gauges, displacement transducers, and pressure gauges. Using seventy pounds of live propellant, a series of tests had also been conducted using subscale hardware but with full-size tang and clevis field-joints at smaller diameters. We had designed these hot-fire tests with intentional flaws in the joint and O-ring areas so as to determine if the joint would leak and, if it did, whether it would reseal, as *Challenger* had, or continue to leak for the entire twenty-second burn-time of the test motor. Our tests couldn't duplicate the dynamics of the joint, and therefore prefabricated flaws in the O-ring, or wires under the O-rings, had to be incorporated to cause leakage.

The results were very revealing, because, as they say, anything that can happen usually does happen. Some of the joints containing the flaws didn't leak at all. Some others duplicated the behavior observed on *Challenger* at ignition by allowing only a puff of smoke to escape the joint, but then sealed up, never to leak again. Others started leaking at ignition and continued leaking, forming a large hole in the joint. Some tests started with a leak in the area of the flaws, eventually progressing all around the joint. Upon disassembly of the joints where only a puff of smoke came out at ignition, our engineers noted that the molten aluminum oxide from the propellant combustion gas did, indeed, solidify in the joint, forming a temporary seal in the flawed area of the O-ring. The small-diameter subscale cases were too stiff to allow any joint rotation like that experienced on the full-size motor, and no external loads simulating wind shears, vibration, and maximum dynamic pressure could be brought into our tests to duplicate the real *Challenger* environment. Even so, it was rather conclusive that the joint could leak and reseal itself as with *Challenger*, with a high probability that the added external loads at Max Q combined with steering the vehicle through high wind shears causing severe vibration and flexing of the vehicle would later break open any brittle aluminum oxide seal that had formed around the burned O-rings. This series of seventy-pound-charge hot-fire tests further substantiated the *Challenger* failure scenario, our test hardware then becoming a very useful tool for testing the tolerance of both the old and redesigned joint to flaws and defects.

In December, I accepted an invitation to speak at Brigham Young University on "Engineering Ethics and the *Challenger* Accident." It was the first time I spoke in public about what had happened on that unforgettable night before the launch. I hadn't realized until this speech how successfully I had immersed myself in the redesign activity and repressed my feelings about the tragic loss of *Challenger*. I had been so involved in how to fix the Shuttle that I had stopped thinking about the failures in the decision-making process leading up to the launch. About halfway through my speech, however, as I started to talk about what had happened at NASA and Morton Thiokol the night before the launch, tears came to my eyes. I had a hard time maintaining my composure. I did my best and managed to finish my presentation.

The year ended with another successful test of the Joint Environment Simulator and receipt of the first redesigned capture-feature case hardware from the Rohr Corporation in San Diego, where the final machining of the case was done. We then put this hardware through successful assembly and proof-pressure tests, a major milestone for the redesign activity. We were well on the way to recovery and reported that to the NRC and the MSFC oversight committees, as well as to a second NRC committee that was looking specifically into failure modes and effects analysis of the new redesigned field-joint. I felt so good about where we were that I finally took some vacation the week before Christmas to be with my family. Nineteen eighty-six had been a very traumatic year for the McDonald family, and the holidays with my wife and children were a blessed relief. I was able to recharge before returning to the hectic schedule of redesigning the solid rocket motors to return the Space Shuttle to safe flight.

It had been a long, hard year financially as well. I had worked over 1,100 hours of noncompensated overtime since January 28, 1986. A portion of my annual compensation came from a management incentive bonus. Because of the *Challenger* accident, all of the management in Space Operations suffered significant cuts in their bonus, even though the company made record profits on the Shuttle program that year. My own bonus was reduced by 90 percent, while our corporate officers received increased bonuses. At the same time, the company gave no annual Christmas bonus at all to its employees—a program that had been in place ever since Thiokol moved to Utah in 1957. CEO Charlie Locke decided to eliminate the program forever. His timing was very poor, and the action greatly irritated the employees, especially when they saw that Locke's already huge bonus had increased.

As much as I enjoyed the holidays, it was impossible to escape frequent flashbacks to the *Challenger* accident. All the "Year in Review" coverage by the newspapers, magazines, and TV stations made it impossible not to think back to the horrors of it all, as *Challenger* was selected by the media as the top story in the world, even over such other major news events as the Chernobyl nuclear power plant disaster in the Soviet Union. I couldn't help but feel a strong camaraderie with all those Russian engineers and managers who had gone through the professional and personal torment of being responsible for their own unmitigated disaster.

32

Boisjoly's $2 Billion Lawsuit

Roger Boisjoly did not return to work from his six-month medical leave of absence in January 1987 as he was supposed to. Instead, he obtained a fancy Washington, D.C., lawyer who convinced him to sue Morton Thiokol for over $2 billion.

Roger called me shortly after his lawsuits were made public. "Roger," I told him, "you're crazy. Two billion dollars is a ridiculous figure."

"Al, my attorney is convinced that it's a legitimate figure for such a high-profile case. With the loss of life of seven American heroes along with a $2 billion Space Shuttle vehicle and its precious cargo on board, this lawsuit is not out of line. Not only am I suing Thiokol for $2 billion under the False Claims Act for selling defective hardware to NASA, I'm also suing the company for $18 million in personal damages for my mental suffering and inability to return to productive work." Under the False Claims Act, Roger was also in a position to recover 25 percent of the $2 billion award if successful in his lawsuit. His attorney also filed a lawsuit against NASA for an additional $10 million.

With Thiokol already dealing with millions of dollars worth of lawsuits from several of the families of the *Challenger* astronauts, I told Roger I thought the company would consider settling with him out of court for maybe as much as $1 million but that $2 *billion* would make them so mad they'd never settle out of court and he would get nothing. "I don't think any judge or jury would award you such a claim, Roger."

Extremely depressed ever since the accident, Roger was emotionally very vulnerable. The lawyer had come to him and had convinced him that this was a legitimate lawsuit and that the principle involved with this case warranted this extreme of an action. Roger bought into that story, partly because the attorney took the case on a contingency fee only. He now owned Roger and told him exactly what he wanted him to do to enhance their case. The attorney clearly wanted to try the case in the press first, taking advantage of the emotions of the public about the national disaster, and make it appear that Morton Thiokol had persecuted Boisjoly wholesale for telling the truth and that the company's persecution of him was what had led to Roger's great mental anguish and post-traumatic stress disorder, which prevented him from working in his profession as an engineer now and maybe forever.

It didn't take long for the media to pick up on the case. On the first anniversary of the *Challenger* accident, the *New York Times* published a feature article on how those few people who had tried to prevent the launch were coping with the fateful experience. The story mainly contrasted Roger and me, the two people who had done their best to stop the launch. Its author, David E. Sanger, explored the dichotomy of my leading the Shuttle redesign activity and working long hours for the same company from which Roger Boisjoly had taken a medical leave of absence in order to get away from being involved in that activity—and was now suing.

It didn't take long after the story broke that I started receiving telephone calls from various magazines and newspapers requesting interviews and from various universities and organizations offering speaking engagements. Roger received many similar calls, the difference being that Roger chose to participate in as many of these interviews as possible because he was seeking publicity for his pending lawsuit, and I refused them all. Roger had extended his medical leave of absence from Morton Thiokol and was supplementing his income with honorariums and stipends from speaking engagements. He was speaking at many colleges and universities on "Ethics" for just traveling expenses and sometimes on his own money, because he believed he had a message to tell. Both he and his attorney appeared on ABC's *Good Morning America* explaining why they had filed their lawsuit. Roger appeared in a feature article in *Life* magazine and received several movie offers for his story, eventually signing a contract for a *Challenger* movie that aired on ABC in February 1990.

I, too, was offered a movie contract. Robert Greenwald Productions, Inc., a well-known studio that had made movies for TV, sent me videotapes of some of their movies to convince me to sign a movie contract with them. Greenwald wanted to produce a human interest story about the *Challenger* accident featuring Boisjoly and myself and was particularly interested in using the events leading up to the decision to launch as the background, with the main part of the story focusing on the diverging paths of Roger and I as we both tried in vain to stop *Challenger* from flying.

I was tempted by the offer, which was generous, but I could not in good conscience help make a movie about *Challenger* when I had accepted the job of leading the redesign to fix the problem that caused the accident in the first place. If I participated in the movie, I would lose considerable support from my colleagues at Morton Thiokol, NASA, and the National Research Council, and I needed to maintain the highest standard of credibility with all of these people to accomplish what still appeared to be an almost impossible task.

So I declined to participate in any movies or even interviews for popular magazines. I only participated in the preparation of articles published by technical magazines and professional journals on the technical aspects of the Shuttle design and the status of the redesign leading to the return of the Shuttle to safe flight.

I was often asked why Roger Boisjoly and I took such different paths after the accident when we were both trying to stop the launch of *Challenger* in the first place. I do not pretend to be a sidewalk psychologist, but the best psychology lesson I ever learned was when I was helping to coach my son's seventy-pound Little League football team in Brigham City, Utah. I never took the coaching job too seriously, like some coaches I knew, because I truly believed the kids (nine to twelve years old) were just out there to have some fun, so I seldom if ever raised my voice during practice sessions. However, at one practice session on blocking, most of the kids were rolling around on the ground wrestling with each other and not paying attention to what I was trying to teach them about blocking. I finally got so irritated that I shouted as loud as I could, "All of you guys need to start paying attention to what I have to say!!!" A few of the kids who had never heard an angry word from me before thought it was a big joke and started laughing. Several of the kids were scared and started to cry. Most of the kids jumped right up and got in line to practice, but a couple kids got angry and gave me the finger, walked off the field, and said they were going home. Now they all heard the same thing, but their emotional responses covered the entire spectrum.

Various congressional committees closely followed the redesign process. In January 1987, the Senate Committee on Space, Science, and Technology requested a briefing on the status of the redesign from NASA and Morton Thiokol, and NASA decided its people would provide the formal briefing with support from Morton Thiokol for responses to anticipated questions. John Thomas conducted the briefing for NASA, with Ed Garrison, Ed Dorsey, and me in support. Garrison was still President of Thiokol's Aerospace Group, while Dorsey was the new Vice President and General Manager of Thiokol's Space Operations.

The House Committee on Science and Technology had supported a bill requiring NASA to consider getting someone other than Morton Thiokol involved in the solid rocket boosters for the Shuttle either by developing a new, more powerful ASRM, recompeting the RSRM, or selecting a second source for the RSRM after it was recertified for flight. NASA had chosen to pursue the ASRM since it had already funded several study contracts with all those in the solid rocket industry, including Morton Thiokol, to examine the possibility.

I helped Dorsey write his prepared statement for the Senate committee, and I also helped NASA's John Thomas prepare his formal presentation material. Though not scheduled to participate actively in the planned testimony, I was there to help respond to potential questions from committee members about the redesign program. I was pretty sure that Thiokol senior management only brought me along to the Senate hearings to show members of the committee that I was still employed by Morton Thiokol and that I was still actively involved in the redesign. Many members of that same Senate committee had been involved in signing the May 1986 letter to NASA Administrator James Fletcher recommending that he consider contracting with someone other than Morton Thiokol if he found that Roger Boisjoly and I had been demoted for our testimony before the Presidential Commission.

When we entered the hearing room in the Russell Senate Building, we were asked to find seats behind the two tables placed in front of the Senate committee. The senators sat in an elevated stadium-like seating area shaped like a horseshoe. All of the NASA management people sat down behind the right-side table below the open end of the "horseshoe" facing the committee, and all of the Morton Thiokol people sat on the left side, starting with the highest-ranking member of our group. Needless to say, I was the last one to find a chair.

Actually, by the time I was to sit down, there were no empty chairs available at either of the two tables, so I turned around to see if I could find a seat close by in the gallery. At that instant, both Senator Al Gore and Senator Fritz Hollings noticed I didn't have a seat at the table. "Mr. McDonald," Senator Gore said, "if you are having difficulty finding a seat down there you are welcome to come up here and sit with us. We're glad to see you here today."

"Fine, I guess I will do that," I said, and started walking up toward the area where the senators were sitting.

Immediately, the entire Morton Thiokol contingency stood up to offer me a seat at the table. Even Ed Garrison commented he would find a seat for me at his table. The last thing Garrison wanted to happen was to have me sitting up in the seats with the Senate committee members looking down on him.

Within a few seconds, a new chair was brought into the room and placed on the end of the Morton Thiokol table so that I would have a place to sit.

Before either NASA or Morton Thiokol senior management could read their prepared statements, several of the Senate members commented on how pleased they were to see me there and to have me actively involved in the redesign process. Everyone could see how uncomfortable Garrison was when these statements were being made.

The hearings lasted several hours. I was not required to either participate or answer questions. Not long after the hearings, Ed Dorsey decided to return to retirement. He had done a good job in reunifying the demoralized workforce at Morton Thiokol and was held in high regard by all of the employees.

Our new Vice President and General Manager of Space Operations, John Thirkill, was also someone I respected and enjoyed working for. Thirkill was a no-nonsense manager on whom you could depend to put all the facts on the table and tell it like it was. John had returned to us from Huntsville, where he had been General Manager at the Redstone Arsenal. I had worked for him for many years when he headed up engineering in Utah, and I knew he could not be intimidated by anyone; if there was anyone in the corporation who would stand up to NASA management, it was John Thirkill.

John did not do anything he liked in moderation, and that included his work or his recreation. He normally worked six or seven ten-hour days a week, and he expected his people to do the same thing if the job required it. When he had time for recreation, he approached it the same way. If it was the ski season, he could be found at Snowbasin or Beaver Mountain as the first man in line waiting for the lift to open, and the last guy on the lift as it closed, and it didn't matter what the weather was like. He wore this old, ugly, dirty yellow ski parka with a sandwich stuffed in the pocket so that he could eat it on the chairlift and not have to waste good skiing time sitting in the lodge and eating. He seldom skied on the groomed trails—he preferred skiing in the powder between the trees. If it was summer, he could be found playing tennis at the crack of dawn until dark or riding his bicycle in the beautiful canyons or over the mountains. On the rare occasion that he took vacation, he would immediately head for the Wind River Mountains of

Wyoming to hike and backpack in the wilderness, or do some serious rock climbing and mountain climbing.

The only thing I didn't like about John was the rotten cigars that he would smoke. I always thought that Fidel Castro sent these to him to try and poison the U.S. rocket effort. However, John was considerate of others, because he never lit up one of those stinking, smoldering old stogies until after the normal quitting time of 4:10 p.m., but you could set your watch on 4:11 p.m. when you could see the first puff of rank smoke coming out of his office every day. I guess I would have been discouraged about this whole assignment because of the politics involved if it hadn't been for John Thirkill and Carver Kennedy. These were two guys I admired and respected and who were known for getting the job done, and I felt good about being part of their team irrespective of how I may have been selected.

It did surprise me, however, that Thirkill was brought back from Huntsville to take over Dorsey's job. John had done a good job solving a lot of problems at Huntsville and turning that division around, but he had the same potential vulnerability that Dorsey did, in that he had been Dorsey's deputy in charge of approving the original SRM design that failed on *Challenger*. I was concerned that the press would eventually find this out and attack both Morton Thiokol and John Thirkill for being given responsibility for the redesign. Fortunately, my fears were unfounded, and the issue was never raised.

A critical moment in the redesign was soon upon us, as the first redesigned development motor (DM-8) was scheduled for testing in the summer of 1987. In May, a full-scale test of the old *Challenger* design modified with a change in the insulation at the segment joints was to occur. This motor, designated Engineering Test Motor No. 1 (ETM-1), included a "U-joint" that was secondarily bonded to the insulation to isolate the field-joint O-rings from the hot gas. Unfortunately, the fabrication of the modified joint didn't work too well. The bonded U-joint experienced several separations that could possibly result in hot gas channeling through the joint and impinging directly on the O-rings.

As a result, we decided to return the ETM motor back to the original pre-*Challenger* design configuration. We redesignated it ETM-1A and tested it by filling the blowholes in the vacuum putty in the field-joints after assembly and heating the joints with the new field-joint heaters. We needed to static test these old pre-*Challenger* motors anyway in order to recover the nearly $7 million worth of metal parts that could be used for future Shuttle flights. There had been eleven of these motors in storage at Morton Thiokol or at KSC at the time of the accident. Everyone at both Thiokol and NASA was very nervous about testing a pre-*Challenger* design. We knew that the redesign program could not tolerate any failure without putting the whole Space Shuttle program at risk.

On the scheduled day of the test, it was cold and rainy. It looked like we might not be able to test the ETM-1A because we could not test in rain due to environmental concerns with acid rain. During a quick break in the dark clouds, we tested the motor. It just completed its firing when the clouds settled over the area again. The huge smoke cloud of hydrochloric acid (HCl) and aluminum oxide formed by the test firing drifted toward the viewing area and rained particles of aluminum oxide coated with HCl down on everyone.

The ETM-1A test was a great success nonetheless. It clearly demonstrated that the Space Shuttle could fly again safely if the vacuum putty blowholes in the field-joints were effectively repaired after assembly and with joint heaters installed until a full redesign could be accomplished. This had been just what General Chuck Yeager had implied in the first executive meeting of the Presidential Commission. However, such a possibility, no matter how rational, was unthinkable back at the time in the emotionally charged atmosphere immediately following the accident.

Still, anything and everything about the Shuttle was making news; even the ETM-1A test had been a media event. In fact, nearly every test, including the Joint Environment Simulators and the Nozzle Joint Environment Simulator, attracted the press. The NASA public relations office didn't have much to do now that the Shuttle was grounded, so they encouraged as much publicity as they could relative to the redesign program in order to maintain the public's interest in the Space Shuttle program.

What surprised me was that NASA was spending more money on the liquid propulsion part of the Space Shuttle—on the ET and SSME—than it was on the solid rocket boosters that were being redesigned and recertified for flight. Nonetheless, the NASA public relations office decided we needed to provide a press center for the media in the test area to witness the test and provide pretest and posttest briefings until we returned the Shuttle to safe flight. So we installed some trailers in our static test viewing area with telephones and restroom accommodations for all future tests.

The approach to testing the full-scale, full-duration, redesigned solid rocket motors became very controversial. The controversy surfaced as a result of some recommendations from the Rogers Commission report and some recent air force decisions about its planned testing program for the Titan SRBs that had failed on the last launch of the Titan 34D in April 1986.

In the past, the Shuttle's solid rocket motors had always been tested in a horizontal position; that had also been the approach in the May 1987 ETM-1A test. However, the Rogers Commission had recommended that consideration be given to testing these motors in a vertical (nozzle down) position to more accurately duplicate the launch configuration in the event that the loads imposed on the field-joint at launch were important to the sealing capability of the O-rings. The air force had already determined this would be more representative of the flight configuration for its Titan IV, and was in the process of constructing a huge new vertical test facility in the desert at the Air Force Rocket Propulsion Laboratory at Edwards Air Force Base in California. Prior testing of the Titan solid rocket motors had also been accomplished in a vertical test stand with the nozzle pointed up. These tests had been conducted at the contractor's facility of Chemical Systems Division of United Technologies in San Jose.

At Morton Thiokol, we had decided earlier that we'd need a new test stand to meet the very ambitious schedule for returning the Shuttle to safe flight as soon as possible. We knew it wasn't possible to "temperature condition" or introduce the launch and flight loads into a full-size Shuttle motor during static test in our current test bay and that it would be necessary to do it in some of the planned static tests. We were going to have to construct a new test bay in any event, so we might as well carefully evaluate the issue of horizontal versus vertical testing.

At first glance, it appeared that vertical testing with the nozzle down was probably the right way to go. The oversight committees with the National Research Council and at NASA Marshall definitely expected us to recommend constructing a new vertical static test facility for the Space Shuttle SRM that would be very similar to the air force's planned new facilities for the Titan IV. Completing our study, however, we concluded that a traditional horizontal test facility was actually far superior to a vertical one. A vertical test stand had one very serious drawback that could not be avoided, and that was its inability to measure the motor thrust with a high degree of accuracy because of the constant change in propellant and insulation weight during motor burn, which affected the thrust measurements. A horizontal test, on the other hand, not only provided the most accurate thrust measurements, it also provided improved launch and flight loading as well as more severe thermal and erosion conditions on the internal insulation and nozzle components (the latter was due to the gravity effects on the aluminum oxide particles in the propellant combustion gases, causing more erosion than would occur in flight motors in some areas). This had been a primary reason why horizontal testing had been so widely used in the solid rocket industry. One of the primary objectives of any static test was to test the motor in a comparable or more severe environment than it experienced when launched. Not testing the Space Shuttle's solid rocket motors in the same condition as they were assembled or in the same environment in which they were to be launched, especially on that very cold morning of January 28, 1986, at the Kennedy Space Center, had been a critical oversight by both NASA and Thiokol.

Concluding that horizontal testing was best for the Space Shuttle SRMs was not a popular decision. It was met with considerable resistance within NASA, by many of those within the SRB design oversight committees, and by many of the staffers who served on the congressional space, science, and technology subcommittees. Their initial reaction was that the decision was self-serving to Morton Thiokol, which had always used horizontal testing and didn't want to admit that it was a mistake. Our critics thought that the decision must also be self-serving to NASA, which had approved the prior horizontal test method and was likely never to accept the same approach as that endorsed by the air force. I knew that none of these factors had entered into our decision. NASA and Thiokol both wanted to make sure that the approach selected provided the best possible simulation of the Shuttle launch conditions.

The decision to continue horizontal testing was a combination of good technical evaluation and serendipity. Launch and aerodynamic flight loads were transmitted to the SRBs through the struts that attached the SRBs to the external tank near the aft field-joint through the ET attach ring that was bolted to the aft segment steel-case attach structure. The axial thrust loads from the SRBs were taken out through a box beam connecting the two SRBs that passed through the intertank structure between the liquid oxygen tank atop the ET and the liquid hydrogen tank on the bottom. When the SRBs were tested horizontally, the axial thrust loads were reacted through a thrust block and measured with a load cell attached to the head-end of the motor, but no external launch or aerodynamic loads were transmitted into the motor during static test. Only those loads induced by the operation of the SRBs were measured and accounted for in a static test.

This was the approach that had been used on all solid rocket motors that had ever been developed in the industry. In analyzing the Space Shuttle system, we decided that we could input the liftoff and maximum dynamic pressure and wind loads with a set of hydraulic actuators attached to the ET attach ring on the aft segment of the SRB. This was much easier to accomplish on a horizontal test, though it could be done in a vertical test. The Space Shuttle was a unique launch vehicle because of the three SSME engines in the back of the orbiter that started six to seven seconds before ignition of the SRBs. These three engines produced just over 1.1 million pounds of sea-level thrust, offset of the launch vehicle's center line, which caused the entire vehicle to bend over more than two feet prior to SRB ignition, while it was still being held down on the launchpad. SRB ignition occurred after the SSME engines were verified to go as the Shuttle stack sprang back to the vertical position. Both SRBs ignited simultaneously with a pyrotechnic separation of the huge nuts on the four huge bolts that protruded through the aluminum aft skirt of the SRB, a skirt that held up the entire 4.5 million pounds of weight of the Shuttle system. These prelaunch or bending loads were extremely difficult to simulate; since they induced a very high bending load into the center and aft field-joints that could affect the joints' ability to seal at ignition, it was imperative that we include the effects of this condition during a static test.

This was not a problem with any other space launch system—and was the main reason why a vertical test stand just wouldn't work nearly as well. To duplicate the same loading condition on a vertical test stand would require some huge hydraulic actuators and load structure to bend the motor in the area of the field-joints. Though possible, this sort of arrangement would be extremely difficult and expensive to do. To externally apply these same loads to a horizontal test stand was itself very difficult. In conducting a normal horizontal test, the 126-foot-long motor was attached by the thrust block to a thrust adapter connected to the forward dome, a semicircular cradle supporting the aft end of the motor. Each of the segments was supported on a set of chocks to aid in assembly; however, after all the segments were mated, these half-moon chocks were removed. The 126-foot-long motor was effectively suspended at both ends; this caused the motor to sag in the middle, making it look like a banana. This sag of 3.5 inches at the center field-joint resulted in a joint loading condition that essentially duplicated the joint bending condition experienced in the aft field-joint during SSME ignition on the launchpad. When the SRB ignited, the motor straightened out, nearly duplicating the operating condition of the motor during launch.

Because of this unique condition of a horizontal test to duplicate passively the effects of the prelaunch bending loads in the joint, it became very clear to us that horizontal testing was much more representative of the Space Shuttle launch than a vertical test.

Our conclusion was not well accepted by many people in NASA and the NRC. Having the test motor in a vertical position just like the SRBs sitting on the launchpad was an easy concept to accept and a difficult one to abandon. Virtually everyone had regarded a decision to test in a vertical manner as a foregone conclusion, especially because of the Rogers Commission recommendation and the air force decision for vertical testing of its Titan SRBs. No one wanted to explain why NASA should take a totally different approach than the air force. It would make it appear that NASA could never agree with the air force on anything, which was generally true, and no one wanted to fight the political ramifications of this perception. Even veteran astronaut John Young was convinced that a vertical test just seemed better, and Young continued to support vertical testing. The fact that all previous motors had used horizontal testing also hurt our cause, because there was a general feeling that what was done in the past prior to the *Challenger* accident somehow was either bad or inadequate. Thiokol and NASA had only gone with horizontal testing because it was the easiest way to static test the motor.

Fortunately, the vast majority of the technical people at NASA and on the oversight committees agreed with our recommendation: to construct a new horizontal test facility that could "temperature condition" the entire motor from 40° to 90° prior to static tests and induce liftoff and flight loads into the ET attach ring during the static test. Prelaunch load effects in the field-joint would be naturally simulated in the vertical sag in the motor in the horizontal test stand. Construction of a new horizontal test facility was authorized to be available for testing within a year, that was, the spring of 1988. It was decided at that time that prior to returning the Space Shuttle to safe flight that the redesigned solid rocket motor must be successfully static tested with all design, liftoff, and flight loads imposed, including thermal conditioning to the temperature extreme most likely to be encountered during the first Shuttle launch with the redesigned motor.

The first solid rocket motor with the redesigned configuration was scheduled for testing in August 1987. The test was to take place in our existing horizontal static test bay, designated T-24. The development motor, designated DM-8 (the previous SRM that had been tested, in May 1985, had been the FWC-SRM, designated DM-7), was to be tested at ambient temperature conditions with no external loads applied. Even though it was a hot, late summer day, the new field-joint heaters (successfully tested on the ETM-1A motor) were also installed on the DM-8 and found to be successful, with the heaters maintaining the temperature of the field-joint at approximately 90°. To maximize our database with this design change prior to launch and make sure that no surprises occurred with this new element on the launchpad prior to the first flight, we decided to test all of our motors henceforth with the heaters operating.

We were in no mood to take any chances.

33

No Mountain Too High

It was eerie crawling through the nozzle into our giant rocket, surrounded as I was by over one million pounds of solid propellant that produced a white-hot flame of nearly 6,000°. Why I had foolishly volunteered to venture inside the DM-8 motor as it rested all assembled in our Utah test bay is beyond me, knowing as I did that if something went wrong and the rocket motor accidentally ignited from a spark of static electricity that my body would crisply become a pile of ashes, not even feeling the nearly 1,000-pounds-per-square-inch pressure that developed within a fraction of a second. But somebody had to take a close look. There were NASA's John Thomas and I inching our way in, crawling along on top of the propellant, meticulously inspecting each field-joint to make sure that the new insulation J-joint had come together during assembly in just the right manner to isolate the O-rings from hot gas as was intended.

The new design looked great, and Thomas and I gave the go-ahead to complete the installation of over 400 channels of instrumentation on the motor in preparation for the static test. This first static firing of the redesigned solid rocket motor, DM-8, was to take place on August 30, 1987—one year, seven months, and two days after the *Challenger* accident.

With every major U.S. space launch vehicle since *Challenger* having experienced at least one failure, including the Titan, Delta, and Atlas, the media showed tremendous interest in the test. NASA was broadcasting the test live on NASA Select TV, and every major newspaper and TV network was on hand to witness what everyone rightfully was regarding as the most important step in returning the United States and the Space Shuttle to safe flight. The space agency sent a legion of its public relations people to take care of the reporters. Not having had much to crow about over the past year and a half, NASA was very interested in regaining the public's attention—and in a much more positive vein. Equally interested in improving its tarnished image, Morton Thiokol fully supported the idea of turning this first test into a major media event. The company had essentially fired its main public relations man, Gil Moore, who had been in charge of handling the press subsequent to the *Challenger* accident. Moore's fundamental flaw as a PR man was he had difficulty in evading the truth. He had never learned how to say "No comment," as he had been instructed to do by his corporate management. Gil was a very sharp physicist and a very personable individual. He really understood the technical and managerial issues surrounding the *Challenger* accident. As such, he was one of my best supporters during the difficult days following the accident, and I suspect that also had something to do with his departure. He was replaced by George "Rocky" Robb, a longtime veteran of the PR business at Cape Canaveral. I considered Rocky also as a good friend, but he had learned from his predecessor's mistakes and knew when not to respond to reporters' questions.

NASA public relations personnel headed up by NASA Marshall's Ed Medal orchestrated the event. They had provided a media center equipped with telephones for the press in a large trailer stationed in the viewing area for the static test. Thiokol provided buses to take the visitors up to the test bay prior to the test to view the motor in the test bay. A press conference was held in the media center prior to the test in which the media were provided with a briefing on the design configuration, test objectives, and predicted performance. This first RSRM static test briefing (DM-8) was conducted jointly by NASA's John Thomas, my boss; Carver Kennedy, Vice President of Space Booster Programs at Morton Thiokol; and myself. I then went directly to the control bunker with Royce Mitchell, NASA Program Manager for the SRM, reviewing prefire checkout data and conditions to provide the final go-ahead for the test. Shortly after the test, Mitchell and I would enter the test bay and conduct a cursory inspection of the postfired hardware. We would then go down to the data reduction center and collect the critical test data to present at another press conference on the results. Mitchell would describe the condition of the test motor and the type of data obtained. I would then present the test data available at the time and compare the data with pretest predictions. After the briefing, we'd respond to questions from those at the press site as well as those who had witnessed the test on closed-circuit television at the various NASA centers.

It was relatively easy to anticipate what the press would ask us. They'd ask: "Are you now ready to fly? If not, what do you have to do to determine that you're ready?" Most predictably, they would ask, "When will the Shuttle go up again?" We would then show the reporters our multicolored chart showing the "network" of testing that needed to be conducted before we would be confident enough to fly, explain how we were building confidence by testing the RSRM, and do our best to avoid projecting specific dates for the Shuttle's first flight. We'd respond, "Sometime next year we'll be ready to launch a Shuttle again," and "The Shuttle won't fly until we conclude that is safe to do so." There would also always be one or two reporters who would ask me something about the *Challenger* accident or its aftermath, and how I felt about it now. It was always evident that it was the media's intent to open old wounds and provide some controversy to whatever story they were about to report. A routinely successful test was not exciting enough for the media, but it sure was for me. I had learned a lot over the past nineteen months and generally sidestepped questions of this sort by saying, "I've been working so long and so hard on the redesign that I haven't had time to reflect on what happened in the past." Most reporters would accept that statement, but there were always some who would continue to probe to the point of being downright rude. At that point I would just pick up my charts and tell them I had nothing more to say. I thanked them for their interest in the Shuttle program and told them I hoped to see them at the next static test. This tactic seemed to work quite well, because I received great support from the press during this recovery period.

After each successful static test, Thiokol would throw a party for the key Thiokol people involved in the redesign along with a host of visiting dignitaries. There would be a cocktail reception and dinner followed by a more detailed briefing of the test results. The affair was held at our Management Training Center out in Ogden Canyon, a beautiful lodge constructed of pine logs and surrounded by steep mountains on both sides with the crystal-clear Ogden River flowing behind. One could literally cast a fly line from the back room into the river below with a good chance of hooking a rainbow trout. If it was summer, cocktails were served on the front porch, with dinner in the main dining hall. In the winter, the reception was held around a large bar inside the lodge, with dinner served around a large fireplace crackling with huge, burning pine logs. Because I was responsible for giving the briefing, I always had to stay in the test area to review the data and prepare a set of briefing charts. As a result, I never got a cocktail and missed all the dinners. I could only grab a small plate, take a few bites, and then make my viewgraph presentation. These were long days, but I really enjoyed them because I could see real progress in the program and was proud to be part of the redesign team that would eventually return the Shuttle to safe flight.

It took several weeks to analyze all the data from our first DM-8 test. Of particular interest was the postfire condition of the new redesigned field-joints and O-ring seals. I remember coming out to the plant site one night to monitor the disassembly of the first field-joint. It was the aft field-joint—the same one that had failed on *Challenger*—and I was very nervous because there was a large contingency of the technical community involved in the redesign that had taken very strong opposition to the design we had selected. Their position was that it would not be possible to isolate the O-rings from hot gases totally on every square inch of surface around the full thirty-six-foot circumference of the rocket motor in all three joints; surely, at some place, the hot gas would jet through and impinge on the capture-feature O-ring.

We had intentionally included the O-ring in the capture-feature for this specific purpose—to prevent any hot gas from reaching the primary or secondary O-ring in the event that this condition in fact occurred. Even though we'd built in this redundant protective device, we expected the insulation J-joint to remain sealed around the full circumference and not allow hot gas into the area of the joint where the O-ring seals were located. To find hot gas penetration in this area would have been very disappointing, indeed, and would have cast serious doubt on the credibility of the design we selected; that's why I was so nervous during the disassembly operations. Since the insulation contained asbestos fibers to improve its erosion and heat resistance, everyone involved in the disassembly operation and on the postfire inspection team had to wear a protective respirator mask to prevent inhaling loose asbestos fibers in the charred insulation. John Thomas from NASA and I were in there with our flashlights inspecting every minor detail of the disassembled joint, looking for anything unusual or unexpected.

What we saw was a joint that was very clean—no hot gas had penetrated at all, and the insulation had enough margin it could have survived another full-duration firing without any problems. I was ecstatic.

It took several days before the motor was totally disassembled, because detailed inspections, video coverage, and numerous still photographs had to be taken in each joint area of the motor. All of the field-joints looked clean and in excellent condition.

The DM-8 had many redesigned areas besides the field-joints. One of the other major areas of redesign was in the nozzle, where numerous changes had been made to improve safety in the basic metal parts, the seal areas, and the ablative sections of the nozzle. One of these areas was the outer-boot-ring (OBR). The OBR was a carbon-phenolic part vulcanized on the end of a rubber flexible boot that protected the flexible bearing allowing the nozzle to move. This movement of the nozzle provided the Shuttle's steering capability during the boost portion of its flight. Like the steel-case segments, all of the nozzle metal parts, as well as the flexible bearing, were recovered and reused on subsequent flights. It was important that the outer-boot-ring and flexible boot remain intact during motor burn to prevent excessive heating of the flexible bearing, so it could be reused. If the boot failed early in flight, it could possibly cause a failure in the nozzle. If the boot did fail, it was most likely to happen late in the burn when it was badly charred; that couldn't cause a failure of the motor, but it might prevent reuse of the bearing. It was an area for redesign because we had noted severe de-lamination of the charred carbon-phenolic outer-boot-ring on several prior flights; in one instance, the de-lamination was severe enough to have totally separated from the nozzle even before the burn was over. Fortunately, it happened so close to the end of the burn that there wasn't any significant heating of the bearing, with no risk to the flight or enough degradation of the bearing to prevent reuse.

We changed the OBR design by making the part thicker and providing a structural wrap of carbon-cloth on the inside of the OBR parallel to the center line of the nozzle (that is, a "zero-degree wrap angle") and nearly perpendicular to the plies of carbon-cloth on the outside of the ring exposed to the hot gas. This was a rather direct engineering solution to the problem to prevent de-laminations of the charred portion of the OBR from propagating all the way through the part, causing separation of the OBR.

Several people at NASA, including most of the agency's consultants—some of them Thiokol's competitors—did not agree with our OBR design. Consequently, NASA suggested we consider a more elegant solution involving a totally new and different fabrication approach. NASA engineering thought our approach would probably work, because it prevented the cracks from propagating all the way through the part. However, they were convinced that the new approach, called an "involute," or "rosette lay up," would prevent the de-laminations of the carbon-cloth in the first place and that this should be the preferred solution to the problem. They felt confident about the approach because it was a technique that had been successfully used by our competitors in making carbon-phenolic parts for smaller solid rocket motor nozzles, even though it was far more expensive and took much more time.

Our nozzle design engineers, led by Jim Lamere, strongly disagreed, because the OBR was totally different than any other nozzle part made by anyone in the industry and was much larger than any part ever fabricated by the involute process. This process required hand "lay-up" of the carbon-cloth fabric cut in patterns, while our process, using a tape-laying machine, was totally automated. We thought our approach had much better quality control.

The OBR was also the only phenolic part not bonded to a metal structure; it provided its own structural capability. All other phenolic parts used in the Shuttle SRM and most other SRMs used carbon-cloth as an ablative nonstructural material in the high-pressure section of the nozzle. We knew that the OBR in the pre-*Challenger* design was not good structurally, and we were fixing that by adding the structural wrap on the inside of the OBR. It was not clear to us that the new involute design would provide sufficient structural capability after it was charred because it was very difficult to model this part. NASA had given a subcontract to an outside analytical modeling firm to analyze the new involute design and provide Thiokol with a detailed structural analysis. Though the contractor concluded that the design met the required safety factors, we questioned the validity of that analysis, which was untested, and again recommended we proceed with the program with the Thiokol design successfully tested in DM-8.

NASA overruled our recommendation, directing us to incorporate the involute design, furnished to us by another contractor, on DM-9, the next static test. We told NASA that we considered the new design a "backup" in the event we observed any future problems with the design tested on DM-8. We also told the agency that we'd agree to test the DM-9 with the involute OBR design, but we wanted to incorporate the DM-8 design on all the remaining static tests as well as on the nozzles for the first flight set of motors scheduled to be shipped to KSC by the end of the year in support of a first flight for the Shuttle in the late spring of 1988—exactly two years from the completion of the Presidential Commission's report and approximately twenty-eight months from the day of the *Challenger* accident.

NASA didn't agree with our recommendation and told us to treat the involute design as a preferred design to be included in all the remaining static tests and on the first flight set of motors; the DM-8 design would be the backup.

We didn't feel comfortable building flight nozzles with a design that we had never tested, and told NASA just that. We were scheduled to ship the initial flight segments to KSC before the DM-9 test—the aft segments, with nozzles bolted, that KSC needed first to begin stacking the first flight, now designated STS-26R. With NASA saying that a successful DM-9 test was a "constraint" to shipping the aft segments, we suggested making a pair of OBRs of the DM-8 design, having one with the old design available in the event that the DM-9 test indicated a problem with the involute design. If that happened, we'd remove the nozzles from the first flight set, cut out the OBR, and replace it with the DM-8 design.

NASA agreed with our recommendation to fabricate two OBRs and have the DM-8 design ready in the event the performance of the involute was not up to snuff in DM-9. This was my first major disagreement with NASA engineering over anything related to the redesign.

It was not unusual to have major disagreements within the technical community on what was the best way to design something, analyze a problem, or conduct a test. There had been a long-running debate over whether the field-joint should be vented, allowing hot gas to pressurize the O-rings uniformly every time, or whether the O-rings should be thermally isolated from ever seeing hot gas. There was also the rather heated debate over horizontal versus vertical testing. These disagreements had involved not only Thiokol engineers but also included NASA and a host of technical consultants, some of whom were from the air force and our competitors in the solid rocket industry.

Many times I felt like pulling out my increasingly gray hair, hoping to arrive at an acceptable decision. On one of these occasions, astronaut Hoot Gibson could see that I strongly disagreed with several people in the room, including some from NASA, and that I disdained the lack of progress. Hoot suggested we break for lunch early and reconvene afterward. He then walked over to Carver Kennedy's office with me. Trying to calm me down, Hoot changed the subject by starting to tell me stories about some of the exciting times he had flying on and off carriers at night in heavy seas in the Vietnam War. I really enjoyed his stories—and enjoyed Hoot Gibson even more. A very congenial, down-to-earth, and fun-loving guy, he really made me relax. He was also a very competent engineer and a terrific pilot, one who had excelled in the navy's first "Top Gun" school.

Based upon all the hair-raising experiences he had had as a pilot in Vietnam, I was curious why Hoot would be so concerned about the Shuttle's safety issues. He replied that on a personal risk basis, he wasn't concerned at all, but that he had more at stake on the Shuttle than just himself. His wife, Dr. Rhea Seddon, M.D., was also an astronaut who flew on the Shuttle, and he had to be convinced that it was safe enough for her to fly. Also, Hoot had lost a lot of very close friends on *Challenger*. For all these personal reasons and more, he wanted to make damn sure the Shuttle was as safe as it could be before anybody flew on it again. Hoot's philosophy was right on target. It was obvious to me why he was selected to be part of the SRM redesign team.

Dr. Rhea Seddon was one of the crew of the STS-51D *Discovery* mission that took Utah Senator Jake Garn into orbit in April 1985. I had the privilege of sitting at the luncheon table with her when Senator Garn brought his crew out to the plant after completing their flight. It was somewhat coincidental that Seddon, along with Rick Hauck and Dick Covey, were the only three astronauts I had personally met prior to the *Challenger* accident. Rick Hauck and Dick Covey, along with Hoot Gibson, would eventually fill very important roles in the return-to-flight program.

When it came time to reconvene our technical interchange meeting, Hoot asked me to explain why I felt so strongly about the positions I had been trying to defend at the meeting before lunch. I told him why I felt the way that I did and why I was against the alternatives being supported by several people in the meeting. Back in the meeting, when I finished with my summary, Hoot stood up and said he had a few comments to make. When an astronaut in uniform stands up in a meeting, everyone knows who he is—it's just like the E. F. Hutton TV ad, "everybody listens." "I've been poring over all of the alternatives that we've been discussing today, and as far as I'm concerned, representing the Astronaut Office on the redesign team, I fully agree with the approach being recommended by Mr. McDonald."

I was shocked, but pleasantly surprised. I then asked if there were any other comments or if anyone was in disagreement with astronaut Gibson. As there were no further comments, much less disagreements, I announced that we would proceed on to the next item on the agenda.

After the meeting, I thanked Hoot for his support. He said he truly agreed with my technical assessment of the issues and that he and the other members of the astronaut corps had a lot of faith in my technical judgment.

It was one of the pinnacles of my career when Hoot told me that because in spite of all the presidential commissions, congressional committees, councils, oversight committees, and all the others involved in the redesign and review process, only the astronauts really counted. I felt really good when I went home that night. I hadn't felt so good about myself since long before the *Challenger* accident. All of a sudden, all the drudgery of those long hours and long weeks didn't seem to matter anymore. I was really pumped up. I felt I was on an authentic mission to make sure the best design was selected and properly tested to ensure the safety of my friend and colleague Hoot Gibson, and his lovely astronaut wife, Dr. Rhea Seddon. The whole redesign activity took on a more personal goal for me after that.

When I was appointed to lead the redesign activity at Morton Thiokol in May 1986, it was envisioned to be a task force–type project much like the famous "Skunk Works" employed by Lockheed to design and fabricate a new aircraft in the shortest time possible. That's why I had been given the title of Director of the Space Shuttle Solid Rocket Motor Verification Task Force. This position rolled the critical engineering and program decisions into one area in order to minimize coordination and enhance critical decision making. The charter I assumed was to develop a reliable design, define the needed testing to adequately verify the design, and structure a program that could return the Space Shuttle to safe flight as soon as possible. Money was not to be a problem; I was to spend whatever was needed to get the job done. In addition to the basic charter, we needed to involve as many outside contractors as possible, improve facilities where needed, and provide the required support to the NASA oversight committees looking over our shoulder.

In order to support this very ambitious redesign program, a huge amount of engineering analysis, test plans, drawings, and reports had to be prepared. Because of this very large engineering workload, Thiokol corporate management decided to hire a new Vice President of Engineering for Space Operations. The previous man in this position, Bob Lund, could not remain in the position because of his involvement in supporting the ill-fated decision to launch *Challenger*. So Ed Garrison hired Dr. Dave Ewing, an old buddy from the air force and a project manager who had worked with Ed on the Minuteman production program back in the 1970s. Dave Ewing was a very personable guy and a good engineer; however, he had been doing program management work most of his career, not directing engineering work. Ewing was coming into our program at a great disadvantage because he knew very little about the Space Shuttle, and even less about how NASA engineering operated. For him, it was kind of like picking up a book and starting to read it in the middle when he hadn't even been exposed to the first twelve chapters.

I felt sorry for Dr. Ewing because he was being put into a no-win position. NASA engineering was a very large community of folks, many of whom didn't have very much to do, so they sat around all day thinking up questions to ask the contractors. This wouldn't have been so bad except many in NASA engineering didn't separate the masses of unimportant questions from the few important ones and expected us to provide detailed answers to all of them. Dave Ewing, like anyone else, had difficulty coping with this even though he had a lot of good help in MTI engineering to help keep the wolves away from his back door. Just keeping track and responding to all of the requests took more of his personal time than it was worth. Dave was not allowed to spend enough time getting up to speed on the first twelve chapters of the book, and it showed. Whenever we went to technical interchange meetings or management meetings with NASA, which was quite often, the NASA folks would ask him numerous questions about the problems we had had in the old design that were making it necessary to change it and how the new design would fix those problems. Since Dave knew very little about the past history of problems encountered in the Shuttle SRMs, I ended up answering most of the questions for him. On several occasions, I was asked by NASA people in hallway conversations why I wasn't the one in charge of engineering. I answered that I was leading the overall technical effort as the Director of the Task Force and at some point in time, probably after the first successful test of the redesigned motor, I would probably return to my pre-*Challenger* position as Director of the SRM Project in Thiokol program management.

It wasn't but a few weeks after the successful August 30, 1987, test of DM-8 that I was promoted to Vice President of Engineering for Space Operations; Dr. Ewing was transferred to a position under my former boss, Carver Kennedy, Vice President of Space Booster Programs. Essentially, Dave and I traded jobs. Even though the first test had been successfully conducted, the real engineering job was now about to begin. Numerous detailed engineering drawings and literally hundreds of engineering analyses and reports needed to be prepared to support a critical design review scheduled for February 1988, followed by a planned design certification review (DCR) that was to take place just prior to the first flight in the late spring of 1988.

The difference between the CDR and DCR was much more than the transposition of the two letters. They were quite different reviews based on the quantity of material available at the different times. The DCR covered all of the CDR information, along with analyses of all of the test data generated in the conduct of the program since the CDR. Both reviews lasted several days and always required a detailed, full-day, stand-up oral presentation before a key NASA management board. If I had thought I was working too much overtime before, this new job required even more. I later learned that several members of NASA management had complained about Dave Ewing's depth of knowledge on the program and that they had suggested how important it was, at this critical stage in the redesign program, to move me into this assignment. Carver Kennedy thought that I should have been in this post from the start. Carver knew I was aggressive enough to give technical direction to the program no matter who our Vice President of Engineering was, and he allowed me to do so by unloading from me all of the cost, schedule, and supplier issues that normally plagued the day of a program manager. Carver and I always made a good team, and he was very influential in getting me appointed as the new Vice President of Engineering as soon as the opportunity arose.

Morton Thiokol Space Operations had not had a major development program since the 1974–1981 time frame just prior to first flight of the Shuttle. All of Morton Thiokol's design and development engineers had been transferred over to Strategic Operations a few years earlier in order to support development of the air force's small, mobile ICBM. Strategic Operations was still in the development phase for the small intercontinental ballistic missile, and, as a result, it did not have an adequate engineering staff to transfer any engineers over to Space Operations to support the redesign program. Numerous contracts had to be let out to various job-shopper companies to support the redesign activities, and many new engineers, most of whom were new college graduates, had to be hired to beef up the engineering staff.

We were still hiring many new engineers when I became Vice President of Engineering, and we were continuing to bring experienced job-shoppers on board to support the heavy workload of developing test plans, analyzing test data, and conducting hundreds of thermal/structural analyses. The need to "train" our new engineers meant an added burden being placed on our supervisors and engineering managers, who themselves were already being overworked. As a result, we subcontracted even more of the engineering work, much more than we desired. We had cold gas and water flow tests of the joint areas being conducted by several different organizations, and several people were trying to develop a very complex, computational fluid dynamics computer model showing the internal flow of the SRM. We were purchasing huge blocks of Cray supercomputing time to analyze the structural dynamics of the field-joint during the ignition transient; in fact, for a while we were the largest user of Cray computer time in the country. NASA engineering was not satisfied with one-dimensional or, in many cases, even two-dimensional computer models anymore. Most every engineering analysis had to be supported by very complex and expensive three-dimensional computer modeling. These computer models were very complex. Not only did they take considerable engineering support, they also required the generation of many new material properties and large databases that were not available. As a result, numerous laboratory testing programs also had to be conducted to support our effort. This was clearly the largest engineering effort ever undertaken in the solid propulsion industry—for that matter, probably in any industry. To say that we had our hands full was a major understatement.

I did receive some very pleasant surprises in the fall of 1987, however. *Design News*, an engineering trade journal with a subscription of about a million engineers and scientists from around the world, announced that I was one of the ten nominees for its first-ever Engineering Achievement Award. The other nominees were Steven Jobs, designer and cofounder of Apple Computer; Dr. Paul Chu of the University of Houston for his breakthrough in superconductivity; Jack Telnack, Vice President of Design at Ford Motor Company responsible for the very successful Tempo and Taurus; oceanographer Robert Ballard for his underwater pictures of the *Titanic*; David Pedrick for his design of the *Stars and Stripes* sailing ship that returned the America's Cup to the United States; Howard Marderness, Chief Engineer at the Jet Propulsion Laboratory, for the Voyager spacecraft; Dr. Omar Bose, of Bose Corporation and MIT, for his numerous patents in the fields of acoustics, electronic, and communications theory; Dr. Robert Jarvik, inventor of the artificial heart; and Burt Rutan, designer and builder of the *Voyager* aircraft that successfully flew nonstop around the world without refueling. Burt Rutan won the award; but, amazingly, I finished a close second and, at a black-tie affair to be held at the Ritz-Carlton Hotel in Chicago in February 1988, I was to receive a "Special Achievement Award."

Just prior to the award ceremony, *Design News* published a feature on Burt Rutan and me. The title of the story on me was "The Making of a Hero," which embarrassed me to no end. However, the article did a great job of accurately quoting what I told the reporter. It renewed some faith for me in the journalistic profession. As for the title, I never considered myself a hero for doing my job in the best manner that I knew how and telling the truth about it. If that is a rare virtue, I'm afraid that doesn't bode well for the moral backbone of this country. I truly believe there are legions of good people out there every day defending their professional opinions and willing to speak the truth at some risk to their own job security. They just haven't been involved in such a high-profile, news-making event like me and therefore they go unrecognized.

About a month after I was nominated for the *Design News* award, I was notified that I was selected as the Distinguished Centennial Alumnus of Montana State University by the National Association of State Universities and Land Grant Colleges in conjunction with the 100th anniversary of the 125 colleges and universities established under the Congressional Land Grant Colleges Act of 1887. This list of Distinguished Centennial Alumni included former presidents, Supreme Court justices, Nobel Laureates, Pulitzer Prize–winning writers and broadcast journalists, congressmen, cabinet members, entertainers, and CEOs of major corporations. I will have to admit that my head swelled a little bit to have my picture appear in the same booklet with these very famous people.

That fall I also received numerous cards and letters from people all over the country and even from Europe congratulating me on the successful test of the first redesigned solid rocket motor for the Space Shuttle. Many of the letters contained first-day-of-issue postcards, postmarked on the day of the DM-8 test, with requests for me to autograph them and return them to the sender. I also received a very nice card signed by all of the members of the Solid Rocket Technical Committee of the American Institute of Aeronautics and Astronautics; the card showed a man climbing a totally vertical face of a high mountain with the caption: "TO SOMEONE WHO'S NEVER SEEN A MOUNTAIN TOO HIGH TO CLIMB."

I suppose growing up in Montana looking at those rugged peaks of the Rocky Mountains had done me some good.

34

The Peacekeeper Explosion

As if I didn't have enough to do moving the redesign along to conclusion, I also spent many hours with the Morton Thiokol corporate attorneys responding to numerous requests for information and interrogatories from various plaintiffs suing Morton Thiokol over the *Challenger* accident. Shortly after the accident, NASA and Thiokol had gotten together and agreed on a fair settlement to be offered to the families of the seven astronauts who were killed in *Challenger*. Reportedly, Thiokol had put up most of the money for the settlement, and the settlements were being made without the admission of any guilt by either NASA or Thiokol. Four of the seven *Challenger* astronaut families accepted the settlement—the Scobees, Jarvises, McAuliffes, and Onizukas—while three had not—the Resniks, McNairs, and Smiths. The actual amounts of the settlement had not been disclosed but apparently had been calculated based upon lost future earnings and the number and age of the dependents in the surviving families. This meant that the parents of Judy Resnik received very little, if anything, in the settlement because she was single and had no dependents. If this was true, it was very stupid of NASA and Thiokol. In any event, her parents initiated a lawsuit against both, leading to an out-of-court settlement that avoided further embarrassing legal action. The families of Ronald McNair and Michael Smith retained personal attorneys who filed lawsuits against NASA and Morton Thiokol that were still pending.

Back in March 1987, I was scheduled for a deposition with the attorneys representing the McNair family. I talked with our corporate attorneys as well as to an outside law firm from Chicago, Adler, Kaplan and Begy, which had been retained by our company to represent them in the litigation. The lawyers gave me a copy of the plaintiff's request for discovery, which included several questions as well as a request for copies of numerous documents containing technical information about what was known about the O-ring problem before the accident, along with contractual and specification documents imposed by NASA on Morton Thiokol. It took me considerable time to prepare my response.

I told our corporate attorneys that I didn't believe any of the plaintiffs had a good case against Morton Thiokol, but I did believe that NASA could be found culpable for all the different reasons I have already covered in this memoir. Thiokol's defense attorneys agreed with my logic and asked me to help provide detailed information supporting this position. It was rather ironic that the company was now going to depend on my actions and conclusions to defend it against lawsuits in the billions of dollars, when I had been punished by the company for revealing these same facts to the Presidential Commission.

As it turned out, I was able to dodge the first bullet, as the McNair lawsuit was settled out of court rather quickly. However, the Michael Smith case was a horse of a different color. Jane Smith, widow of *Challenger* Pilot Michael Smith, was very upset by what she had observed from the publicly televised hearings of the Rogers Commission. She had also learned that the astronauts really had survived the explosion and had not been killed instantly as NASA first reported. Close examination of the flight films revealed the crew cabin separating intact from the rest of the orbiter after the explosion. Some of the crew members, one of them being Michael Smith, had had time to hook up their emergency oxygen. NASA's recovery of audiotapes inside the crew cabin indicated that several members of the crew, if not all of them, had been alive until water impact (around 200 miles per hour) several minutes after the explosion. The last recording released to the press was a statement from Smith, "Oh! Oh!" just prior to the explosion, indicating that he knew something was wrong. Realizing this and hearing the evasive and arrogant testimony of many of the NASA officials and of Morton Thiokol senior management had outraged Jane Smith. She bore no ill feelings toward either Roger Boisjoly or me; in fact, she had made a statement to the *Washington Post* that she admired our honesty and courage before the Presidential Commission. She was particularly upset with what she had heard from Larry Mulloy—so upset that she had his name added to the list of defendants, the only individual name in her lawsuit.

As it turned out, Jane Smith filed her civil suit in Florida. Because of a large backlog of civil and criminal actions—mostly drug-related cases that the new governor had promised to clear off the docket—her case was put on the back burner. Seeing no action on her case for over a year, the attorneys for the Smith family filed another lawsuit, this time in the Commonwealth of Virginia, where Smith and her children now lived. As soon as that happened, lo and behold, the state of Florida, to avoid any further embarrassment, moved her case up. At the preliminary hearing in Florida, the government's defense attorneys, representing NASA, were successful in having NASA removed as a defendant, because Michael Smith was still a captain in the U.S. Navy and thus a government employee; his accepting the risk of flying the Shuttle was just like any other military pilot accepting the risk of flying military aircraft in the service. On the other hand, Larry Mulloy, a government employee, remained as a codefendant in the case with Morton Thiokol, Inc.

In October 1987, our attorneys received a set of interrogatories from Smith's attorneys relative to the lawsuit in Virginia. The interrogatories were basically the same as those I had received earlier in the year in the McNair litigation. Smith's attorneys also requested all of my records concerning my personal and office files, calendars, work assignments, and appointment logs spanning the period October 1, 1985, through March 30, 1986. It was a difficult and time-consuming task to retrieve all of this information. I finally managed to assemble it in a package for Smith's attorneys, including an update to the material I had supplied for the McNair litigation. Generating all this information brought back a lot of bad memories about *Challenger* that I thought I had finally gotten over. It wasn't long, however, before I was back to work on the redesign and feeling very good about the progress we were making.

My spirits were high as we moved toward the DM-9 test in December 1987. But as it seemed to happen all too frequently for me in recent years, the euphoria was short-lived.

The day of the DM-9 test dawned very cold. It was two days before Christmas, and a strong north wind was blowing across northern Utah. The temperature was in the low twenties, with a wind chill well below zero. We had attempted to test the motor the day before, but during the countdown the electromechanical safe-and-arm device failed to rotate over from the safe to the arm position in the required time. In the bone-chilling cold, foolishly wearing just my suit without a topcoat, I attended the pretest press briefing, held in the media trailer inside the viewing area. I sat in the back with a hot cup of coffee while Carver Kennedy and John Thomas explained what happened the day before to delay the test. I left before the press conference ended to drive up to the test area. I had to be in the control room during the final phases of the countdown and the test itself in the event anything went wrong.

We planned to test the DM-9 motor at 1:00 p.m. A large crowd of people nearly froze to death in the viewing area. Not nearly as many showed up as for the first RSRM test in August, but far more than I'd expected in this weather. Perhaps interest in the test had been heightened by the cold temperatures, which caused people to reflect on the cold *Challenger* launch, and now we were going to test our redesigned motor in a far more severe environment. It was going to be a good test for the new field-joint heaters. In the control room, everything looked good, so NASA's Royce Mitchell and I gave the go-ahead to proceed with the terminal phase of the countdown.

The motor fired right on schedule, and everything looked good. Immediately after the test, Mitchell and I went up to the test bay to inspect the motor in the test stand. I remember getting out of my car and walking around the motor, inspecting all the joints for any soot blowby and examining the case and nozzle for any heat-affected areas. Still without a topcoat, I was freezing to death, but as far as we could see the motor looked in good shape. We drove to another building to take a quick look at some of the data traces from the instrumentation. The data also looked good, so Royce and I went down to the trailer in the viewing area to brief the press. So elated were we over the results that it seemed to us like Christmas morning. Morton Thiokol always closed the plant down from Christmas through New Year's Day, and with the successful test behind us, we all left to enjoy a long holiday. Normally, we'd have someone put on a full respirator and go inside the motor a day after the test to inspect the interior of the motor and nozzle for any anomalous conditions. But since that was going to be Christmas Eve, we told our people to delay the interior inspections until sometime convenient during the week after Christmas.

On December 28, I received a phone call from John Thirkill, our General Manager. The engineering folks had decided to go out and inspect the interior of the motor. A workaholic, Thirkill had gone over to the test area and talked to one of the engineers who had been inside the motor. The engineer told John that the nozzle looked really bad: there was a large piece of the nozzle lying in the bottom of the motor. It appeared that we had lost a significant portion of the carbon-phenolic outer-boot-ring and that the portion still remaining on the nozzle was severely de-laminated. Thirkill himself briefly entered the motor along with Ross Bowman, our Vice President of Safety, Reliability, and Quality Assurance. The two of them confirmed the sad state of affairs. John told me to get to the plant right away, because we needed to decide what to do about the two aft segments we had ready to ship to KSC for the first flight, which had similar nozzles already installed.

In order to carefully inspect all other areas of the motor for any possible anomalies, we decided to expedite the disassembly of the DM-9 motor. To NASA, we recommended removing the nozzles from the first flight set and replacing the OBR in these nozzles with the design successfully tested in DM-8, which meant it was a good thing we had fabricated two extra OBRs of the DM-8 design in the event something happened to DM-9. NASA was reluctant to accept our recommendation until we had a chance to review all the test data, completely disassemble the fired DM-9 nozzle, and understand why it had failed. Immediately, we put together a DM-9 failure analysis task force.

I went home that night very dejected. Our holidays were cut short, and all of a sudden we were back in the soup again. This was the first major technical setback for the redesign, and it occurred on a component of the nozzle that we hadn't wanted to change in any way from the successful DM-8 test.

I was in my office for only a very short time the next morning when someone came in and asked me if I had heard what had just happened over in the Strategic Operations area of the plant. "No, what happened?" "There was a fire in a building where the solid propellant is cast for the air force's Stage 1 Peacekeeper (MX) motors, and the word is that five or six people were killed."

A friend over in Strategic Operations confirmed the story. "Yes, Al, a crew of people was in the casting building trying to loosen a mandrel from a fully loaded Peacekeeper motor." I knew that a mandrel was a Teflon-coated aluminum pipe used to form the geometry of the propellant grain. The propellant was cast around a mandrel, or core, with the propellant being cured at a temperature of some 135°. After it was cured, the motor was cooled to ambient temperature and the core removed remotely by using a crane to pull it out of the motor. Apparently, there had been some problem with this removal operation, because the mandrel wouldn't break loose from the propellant grain. A group of people went into the building to see what caused the problem. While they were in there trying to loosen the core, the motor ignited, causing an explosion. A 6,000° flame spewed out, killing all five people in the building.

All of a sudden, our problem with the DM-9 outer-boot-ring didn't seem significant anymore.

I really wondered whether this company I worked for could survive this latest disaster. At Congress's direction, NASA was already funding a study with our competitors for an advanced solid rocket motor that would eventually replace the motor we were redesigning; maybe NASA would just decide to go ahead and accelerate that approach and avoid Morton Thiokol altogether. After all, the newspapers were certain to pick up on our most recent woes and remind everyone that the same company responsible for killing seven American astronauts had now, less than two years later, killed five of its own employees. How long should the American public allow these yo-yos in Utah to continue? At best, the bad publicity was going to result in a further postponement of the Shuttle's next flight. Most likely the government would shut down the whole MTI plant until this latest catastrophe was understood and corrective action taken. The public outcry might not only further delay the next Shuttle flight but force NASA to fix the SRB problem with a new contractor, thereby stopping Thiokol from killing any more people.

I didn't know any of the people personally who were killed in the Peacekeeper incident, but I did know several of the supervisors and managers who were involved in this program. They were going to be under siege, and I felt very sorry for them. I had been there myself and had walked that mile in the same shoes, carrying guilt that I knew they, too, would be carrying for a long time. I had heard that four of the people had been killed instantly, but one survived for several hours. The fire and ambulance crew that removed the bodies indicated that that fellow had been able to speak and communicate for a few hours before he died. The paramedics prevented his wife from talking to him in the condition that he was in, afraid his wife would go into shock if she saw him. The 6,000° propellant fire totally burned off his nose, and his ears and face were unrecognizable; he had apparently been standing over the casting pit looking down into the motor when the motor ignited. At the bottom of the casting pit they later recovered some body parts from some of the victims of this horrible explosion.

After that horrendous day, I couldn't get enthused about my own problem with the DM-9 outer-boot-ring failure. I was more interested in finding out what had happened in the Peacekeeper accident and why it occurred. We had several Shuttle segments in propellant cure awaiting core removal. We obviously wouldn't risk removing the cores from the segments until we fully understood what had happened in the Peacekeeper incident. I just stared at the walls in my office, wondering how we were going to be able to recover from all this and whether the Shuttle would ever return to flight. After all, catastrophes such as these sometimes happen in threes. I couldn't help thinking, When is the next one going to happen? What will it be? Will I be responsible or accountable for it? In any event, this will surely be the straw that will break the camel's back. We'll probably be out of business at best, and under some grand jury indictment at worst.

It was a long forty-mile drive home that day with all these horrible things that kept circulating in my head. I couldn't help but think about what happened to those five guys in the Peacekeeper fire and how that could have happened to me when I had gone "inside" the propellant grain of the RSRM to inspect the field-joints after motor assembly in the test area. I had done that for both DM-8 and DM-9 and had intended on doing that for the remainder of the RSRM tests. Maybe I should rethink doing that. It was later concluded that the Peacekeeper accident occurred because of a static charge buildup from the people and tooling that were improperly grounded. Crawling inside the rocket motor on a black velostat sheet (a conductive plastic film) on the surface of the propellant also created a static charge buildup if it was not properly grounded or the velostat was not as conductive as it should be or the wrong black plastic sheet was used. At home, I looked so depressed that my wife told me I should stay home the next day. "I can't do that, Linda. I really need to get back out to the plant tomorrow and help develop a plan to get us back on track for returning the Shuttle to safe flight." Even though I felt like hanging it up, I had invested way too much time and had suffered way too much agony to give up now.

At the plant the next day I met with John Thirkill and Dick Davis, and we put together a recovery plan. Davis had replaced Carver Kennedy as the Vice President of Space Booster Programs, with Carver returning to KSC in his old job, Vice President of Space Services responsible for SRB and ET assembly operations in the Vehicle Assembly Building. We had just sent to KSC an assembly test article consisting of two loaded center segments, to be used for checking out the new field-joint mating procedures and new ground support equipment prior to conducting these same operations on the first flight set. Dick Davis had been Vice President and General Manager of Thiokol's Elkton Division in Maryland. He recognized that our Aerospace Group could not survive if we weren't successful in returning the Shuttle to safe flight, so he told Ed Garrison he'd be glad to come out to Utah and help in the Shuttle recovery activity. Garrison accepted his offer when Carver returned to KSC. Not intended to be a temporary assignment, the company also made him the Assistant General Manager reporting to Thirkill, who expected Davis to replace him when he left the General Manager's job.

When Thirkill, Davis, and I met with John Thomas and Royce Mitchell, the NASA guys basically agreed to our approach. We recommended removing the nozzles from the first two flight motors and cutting out the flexible boot and outer-boot-ring and replacing those parts with the DM-8 design we had already fabricated as backup parts. We wanted to go ahead and install the nozzle with the involute (or rosette) OBR design on the next planned test motor (QM-6), then remove the nozzle, cut out the involute OBR, and replace it with a newly fabricated DM-8 OBR design just like we were doing on the two flight motors. We would delay the shipment of the first two flight motors until we tested QM-6, which was now delayed to April 1988.

I felt good about the recovery plan, but it quickly became apparent that issues related to the Peacekeeper fire were going to get in the way. The forward segment for the QM-6 motor was still in the casting pit with the core mandrel in place. We needed this segment to assemble the static test motor in the test bay. Assembly sequence of the static test motors was exactly opposite to the assembly sequence of the flight motors. The flight motors needed the aft segment to attach the aft skirt, which bolted on to the mobile launch platform to hold up the entire Shuttle stack. The remaining segments were then stacked vertically, with the forward segment placed on top of the stack last. In the horizontal test arrangement, the forward segment was needed first to bolt into the forward thrust adapter. The center segments were then joined to the motor, with the aft segment last.

Murphy's Law had struck again. We needed to get the aft segments for the flight motors to KSC as soon as possible, but they contained the nozzle that now posed a big problem in terms of needing to be removed, repaired, and replaced. We needed to get the forward segment of QM-6 over to the test area as soon as possible, but now we had a big problem with that because we couldn't remove the core tooling from the propellant grain until the Peacekeeper investigation had been completed. We couldn't ship the flight motors to KSC until the QM-6 static test was conducted successfully.

My hair was turning grayer by the minute.

It didn't take long for the press to spread all of the bad news. NASA fell under great pressure to assess the impact of our explosion to the planned launch of the Shuttle, and the agency reported that the failure of a critical part of the nozzle on DM-9 would, indeed, result in a further delay to the

Shuttle program, postponing the first flight of the redesigned Shuttle from the spring of 1988 to late summer or early fall of 1988.

And another clunky shoe was about to drop. During the detailed inspections of other areas of the DM-9 motor, it was noticed that one of the operational pressure transducers (OPTs), which measured the pressure in the motor, was loose on the motor. The motor held three OPTs, and they provided pressure data to determine the motor's ballistic performance and provided data to separate the boosters from the ET at the end of SRM burnout after launch. This pressure data was transmitted to the orbiter, and when the last of the three pressure transducers reached fifty-pounds-per-square-inch pressure at burnout, the motors then separated from the ET.

The Shuttle could not achieve orbit with the empty motors attached. When the safe-and-arm device was removed after the aborted static test attempt of DM-9, one of the special bolts that contained one of the OPTs was also removed to allow better access to the safe-and-arm. After the new safe-and-arm was installed, this OPT was reinstalled, but the technician forgot to retorque the OPT special bolt to the proper load, and therefore the OPT assembly was only finger tight on the motor and loosened somewhat from the vibration during static test.

This problem, which had been basically overlooked because of the OBR failure, was actually far more critical and came much closer to a motor failure than the loss of a portion of the OBR. If the OPT assembly came out or failed to provide an adequate seal, hot gas would leak from the forward dome of the motor, causing a catastrophic failure in just a few seconds.

I was far more concerned about this problem than about the OBR, because I knew we had a sound backup design for the OBR and an OBR failure was not as critical as a loss of an OPT. Not tightening the OPT special bolt was a symptom either of lack of attention, worker fatigue, or downright sabotage. The last possibility really scared me. Sabotage was certainly not out of the question. We had recently found some O-rings that were to be installed in a motor that clearly had been cut intentionally. The FBI had been notified, and an investigation was in process.

There were plenty of things to worry about as we closed out 1987, and thinking about the upcoming second anniversary of the *Challenger* accident didn't help.

35

Witness for the Defense . . .
and Prosecution

When NASA declared a slip in the planned launch of STS-26R, several journalists conducted interviews with the astronaut crew selected to be the first to fly the Shuttle following *Challenger*, a catastrophe all of them had witnessed firsthand. Being in this position was newsworthy in itself, but now that there was a delay attributed to a test failure in the SRB made their situation all that more ominous. The reporters asked the astronauts about the Thiokol failure and how they felt about the delay. Basically, the Shuttle crew responded by saying that, during a development program, one expects things like this to happen, and they were confident the problems would be fixed.

Up till now I had been too engrossed in my work to recognize the significance of the crew selected for *Discovery*, STS-26R. Comprised of Commander Navy Captain Rick Hauck, Pilot Air Force Major Dick Covey, and Mission Specialists Pinky Nelson, Dave Hilmers, and Mike Lounge, the entire crew was military. Ironically, their prime mission was the same as the *Challenger* crew's: to deploy NASA's Tracking and Data Relay Satellite. For me personally, there was the additional irony that the two men in charge of flying the machine were two of the three astronauts I had personally known before the *Challenger* accident occurred. I had met Rick Hauck in Israel in February 1985, and had even met his wife shortly after *Challenger* when she toured our plant with Congressman Bill Nelson from the House Committee on Science and Technology. I had also spent a considerable amount of time with Pilot Dick Covey during his visit to our plant at the time of the first static test of the FWC-SRM (DM-6) in October 1984. The other astronaut that I had met before the *Challenger* accident was Dr. Rhea Seddon, M.D., Hoot Gibson's wife; Hoot was scheduled to be commander of the following flight, STS-27R *Atlantis*.

With a near-panic mode having set in and a critical design review coming up in February 1988, I began working tremendously long hours and weekends once more. We had to recover from the Peacekeeper accident so that we could start up our propellant operations once again to support the program and that involved, among other things, removing the nozzles from the first flight set of motors to replace the outer-boot-ring with the DM-8 design. We were also preparing to complete this same identical operation on the QM-6 static test motor when it became available. If that wasn't enough, NASA engineering at MSFC insisted that we conduct a very detailed failure analysis of the failed OBR design on the DM-9 motor.

We were already short-handed for engineering support, as we were also in the process of testing a series of short-duration Joint Environment Simulators at both Thiokol and Huntsville. Our first successful test of the new Transient Pressure Test Article (TPTA) had just been completed at MSFC in November 1987. This was a very sophisticated piece of test hardware that included two full-scale field-joints, a nozzle-to-case joint, and a full-scale igniter. We tested the TPTA with a full Shuttle mass simulator on top of the assembly, an aft skirt attached to the bottom, and liftoff strut loads applied to the ET attach section during ignition. With all of this activity going on simultaneously, we resisted as best we could conducting a detailed failure analysis of the DM-9 OBR, because we had no plans to use it in the future. The head of the Propulsion and Structures Directorate at MSFC, Dr. John McCarty, formed a task force on the DM-9 OBR failure, wanting further analyses and laboratory structural tests on the failed involute OBR design. We tried to convince McCarty to put this activity on the back burner—or better yet, shove it off the stove—because we weren't going to use that design anyway. There were too many more important things to accomplish in a very short period of time. McCarty didn't agree, insisting on support to the task force from an already overworked Morton Thiokol engineering staff.

Getting ready for the critical design review was itself a monumental task. The engineering documentation required to support the CDR was mind-boggling. Over 35,000 pages of engineering drawings, test plans, test data, and analysis had to be assembled. We also had to prepare a full eight-hour stand-up presentation summarizing the key engineering data. The CDR board would meet with scores of subordinates for nearly two weeks, poring over the details of the engineering documentation we submitted. Any concerns with anything in our CDR package would be written up as review item discrepancies (RIDs), with each RID requiring a response involving an analysis or test data to close it out. In all, we had to resolve over a thousand RIDs.

Somehow we made it through all of them, but several hundred RIDs simply could not be closed out and had to be carried over to the design certification review, scheduled for a couple months before the planned launch.

The following month was hectic. We had to complete the nozzle rework of the aft segments for the first flight as well as the fabrication of the remaining segments for the first flight of the redesigned solid rocket motor. We also were preparing the segments for shipment to KSC pending the outcome of the first qualification motor test (QM-6) planned for late April. Everyone was very tense, because we all knew that this test had to be perfect or we wouldn't return the Shuttle to flight by the end of the summer—and probably not this year.

As it turned out, the static test of QM-6 went off without a hitch. The original Morton Thiokol design of the outer-boot-ring, the one successfully tested on DM-8, worked well again, and the decision was made to ship the first flight set of motors for STS-26R to Kennedy. The Shuttle's return-to-flight was now set for the last half of September, only five months away.

It never seemed to fail; the enthusiasm of passing the critical test of QM-6 was quickly dampened by negative news surrounding ongoing controversy over the *Challenger* accident. In an interview with *Business Week* about the recovery activity at Morton Thiokol, Charlie Locke had stated that the responsibility for launching *Challenger* had rested with NASA because their people launched the vehicle in spite of our objections and launched it in a condition that was outside of the Shuttle's specification limits—exactly the position I had outlined for our corporate lawyers. The article appeared in the magazine's March 14, 1988, issue. Larry Mulloy read the article and submitted a vindictive and defensive response as a letter to the editor, which *Business Week* published in its "Readers Report" on April 18. Mulloy's response appeared under the title, "Morton Thiokol: The Wrong Kind of Industry Leader." It read:

> In "Morton Thiokol: Reflections on the Shuttle Disaster" ("The Corporation," March 14), Chairman and CEO Charles S. Locke's arguments regarding Morton Thiokol Incorporation's lack of approval to launch *Challenger* were specious. No matter how many Thiokol executives try to dupe investigating commissions, committees, business reporters, the public, and ultimately a jury, the truth is that the *Challenger* was

launched within specification limits with the unqualified approval of
senior management. The 40° F. temperature specification is the lower
limit on the operating range for the mass of the propellant in the motor. That temperature was determined to be 54° F. at the launch of
Challenger—well within limits. The lower temperature limit for the
surrounding air at launch is 31° F. . . . It was 36° F. at the launch. It is a
small wonder that Morton Thiokol seems to lead the solid rocket motor industry in the destruction of lives and property. The combination
of arrogance and apparent ignorance their management continues to
display is lethal.

Larry Mulloy,
Arlington, Va.

The letter was vintage Larry Mulloy, still a defendant in the Michael
Smith litigation and still in denial for accepting any responsibility, much
less blame, for the decision to launch *Challenger*. What Larry failed to mention in his cryptic reply was that he personally didn't agree with Thiokol
engineering's recommendation not to launch *Challenger* the night before
the launch, and it was only after his caustic remarks during the telecom that
he forced the "unqualified approval of senior management." He also failed
to mention that the approval he finally received, which he quickly accepted
without reservation, was based on launching at an expected surrounding air
temperature of 26° to 29°, clearly below even his postlaunch interpretation
of the Shuttle's low temperature limit of 31°. Mulloy not only accepted a recommendation that was outside anyone's interpretation of the Shuttle launch
constraints, which should have required a waiver, but he chose to intentionally withhold this information from the NASA Mission Management Team
during his discussions with them on the morning of the launch.

There was no love lost between Charlie Locke and me, and Locke was, in-
deed, pretty high up on the arrogance scale, but, in my opinion, Mulloy took
the title away from "Commodore Locke" for the insensitivity of his *Business
Week* letter. It was an insult not only to me but to all of the good people at
Morton Thiokol who were presently working long and hard to return the
Shuttle to safe flight.

Other events rained on our parade besides Mulloy's letter. To expedite
the shipment of the Shuttle's segments to KSC, dedicated trains from the
Union Pacific Railroad were used to transport the segments from Utah to
Florida. Because the segments were transported on specially designed rail-

cars that approach the maximum height and weight for rail transport, the train followed a special route to provide sufficient clearance through tunnels, bridges, overpasses, and rail yards.

As Murphy's Law would have it, the train carrying the last three segments was involved in an accident at a railroad crossing in Mississippi. The train hit an automobile that was attempting to cross the tracks before the train got there. The huge locomotives with these two large railcars carrying over 450 tons of rocket motors hardly slowed down, literally flattening the automobile and instantly killing the car's two occupants, although the train didn't derail or suffer any significant damage.

It wasn't the only train incident. Because the shipment of our segments was such a significant event to the local railroad community back in Utah, the site of the completion of this nation's first transcontinental railroad, the Union Pacific Railroad hooked up one of their old historic dining and club cars to accommodate the Thiokol people accompanying the segments to KSC. During the train's journey through the Deep South, two young renegades used the moving train for target practice. With their hunting rifles, they shot at the large yellow fiberglass railcar covers that were placed over the segments, but the bullets ended up coming through the glass windows of the dining car. Fortunately, no one was struck. The conductor called ahead to the nearest station, which then made a call to the local sheriff's office, and the two perpetrators were apprehended.

The job of returning the Space Shuttle to safe flight seemed to be under some kind of curse. On May 4, 1988, a terrific explosion occurred in Henderson, Nevada, at the plant site of Pacific Engineering & Production Company, the manufacturer of the most important ingredient in our solid propellant—the oxidizer, ammonium perchlorate. Pacific Engineering supplied about 75 percent of the ammonium perchlorate to Morton Thiokol for the Shuttle program. Fortunately, there was a second source for this critical material, the Kerr-McGee Corporation, also located in Henderson, Nevada, which supplied most of the ammonium perchlorate for DOD solid rockets. The blast from this explosion in Henderson was the strongest recorded in the U.S. since nuclear testing had been terminated. The shock wave shattered windows for miles, blew automobiles into ditches, knocked doors off hinges, and broke gas lines miles away. The earthquake center on the campus of Caltech in Pasadena recorded the explosion at 3.5 on the Richter scale. Two people eventually died, and 326 were injured.

There was other controversy not associated with *Challenger* or the transport of segments to KSC. During the rapid buildup of employment at Morton Thiokol to support the ambitious redesign activity, the company contracted with several engineering service job-shoppers to provide experienced engineering help in different areas, particularly in systems engineering, test planning, and system safety. As a result, a large number of support people were hired under the engineering services contract. This contractor was responsible for verifying the academic and professional credentials of the people supplied to Thiokol under the contract, over 100 of them at one point. For the most part, these people were well trained and provided valuable help to us during a very critical time; however, there were a few who not only lacked the proper credentials but came with their own hidden agenda.

One of these, Dr. Anthony Lane, spent an inordinate amount of time in the technical library trying to access classified data from our ballistic missile program. Lane claimed he was going to provide nuclear "hardening" to the redesigned boosters. As there was no reason for that with the Shuttle, his Thiokol supervisor called him in and told him to stop looking into that. Lane then started pursuing what he thought were other failure scenarios for the *Challenger*. He was convinced that plasma generated from the rocket plume had somehow arced to the external tank, causing the failure. He spent a good deal of time researching this in the technical library, and during one of his library visits he again ordered some classified documents concerning nuclear hardening of missiles. When the library checked to see if Dr. Lane had a "need to know" for these classified documents, the supervisor withdrew Lane's library request and escalated the problem to John Thirkill, Vice President and General Manager of Space Operations, who decided to release Lane from his work for the company. Because Lane was not an employee of Morton Thiokol, the company could not fire him, but it did contact the engineering services company that had supplied him to us and told the contractor that we no longer needed or desired Lane's services.

Shortly after his termination, Lane hired an attorney to sue Morton Thiokol on the grounds of unjustifiable termination of his employment. Lane claimed he had uncovered safety problems with our SRM redesign and that he had been fired when he tried to make these problems known to senior management. What a crock! When our attorneys checked on his background, they found out Lane not only had been dismissed from prior jobs but that he didn't even have the Ph.D. degree that he claimed. I was told that he had received a doctorate from some mail-order operation. His case was quickly dismissed at the initial hearing.

Shortly after this incident and about the same time that Boisjoly's multibillion dollar lawsuit against Thiokol and NASA was getting considerable press, another one of our job-shoppers quit his job and sued Thiokol for undisclosed safety concerns with the redesigned SRM. This opportunistic fellow, Steven Agee, made several preposterous claims about the lack of response by Thiokol management concerning safety issues he had raised. While working as a job-shopper at Thiokol, Agee had gone to the FBI and voiced his concerns. At the request of the FBI, Agee wore a hidden tape recorder and turned over copies of hundreds of documents related to our booster project; he also recruited others to act as informants. Robert Bryant, one of the FBI agents involved in the case, was later quoted as saying, "If any information came to our attention that would endanger the safety of the launch, we would have passed it on to NASA." Agee later appeared on an investigative TV program alleging that the SRMs were being redesigned without proper attention to safety issues. He sued Morton Thiokol, hoping to get an out-of-court settlement to keep quiet, but again the company challenged his allegations, which were totally baseless, and his case, too, was quickly dismissed. The company continued to receive a lot of bad press over all of this. It was hurting morale during a very critical time in the recovery program.

It seemed like I was spending more time with the lawyers than I was getting prepared to return the Shuttle to safe flight. Earlier in the year, the newspapers printed a story saying that the parents of *Challenger* astronaut Judith Resnik had reached a multimillion dollar settlement with Thiokol over their daughter's death. The article further stated that the Resnik family placed the amount at between $2 million and $3.5 million, which was the range of settlements received by other families of the seven-member crew. These lawsuits really hurt Morton Thiokol's image, but not its pocketbook, because the company had a $500-million insurance policy covering such contingencies.

But Roger Boisjoly's lawsuit for over $2 billion was still pending, and the Michael Smith lawsuit for $1.5 billion was starting to heat up. I was scheduled for a deposition by Jane Smith's attorney, a Winston-Salem, North Carolina, aviation lawyer named Bill Maready who had been a friend of Mike Smith's and whose family had vacationed together with the Smith family the week before Smith's death. The deposition took place in a room in the Ogden Park Hotel in Utah. Several people from Thiokol, as well as some ex-Thiokol employees, including Roger Boisjoly, were also scheduled for depositions that week.

Before the depositions were given, the attorneys that Thiokol had hired from the Chicago law firm of Adler, Kaplan and Begy met with each of us to prepare us for the depositions and answer any questions we might have about the process. I was scheduled for a deposition on the same day as Bob Ebeling, who was working for me at the time of *Challenger* as the Manager of Ignition Systems and Motor Assembly and had the O-ring seals under his direction. Suffering from a serious bout of depression, he had retired from Thiokol in his late fifties in the summer of 1986. At the question-and-answer period with Thiokol's attorneys, one of them, John Adler, asked if there were any questions that we thought we might be asked that would be uncomfortable for us to answer. Ebeling said he did not know how to respond to the question about the supplemental money he was receiving from Thiokol in conjunction with his early retirement income. Bob said that he had visited Ed Garrison prior to his retirement, and Garrison had agreed to supplement his early retirement income with $100 per month for each year of his service to the company. This supplemental income was to be provided for ten years. Adler was already aware of this situation and acknowledged that Garrison had authorized those payments as a supplement to the pensions for those who had been intimately involved in the *Challenger* accident and aftermath. Ebeling was a longtime employee with the company, so this amounted to a sizable sum of money over ten years.

This was the first time I had heard anything about this; apparently Ebeling had been sworn to secrecy when given this lucrative early retirement bonus. Adler told him that if the plaintiff's lawyers asked him if he had received any additional payments from the company as a result of the *Challenger* accident, that he didn't have to consider his supplement as any kind of a payment that could influence his testimony in this case. Adler then stated it was his understanding that Garrison intended to provide similar payments at the time of their retirement to others who had been central figures and borne the brunt of the *Challenger* investigation. Jerry Mason was the only other one who was at the early retirement age of fifty-five or older, and it was rather general knowledge that he had received a lucrative golden parachute to take early retirement.

After this briefing session, I asked Ebeling what he knew about Adler's statement that others were to be offered similar bonuses to come at the time of their retirement. Bob told me what he knew and that it was his understanding I would receive considerably more than any of the others because of my long service and willingness to head up the redesign activity and help defend the company from the pending lawsuits. That sounded pretty great to me because that would mean I could retire in four years at age fifty-five with nearly a $400,000 premium or maybe even more spread over ten years for my thirty-three years of service by that time.

My deposition with attorneys for Michael Smith lasted for several hours and was not as stressful as I had anticipated. I was very concerned because of the large volume of information I had supplied in response to interrogatories from these same attorneys in October 1987. I had given them a huge three-ring notebook full of material. Maready's demeanor was clearly influenced by the fact Jane Smith had expressed to him her admiration for what I had done in trying to prevent the launch of *Challenger* and for my testifying before the Rogers Commission. It was unfortunate, Maready said, that Thiokol alone was now holding the bag, as preliminary hearings in Florida and Alabama had removed NASA and Larry Mulloy as defendants in the case. Jane Smith was so upset over Mulloy being removed as a defendant, Maready indicated, that she insisted her attorneys find a way to reinstate him as a defendant in this case. Her lawyers told her that they didn't feel that was possible, but she insisted they pursue it anyway.

Maready's questions were primarily related to my testimony before the Rogers Commission and did not delve into the difficulties I had had subsequently with Thiokol management. I was particularly thankful for that because I didn't want to open old wounds, especially now, when it appeared that senior management had finally recognized that I was really helping the company, not hurting it, by my actions during and after the *Challenger* accident. What really surprised me was that Smith's attorneys were planning to use me as their chief witness. As I was already the chief witness for the defense, I found myself in an unusually awkward position acting as the chief witness for both the prosecution and the defense.

It naturally took a while before Jane Smith's lawsuit went to trial. There was a long series of negotiations between her attorneys and ours looking for an out-of-court settlement, but to no avail. The situation really upset me, because I was in a no-win position. The plaintiff was claiming that Thiokol management had acted irresponsibly and negligently by not supporting me or consulting with me relative to my objections to the launch when I had been the company's senior management person at KSC responsible for approving any launch decision and, furthermore, that Thiokol had improperly deviated from government specifications. (At about this same time, the U.S. Supreme Court decided five to four that when a manufacturer complied with government specifications, even if the specs resulted in a defective product, the company got the same immunity as the government.)

Thiokol's attorneys, on the other hand, claimed that I had, indeed, exercised such authority even after Thiokol management changed their minds and gave approval to launch. I had passed on my concerns to NASA management, including Stan Reinartz, who was a member of the Mission Management Team. NASA management chose to ignore my warnings and furthermore had not even conveyed Thiokol's concern with the O-rings in cold weather to the MMT on the morning of the launch. Had it done so, the launch likely would have been canceled when temperatures of 7° to 9° were recorded on the right SRB aft field-joint. If the discussions I had had with the NASA management had been revealed to the MMT, they would have been sensitized to the cold weather issue and canceled the launch when the IR temperature data was obtained.

I had outlined this defense to our attorneys, and they'd agreed it would be a good defense. I was not looking forward to several days of ugly testimony in the courtroom, especially when we were now within just a few weeks of the Shuttle flying again.

36

Confirming the Silver Bullet

My presentation at the design certification review took over eight hours. Supporting it was over 50,000 pages of engineering reports and over 200 viewgraphs depicting the design changes made after *Challenger*, why the changes were made, test results, and mountains of analysis verifying that the redesigned motor met all of the design requirements. My engineering staff and I spent literally hundreds of hours preparing for this DCR, and we were all greatly relieved when it was over, especially when Admiral Richard Truly, the NASA Associate Administrator for Space Flight, remarked to everyone that it was the best-prepared presentation he had heard.

But not everything was going smoothly. In reassessing the launch commit criteria for the Shuttle, NASA had asked Morton Thiokol if we had any new requirements for the RSRM. I had always been concerned about the extremely low temperatures of 7° to 9° that had been measured just hours before liftoff in the area of the aft field-joint that failed on *Challenger*. NASA had concluded that the infrared gun used for obtaining this data was not calibrated properly, and that this data was in error, so it corrected the measurement upward to reflect higher temperatures. This had always bothered me because these extremely low temperatures had been measured *only* on the lower part of the right-hand SRB in the vicinity of the *only* field-joint that failed on *Challenger*. Furthermore, after all the IR data had been corrected, some of the temperatures measured on the left-hand SRB—ones that were basically reflecting local ambient temperature—now measured above the local ambient temperature, and ice was still forming in the antifreeze troughs below the nozzles. This correction business just didn't make any sense.

In the fall of 1986, I had asked one of our heat transfer experts, Stan Boras, to find out where all of the cryogenic boil-off and purge gas vents were located on the Shuttle launchpad; I wanted to determine if these vent gases could have caused some local cooling on the SRB. I knew that the boil-off from the liquid oxygen that was contained in the upper portion of the external tank was vented to the "beanie cap" on the top of the ET, and that the hydrogen boil-off was ducted away from the pad area, where it was burned. There were several other cold purge gases, such as nitrogen and helium, which were also used in various areas of the vehicle. I told Boras that I was particularly concerned about the gaseous oxygen (GOX) that was ducted away from the vehicle through the GOX vent arm attached to the beanie cap and dumped back into the atmosphere at the fixed service structure. It appeared from the prelaunch films of *Challenger* that I had reviewed on the morning of the launch that the GOX vapors drifted back toward the vehicle. I wondered if those GOX vapors could supercool the surrounding air; if the denser cold air mixed with GOX and migrated down between the ET and the SRB, it could very well be that the bottom of the SRB would be colder than the top. After finding out what he could about the prelaunch Shuttle configuration and the venting of cryogenic vapors, Boras informed me that it was possible to model the GOX venting effect analytically, but that the modeling involved a very complex and difficult flow analysis and heat transfer problem. But Boras had a man working for him, Dr. Rashid Ahmad, who had completed his Ph.D. in an area of convective flow computer modeling, and he would solicit Dr. Ahmad's help.

With his heat transfer section already working overtime in support of the redesign activity, Stan Boras wanted to know the priority of this new task; I told him it was a lot lower than the redesign activity and that he should only work on the cryogenic vapor venting problem when he found time to do so. Boras began collecting data on the venting of cryogenic gases from the Shuttle in the prelaunch configuration, but he didn't have much time to start analyzing the problem until the following year. During the previous year, however, he and Ahmad had developed a convective flow and heat transfer computer model that simulated the prelaunch configuration of the GOX venting from the ET. They had obtained some cryogenic vent gas flow rates from NASA and then modeled the *Challenger* wind and local ambient conditions as best they could. Their preliminary results indicated that the GOX vented from the GOX vent arm during the morning of the *Challenger* launch did, indeed, tend to flow back toward the vehicle around the ET, causing cooling of the local ambient air sinking down between the ET and the right-hand SRB and significantly depressing the temperatures of the lower part of the SRB. In fact, the computer modeling results basically duplicated the measured *uncorrected* IR gun temperatures that had been recorded on *Challenger* several hours before launch.

I told Boras he needed to present the results of this analysis to NASA as soon as possible. Even though the new RSRM field-joint design would be unaffected by this phenomenon because of the joint heaters, the "acreage" of the case should not be at such low temperatures because it decreased the fracture toughness of the steel-case material, drastically decreasing the ability of the case to tolerate undetected flaws in the case when it was pressurized. Stan Boras met with NASA engineers from MSFC and JSC on this issue, and they did not believe Thiokol's analysis, nor did their systems engineering contractor, Rockwell International. Rockwell was responsible for defining prelaunch environments and claimed that its own analysis did not indicate any problem with GOX venting. We were basically told by NASA that we should stop our effort in this area because it wasn't our responsibility area and, furthermore, the heat transfer experts from Marshall, Johnson, and Rockwell all thought our analysis of the problem was not correct.

I believed Boras was on the right track, and I told him so. To me, it couldn't be purely coincidental that the results of Stan's thermal model so accurately portrayed the actual temperatures measured on *Challenger* in the early morning hours prior to launch. I told him to continue analyzing different sets of ambient conditions of temperature, wind direction and velocity, and so forth, so that we could ultimately make a recommendation for defining acceptable and unacceptable launch conditions as part of the Shuttle launch commit criteria. No sooner had Boras initiated these additional studies when we received a formal letter from the NASA/RSRM Program Manager at MSFC, Royce Mitchell, telling us to discontinue our launchpad environment studies immediately. Mitchell wrote that GOX venting had been determined to be "insignificant," that it was not even to be an issue for the upcoming design certification review, and that we already should have adequate ground environmental instrumentation on the RSRM to verify the environmental conditions prior to flight.

Consulting with our legal people, they supported our engineering recommendation to continue working in this area at company expense, so as to strengthen our position for establishing acceptable launch commit criteria. I told our attorneys that we could also use the results of our flow/thermal model to support the litigation against us from Michael Smith's widow, which was then still pending. These results would further enhance Thiokol's defense at the expense of NASA's defense, because NASA had ignored the prelaunch IR data, which now looked to have been correct—data that had not even been made available to Thiokol before the launch. Furthermore, if the GOX venting phenomenon we were modeling proved to be real, then NASA could rightfully be considered negligent for not properly analyzing the situation or providing accurate prelaunch environmental criteria to Thiokol as part of its contractor-end-item specifications. NASA and Rockwell both claimed that not only was there no impact of cryogenic venting on the launch vehicle but their statistical analysis of prelaunch weather conditions that we claimed caused the STS-51L problem had a probability of occurrence of only 0.52 percent. Stan Boras, who conducted our analysis, later found out they misrepresented the historical statistical weather data (temperature and wind direction, velocity, and duration) to come up with such a low probability; they did this by using seven months of KSC weather data from October through April. If they would have used only the month of January or February like they should have, the probability of these unacceptable conditions occurring simultaneously on one day of each of these months would have increased by a factor of fifteen, to 7 to 9 percent.

Ultimately, NASA and Rockwell finally conducted a computational fluid dynamics analysis that confirmed the Thiokol analysis that this issue could indeed be a problem, so that a new set of LCC was finally developed, but not for months after returning the Shuttle to flight status. The criteria recognized that low wind speeds (below five knots) could result in cryogenic vapors accumulating around the vehicle on the pad, causing significant depression in temperature in local areas on the vehicle. As a result, NASA raised the ambient temperature prior to launch to 42° if the wind speed was below five knots, while temperatures down to 33° would be acceptable for ground winds at the pad in excess of five knots.

Additional new launch commit criteria pertaining to the ambient environment were based on a premature failure that occurred in the aluminum aft skirt of the SRB during a structural test, limiting ground winds from a southerly direction to less than twenty-four knots. This launch criterion was implemented because the ignition of the SSMEs prior to SRB ignition would always bend the whole Shuttle vehicle over toward the north, resulting in high bending loads on the aft skirt that supported the entire weight of the vehicle while it was bolted to the mobile launch platform. The premature failure of the structural test article indicated that this element's safety factor was only 1.28 instead of the required 1.4. To fix this problem, reinforcement plates were added to the SRB skirts, but this redesign by United Space Boosters Production Company did not result in any improvement of the skirt's capability, meaning that the Shuttle had to return-to-flight with a permanent waiver against the skirt for not meeting the required structural safety factor.

In general, higher wind velocities quickly removed any cryogenic vapors from the launchpad area, and higher wind speeds from the north reduced the loads on the aft skirt during SSME ignition. The new LCC established for ground winds and ambient temperatures recognized the potential of cryogenic-vapor accumulation. Limited three-dimensional modeling of the pad environment confirmed our original Thiokol one-dimensional thermal analysis, with the results later published by Boras and Ahmad in the technical literature long after the Shuttle successfully returned to flight.

Next to the spring of 1986 that followed *Challenger*, the summer of 1988 was the most hectic time of my life. The successful test of the first qualification motor (QM-6) in April 1988 had verified the integrity of our selected nozzle outer-boot-ring design. The first flight set of motors had been modified and sent to KSC, and NASA had announced that it planned to launch the first Shuttle since *Challenger* in late September 1988. As Vice President of Engineering, my plate was overflowing. We had several critical tests, numerous important meetings requiring my participation, and several activities at KSC to support. Before the first flight could occur, we had to test in our new test bay the second qualification motor (QM-7) at high temperature with simulated liftoff and maximum dynamic pressure flight loads applied during the static test, conduct short burn tests of both the Joint Environment Simulator at Thiokol and the Transient Pressure Test Article at Marshall (with intentional flaws induced in the insulation and seals in the joint areas), conduct a hydro-burst of the previously fired Nozzle Joint Environment Simulator to verify the structural integrity of the new radial bolted case-to-nozzle joint, and conduct a final all-up, full-scale, full-duration static test (referred to as Production Verification Motor No. 1 [PV-1]) with prefabricated intentional flaws in all of the joint areas of the rocket motor to demonstrate the fail-safe capability of the redesigned solid rocket motor.

In addition to these critical tests, we had to complete the design certification review and close out literally hundreds of review item discrepancies that were generated during the various design reviews. We also had to meet with the National Research Council on a monthly basis, review our progress with the NASA Marshall oversight committee (headed by Al Norton), and support a series of Level I, II, and III flight readiness and simulated countdown reviews for a planned flight readiness firing (FRF) at KSC, all heading to the launch of STS-26R.

If this wasn't enough, I was told to spend some time preparing my testimony in conjunction with the Jane Smith litigation. Thiokol had been unsuccessful in settling this lawsuit out of court, and a starting trial date of September 12, 1988, was on the docket in Richmond, Virginia. It was anticipated that the trial could take as long as two weeks. Because of the heavy workload leading up to the scheduled Shuttle launch in late September, Thiokol's attorneys had filed a request for a change of venue to move the trial from Richmond to Salt Lake City, which would make it easier for those who were testifying to accommodate their work schedule. In addition to the Smith litigation, I was also providing information to our attorneys in response to questions being submitted by the attorney for Roger Boisjoly with regards to his $2 billion lawsuit against the company.

My days were very long and nights very short. The pressure was building, and with an official new launch date for the Shuttle announced by NASA, the press became reenergized. The media were once again present at every test, scrutinizing the results with the hope of uncovering some problem that would surely delay the launch and cast further doubt on whether Thiokol and NASA were capable of ever recovering from the *Challenger* accident.

When NASA had announced its new launch date for September, it was based on an ambitious 100 percent success-oriented schedule of tests that must be successfully conducted before we could launch. The QM-7 motor had been in high temperature conditioning with 100° air for over a month in order to bring the propellant to a minimum propellant mean bulk temperature (PMBT) of 90°—the actual PMBT being 92°, with some areas of the outer skin at 102° at time of the test. This motor also had simulated liftoff and flight loads induced through a set of hydraulic actuators into the external tank attach ring just below the aft field-joint. We felt some apprehension about the test because our new test bay (T-97) stood much closer to the mountain than did our old one. With the rocket's 6,000° plume shooting out from T-97 directly into the mountain, there was a possibility of higher acoustical reflections and added debris from tons of dirt and rock blasting free. Fortunately, the QM-7 test went off without a hitch on June 14, 1988, with the test results right on our predictions. Like most posttest press briefings, I was asked if this test now cleared the Shuttle for flight, and if not, what were we still worried about. My response was, "I spend most of my time worrying about what we ought to be worrying about and when enough is enough. We are a little over three months away from the planned launch; we still have several critical tests and a huge pile of data reduction and analysis to be completed, along with a very critical design certification review with NASA."

The final design certification review for the RSRM was conducted in July. In the midst of this heavy workload, NASA added even more to our burden. Under pressure from Congress, the agency decided to go into a full-scale engineering development of the ASRM at an abandoned nuclear power plant site that the government owned at Yellow Creek just outside of Iuka, Mississippi. More than a year earlier, NASA had issued study contracts on the ASRM to everyone in the solid rocket industry who was interested. Thiokol had bid, won one of the study contracts, and put a small group of people looking into alternative designs. The emphasis was to further enhance the reliability of the SRMs used on the Shuttle, but it was becoming very obvious that the ASRM program was totally political in nature. The program was being conducted to punish Thiokol and bring jobs into the district of one of the most powerful men in the House of Representatives and chair of the critically important House Appropriations Committee, Mississippi Congressman Jamie Whitten.

Thiokol management was in a quandary over how to deal with this issue. They thought that we could not just walk away from a system that might eventually replace our RSRM business, which represented over one-third of our total corporate activity. I told my boss, John Thirkill, "Space Engineering cannot support such a proposal at this time: we can't afford to place any new attention on anything prior to returning the Shuttle to safe flight. The ASRM work will be a tremendous distraction. Furthermore, I do not believe there is a chance in hell that we can win the ASRM program because the program has been instigated in the first place by Congress to get someone other than Thiokol involved in making the Shuttle's solid rocket boosters. I feel comfortable that the ASRM program will likely die after we have several successful flights of the RSRM under our belt; NASA will not be able to continue justifying the program or support it financially."

Thirkill agreed and went to the corporate office and met with Phil Dykstra, our Vice President of Strategic Planning and the individual who had been responsible for bringing me into the Shuttle program in 1984. Dykstra was a very intelligent guy but with a large ego that could be very intimidating; he was alleged to be one of the original "whiz kids" in the 1950s. He agreed with Thirkill that Thiokol should "no bid" the ASRM program and would take that message to Charlie Locke and Ed Garrison. Ironically, both of these men had been born, raised, and educated in Mississippi, where the

ASRM work would be conducted. Dykstra was successful in convincing our company's two top men that the best course of action was for Morton Thiokol to focus all our resources on returning the Shuttle to safe flight and not responding to the request for proposal (RFP) for the ASRM.

Actually, the primary instigator of the ASRM program had been our friends from the Hercules group down at the other end of the Great Salt Lake. Hercules had teamed with Atlantic Research Corporation and Martin Marietta to leverage its political position. Atlantic Research Corporation was the study contractor that had located and recommended the Iuka, Mississippi, site for the ASRM project. The RFP required all bidders to use the Yellow Creek site for the ASRM. Hercules had also made a deal with the only two independent large solid rocket motor nozzle manufacturers in the United States, Hitco and Kaiser, to exclusively support its ASRM proposal. Chemical Systems Division of United Technologies, the manufacturer of the Titan SRM, was also bidding with its Titan nozzle vender, Rohr Corporation. A team from Lockheed and Aerojet came together also to bid the ASRM, but the combination had no nozzle fabrication capability. Lockheed asked us if we would join their team by supplying the nozzle for the ASRM if they should win. Since the RFP required that the ASRM be manufactured in a government-owned-company-operated (GOCO) plant, the Lockheed-Aerojet team recommended putting the ASRM nozzle fabrication into the NASA Michoud facility just outside New Orleans, where the Shuttle's external tank was manufactured. This would give them added political clout from the Louisiana delegation, namely Senator Bennett Johnston. If Thiokol refused to support the proposal, the Lockheed-Aerojet team wouldn't be able to submit a credible bid.

As Lockheed was the prime contractor for the Trident II (D-5) fleet ballistic missile and a good customer for Morton Thiokol, we decided to support its ASRM bid. Thiokol agreed to operate a new government-owned nozzle manufacturing facility at Michoud with the proviso that Thiokol would not have to invest a single dime in the operation if the Lockheed-Aerojet team was successful in its bid. Furthermore, Lockheed and Aerojet would have to provide most of the proposal preparation work with minimal support from us.

Early in the program, we decided to hydro-burst the Nozzle Joint Environment Simulator test motor after the hot firing tests were completed with this hardware. A hydro-burst was a test-to-failure made in order to determine the actual safety factors of the structural elements. This was achieved by pressurizing the test configuration with water until it broke. In the technical community, there had been some concern that the incorporation of the radial bolt holes in the nozzle-to-case joint had weakened the area below the required safety factor of 1.4. We had beefed up the nozzle fixed housing in this area to accommodate the redesign, but the steel aft dome of the motor was still using the old hardware manufactured before the *Challenger* accident to save schedule time and costs. The engineering analyses indicated that 100 holes (1¼ inches in diameter) could be drilled into the aft dome area and still exceed the 1.4 safety factor requirement. In fact, the analysis predicted that the aft dome membrane (that is, the thinnest section without holes) away from the joint would remain as the failure point with a safety factor in excess of 1.6. In the press, Roger Boisjoly had strongly criticized Thiokol's radial bolted nozzle-to-case joint design, stating that the radial bolt holes in the aft dome had significantly degraded the structural capability of the motor. So we were relieved when the Nozzle Joint Environment Simulator hydro-burst test went as planned: the motor failed in the membrane area of the domes at precisely the pressure predicted by engineering, demonstrating a safety factor of 1.62, well above the 1.4 requirement.

The National Research Council suggested that we verify that not even severe flaws introduced into the joint areas of the short-burn (one second) JES and TPTA test motors would significantly damage the rocket. At the request of the NRC, we introduced such flaws into all the joint areas, including the field-joints, igniter joint, and nozzle-to-case joint. We punched holes through the insulation and joint filler materials to allow hot gas jets to impinge on the new O-ring seal design. We cut away sections of the barrier seals and sections of the primary O-rings. In all of these areas, we allowed hot gas to reach the last line of defense, the secondary O-rings. We even intentionally unbonded the case insulation on the edge of many of these areas to determine if hot gas would penetrate and promote further unbonding that could eventually burn through the steel case in a full-duration motor. Our structural, thermal, and flow analyses indicated that the motor would work just fine with all the induced flaws. The results verified our engineering analysis, and we regarded the tests as a roaring success.

But the real proof of the pudding was to incorporate these very severe flaws into a full-scale, full-duration, two-minute static test—a subject of great debate among the large technical community involved in the redesign of the solid rocket motor. The idea to conduct such a test, which was unprecedented in the industry, originated in the approach to the joint design that had been selected for the motor. As indicated earlier, a bitter dispute had erupted over the selection of the unvented (or isolated) O-ring design. Because we could not conclusively prove through prelaunch testing of the assembled joints that they were indeed leak proof and would not let hot gas through into local defective areas of the joint insulation, many in the technical community concluded that local hot gas jets could still occur, albeit infrequently, and impinge on the primary O-ring. Their concern was that Thiokol might at some time assemble a defective joint that had a flaw in the capture-feature of its O-ring and in its insulation J-joint, thereby allowing a hot gas jet to penetrate the joint and impinge on the primary O-ring. For this reason, the National Research Council suggested that the new RSRM design be tested with intentional flaws in the joint areas to show that it was tolerant to these conditions.

I did not disagree with this recommendation, nor did Carver. We both strongly endorsed such testing with joints that incorporated worst-case manufacturing defects, going so far even as conducting tests with moths and other insects stuck on the contact adhesive and creating wave defects in the joint insulation that would prevent 100 percent sealing of the insulation joint at assembly. But we thought that should be the end of it. It would be stupid to intentionally incorporate severe flaws into *all* of the new design features of the joint and then require that joint to meet the new, more stringent CEI specification for the motor. But some people wanted to do just that. With our engineering analysis and results from the Joint Environment Simulator tests indicating that we could tolerate even more severe flaws in the joint without eroding or blowing by the primary O-ring, a recommendation to do just that came alive. Several people at NASA and some members of the National Research Council and NASA MSFC oversight committees suggested that we conduct a full-duration static test with every possible severe intentional flaw and that the criteria for success be based upon meeting the CEI specification requirements for the motor.

I just about went ballistic. Even though it was most probable that the motor would meet the requirement, there was a reasonable possibility that it might not. I felt reasonably comfortable that the motor wouldn't fail or allow hot gas past a secondary O-ring, but I didn't feel comfortable that it would meet the CEI specification requirement of no blowby or thermal distress to the primary O-ring. Carver agreed, and we both strongly recommended that such a severe flaw test should only be required to meet fail-safe criteria, which meant the motor remained intact during the entire test and did not allow hot gas to leak outside the motor from the joint area. We both thought that the proposed flaw test was too severe and that we should test the motor with the worst-case manufacturing and assembly flaws. With these types of flaws, we would accept meeting the CEI specification requirements as the pass/fail criteria for the test.

J. R. Thompson, Center Director for NASA Marshall, felt very strongly otherwise and believed that the pass/fail criteria should be that the motor meet the new CEI specification requirements. I strongly disagreed with that logic and told him so, because there was more than a reasonable probability that the motor would not meet that criteria, and we'd have to shut down the Shuttle program for God only knows how much longer because of this crazy "9 sigma" test. "The motor won't fail," I told Thompson, "but there is more than a remote possibility that the motor will allow some hot gas to leak out of the joint during the test. If this happens the entire Shuttle program may be canceled by Congress!"

I recommended that we not conduct such a test. The conditions were so unrealistic, and so far out of the design envelope, that it was not worth risking the Shuttle program on such a test. It was akin to testing the SSMEs at a 200 percent-rated power level for full duration when they were designed for a maximum power level of 109 percent. No one would think of doing such a thing.

Thompson didn't agree. "There are several members of the NRC panel on the redesign of the solid rocket boosters that agree with my position, Al."

"The risk of stopping the Shuttle program is too high to do this," I countered. "It's somewhat akin to placing a bet on the roulette wheel in Las Vegas and the casino changing its house rules to allow you to bet $10 on every number except double zero and let you win *all* your bets if the marble landed on *any* one of the numbers you played, and the casino can only win if the marble stops on double zero. Then the odds of winning would be so good that anyone would be willing to make a bet, anytime. The odds for our motor to *not* allow hot gas out of the joint with these severe intentional flaws are quite similar; but there is some slight finite possibility the motor could fail in such a test. Now if the motor fails, or even allows hot gas or soot out of the joint, it will not be just the loss of that particular motor to be replaced by another test; it will most likely be the death of the entire Shuttle program, with the *Challenger* accident as vivid in the minds of Congress, the public, and the media as it still is.

"But now let's assume that the casino wants to change one of the house rules. The house still lets you win every one of your bets if the marble falls on any number but double zero; however, if the marble *does* fall on double zero, the house not only takes every one of your bets but you also lose your car, your home, your family, and you go to prison for twenty years! The odds of you winning haven't changed one bit—and they're still extremely good—but I wouldn't even consider taking such a bet. There's far too much to lose versus what there is to gain!"

Well, I lost the argument, but I did win agreement on what the pass/fail criteria would be: no hot gas could exit the joint though it could blow by the primary O-ring or erode through the primary O-ring and part of the secondary O-ring. John Thomas and Royce Mitchell from Marshall fully supported the pass/fail criteria being based upon meeting a fail-safe condition rather than meeting the CEI specification requirement.

The flawed test motor, designated PV-1, was scheduled for testing in mid-August 1988 and was the last major hurdle to clear before the Shuttle could be given the go-ahead for launch. About one week before the PV-1 test, the Shuttle *Discovery*, designated STS-26R, successfully passed a flight readiness firing on Pad 39B at KSC. The FRF consisted of a full dress rehearsal countdown wherein all systems were activated with the exception of ignition of the SRBs. The Space Shuttle main engines were ignited in the proper sequence and continued to burn for approximately twenty seconds while the Shuttle remained bolted down through the SRB skirt to the mobile launch platform. This was also a critical test for returning the Shuttle to flight, because it had been two and a half years since a Shuttle launch had been conducted at KSC and was a good training exercise for the new Shuttle launch team as well as for Launch Control Center management and support.

I sat at the same console for the FRF as I had on the day of *Challenger* two and a half years earlier; it seemed like just yesterday and brought back a lot of sad memories. Although the FRF was primarily focused on the SSMEs, the test was also very important for clearing the SRBs for flight. Because the SRB aft skirt had failed prematurely in a structural test article, it had been heavily instrumented to verify that the prelaunch loading and behavior of the skirt were as predicted, because the aft skirt was approved to fly with a waiver for not meeting the required structural factor of safety of 1.4, whereas the waiver allowed the skirt to be used with a minimum safety factor of 1.28. The FRF verified the skirt would indeed meet the lower safety factor of the approved waiver, which meant that the PV-1 test then became the only major constraint to the planned launch, returning the Shuttle to flight after thirty-two months of agonizing redesigns, testing, and endless reviews.

Finishing assembly of the flawed PV-1 motor in our static test bay, we had to conduct a leak-check on the field-joints to verify that those that were not flawed were acceptable and those that were flawed were in the planned pretest condition. During the field-joint leak-check operation, we detected a problem when checking the cavity between the capture-feature O-ring and the primary O-ring in one of the joints wherein we had not "flawed" the capture-feature O-ring. The test indicated that the capture-feature O-ring was leaking. The problem was that the cavity was pressurized to 1,000 psia rather than the planned 100 psia. The maximum operating pressure of 1,000 psia was used as a leak-check pressure between the primary and secondary O-rings but never between the capture-feature O-ring and the primary O-ring. We'd learned early in the program that high pressures would extrude the O-ring out of the shallow capture-feature O-ring groove and up into the insulation, and that is exactly what had happened. It wasn't a problem with the primary and secondary O-rings because they had a much deeper groove and smaller extrusion gap behind the O-ring groove. What had happened was that the pressurizing line connected to the motor was also connected to the high-pressure outlet on the leak-check console; there was no differ-ence in the fittings or the pressure lines between the low-pressure and high-pressure connections. The test technician had simply made a mistake that unfortunately was not caught by the quality inspector who had signed off that the leak test set up was proper.

This leak-check mistake caused a day slippage in the schedule to de-mate the field-joint, replace the O-rings, and redo the leak-check. Because the PV-1 test was constraining to the scheduled launch date in September, it wasn't surprising that NASA was very upset over this mistake. Thiokol's President, Ed Garrison, was also very upset and insisted that we find out who was responsible for the error and fire him. My boss, John Thirkill, called me in on the situation. "I don't agree with firing the technician who made this mistake," I told him. "Our engineers, who work for me, should have designed this leak-check system to be more foolproof and user-friendly by having different size fittings and connections for the various pressure lines so that this could not occur. Mistakes do happen, and we must accept that. We were able to recover from this mistake without a loss of hardware. All it cost us was a day or two on the schedule."

"Look, Al, Garrison wants all those associated with this problem disciplined, and those most responsible, including the test technician and quality inspector, fired."

"I disagree, John. Even though some of my people in engineering were responsible for designing the leak-check system, I don't plan to discipline them, but I'll have them participate in a redesign of the system to avoid such errors in the future."

Ross Bowman, our Vice President of Safety, Reliability, and Quality Assurance, agreed with me and didn't want to fire his quality assurance inspector who had also made a mistake. As it turned out, both the technician and the Quality Assurance inspector were fired. The reason given to them was pressure from NASA. The firings did not sit well with the rest of the technicians or the people working in the test area, who had been busting their butts on this program for the past two years to get us to this point.

A few days later, one of the NASA people who had been recently transferred from MSFC to Utah to provide oversight in the test area reported that he was intentionally forced off the road by a disgruntled Thiokol employee. This employee apparently thought the NASA man was responsible for the dismissal of the two Thiokol employees. Relations with NASA, which were already strained at best, increased the pressure and stress to a new level not only because of the importance of this test, but also because of the added concern that some disgruntled employee might potentially sabotage the test motor or, worse yet, destroy a flight motor. It was a worry we didn't need at this very critical time.

Our fear of sabotage was real because we had found cut O-rings just prior to installing in one of the earlier test motors. We thought the O-rings had been intentionally cut and brought the FBI in to investigate. The FBI concluded that the O-rings had in fact been cut intentionally at our O-ring vendor, Hydra-Pak in Salt Lake City, which was responsible for splicing the RSRM O-rings. However, the criminals were never identified. With this danger in mind, we introduced stringent precautions into the inspection and packaging of the O-rings, adding cost to the program since two people were now required in all operations associated with O-rings.

To no one's surprise, a legion of reporters and news media showed up for the PV-1 test on August 18, 1988. I was very nervous over this test and was particularly thankful that Carver Kennedy and John Thomas were providing the morning's pretest briefing, which took place in a trailer that had been parked near the highway for the past year just west of the test site. The briefing included a description of the flawed PV-1 test motor, the test objectives, and the pretest predictions for the motor. The press was told that upon successful completion of the test, the Shuttle would be cleared for launch in about a month. Two to three weeks would be needed to analyze the nearly 500 channels of data, along with very detailed postfire disassembly inspections of the motor and its components. I listened to the briefing in the back of the press room. As I was scheduled to provide the briefing about thirty minutes after the test, I found it particularly beneficial to hear what questions were being asked by the media prior to the test.

I left the trailer immediately after the press conference and headed for the static test control center, or T-22. I wanted to make sure all the pretest checkouts of the instrumentation and firing circuits were OK and find out if any anomalous data had surfaced. When I got to T-22, I found that everything was fine, and the countdown going as planned. "Could you please quickly scan all of the TV cameras for me," I asked, "so I can get a good feel for what we might be able to see during the test?" Normally, the only TV coverage seen in the control center during the test came from a camera focused on the nozzle exit plane. This was usually the most important area of a solid rocket motor to watch, because if anything abnormal happened, it usually showed up as a disturbance in the exhaust plume of the rocket right behind the end of the nozzle. I told them to leave at least one of the cameras focused on the most severely flawed field-joint appearing on the TV screen, because I absolutely needed to know during the test if any hot gas or smoke was escaping from the motor through that field-joint. This happened to be the aft field-joint where a 0.1-inch diameter hole had been formed through the case joint insulation design. We had established this hole size to provide the worst-case hot gas jet that could impinge on the capture-feature O-ring below the J-joint insulation. The hole through the insulation was placed in-line with the capture-feature O-ring that had had a section ground out of it, which effectively removed 80 percent of the O-ring's cross section. The vent port plug that filled the leak-check port between the capture-feature O-ring and the primary O-ring had been replaced with a transducer to measure pressure and temperature in the area since hot gas would immediately fill it at motor ignition. A similar set of transducers had been installed in the leak-check port between the primary and secondary O-rings. From pretest predictions of the pressure and temperature to be measured by transducers in front of the primary O-ring, we expected that the primary O-ring would not erode or leak hot gas—in other words, that the transducers behind this O-ring would show no response. However, these transducers would prove extremely useful if the primary O-ring were to leak or if the motor failed. All of the instrumentation was working properly, and the countdown proceeded on schedule for the 1:00 p.m. static test.

Around 11:30 a.m. I decided to go back to the A-1 administration building to grab a quick lunch. Picking up a sandwich and a cola in the cafeteria, I was so nervous I took one bite of the sandwich and lost my appetite. I left my lunch uneaten and drove back to the control center. You could feel the tension building in the blockhouse. This was clearly the biggest, most important, most unusual test ever conducted at Morton Thiokol or, for that matter, in the entire solid rocket industry. Here we were, just one month away from the planned return of the Space Shuttle to flight, testing the largest single rocket ever to be flown and the only man-rated solid rocket ever built, with flaws intentionally introduced into the most critical areas of the rocket motor, to prove that our new motor design was safe. Hot gas was being purposefully channeled through the new insulation and through the O-rings with the intention of proving that this new motor was much safer than the pre-*Challenger* design.

If that wasn't enough, a short section of the insulation had been intentionally unbonded from its steel case on both sides of the field-joint, where the hot gas came through, to demonstrate that this would not propagate a case insulation failure that would be catastrophic. These additional insulation unbond flaws had been incorporated into the test motor because of the Titan 34D that failed catastrophically in the April 1986 launch from Vandenberg Air Force Base. With RSRM design having far superior insulation, the decision was made to prove that the RSRM was tolerant to small insulation unbonds throughout the motor, including on the ends of the segments in the field-joints, at the case-to-nozzle joint, and in the igniter area.

After exhaustive investigation of the Titan 34D accident, the air force had concluded that the failure was due to a case insulation unbond in the area of a field-joint in the SRM. Inspection of other Titan SRMs revealed many segments with similar problems. This slowed down the seven-segment Titan IV SRM development program that had started in 1984, and had totally shut down the Titan flight program. The air force's "assured-access-to-space" program was actually in far worse shape than the Shuttle. It took the air force one and a half years to build up sufficient courage to fly another Titan 34D system without introducing any new design changes, and its Titan IV SRM program was way behind schedule. The Titan IV SRM was basically identical to the seven-segment SRM tested by the air force some twenty years earlier, had been in development nearly twice as long as the RSRM, had fewer design improvements being incorporated, was not man-rated, and still had not flown a single mission. In fact, only one static test had been accomplished in the new vertical test stand constructed for the Titan IV at Edwards AFB, whereas we had tested six full-scale Shuttle motors during the same time frame. The fact that the only other large solid rocket motor that had flown had also failed shortly after *Challenger*, and that the development of the Titan IV with a twenty-year-old rocket design was progressing at a snail's pace, provided added pressure and visibility to the RSRM for returning the Shuttle to safe flight for both NASA and DOD payloads.

At 1:00 p.m., when the countdown went to zero, both of my eyes were glued on the TV monitor and sternly focused on the flawed aft field-joint. It felt like I held my breath for a full two minutes until the motor burned out. To my great relief, no smoke came out of the joint at ignition. But I knew there was always the chance of invisible damage that could materialize in a failure later on. The flawed joint still had to withstand motor pressures of over 900 pounds per square inch and gas temperatures near 6,000° for another two minutes. Everyone in the control room watched the TV in absolute silence until the motor burned out, when the test conductor called for activation of the carbon dioxide quench system to eliminate the post-fire burning of the insulation and cool down the motor. The whole room erupted in a tremendous cheer. We had had a very successful test.

John Thomas, Royce Mitchell, and I immediately went out to inspect the motor in the test bay. We made sure there was neither any trace of hot gas exiting the motor from any area nor any "hot spots" on the motor. The safety folks held us up just outside the test bay until they were sure there were no hazardous conditions to encounter during the postfire inspection process. The motor was in excellent condition, with no anomalous indications anywhere. We then proceeded down to the area where the test data was recorded and being reduced. I had requested that all the data channels that were located downstream of primary O-rings be immediately examined as a part of the "quick look" report we'd be presenting to the media thirty minutes after the test. Inspection of the motor in the test bay clearly established that we had succeeded in meeting the "fail-safe" test objectives of no hot gas exiting from any place in the motor. The "quick look" data behind primary pressure seals gave us a good indication of how close we came to having hot gas escape the joints. All these instruments recorded no pressure and no temperature rise, indicating that all primary pressure seals worked as designed.

This was great news, and we proceeded to the briefing room in a state of absolute euphoria. It felt like the whole world had been lifted from my shoulders. Royce Mitchell and I gave the posttest briefing, saying that all indications were that the PV-1 motor test had been a huge success. It would take another week to fully disassemble the motor and inspect the critical areas, and another two to three weeks to analyze all 500 channels of test data fully and confirm our preliminary conclusions. Following that, the Shuttle would be ready to fly. Reporters barraged us with questions. I answered those associated with the technical performance of the motor, and Mitchell responded to those dealing with the Shuttle launch schedule and other activities being conducted by NASA in preparation for returning the Shuttle to flight.

There was a huge celebration at the Thiokol management training center up in Ogden Canyon that evening. As with the past motor tests, I stayed out at the plant to review the "quick look" data and prepare some viewgraphs for presentation to corporate management as well as guests from NASA and the NRC. I arrived a bit late again for cocktails and dinner, but didn't really care. I floated on cloud nine as I delivered all the excellent data we'd obtained from the PV-1 test. It was a pleasant surprise to receive a "Silver Snoopy" award from the Shuttle astronauts attending our event, a lapel pin that had flown in space. It was an award that could only be presented by an astronaut—in this case, by Dan Brandenstein, then a veteran of two Shuttle flights (STS-8 and STS-51G), with two more to come (STS-32 and STS-49), and chief of the astronaut corps. Commander Brandenstein, along with fellow rookie astronaut Steve Oswald, who later flew on STS-42, STS-56, and STS-67, presented it to me for being someone whom they had considered to have done exceptional work to improve the quality and safety of manned space flight. I was very honored.

Shortly after I received the Silver Snoopy award, veteran astronaut John Young sent me two autographed photographs, one as an *Apollo* astronaut on the Moon and one as a Shuttle astronaut, congratulating me for the fine job I had done in leading the redesign of the solid rocket motors. John was a member of the Norton Committee, which was providing oversight for NASA along with the National Research Council on the SRM redesign program. U.S. Navy Captain John Young was the pilot of *Gemini III* and commander of *Gemini X*, commander of *Apollo 10* and *16*, and commander of Shuttle STS-1 and STS-9. By far the world's most experienced and versatile astronaut, John flew on three different spacecraft twice each—that is, *Gemini*, *Apollo*, and the Space Shuttle; flew once on the Lunar Excursion Module; drove the lunar rover on the Moon; and was on the first manned flight of both *Gemini* and the Shuttle (which was the only spacecraft and launch vehicle ever to carry astronauts on its maiden flight). John was also the most underappreciated astronaut of all time. After his last flight on the Shuttle in 1983, he continued to serve NASA for more than two decades in the Astronaut Office and in various support roles for engineering and safety at the Johnson Space Center. John was never bashful in voicing his opinions, which didn't make him too popular in many circles within the agency. He wasn't always right, but you always knew where he stood, and I always admired him for that.

I felt extraordinarily good about how far we had come and anxiously awaited preparations for the Shuttle's first flight following *Challenger*. During the extensive testing program, in response to the question, "When will the Shuttle fly again?" I always responded that it will when we feel we have done sufficient testing to make it safe enough to do so. Once I was asked how I would ever know when we reached that point. My response was, "It is safe enough when I would personally be willing to fly in the backseat, but there would be no way in hell that I would allow my wife or children to fly in it." I no longer had any qualms at all about personally flying the redesigned motors into space. It was time to stop testing. Time to fly!

Sir! We have some Visitors
from NASA!!

It was easy to lampoon the type of technical interchange meetings that took place between NASA and Morton Thiokol occurring on the Shuttle booster redesign. (Noncopyrighted cartoon, cartoonist unknown; a Morton Thiokol employee gave this cartoon to me.)

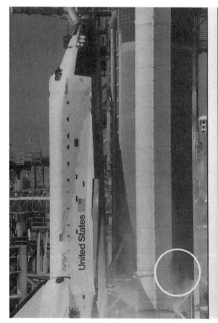

Photographs of the smoke emanating from the joint that failed on *Challenger* at liftoff (T+0.678 second), and similar smoke (at T+0.600 second) coming from a test joint in JES-1B that simulated the actual *Challenger* joint conditions and temperature at ignition. (Courtesy of NASA/Morton Thiokol.)

The monstrous Space Shuttle transporter with a full Shuttle stack preparing to make a left turn to Pad 39B while creeping down the road at less than one mile per hour. (Courtesy of NASA.)

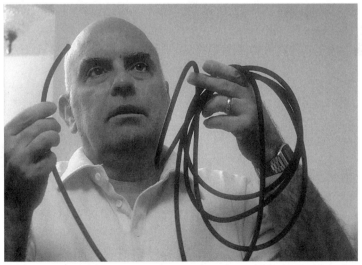

To demonstrate his point, Roger Boisjoly held up sections of typical O-rings used in the field-joints of the Space Shuttle solid rocket motors. (Photograph by Phillip Jones Griffiths, 1988, Magnum Photos, reprinted with permission.)

In August 1987, NASA's John Thomas and I answered questions at a NASA press conference held in Utah just prior to the first test of the redesigned solid rocket motor (DM-8). (Courtesy of NASA/Morton Thiokol.)

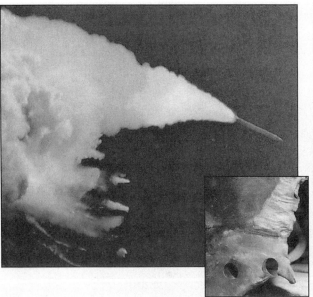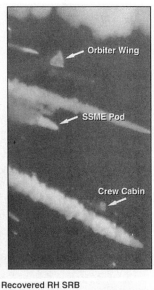

Orbiter Wing

SSME Pod

Crew Cabin

Recovered RH SRB
Aft Field Joint

Debris falling from the sky at T+78 seconds (approximately 5 seconds after the *Challenger* explosion) clearly showed a large section of the orbiter, SSME pod, and the entire crew cabin intact. The inset photo of the recovered joint that failed on *Challenger* exhibited burn-through of the steel case at that location. (*Report of the Presidential Commission on the Space Shuttle Challenger Accident*, Vol. 1, June 6, 1986.)

A horrific propellant fire during core removal in December 1987 destroyed the Peacekeeper casting building at Morton Thiokol, resulting in the deaths of five Morton Thiokol employees. (Photograph by Gary Clark, 1987, *Standard Examiner*, reprinted with permission.)

A May 1988 explosion of the main ingredient in solid propellant rockets—ammonium perchlorate—at the Pacific Engineering and Production Company in Nevada killed two people, injured 326, and was likened to the nuclear bomb blasts once conducted in Nevada. (Photograph courtesy of the 1988 Associated Press and AP Images, reprinted with permission.)

In the early morning hours before the *Challenger* launch, cryogenic gaseous oxygen vapors, which had been dumped into the atmosphere near the fixed service structure, drifted back onto the Shuttle's external tank. (Courtesy of NASA.)

PART VII

Return-to-Flight

Judge a man by his questions rather than by his answers.
—*Voltaire*

I don't think much of a man who is not wiser today than he was yesterday.
—*Abraham Lincoln*

37

Flight of the Phoenix

The postfire inspection of the disassembled PV-1 motor hardware provided no surprises—the motor was in excellent condition, as predicted. There was some minor erosion (0.003 to 0.004 inch worth) on the nozzle-to-case joint primary O-ring, located just downstream of a flaw in the polysulfide joint adhesive (and in-line with a flaw through the wiper barrier O-ring), but this was expected and so noted in the pretest predictions. Those predictions indicated that the primary O-ring in this joint might (or might not) experience some erosion from the hot gas jet flowing through the intentional flaws created during the joint's fabrication. Within the accuracy of our thermal analysis, we concluded that if we did observe any erosion of the primary O-ring, it would be less than 0.006 of an inch—not enough to even come close to compromising the sealing capability of the O-ring.

We had convinced NASA and the NRC to accept "fail-safe" criteria rather than CEI specification criteria for determining whether the PV-1 test was successful. The CEI specification did not allow any erosion of primary O-rings under *any circumstances*; the PV-1 test, as great as it was, would have failed this criteria. Had that happened, there would have been a lot of gas pains and gnashing of teeth by several very important individuals before the Shuttle could have been cleared for flight. At best, the Shuttle would have gotten a conditional go-ahead with a waiver for not fully meeting the PV-1 test objectives. At worst, the launch would have been delayed until another static test could be successfully conducted, delaying the launch at least six months.

It seemed like launch day would never get here. The five weeks between the PV-1 static test and the planned flight of STS-26R crept by very slowly. Even though I had a great deal of confidence in the RSRM as I waited to leave for Florida, I was still very nervous. I still had occasional flashbacks of the *Challenger* accident, and even though I thought the field-joints would never fail again, I knew there were plenty of other potential failure modes in the nozzle and that the propellant and insulation systems did not have nearly the safety margins or redundancy features that our field-joints had. I was even more concerned for other elements of the Shuttle, including the SSMEs, the ET, the thermal tiles on the orbiter, the electrical and computer systems, and the flight software that had not flown or had been subjected to as much scrutiny as the SRBs over the past two and a half years. Any primary system failure on the Shuttle could be a disaster during launch or reentry—and thus a disaster for the entire Shuttle program.

We gave our posttest analysis of the PV-1 hardware and static test data a very high priority in support of the various flight readiness reviews prior to launch. I worked long hours preparing viewgraphs for these critical meetings, and our entire engineering and test operations staffs worked weekends and nights to support me. Meanwhile, the manufacturing operations were also given the go-ahead to resume routine manufacturing of production motors. The first three Shuttle flights were being considered as development flights and as such were heavily instrumented. Each flight readiness review lasted many hours, with so many questions from the various FRR boards needing to be answered. We had generated lots of good data to support our recommendation to fly, and we passed all these reviews with flying colors.

It was still not a time for all joy and happiness though. Jane Smith's lawsuit against Morton Thiokol was still pending and scheduled to go to trial just two weeks prior to the launch. I was told to start preparing for this trial. There had been a long series of negotiations between the attorneys representing her and those representing Morton Thiokol for an out-of-court settlement, to no avail. This really upset me, because I knew that I was in a no-win position. I was effectively being used as the chief witness both for the plaintiff and for the defense in this case. As it turned out, I was able to dodge another bullet, because the case was satisfactorily settled out of court at the eleventh hour, avoiding a well-publicized courtroom trial just days before the planned launch of the Space Shuttle, the first since the *Challenger* accident.

Not everything was coming up roses, however. Roger Boisjoly was still running around the country making speeches claiming that our redesign was "lousy" and, in his opinion, had only a fifty-fifty chance of being successful. In one of his speeches, Boisjoly said that the redesign of the safety equipment on our SRB was "an absolutely, unbelievably lousy design and even less reliable than the equipment that failed on *Challenger*." Roger further stated, "If the design is not changed, we're going to have another disaster on our hands. We'll see the loss of more lives, the loss of a $2.5 billion vehicle, and maybe even the loss of the entire manned space flight program." "The 'inside word' among experts even at the Johnson Space Center," Roger claimed, "was that there was a one-in-twenty chance that the Shuttle would fail during its next launch." Earlier in the year, Roger had stated that it was very unlikely that the Shuttle would be launched in 1988. NASA had repeatedly set unrealistic launch dates that were postponed at the last minute. Roger claimed that after he left the Thiokol redesign team, the company had ignored the very best proposal—a metal ring offered by Vetco Gray, Inc., of Houston, Texas—and instead, for "political reasons," chose "a severely flawed design" put forward by NASA Marshall.

I was deeply disappointed by my friend's statements because I had supported him. I knew he was a better engineer than that and clearly would not be making these comments had he been more involved in understanding the redesign and the test program results. Earlier, I had told him that he was selling his soul to his attorney when he filed his outrageous $2 billion lawsuit. I was especially concerned how his rhetoric was poisoning young minds when he spoke on college campuses.

Three days before the launch, I flew to Florida for the L-2 FRR. Tension was building as the Shuttle's return-to-flight generated tremendous public interest. Some in the media predicted that more people would witness the launch of STS-26R *Discovery* than had been there for the Shuttle's first flight, by STS-1 *Columbia*, in April 1981: possibly as many as one million people. Over 3,000 members of the press would attend the launch, and, because it was such a newsworthy event and a critical moment for the Shuttle program, most of the key executives from the Shuttle contractors would also be at the Cape, with their spouses, to witness the historic liftoff.

Morton Thiokol was no exception. The company even invited my wife, Linda, to come along to Florida on the company jet. While I was in the Launch Control Center, she would be in the new VIP viewing area just northwest of the Vehicle Assembly Building, approximately seven miles due west of Pad 39B, where *Discovery* sat. This would be only the second launch attempt from Pad 39B—the only other launch from it was *Challenger*'s. Charlie Locke was so afraid that MTI personnel at the Cape would be sought after by the press that he made arrangements for all our employees to stay in the Marriott Hotel near the Orlando airport. All the other contractors and NASA people stayed in Cocoa Beach, where all of the action was. Locke was absolutely terrified of the press; following *Challenger*, he had had far more exposure to the media than he had ever wanted, and it was all bad. Realizing it would be impossible for all our key people to avoid the media totally during these last days leading up to the launch, Locke had everyone attend a seminar on what to expect and how to best respond to the media. He also hired Nancy Hobart, a young woman from our corporate office in Chicago, to prepare everyone for media questions. Hobart was to handle all of the requests for information and involve the appropriate MTI people when required. She had prepared a set of anticipated questions and responses for everyone concerning the corporation's recovery from the *Challenger* accident, and she advised everyone to coordinate any request for an interview with her. She also prepared several brochures on the corporation with detailed background information on the redesign (much of it contributed by me) to be handed out at the press site, located at KSC just across the street from the Vehicle Assembly Building.

My wife was very excited to be at the launch and particularly happy to visit again with our dear friends Martha and Carver Kennedy, who lived in Titusville. It was not Linda's first Shuttle launch; she had witnessed a launch before the *Challenger* accident, but it was nothing like this. This was not an event; it was a spectacle.

I presented the Thiokol briefing at the L-2 FRR, making the final recommendation on our RSRMs as "go-for-launch." Following this very successful FRR, we enjoyed a splendid dinner with the Kennedys at the Mango Tree Restaurant in Cocoa Beach. As we were finishing dessert, Dan Brandenstein, chief of the astronaut corps, sat down with his guest at the table next to ours; with him was Lynn Sherr of *ABC News*. Sherr was the reporter who first broke the news on the closed hearing of the Rogers Commission during which I had revealed what really transpired between Morton Thiokol and NASA the night before the *Challenger* launch. Seeing me, Lynn came over and asked if I would be available for an interview by Peter Jennings, the *ABC News* anchor, on the morning of the launch. I told Lynn that I wouldn't be available because I'd be at a console in the Launch Control Center. As Vice President of Engineering, I had to be there to represent Thiokol and give the final go-ahead for launch as part of the engineering launch support team. Lynn wondered if I could be available sometime tomorrow, the day before the launch, for an interview with Jennings. She knew her boss had a very busy schedule for tomorrow, but thought Jennings would want to talk to me. I had to participate in the final L-1 FRR in the morning, but if everything went OK, I told her I should have some free time in the afternoon. But I quickly added that she would have to coordinate this request through MTI's public relations director, Nancy Hobart, at the press site.

On the drive back to our hotel—which, unfortunately, was nearly fifty miles back to Orlando—my wife expressed her nervousness about me talking to the press, much less Peter Jennings, without first getting it cleared with Charlie Locke or Ed Garrison. She knew how they absolutely hated any publicity and had taken great steps to avoid the media whenever possible. Linda was concerned that it would open old wounds between senior corporate management and me, and that most likely I would get into hot water for even agreeing to an interview.

"I think I played it straight," I told my wife, "by telling Lynn Sherr to contact our Public Relations Director first. If Nancy wants to cancel the interview, that's fine; at least, I won't have appeared uncooperative with the press. But I don't think Nancy will cancel the interview. She'll talk to Mr. Locke and/or Garrison first, but I suspect she'll recommend to them that I go ahead with the interview. Thiokol certainly doesn't need any more bad press, which they would be assured of getting if they declined this opportunity. What's more, Nancy's new job as Director of Public Relations will suffer irreparable damage if she says no. She seems to have a good head on her shoulders, and I think she'll be very convincing to Locke, who is also a very intelligent person. He's always interested in damage control when it comes to the media, and he'll most likely conclude that granting the interview is much better than the alternative. The only real damage will likely be to Locke's and Garrison's egos, because they weren't asked to be interviewed by Peter Jennings!"

As it turned out, Nancy Hobart must have been successful in convincing Locke to agree to the interview, because I had a note from her the next morning to meet her at the Thiokol booth at the press site in the early afternoon after the L-1 FRR was over. The review went well, and after a lunch in the KSC cafeteria, I strolled over to the press site. It was like a circus! TV and press interviews were being conducted all over the place, and various media personnel were going around to the booths of all of the Shuttle contractors picking up armloads of press material, brochures, videos, and other information they might be able to use in their newscasts, newspapers, or magazine articles. We had staffed our own booth with our best human resources personnel at KSC, and they were glad to see me, as they were being inundated with questions concerning the redesign, the qualification testing, and readiness for this flight. The PR folks really didn't know how to answer most of these questions, so they quickly glommed on to me to provide answers.

Soon Nancy Hobart showed up, and we talked for a while about the upcoming interview with Peter Jennings. Nancy thought it was a great opportunity for Thiokol to shed its "black hat image" and reveal to the world all the good things we had done in the redesign to get ready for this launch. She had talked to Lynn Sherr and had gotten directions how to find the *ABC News* set. Once there, we waited until Jennings was finished with his evening broadcast. When he was finished, Lynn returned to our waiting room and told us, "I'm very sorry but Mr. Jennings is totally occupied at this time, and with the other commitments that he already has today, it doesn't appear he'll have the time to interview Mr. McDonald today. Couldn't you be available in the morning?"

"No," I answered. "I really can't. I'll be in the Launch Control Center."

Leaving the studio, I had really mixed emotions. I was relieved not to have to do the interview because it eliminated the possibility of opening old wounds concerning the *Challenger* accident, yet at the same time I was disappointed in not being given the opportunity to tell the world what a terrific job we had done in the redesign to enable us to finally resume Space Shuttle flights tomorrow with the highest degree of confidence.

We went to bed early in our Orlando hotel that night, but I couldn't sleep. I had tremendous butterflies in my stomach about next morning's launch and just couldn't keep pictures of *Challenger* from flashing through my brain. I got up extra early that morning, September 29, 1988, and drove back out to KSC.

Arriving at the Launch Control Center, I saw numerous people that I knew. Many were new friends I had made as a result of the redesign; many others were people I'd known at the time of the *Challenger* accident, and I didn't know whether they were still my friends or not. Because of this, my anxiety waxed as I waited out the several hours until launch. Sitting at the console with my headset on, I scanned all the TV monitors on the console and carefully reviewed the SRB data displayed on the computer screen. I had to be ready to respond promptly to the engineering launch support "poll" that would be taken on the communication network during the terminal countdown. At the planned "hold" in the count at T-9 minutes, I had to be totally assured that I could say that Morton Thiokol engineering was working no problems and was "go-for-launch." It seemed like that time would never come.

The countdown proceeded smoothly, except for a short hold: the weather was beautiful, but the winds aloft were too low. Too low? That surprised me also, so I listened carefully to the network. The flight profile fed into the computers on the orbiter was based upon "seasonal wind profiles," but the winds aloft that morning were much lower. The count was held for about ninety minutes until the winds increased to values closer to nominal.

The poll came just before we came out of that hold.

"Morton Thiokol engineering is working no problems," I stated proudly, "and we are go-for-launch!"

After my say-so came the voices of the Mission Management Team in the Launch Control Center's firing room No. 1. The MMT had been expanded to include the chief program manager from each Shuttle contractor as well as the element program managers from NASA. Representing Morton Thiokol was Dick Davis, while Royce Mitchell spoke for NASA Marshall. My boss, John Thirkill, was sitting in a VIP room for executives from the other Shuttle contractors and from NASA in a room adjacent to the firing rooms.

"10, 9, 8, 7, 6, we have a go for main engine start, 3, 2, 1 and we have lift-off!"

I held my breath just as I did during the PV-1 test. Actually, I felt like I kept it in until approximately T+70 seconds, when the launch conductor said, "*Discovery*, go for throttle up!"

A chill went through my spine, because I'll never forget that those were the last words I'd heard over the network transmitted to *Challenger* prior to its horrible explosion. I had one eye glued on the TV focusing on the SRB plume and the other eye glued on the real-time pressure trace of the two boosters displayed on my console. I was straining to see if anything was abnormal in the plume or if the pressure traces departed in any way from our prelaunch predictions. Considerable reverse-flow after-burning of the fuel-rich exhaust gases from the SRB and SSME always occurred at the back end of the SRB up along the aft skirt and onto the boosters. Although it seemed to be perfectly normal, it always looked like the whole aft end of the boosters was on fire. The flames appeared to be licking at the elevons on the back of the orbiter's wings and dancing around the entire circumference of the aft skirt of the SRB. This phenomenon, which also occurred on *Challenger*, may well have been what several people including Ali Abu Taha interpreted as a "fire between the *Challenger* and the ET" long before NASA reported a fire emanating from the aft field-joint of the right-hand SRB at T+59 seconds. This aerial pyrotechnic display always makes me nervous.

A big sigh of relief came out of me after *Discovery* successfully passed the "throttle up" point. The pressure trace for both boosters tracked preflight predictions to a tee, and the boosters burned out and separated from the external tank exactly as planned. Everyone in the Launch Control Center stood up and cheered—something that was not only unprecedented but a clear violation of Launch Control Center protocol. Everyone there is supposed to cease any nonessential conversation during the final minutes of the terminal countdown through the entirety of launch and rocket ascent. Following the great cheer, the word came over the network that booster separation had occurred as planned and that performance was nominal.

We all sat back down at the console, where I had the console director switch our SRB screen over to the data being obtained on the SSMEs. The SSME guys from Rocketdyne were still propped on the edge of their chairs, holding their breath, though it was impossible for them because the SSMEs burned for some eight and a half minutes. The main engines had another six and a half minutes to go before cutoff. I watched the performance data for the three SSMEs with great anxiety as well. Launching a Shuttle was a real team effort, and even though our redesigned SRBs worked perfectly, it wouldn't be a good day if the rest of the system didn't perform equally well.

As it turned out, all elements performed well, and *Discovery* reached orbit as planned. When the astronauts acknowledged main engine cut off (MECO), everyone in the Launch Control Center stood up once more and cheered. What a great feeling that moment was, for me and all the others. Several people came over to shake my hand and congratulate me for a job well done. I still had my headset on and was trying to switch to another channel to hear if the SRBs had been sighted in the water. It was very important to retrieve this set of motors so that we could carefully inspect their flight performance and verify our static test results. I finally found the right channel and was told that both SRBs had been observed under main parachutes impacting the water very close to the recovery ships. Both appeared to be in good shape floating in the water—more good news.

NASA Administrator Dr. James Fletcher then came on the network and congratulated everyone on the launch team. The Associate Administrator for Space Flight, Admiral Richard Truly, did the same. Goose bumps tickled my skin, I was so happy to have played an important role in this wonderful accomplishment of returning the Shuttle to safe flight. It didn't erase the memory of *Challenger*, but it placed a new picture in my memory bank that helped me feel much better about myself. I finally felt, after what seemed like many years, that the *Challenger* astronauts had not died in vain.

After exchanging many rounds of congratulations with numerous NASA and contractor personnel in the Launch Control Center, I went next door to the engineering data room to obtain a printed copy of the SRB pressure traces. They were both absolute overlays on one another just like the pre-flight predictions. A good friend of mine, Hoyt Sherard, had transferred down to KSC from Utah to be the engineering representative in our launch support services office. Sherard was as excited as I was, and he retrieved what data he could for me. The remainder of the engineering data had been recorded at the Huntsville Operations and Support Center; we had several people from Thiokol engineering there, and they faxed down a set of "quick look" data on the performance of the joint heaters, the measured tempera-tures on the motor, and the measured chamber pressure data for an L+2-hour report that I had to prepare for KSC. All the data looked great.

It was one big party in Cocoa Beach that evening. All the Shuttle con-tractors and many of their support contractors hosted parties and hos-pitality suites all along the strip—all except Morton Thiokol, that is. The high poobahs of our corporation held a sedate dinner and reception at the Marriott Hotel near the Orlando airport where we were all staying. Over half a million people had witnessed the launch, and even five hours after launch there was a massive traffic jam on the Beeline Expressway heading for Orlando. The MTI cocktail reception and dinner were both very elegant. The top dogs in corporate management each took a turn congratulating Space Operations for its dedication to success. It was a very nice evening, but rather subdued and quiet compared to the wild celebrations going on in Cocoa Beach. As nice as it was, I would much rather have been in Cocoa Beach celebrating with the rest of the Shuttle team.

I had planned to stay in Florida to see the boosters towed into Port Ca-naveral and docked at Hangar AF. I was really interested in inspecting the joints from the disassembled motors: all of the redesigned field-joints and nozzle-to-case joints. I was told it would be several days before both boost-ers would be completely disassembled for detailed inspections. The boost-ers first had to be lifted out of the water into a sling, where they would be made safe for future operations. All of the pyrotechnics systems had to be checked to make sure they had been fired and were in a safe condition. The large hydraulic actuators that moved the nozzle that steers the Space Shuttle were powered by a hydraulic system that was pressurized by a set of redundant hydrazine-powered gas generators. Hydrazine is a very toxic liquid, and the residual hydrazine needed to be carefully drained from the system before it was safe to begin disassembly and postflight inspection operations. The boosters were then sprayed with fresh water, much like a car wash, to remove the saltwater to minimize corrosion. Then the boosters were moved into the hangar for a detailed walk-around inspection before the disassembly operations began. All of this activity took several days, so I decided to return home first; I still had a ton of work to do. We had a lot of development flight instrumentation on *Discovery*. This instrumentation was an exact subset of the most critical instrumentation that we'd used on the ground tests in the development and qualification static tests of the RSRM. All of this data had to be reduced, analyzed, and compared to the static tests to verify that the flight motors had behaved exactly as predicted and that the response had been the same as the ground test motors.

When Linda and I got home to Utah the next day, the successful launch of *Discovery* was still headline news. The local NBC affiliate showed a rerun of the launch, along with scenes of the huge crowds witnessing the event. It also showed a short clip of the Utah delegation in the VIP Shuttle viewing area; there was Linda standing in the bleachers cheering the liftoff. The news also explained that the *Discovery* crew had been able to place the Tracking and Data Relay System satellite successfully. This was the very same payload that had been lost on board *Challenger*. It made me very proud, and we were both very happy to finally enjoy this wonderful moment that had totally consumed our lives for the past thirty-two months and one day.

I was particularly pleased with the coverage of the launch from our local newspaper, the *Standard Examiner*. The front page carried a photo of *Discovery* lifting off the pad with a subheadline, "Liftoff a White-Knuckle Time for Thiokol Workers and Spouses." Excerpts from the article:

> CAPE CANAVERAL, Fla.—Linda McDonald was next in line for a shoulder rub, as loudspeakers crackled updates on the *Discovery*, which was about 45 minutes from launch. She stood in front of Peter Evanoff, who works for Morton Thiokol in Washington, D.C.... As he kneaded her shoulders, he looked beyond the cattle-crowd of guests to the faraway Shuttle. Sun beat through Evanoff's straw hat and threw a checkered pattern on his cheeks. He'd just dabbed a bit of suntan lo-

tion on his nose. Martha Kennedy, whose husband works for Thiokol at the Cape, had doled the lotion out to Evanoff and Tom Russell, of the corporation's headquarters in Chicago. Yellow rope separated the news media from the Very Important Guests. But they were shoulder-to-shoulder anyway, shoved even nearer as a crowd of 1,500 nudged closer and closer to the perimeter of the rope.

McDonald, whose husband Allan is Thiokol's Vice President of Engineering, was quietly trying to keep a little space around her and Kennedy.

"You'll hear heartbeats the last five minutes before liftoff," Kennedy said. Final countdown began.

Linda McDonald agreed. "From one minute on down . . . " she didn't finish. Her words fell into a sigh. She peered through binoculars as the *Discovery* was about to go. Thirty-one seconds away from blastoff, NASA held the count a moment—for a heartbeat. They cleared the way again. The crowd broke out cheering.

"They're going," Kennedy shouted.

"Yeah, they're going," McDonald said quietly. Beyond the swarm of heads and shoulders, the Shuttle was lifting off.

McDonald strained to see it, and Kennedy said, "Where is it? Where is it?" Suddenly a black trail of metallic smoke, tipped in neon peach, rose above the heads and shoulders. The tip disappeared into a cloud. It shot into blue sky and was swallowed a moment later by another cloud.

"Martha, we did it," McDonald said.

"Not yet, Linda, not yet," Kennedy answered. Their skins smoothed over their knuckles as they grabbed each other's hands. Their hands shook. They were waiting for the Thiokol-made solid rocket boosters to separate from the Shuttle, tumble by parachute into the ocean to be recovered.

They suddenly saw that the boosters were separating as they should, two minutes into flight. Only then did Linda McDonald bend her face into her hands and turn away from the sky. And only for a moment. *Discovery*'s booster rockets—redesigned after their failure caused the *Challenger* explosion—did their job flawlessly today, powering the Shuttle toward orbit for 2 minutes, 4 seconds before dropping off into the Atlantic Ocean.

At work the next day, I congratulated all the folks who worked for me. Several of them had witnessed the launch on a big screen projection of

NASA Select TV in our Management Information Center, which was tied in with the launch communications network at KSC. Several of my people told me it had been good to hear my voice on the network once again; it had been a long and hard struggle for them since *Challenger* as well. All of *Discovery*'s preliminary flight data looked really good, and preparations for the upcoming FRRs were already under way for STS-27R *Atlantis*, scheduled for late November or early December. This was a secret DOD mission to deploy a classified military satellite into orbit.

At one time, I had seriously considered requesting a transfer out of the Shuttle program as soon as we returned the Shuttle to safe flight. However, my friend Hoot Gibson was scheduled to command *Atlantis*, and he had asked me some time back to make sure I was in the Launch Control Center when he was launched. I had told Hoot, "I don't intend on doing this job for the rest of my career, but I promise I'll be there for your launch." As a result, preparing for the next launch was as significant to me as preparing for the first.

The pain of *Challenger* never went away, nor did all of my company's antipathy toward me because of my role in the *Challenger* investigation. Shortly after returning to Utah from the successful *Discovery* launch, John Thirkill hosted a reception at the Ogden Golf and Country Club. As I was walking down one side of the buffet line, a fellow employee commented that it was sure nice seeing those boosters strapped to the Shuttle once again. As he said it, the wife of one of my former bosses leaned over the table and nearly spat at me, "It's too bad you weren't strapped to one of those boosters."

I was shocked and deeply hurt, and it quickly spoiled what should have been a great celebration. On the brighter side, John Thirkill later received a poem in the mail that was very appropriate for the Shuttle's happy return-to-flight. It read:

Two and one-half years ago
You stood, listened, and you took it.
The blame was placed entirely on your shoulders,
And the whole world stood by to see it.
At you, stones were thrown, some as big as boulders.
Yet, you stood, with heads bowed with grief for those seven.
Paid your respects, then lifted your heads to the heavens.
For now you must strive toward a new beginning,
Knowing well you could not accept the challenge without winning.

Redesign was the order; you took it without delay,
Knowing that the challenge would be upon you each and every day.
Weeks and months passed, you worked hard on each motor section,
The redesign was going well when you received new directions.
Frustrations set in on some of you, but falter you didn't do.
The task seemed impossible but was weathered by each of you.
Management somehow gathered the strength to go that last mile,
Urging all the workers that the lights at the end of the tunnel would bring
    a smile.
There were those that did not believe in you,
But you stood, you tested, you did all that you could do.
Beautifully you stood, redesigned, and you did it all together;
Now I believe there is no storm big enough, that you cannot weather.
Now that the return to space is mostly all completed,
Hold your heads high because each of you have succeeded.
Set your goals to continue, I wish the best to you all.
Put on your Distinguished Service Medal, you did beckon to the call.

I dedicate this writing to each of you, at Morton Thiokol, for your
devotion and hard work in returning America to space.

Thank you,
Shawn Ferd

38

Premonitions

I did go back to KSC after the boosters were all disassembled to inspect the returned hardware. All the joints were in excellent condition, and the data from the flight instrumentation was very close to preflight predictions and previous data obtained from the static tests. However, there were two major anomalies in the returned flight hardware. The first involved a series of small longitudinal scratches in the steel cases in the area of the interference fit, where the capture-feature on the tang in the field-joint contacted the inner clevis leg. We saw scratches to some degree in every field-joint—a few tenths of an inch long and ranging from 0.001 to 0.013 inch deep. As the scratches had not been noted in any of the ground test hardware, we concluded that they'd been formed while in one of several dynamic environments experienced by the motor in the assembled condition. Detailed examination of the scratches indicated that small pieces of metal had been removed, and the presence of this removed metal was observed on the mating parts. The phenomenon, known as "fretting," had been observed in various contact metal parts when subjected to cyclic loading during transportation and other vibration. It wasn't possible to determine where the fretting had occurred—whether it was during rollout to the pad, in vibration due to aerodynamic loading during ascent flight, or during reentry, splashdown, or tow-back from the ocean. The area where the fretting occurred was in a low-stress region of the case and was very minor in terms of depth; thus, no one considered it a flight safety concern.

The other anomaly noted on the SRM was the loss of a few pieces of cork insulation that had been bonded over the top of the joint heater cables that were placed next to the systems tunnel that ran down along the side of the case. In any other rocket system, this would not be of concern, because the heater cables were inoperative at time of launch. However, the Space Shuttle was very sensitive to sources of debris during ascent that could potentially impact the wing and the delicate thermal tiles on the orbiter. This had become a very serious issue on STS-26R when *Discovery* landed at Edwards Air Force Base in California. The damage to the thermal tiles on *Discovery* was very extensive and severe—the worst condition of any orbiter ever flown in the Shuttle program. Naturally, we immediately put together a task force to determine if the cork on the RSRM had come off at splashdown or in flight, and to find how to prevent losing cork insulation on future flights.

The ecstasy of returning the Shuttle to safe flight quickly dissipated. We were back in the soup again solving a potentially new life-threatening problem. The next launch, STS-27R *Atlantis*, was already stacked in the VAB with the ET attached awaiting orbiter mating and rollout. Once again we were under intense pressure to resolve the cork insulation problem and provide a detailed engineering analysis that confirmed that the fretting observed in the field-joint region did not compromise the sealing capability of the joint or any structural margins of safety. My engineering staff was back under the gun to provide answers at the upcoming flight readiness reviews.

In a fairly short time, we resolved both of these issues to our satisfaction and were prepared to present our findings to the Level III FRR Board at Marshall. Completing our analysis of the fretting anomaly, we confirmed that it was, indeed, restricted to the interference fit location, which was outside of any O-ring sealing area and located in a low-stress area of the joint, where it did not pose any threat to structural margins of safety. As far as the loss of external insulation was concerned, we noted several areas where small pieces of cork were missing. Based upon detailed examination of the failed cork surfaces, we believed that most of the cork had been lost at water impact; however, the largest pieces of cork, located in the area of the forward field-joint, apparently had come off some time during powered flight. We based this conclusion on our observation of soot from thermally decomposed cork on the leading edge of the failed cork surface, which indicated that the surfaces had been subjected to aerodynamic heating during booster burn. Inspection of the cork adhesive that was still attached to the case indicated several large voids where the cork was lost. Tap-testing and ultrasonic inspection of the remaining cork also indicated several large areas of voids in the adhesive, apparently caused by trapped air during cork installation and closeout. We recommended ultrasonically inspecting the cork on STS-27R and filling all of the voids with an epoxy adhesive from a hypodermic needle. We also recommended incorporating a vacuum bag over future cork installations to eliminate the problem. This we accomplished on some cork runs that had been installed at our plant, a technique that appeared to work quite well.

We presented these findings to the Level III FRR Board at Marshall. The board agreed with our findings and approved the inspection and repair of the cork areas on STS-27R as we proposed. Immediately we sent people down to KSC to inspect and repair the cork on the assembled motors of STS-27R. Done on the assembled motors, postrepair inspection verified that the technique worked well. Our plan was to go to the Level I/II FRR Board with a recommendation to proceed with the STS-27R launch based upon our satisfactory resolution of the two in-flight anomalies from STS-26R. Successfully completing the cork repair, we conducted a pull test on various repaired sections to verify that the fix was acceptable and provided adequate margins of safety during ascent.

I felt really good about our resolution of these problems and was ready to make our presentation to the Level I/II FRR Board at KSC chaired by Admiral Truly.

The FRR was scheduled to last all day—a marked improvement over the previous FRR, which lasted two full days. Our presentation came near the middle of the day, after the systems engineering presentation from JSC, from Rockwell on the orbiter, from USBI on their SRB components, and from Martin Marietta on the ET. Each speaker addressed concerns with his as-built hardware on STS-27R and with the flight results from STS-26R, with particular attention to anomalies from *Discovery* that could affect the reliability or risks associated with the next flight.

I was absolutely shocked at what I had heard from Rockwell on the orbiter. Its engineers revealed that the orbiter had experienced severe damage to its thermal protection tiles and that one area on the orbiter wing had experienced such severe damage during ascent that the tile had eroded during reentry down nearly to the wing's aluminum structure. This severe area of damage was of great concern because it had come very close to damaging the structural integrity of the wing, a problem that would have resulted in catastrophic failure of the orbiter during reentry. Rockwell had carefully analyzed the problem and concluded that the damage to the tile had occurred early in ascent as a result of loss of external cork insulation from the solid rocket motors during their burn.

After running numerous computer simulations, Rockwell engineers had concluded that the size and pattern of the damaged thermal tiles matched the location, size, and weight of the largest piece of cork lost on the RSRM in the area of the forward field-joint on the right-hand SRB. Rockwell indicated that, due to the aerodynamic flow stream during ascent, the location of the piece of cork would normally not be able to impact the tile on the orbiter, unless one of the following conditions existed: the piece of cork didn't tumble as expected, but had sufficient rotational lift to carry it over to the impact area on the orbiter wing; there was sufficient void-volume in the adhesive in the cork area to provide sufficient elevated temperature to increase the pressure under the cork (from sea level ambient pressure to altitude) from aerodynamic heating, thereby bursting the cork and ejecting it across the free stream to impact the orbiter wing; or a piece of falling cork hit a protrusion on the RSRM case and ricocheted across the aerodynamic free stream several feet away to impact the underside of the orbiter wing.

In any event, Rockwell concluded that we were very lucky on STS-26R *Discovery* because it wouldn't have taken much more damage to destroy the vehicle during reentry. Rockwell's analysis also showed that we were lucky that it was *Discovery* that was flying, and not one of the other orbiters in the fleet, because its wing surface roughness, though significant, was considerably less than that of one of the other orbiters. That orbiter turned out to be *Columbia*, a fact Rockwell did not mention at the time. But its analysis showed that the coefficient of surface roughness for the left wing of this unnamed orbiter (*Columbia*) was 50 percent higher than with *Discovery*. If this vehicle this time had been *Columbia*, it most likely would have experienced a burn-through of the left wing and suffered a catastrophic reentry.

At the Level I/II FRR Board, Rockwell matched the final eroded profile of the tile with the piece of cork that had been lost from Thiokol's SRB. All of a sudden, it sounded like *Discovery* was not really the successful return to safe flight that we had thought it was. It was dangerously close to a catastrophic failure, again as a result of Thiokol's solid rocket motors, and just a hair away from another *Challenger* accident.

My heart sank into my stomach once again.

Prior to the Rockwell presentation, I had felt good about the launch of *Discovery* and our ability to respond quickly to the in-flight anomalies from the flight, but I was astounded by how critical the loss was of such a small piece of cork (approximately 2.5 inches wide, 12.0 inches long, and 0.25-inch thick, weighing only 2.1 ounces). In its analysis, Rockwell had assumed a maximum impact velocity of 1,180 feet per second, while not being able to determine exactly when the cork came off during ascent. At this speed, the impact energy (assuming a direct hit) could have been slightly higher than a 30.06 rifle bullet at 100 yards, which was about 2,400 foot-pounds!

I was depressed, to say the least, but I was also very angry because Rockwell was assuming that the problem was caused by Thiokol's boosters and didn't even discuss any other possibilities as to the source of the observed tile damage on the orbiter. I had sat in on several telecoms on the subject and saw photos taken at the landing site at Edwards AFB showing numerous areas of damage on the orbiter's tiles. When we'd found the large voids in the adhesive, we, too, had been concerned that there could be sufficient pressure being generated under the cork to fracture the cork cohesively and eject it from the surface rather than having the aerodynamic forces from the high-velocity air-stream tear off cork poorly bonded to the case. We'd run tests in our lab to determine how much pressure was needed to fail the cork cohesively and "shoot" it off the case. The minimum pressure required was above 60 psia—or three times higher than the very conservative maximum pressure of 20 psia achievable in a void area with 14.7 psia trapped-air warmed up by aerodynamic heating through a thin section of the cork during flight.

"Al, how are you going to respond to the Rockwell conclusions when you get up there?" my boss, John Thirkill, turned to me and asked.

"By saying that I don't agree with Rockwell's analysis, that's how," I replied. When it was finally time for the RSRM on the FRR agenda, Gerald Smith, NASA's Project Manager for the SRB, introduced me as the speaker. Located in the Shuttle Operations Building at KSC, the FRR presentation room was a rather large auditorium with three large screens in the front with backside projection of viewgraphs. The room was packed with people—all the chairs were filled, and several people stood along the side and back walls. Admiral Truly and his board members sat at a huge U-shaped conference table directly in front of the speaker's podium. Generally, I used all three screens simultaneously in my presentations in order to cover all the material I had to present in the time allotted. Only the title page of my presentation was being projected on the middle screen when I came up to the podium.

"Before I start, I'd like to make a few remarks about the previous presentations, and in particular the presentation made by Rockwell on the condition of the orbiter *Discovery* from the last flight. Personally, I do not agree with Rockwell's conclusions that the severe damage to the orbiter tile was caused by a loss of a single piece of cork from our SRM."

You could hear a pin drop in the room. "If I was an astronaut flying on the next mission, I certainly wouldn't be comfortable with the explanation for the severe orbiter damage that I heard here this morning."

From all around the room and from the great U-shaped table, I could see glaring stares of utter amazement ranging from "who in the hell does this guy think he is" from those who didn't know me, to "here he goes again—Al McDonald, the pest who raised all the hell about the launch of *Challenger*, is up on his soapbox again."

"Am I to understand, Mr. McDonald," boomed the voice of USAF General Forrest McCartney, the current Director of Kennedy Space Center, "that you do not believe the detailed analysis presented by Rockwell this morning, which to us certainly appeared to indicate that the loss of the cork on the RSRM was indeed the source of the problem?"

"That's right, General McCartney."

"Well, if that wasn't the cause of the problem, could you please tell us what was?"

"I'm not an aerodynamicist," I answered, "but I'm surprised that Rockwell didn't consider other sources for the debris, sources like the external tank and other SRB components like the frustum and nose cone. I personally feel that these are more likely sources of debris that could impact the tile on the orbiter and cause the problem that was observed than was the loss of a few pieces of cork on our SRM."

"What is the basis for your conclusion, Mr. McDonald?" McCartney wanted to know.

"It's my understanding that the orbiter experienced damage in over 400 different areas in the tile system. Clearly all these couldn't have been caused by a few pieces of cork falling off the RSRM. Isn't it quite possible that the source of the severe damage at the location of concern might also be the same as the source that caused less severe damage at the other 400-some locations? I'm not a statistician, but it seems reasonable that this may well be the case, and I'm surprised that Rockwell didn't even discuss this is a possibility."

"Your numbers are wrong," the fellow from Rockwell who'd made the presentation said as he jumped to his feet. "There wasn't anything close to 400 areas of tile damage on *Discovery*."

"Well, that's the figure that I remember from the telecoms I participated in on this subject." I could tell from the expressions on the faces of many of the FRR Board members and many in the audience that they were think-ing, "There goes McDonald again, exaggerating the situation to support his position."

"If you're so smart, Mr. McDonald," McCartney countered, "then where did all of this debris come from that caused the problem?"

"I would be very suspect about the external tank and the frustum or nose cone of the SRB. All of these components are insulated with an external in-sulation that would most certainly impact the underbody of the orbiter if it came off on the inboard side, and do so without the need for any deflections, Frisbee-type aerodynamic maneuvers, or puffed-wheat-action of being shot from guns. Rockwell's analysis clearly shows that any debris in the aero-dynamic streamlines from the inboard side of the ET's ogive and the nose cone of the SRB most certainly would impact the tile on the orbiter. Both the ET and the nose cap of the SRB are not recovered, so we can't say with any degree of confidence that they were *not* the source of the problem."

"I'd like to defend the ET," someone from Martin Marietta spoke up. "We noted this sort of problem early in the Shuttle program, but we solved the problem by making some process changes to prevent any loss of the insula-tion. Furthermore, we incorporated a camera into one of the earlier flights; it focused on the ogive and confirmed that these changes had indeed fixed the problem. We haven't experienced any problems since."

"Sorry," I said, "but I'm not impressed when a single camera on a single flight indicates there could never be a problem on any other flight, espe-cially since we never recover any of that flight hardware. Furthermore, I've learned from our management at KSC that the ET used on STS-26R was the same ET that was shipped to Vandenberg through the Panama Canal for use on the first launch from Space Launch Complex 6 at Vandenberg prior to the *Challenger* accident. This ET was later sent back to KSC through the Panama Canal again and assigned to STS-26R. It was supposedly the old-est ET ever flown and required the most repairs on the external insulation prior to flight of any ET previously used. These facts certainly make me very suspicious about the ET as a source of debris on STS-26R."

The Martin Marietta fellow sat down.

"I'm equally concerned about the forward section of the SRB," I contin-ued, referring to the USBI-supplied frustum that housed the main para-chutes and the nose cap containing the pilot and drogue chutes. "I've per-sonally inspected the returned frustums, and when I did I noticed several areas of missing external insulation."

Now a fellow from USBI jumped to his feet to defend his components: "We've examined the frustums carefully and concluded that all of the missing insulation occurred as a result of splashdown in the ocean. If it had been lost during ascent, we would have been able to tell by the nature of the fracture of the insulation and by the fact that we couldn't find any area of soot or any indication of aerodynamic heating on the exposed surfaces of the missing insulation."

"That may be the case," I responded, "but you don't know the condition of the external insulation on the SRB nose cones because they, like the ET, are never recovered, so you cannot discount these as a source of debris. Nor do you know what the other surface of the missing insulation on the frustums may have looked like."

Now that I indicted both the ET and SRBs, Bob Marshall, Director of the Space Shuttle Projects Office at Marshall and a member of Admiral Truly's FRR Board, also became defensive. Marshall was the NASA Program Manager responsible for all of the Shuttle propulsion elements at MSFC, including the SSMEs, ET, RSRM, and SRB. He challenged my allegations relative to the ET and the SRB parachute hardware and my disagreements with the Rockwell analysis.

Dick Davis, Thiokol's Vice President of Space Booster Programs and a member of the board, got up and defended me: "We are not here to tell you necessarily what you want to hear. I'm rather new on the program, but if that is what you expect, you had better find someone else to take my job."

The room became very quiet. This had quickly become a very heated and emotional discussion that could consume the rest of the day. Admiral Truly intervened: "This is a serious enough disagreement that I want Rockwell to reassess its scenario and report back tomorrow in a special FRR meeting on the exact extent of damage to the orbiter tiles and whether these other potential causes that Mr. McDonald has raised could explain the severe tile damage observed on *Discovery*. I'd like to table further discussion on this issue until tomorrow so that we could get through the remaining FRR agenda. Mr. McDonald, would you now proceed with your planned presentation?"

I did, and the presentation went well. Everyone seemed comfortable that, even if the cork was the source of the problem, we had taken the proper corrective action to prevent it from becoming a problem on STS-27R.

The rest of the FRR was rather uneventful and boring.

The next morning, Rockwell made the presentation on the debris issue that Admiral Truly had requested. Its first statement was that there were over 400 areas of impact damage noted on the underside of the orbiter when it was inspected after landing at Edwards AFB. Someone in the audience spoke up, "That seems to be consistent with the number we had heard yesterday in this room that Rockwell had disagreed with!" The Rockwell fellow, who could only agree, turned red-faced. Describing the extent of damage in detail, he quickly focused on the area of the most severe damage—all of the other areas certainly did not reflect the most desirable conditions, because they needed repair, but they posed no threat to the safety of the orbiter or crew.

The Rockwell folks then explained the basis for their aerodynamic analysis, impact damage analysis, and the reentry thermal analysis of the damaged tile. They reiterated that an object about the size of the missing cork on the RSRM, with the approximate density of cork, was required to produce sufficient damage to the orbiter tile and cause the eroded tile profile during reentry that matched what was observed in the area in question. So they still believed that the loss of cork on the RSRM was the most probable cause of the problem. However, they did believe that the corrective actions taken by Thiokol were appropriate and should eliminate cork as a source of debris from the RSRM for the next flight.

Rockwell recommended proceeding with the launch. As far as Admiral Truly was concerned, that was the most important conclusion. I listened to the discussion and had nothing more to say other than I still didn't agree with Rockwell's conclusion that the cork was the most probable source of debris causing the problem. The meeting ended with an agreement to disagree. All parties did agree that the repairs on the cork of the RSRMs should fix the problem if it did exist; therefore, the RSRMs were safe to fly on the next flight.

I walked out of the room muttering to myself that I sure hoped that the next flight would come back safely with even more damage to the orbiter tile and that the RSRMs would be recovered with no cork missing in any area. I was fully convinced that it would take such a condition to get all of these people to take their heads out of the ground and face up to the real nature of the debris problem and not feel so comfortable that they had solved the problem by just fixing the cork on the RSRM.

A few days after I got home, I got a phone call from Captain Hoot Gibson, commander of the next flight, STS-27R. Hoot had asked me some time back if I had ever witnessed a Shuttle landing. I hadn't, and he invited me and my family to be his guests at Edwards AFB to witness the landing of

Atlantis. As STS-27R was a classified DOD mission, the VIP viewing area for the landing at Edwards wouldn't be open to the public but only to invited guests of high NASA officials, air force brass, and the astronaut crew. I knew my family would be excited to see a Shuttle land, and I was, too. Hoot also complimented me on my stand at the recent FRR concerning the orbiter tile damage, because he also thought that this problem was too widespread and too severe to focus only on the cork from our RSRM as a source of significant debris.

Linda was ecstatic about the chance to see a Shuttle landing. When we received the formal invitation in the mail, it also included the rest of my family, including our son Greg, who was living in Salt Lake City and studying airline operations and management with a pilot's option at Westminster College. Greg wanted to be a commercial airline pilot, had most of his licenses, and was really excited about seeing a Shuttle landing. My youngest daughter, Meghan, who was six years old, was also very excited. A great fan of her big brother, she had decided, too, she wanted to become an airline pilot when she grew up. Our middle daughter, Lora, who was going to school at Loyola Marymount University near the Los Angeles International Airport (LAX), also wanted to attend. Thanks to John Thirkill approving use of the company's corporate jet, we were able to pick Lora up at LAX on the way to Edwards. The only family member who was not able to see the Shuttle land was our oldest daughter, Lisa, who was a senior at Boston College.

On November 30, 1988, I returned to the Cape to participate in the L-2 FRR for *Atlantis.* Launch preparations for STS-27R were proceeding very smoothly, and the L-2 FRR did not uncover any significant issues or concerns. The countdown started, and all indications were a "go-for-launch" as planned on the morning of December 2. I sat at the console in the Launch Control Center on the morning of the launch and was as nervous as I had been for the previous launch. The return-to-flight of the Shuttle in September had been important to me, but this second flight was very special to me as well because Hoot Gibson, whom I had gotten to know very well on the SRM redesign team, commanded this flight.

I gave my "go-for-launch" during the engineering poll taken during the T-9 minute hold but became increasingly nervous as the clock started into the terminal count. My hands and arms were cold and clammy by the time the final seconds ticked off to SRB ignition and liftoff.

Everything went well: the SRB's performance was right on target, and the rockets burned out and separated from the ET just as planned. Normally, I breathed a great sigh of relief at this point when our boosters had successfully done their job; this time I felt differently. I knew that the SSMEs had to operate another six and a half minutes before the Shuttle was safely in orbit, and as far as I was concerned that was now a more hazardous part of the ride than on the redesigned solid rocket motors. I thought there was now more margin of safety in the RSRMs as a result of our redesign than there was with the SSMEs.

Quickly I switched the console's screen to display the SSME data, loaded as it was with pressures, temperatures, turbine speeds, voltages, and all kinds of engineering data. The more I looked at the amazing amount of data being monitored, the more I was convinced that our simple RSRMs were indeed more reliable; the safest part of the flight was over.

As I kept my eyes glued on the monitor, one of the three columns of data on the screen went totally red. This indicated a potential Criticality 1 failure on one of the engines. My heart just about stopped; I couldn't swallow. One of the engine temperatures had exceeded the redline and a decision had to be made quickly whether to shut down the engine, forcing a possible transatlantic abort, or ignore the temperature measurement and proceed to orbit as planned.

This data was also displayed, I knew, in the cockpit of the orbiter; I was surprised how calm Commander Hoot Gibson was being about the situation. He opted to proceed to orbit on the basis that a redundant thermocouple on that same engine indicated it was operating OK. The folks from Rocketdyne and those monitoring the situation concurred, and the decision was made to proceed to orbit on the basis that the temperature measurement causing the redline violation was probably in error.

Wow! What a sweaty palms operation this was now, with that big red column of data remaining on the screen and several minutes of SSME operation yet to go. If the judgment was wrong, the turbine blades running at nearly 40,000 revolutions per minute could overheat and fail, causing loss of mission and crew.

Thank God, it turned out that the decision was correct. *Atlantis* made it safely to orbit without incident, but I knew I could never take another one of these. It was time for me to get off this hot seat and turn it over to someone else who would not take this job so personally. I was ready to go home to see my family and get ready to witness *Atlantis*'s landing at Edwards. It was hard to relax, though. I was very anxious to find out whether the RSRMs had lost any cork on the flight and whether the joints performed the same as on the previous flight, including the fretting.

Arriving at Edwards on the morning of the Shuttle's landing, my family was given badges and escorted to the VIP room, where there was an ample supply of snacks and soft drinks. Hoot Gibson's mother, Mrs. Paul A. Gibson, was there. She was a very charming woman, and Linda and I enjoyed talking with her. Hoot's wife, astronaut Rhea Seddon, was not there; she was in training at JSC for an upcoming Shuttle mission. There was a large television screen monitoring NASA Select TV.

Never having had to focus on the requirements for landing a Shuttle, I had never really given much thought to the return to Earth part of the mission. However, I quickly got butterflies in my stomach when they announced over the loudspeaker that *Atlantis* was preparing to fire the onboard orbital maneuvering system to begin its fiery descent to Earth. It had just struck me about the tile damage on the previous flight and how close it had come to causing a disaster during reentry. Worse yet, though I didn't particularly believe the Rockwell analysis, it certainly was a possibility that the cork on our RSRMs was the problem. I broke out into a cold sweat thinking about that, hoping that the repair we had made to the cork was really going to work. I worried even more about the things they didn't fix on the ET and SRB nose cap. I listened very intently to the communications between *Atlantis* and the ground and became very nervous during the peak heating portion of the reentry when the ionized plasma forming around the orbiter blacks out the communication with the orbiter for a period of time. Knowing that *Atlantis* was speeding home somewhere high over the Pacific Ocean literally gave me the chills.

As the Shuttle *Atlantis* approached the West Coast, we went on top onto the roof to the VIP viewing area. It wasn't long before we could see a little black dot high in the sky approaching from the northwest. As the Shuttle passed overhead, two very strong sonic booms, one right after the other, startled everyone. You could feel them from head to toe. The Shuttle then made a spectacular banking U-turn to approach the long, sandy landing strip on the dry lake bed surrounding Edwards. The landing was right on target, and everyone on the roof gave out a big cheer. I had the same feeling of relief that I'd had when I sat in the Launch Control Center and heard the word "MECO" on the network, because then I knew that the Shuttle had safely made it to orbit. Now it had safely returned home. I was very proud to be part of this memorable occasion, and my family was overjoyed for having the opportunity to witness our spaceship return to Earth with one of my friends aboard.

As the astronauts came out of the orbiter, they were picked up in a blue van. Still in the VIP room, I was hoping we'd have the opportunity to talk to Hoot in about an hour or so and thank him for inviting my family to this exciting event. We waited around for over an hour when one of the hosts informed us that the crew was in a debriefing room and would be there for some time. So we decided to head for home. After dropping daughter Lora off at LAX, our jet headed back to Ogden. The Thiokol pilots knew that my son, Greg, was training to be a pilot, and they let him sit in the right seat and fly the plane for a while. Six-year-old Meghan also got to sit in the right seat and fly the plane for a few minutes in the pilot's lap. What a thrill that was for her. We got into Ogden very late that night totally exhausted, but it had been a wonderful day.

On my desk the next day at work, I had a pile of good news. The STS-27R boosters had performed as expected, with no loss of cork insulation during ascent (a three-inch-square piece broke off the right-hand center field-joint at water impact at sixty miles per hour), and the joints were in excellent condition; however, some fretting was noted. Even that was good news, because now we knew that fretting was "normal" behavior and that our analysis had already confirmed it was not a flight safety issue.

Detailed inspection of the orbiter *Atlantis*, however, revealed some severe tile damage that could not be blamed on the RSRM and would require a detailed investigation before the next Shuttle flight. What I didn't know at the time was how severe the damage was due to debris hitting the tiles. I didn't realize that it was by far the worst impact damage to an orbiter prior to the loss of *Columbia* (STS-107) over fourteen years and eighty-six flights later, in February 2003. If I'd known what had really happened during ascent, when it was noticed that *Atlantis* had a severe debris impact, I would have been even more nervous watching Hoot's landing.

In 2003, the *Columbia* Accident Investigation Board report clearly indicated how severe the problem had been on STS-27R, noting that it was through sheer luck that *Atlantis* had been able to reenter the Earth's atmosphere and land safely. The report stated: "One debris strike in particular foreshadows the STS-107 *Columbia* event." When *Atlantis* was launched on STS-27R on December 2, 1988, the largest debris event up to that time significantly damaged the orbiter. Postlaunch analysis of tracking camera imagery by the Inter-center Photo Working Group identified a large piece of debris that struck the thermal protection system tile at approximately eighty-five seconds into the flight (this was within three seconds of a similar impact on *Columbia*, STS-107, launched on January 16, 2003). On Flight Day Two, Mission Control asked the flight crew to inspect *Atlantis* with a camera mounted on the remote manipulator arm, a robotic device that was not installed on *Columbia* for STS-107. Mission Commander R. L. "Hoot" Gibson later stated that *Atlantis* "looked like it had been blasted by a shotgun." Concerned that the orbiter's thermal protection system had been breached, Gibson ordered that the video be transferred to Mission Control so that NASA engineers could evaluate the damage.

When *Atlantis* landed, engineers were surprised by the extent of the damage. Postmission inspections deemed it "the most severe of any mission yet flown." The orbiter had 707 dings, 298 of which were greater than an inch in one dimension. Damage was concentrated outboard of a line right of the bipod attachment to the liquid oxygen umbilical line. Even more worrisome, *the debris had knocked off a tile*, exposing the orbiter's skin to the heat of reentry. Postflight analysis concluded that structural damage was confined to the exposed cavity left by the missing tile, which happened to be at the location of a thick aluminum plate covering an L-band navigation antenna. "Were it not for the thick aluminum plate," Gibson stated during a presentation to the Board, "a burn-through may have occurred." After the STS-27R damage was evaluated during a postflight inspection, the Program Requirements Control Board assigned in-flight anomalies to the orbiter and solid rocket booster projects. Marshall Sprayable Ablator (MSA-1) material was found embedded in an insulation blanket on the right orbital maneuvering system pod confirming that the ablator on the right solid rocket booster "nose cap" was the most likely source of debris.

Back in the aftermath of the *Atlantis* flight, I was really glad to hear this, because we had not lost any cork on the RSRMs during ascent, so this severe debris problem couldn't be blamed on Thiokol again. NASA and Rockwell had to take the problem more seriously—and they did, forming a big Debris Task Force headed up by Marshall's John Thomas. This was the same John Thomas who had headed up the redesign of the SRMs for MSFC. He was a good choice, as John was a top-notch engineer and a very hard worker. If anybody could flush out the real culprit, John was the guy to do it. The Shuttle was temporarily grounded until the problem could be resolved.

In late 1988 I had some additional good news. I was being awarded the NASA Public Service Medal for my contribution in leading the engineering effort in redesigning the SRMs to return the Space Shuttle to safe flight. The certificate read, "For exceptional leadership demonstrated through the successful management of the engineering activities associated with the redesign of the solid rocket motor at Morton Thiokol." As honored as I was to receive this award, I was happier that three other Morton Thiokol Space Operations engineers, Charlie Bown, Stan Graves, and Scott Stein, who had worked for me on the redesign, were also being honored with special awards for their outstanding contributions to the RSRM redesign activity.

Like many other experiences I had had over the past three years, the awards ceremony, which was held at NASA Marshall, turned out to be a bittersweet occasion. No one in government or industry had carried a heavier burden or worked harder for the safe return-to-flight of the Shuttle than Carver Kennedy and myself. Without question, John Thomas, Royce Mitchell, and Gerald Smith from the NASA/MSFC SRB Program Office had also put in many long days and had been very instrumental in the successful redesign activity, and I believed they deserved much of the credit as well. However, as it turned out, politics and rank once again entered the scene as Carver, Royce, Gerald, and I were presented with the NASA Public Service Medal that night in Huntsville. I had some problem with NASA's John Thomas receiving a higher award—NASA's "Distinguished" Public Service Medal—but it was even more difficult to see Joseph Pelham, Chief Scientist at Morton Thiokol; Dick Davis, Vice President of Space Booster Programs; and John Thirkill, Vice President and General Manager of Space Operations, also receiving this higher award. Joe Pelham probably deserved the award for making a tremendous contribution to the redesign activity by inventing the Joint Environment Simulator, but even he had retired from the company over a year before the job was completed. John Thirkill didn't even come to Utah until a year after the *Challenger* accident. Worse yet, Dick Davis came into his position less than a year after the actual design activity was totally over. Adding insult to injury, NASA presented Davis with a "Team Award" for the Morton Thiokol Space Engineering Team for its outstanding contribution to the recovery activity—and I was the Vice President of Morton Thiokol Space Engineering at the time.

I was flabbergasted. I guess NASA thought it was more important to give its highest awards to those with the highest rank in the company rather than for their personal contributions to the successful recovery activity.

Several members of the National Research Council on the SRB Redesign told me they could not believe I did not receive the highest award possible. Many of the other Shuttle contractors received the same award that I did, but I had learned a long time ago that life was not fair.

Meanwhile, after several weeks and several thousand man-hours of detailed analysis and laboratory testing, the Shuttle Debris Task Force confirmed that the principal source of debris causing the severe damage to the orbiter tiles was from loss of insulation on the nose cap of the SRBs. A change made in the external insulation material and processing for the nose cap had degraded the capability of this material.

The nose cap insulation material was changed before the next scheduled flight of the Shuttle, which was STS-29—a reflight of the repaired *Discovery* orbiter, planned for March 1989. The pilot for this flight was navy Captain John Blaha. Blaha, along with astronaut Steve Oswald, had gotten involved with the RSRM program as a representative of the astronaut corps when Hoot Gibson had begun his training for flying again, so I had also gotten to know these two men quite well. After I told Blaha that my six-year-old daughter, Meghan, wanted to be a Delta pilot, he had given her a signed photograph of him with the message, "See you on the Space Station, Meghan."

At the flight readiness review prior to the next flight of *Discovery*, John Thomas gave a very lucid presentation on the results of the Shuttle debris task force. He and his team had done a very thorough job. Even though his data indicated quite clearly that the primary source of the debris problem was the SRB nose cap insulation material, the investigation team rightfully chose to implement some corrective actions in all areas of the Shuttle, aiming to eliminate the possibility of debris from all sources. This was as it should be, and everyone on the FRR board felt comfortable that the problem had been properly resolved and that it was safe to continue flying the Shuttle—although I don't recall anyone addressing the large piece of debris seen coming off the ET at eighty-five seconds into the flight of *Atlantis*.

It was a little disappointing to me that no one on the FRR Board mentioned the discussion about the debris problem that I had raised prior to the previous flight, when this whole problem was blamed on Thiokol's RSRMs. All that Thomas said was that the corrective action taken by Thiokol to eliminate cork debris from the RSRM was appropriate, as indicated by the excellent condition of the RSRMs from the previous flight. I was proud of my contribution, even though no one from NASA, Rockwell, or the Shuttle contractors ever acknowledged it.

We still had the last qualification motor (QM-8) in the redesign program to test. It was the sixth ground test of the RSRM design and a very important test in completing the redesign program.

Prior to this test, the Space Shuttle was restricted to high-temperature launch conditions only, even though the motors had electrical heaters on the field-joints and a warm gas nitrogen purge in the aft skirt region for temperature control of these areas. An electrical heater was also added to the igniter joint on the second flight and subsequent vehicles. The RSRM specification still required that the motor operate successfully at a propellant mean bulk temperature as low as 40° under all flight loads, something that had not been accomplished until QM-8. It took over a month at conditioning temperatures around 30° to condition the motor to a PMBT of 40°. Furthermore, we wanted to test the motor in the wintertime, so we could maintain the low temperatures on the motor after the conditioning house was removed from the motor prior to test. The joint heaters and warm gas purge operated to keep the joints warm, but the PMBT was actually 39°, and there were some locations on the skin of the motor that were as low as 29° at the time of firing. The motor was static tested in our snow-covered T-97 facility with design launch and aerodynamic loads applied during the test.

The QM-8 static test of January 1989 was an unqualified success. When it was over, we had completed the most extensive and successful solid rocket motor development and qualification program in history. Over 11,000 channels of data had been obtained in the redesign program, verifying the performance and reliability of the motor. I was proud to have played a significant role in the most scrutinized engineering activity in history. The ad hoc NRC Committee on the Redesign of the Space Shuttle Solid Rocket Boosters met 22 times during their 30-month tenure in addition to sending individual members to participate in 77 events (site visits, component suppliers, reviews, tests, etc.). The level of dedication of the NRC committee members was well beyond anything the NRC had ever seen before or since that time, and all of these members served pro bono on their own time and were compensated only for their travel expenses. The NRC made a significant contribution to the success of the redesign program, and both NASA and Morton Thiokol owed them a sincere debt of gratitude for their service.

I was particularly pleased by comments about the redesign program made by the National Research Council in its final report to NASA Administrator Dr. James Fletcher in December 1988 and in a presentation that the chairman of the NRC Panel on the Redesign of the SRBs, Dr. H. Guyford

Stever, made at an AIAA conference in the summer of 1989. In a final report to Fletcher, Chairman Stever wrote:

Finally, Jim, all of the members of the NASA/contractor team responsible for the redesign of the solid rocket boosters have earned the appreciation and congratulations of the nation for their tireless dedication leading to the return-to-flight of the Shuttle in the face of intense public scrutiny. We join in expressing our own thanks, particularly because we could not have discharged our responsibilities without their earnest cooperation and support. We are especially grateful to John W. Thomas, of NASA's Marshall Space Flight Center, and Allan J. McDonald, of Morton Thiokol, Inc., who bore a primary responsibility for making the engineering judgments required in the redesign program and the burden of convincing us that they were right—*or accepting our recommendations.*

Later that summer of 1989, Dr. Stever and a member of his staff, Dr. Myron F. Uman, presented a paper at an AIAA conference entitled "Some Lessons Learned from the Redesign of the Space Shuttle Solid Rocket Boosters." Some of their key bullet points bear repeating:

- "Human and institutional failure to fix a weak design when strong evidence of weakness was mounting and a similar failure to observe redline constraints in operating a complex technical system led to the accident."
- "Redesign of the SRBs remarkably focused on four interdependent concerns, i.e. safety, schedule, cost, and performance. Safety was prime consideration although critical national need for the launch capability of the Shuttle made schedule a close second."
- "The redesign team convinced us that the principal line of attack should be modifying the original design to correct its deficiencies."
- "Our panel concluded that the program was well conceived and executed."
- "We found that the team was conscientious in establishing specific pretest criteria for success and predictions of results."
- "The redesign program has set new standards for control of materials and processes and inspection of components."
- "The redesign was a difficult and sometimes a thankless task. Car-

ried out in the blinding lights of the Presidential Commission, congressional hearings, oversight committees from both within and outside NASA, thorough professional society reviews, a disturbed and fascinated public, and a hyperactive media, it was remarkably well done, albeit with considerable grief."

I had decided that the next flight was going to be my last in the program. It was the last flight with development flight instrumentation, and we had hopefully solved all of the major problems so the Shuttle could return to routine operations. Besides that, I found myself getting too personally involved again because of my acquaintance with John Blaha; STS-29 *Discovery*, with John Blaha as the commander, launched successfully on March 13, 1989. All systems operated normally, and the orbiter returned from space with minimal tile damage—all of the fixes recommended by the Debris Team had apparently worked. Inspection of the recovered RSRMs again confirmed their excellent condition. It was an opportune time for me to give up the hot seat.

I had achieved my goals and been given my share of credit and awards, including being the American Institute of Aeronautics and Astronautics' nominee for the first ever prestigious $350,000 Charles Stark Draper Prize, for which my supporting nomination letters were written by William Rogers, Admiral Richard Truly, and Hoot Gibson. Excerpts from those letters included:

McDonald's courage and willingness to speak the truth contributed greatly to the successful conclusion of the Commission's work. Although there was considerable unhappiness on the part of Morton Thiokol for McDonald's forthright testimony, because of his experience and engineering skill, he was asked to return to work on the *Discovery* Shuttle which was successfully launched. I think that the Nation owes a debt of gratitude to Mr. McDonald.

William P. Rogers

The success of the endeavor to return the Space Shuttle to safe flight following the *Challenger* accident was critical to this nation's reclamation of spirit and prestige. Most important, at a time of crisis, Mr. McDonald instilled in the Morton Thiokol engineering team a renewed sense of confidence and purpose. For the engineering excellence and leadership personally contributed by Allan J. McDonald to this endeavor of paramount national importance, I, without reservation, endorse his nomination.

Richard H. Truly

I worked closely with Al McDonald and had the opportunity to observe him under a wide variety of situations over the past 2½ years. As an engineer and a manager, Al is unsurpassed. His task of rebuilding the SRBs and reestablishing confidence in Morton Thiokol was a nearly insurmountable task. Al approached it with the leadership and engineering judgment he is well known for, and was instrumental in guiding the team to the eventual selection of the redesign concept. I have again had the thrill of riding two of Al's booster rockets in December 1988, as the Mission Commander of *Atlantis* on STS-27, and I can personally confirm that the confidence that we place in the booster rockets is in no small part due to the confidence we have in the talent and ability of Allan McDonald and the company he helped rebuild.

Robert L. "Hoot" Gibson

But nothing related to *Challenger* could ever come up that could make me, or anyone else, really happy. As recounted earlier, shortly after the congressional hearings into *Challenger*, I had been contacted by a Hollywood movie producer to participate in a film about the events leading up to and following the accident. The movie was to center around Roger Boisjoly and myself—"two Morton Thiokol rocket engineers who both opposed the launch of the *Challenger*, one ended up quitting his job and suing the company for over a billion dollars and the other stayed with the company to lead the redesign activity to return the Shuttle to safe flight." The movie's theme was to focus on our two contrasting personalities: both men were right, both were mistreated by their company, but each man responded to the situation in a totally different manner.

Even after I declined to participate in the "*Challenger*" movie, the film studio chose to make the movie anyway, changing the theme to focus on the families of the *Challenger* astronauts. Roger Boisjoly decided he still wanted to play an important role. When Thiokol had been contacted by the movie company about my participation, it had not responded. The movie studio also contacted Larry Mulloy, and he agreed to participate. The movie did not turn out to be very good, but it made Larry Mulloy look better than he should have. I was portrayed in one scene smoking a cigarette (which I have never done), wearing glasses (I wear contact lenses), and commenting to Mulloy that we need to remove the O-ring problem from the Problem Assessment System Report. Mulloy was portrayed more like a hero and I as the risk taker. It was a picture far from the truth. Both Boisjoly and Mulloy were paid for their contribution to the movie, as they were central figures in it. The made-for-TV movie appeared on ABC in January 1989. It was not at all a good representation of what happened prior to the launch of *Challenger*, and it opened old wounds for me.

It reinforced my desire to get off the Shuttle program to find some peace and quiet.

39

Dissolution

My timing for leaving the Shuttle program was perfect because the ASRM contract had recently been let to a surprising winner, the Lockheed-Aerojet team. Not only had Thiokol agreed to fabricate the nozzle for Lockheed-Aerojet if it was successful in its bid, which no one expected, the really bad news was that the nozzle had to be fabricated in a plant in Michoud, Louisiana, a plant that didn't yet exist! Though Thiokol didn't have to provide any capital funding for the project, we had to provide the physical layout for the plant and define all the needed manufacturing equipment, not to mention the design of the ASRM nozzle, which was under subcontract to the Aerojet Solid Propulsion Company, one of our major competitors. As we already had all of the Shuttle knowledge, Aerojet expanded the scope of our effort on the ASRM program to help them in other areas of the motor design and testing requirements.

The complications of this business were many and varied. The Lockheed-Aerojet team had subcontracted the Michoud plant design to Rust International in Birmingham, Alabama, while at the same time being responsible for the ASRM manufacturing plant in Iuka, Mississippi, to be operated by Aerojet. The nozzle manufacturing plant in Michoud was to be operated by Thiokol under the direction of John Sucher, who had been our Chief Program Manager on the RSRM nozzle. Testing of the ASRM was to be directed by NASA at a newly constructed solid rocket motor testing facility located at the NASA Stennis Space Center, also in Mississippi. NASA's plan was to have the ASRM flying by 1994, so that production of the RSRM could be discontinued in Utah by 1996. Morton Thiokol would be left with only ASRM nozzle production in a GOCO facility in Michoud.

One of the strong selling points for the GOCO facility made by NASA to Congress was that such a facility would allow the space agency to recompete the ASRM, or throw out the existing contractor at any time, if NASA thought it was in the government's best interest to do so. NASA had used this argument after the *Challenger* accident by saying if it had had a GOCO solid rocket motor facility at the time of *Challenger*, it could have thrown out Morton Thiokol and gotten a new contractor to do the RSRM. The way it had been, NASA didn't have any choice but Thiokol for the RSRM, because MTI owned the only facility in the world capable of manufacturing such large solid rocket motors. NASA would have had to shut down the Shuttle program for an additional four or five years if the RSRM contract had been given to anyone other than Thiokol.

With the ASRM program now posing such a threat to Morton Thiokol's future business, the corporation decided to examine what it would take to modernize its facilities in Utah to do the ASRM job and at the same time provide more incentive for NASA to extend production of the RSRM. Thiokol had been successful in winning every large solid rocket motor project but one (the Titan III) since the late 1950s, because MTI had invested its own capital and built the only facility in the world for doing such a job. Now the government was going to pay for a totally new facility that anyone could bid on to compete with Thiokol.

NASA had sold the ASRM program and GOCO facility approach to Congress on the basis that the new motor would have higher reliability at lower cost than the RSRM. It would do this by changing the steel-case design to allow for a bolted flange design for the field-joints and having a more modern computer-operated and more robotic factory for better control of the manufacturing processes. Thiokol had initiated a project called Advanced Launch Vehicle Organization (ALVO) to define what it would take to upgrade our current RSRM facility for better control of the manufacturing processes. The difference would be that MTI would provide the capital for such an expansion if we could show a net cost savings to the RSRM Program within completion of our current Buy III contract, scheduled for completion by 1996. In 1989, NASA was still hoping to fly as many as ten to twelve Shuttle flights per year.

Phil Dykstra had been put in charge of our ALVO Project. Phil had previously been the Vice President and General Manager of Strategic Operations, which produced the first stage of the Peacekeeper ICBM for the air force, our joint venture with Hercules on the navy's Trident II (D-5) Fleet Ballistic Missile, and the full-scale engineering development program for the first stage of the air force's new mobile small ICBM. Strategic Operations had suffered a serious accident in December 1987 during the manufacture of a Peacekeeper motor, resulting in the horrible fire that killed five Thiokol employees. The air force general in charge of this program had not been very happy with Thiokol management's response to this accident, and the company was being sued by some of the relatives of those who were killed. Corporate management, under pressure from the air force, had removed Dykstra from his Strategic Operations job and put him in charge of the ALVO Project as part of his newly created job as Corporate Vice President of Strategic Planning. When I left the Shuttle program, I was selected to take over as Vice President of the ALVO Project, replacing Dykstra. I was also asked to form an advanced technology design group for examining new launch vehicle system programs—an organization simply called Advanced Programs and Technology.

My first assignment as Vice President of Advanced Programs and Technology was to complete the modernization plan for manufacturing the RSRM and selling it to NASA. Shortly after accepting the job, we submitted a $200-million proposal, called the RSRM Enhancement, to NASA. Thiokol would fund the entire project using MTI capital, with NASA agreeing to let MTI recover its investment with the savings projected in the remaining portion of the current production contract on the RSRM. If the projected savings were not realized, then MTI would end up with unamortized facility costs at the end of the scheduled RSRM production program.

NASA ended up agreeing to portions of the proposed modernization project, because it was clearly a win-win situation for the Shuttle program; however, the agency would not agree to approve it all, because it was viewed as a political threat to the ASRM program.

The estimated cost of the ASRM program at the time was over $1 billion, whereas the RSRM Enhancement program would not cost the taxpayers anything; in fact, it would save the taxpayers money. Since we were already the selected ASRM nozzle contractor, we proposed coproducing the ASRM nozzle with the RSRM nozzle in a new nozzle manufacturing facility just outside of Ogden, with a provision to turn the deed of this $70 million facility over to NASA as a GOCO facility after MTI recovered its investment on the current $3 billion RSRM production contract.

NASA liked the idea until Senator Bennett Johnston from Louisiana found out about it and rebuked NASA for reneging on the ASRM nozzle factory to be put into Michoud and employing 350–400 people. If we wanted to do the nozzle production facility in Utah, NASA wanted us to find another project to put into Michoud that would employ the same number of people. We couldn't find one, so the more modernized nozzle facility was dropped.

Following the *Challenger* accident, we had upgraded our nozzle manufacturing facility, so it was in much better shape than some of our other facilities. We proceeded with the improvements to our propellant mixing and casting areas and final assembly area, but NASA rejected our plans for improving the case refurbishment and case insulation and lining operations because these areas were targeted for major changes in the ASRM program at Yellow Creek in Mississippi.

In my new position, one of my first suggestions to MTI management was to convince NASA that the "R" in RSRM should stand for "reusable" rather than "redesigned" solid rocket motor. The title change, which would put the RSRM in a better light, had been very successful for the air force when it changed the meaning of its Interim Upper Stage (IUS) from "Interim" to "Inertial" Upper Stage. The IUS was intended to be an "interim" solution until a new, more expensive reusable liquid space tug could be developed. This never happened because it was far too expensive, so the air force renamed the system to an Inertial Upper Stage to fit the already well-recognized acronym of IUS.

NASA and Thiokol senior management initially rejected my suggestion because they thought the NASA hierarchy in Washington might consider it a threat to ASRM. It sounded better for the ASRM to replace a "redesigned" rocket motor, which sounded kind of negative, rather than replace a "reusable" rocket motor, which sounded positive. NASA did consider the RSRM as an "interim" solution at the time, with the RSRM eventually to be replaced with the ASRM.

It was also clear Charlie Locke couldn't stand the heat of the Space Shuttle program and wanted to wash his hands of the whole deal if he could. Locke was a very keen businessman, and the advent of the ASRM program made it quite evident to him that the "cash cow" called the RSRM would disappear from his books in the next seven to eight years. With even greater financial pressures on the defense side of the aerospace business, Locke had decided some time earlier that it would be in the best interests of the company to sell off the Aerospace Group. At the time it had merged with Thiokol Chemical Corporation some seven years earlier, Thiokol had had two product groups within the company: Aerospace and Specialty Chemicals. The latter had been integrated into Morton's chemical operations, while the Aerospace part of the company was maintained as a separate Aerospace Group within MTI.

However, one of the fledgling operations within the Aerospace Group was a product line called "air bags." Thiokol had been nurturing this fledgling business for over twenty years, and it was finally starting to produce some revenue through a small production contract with Mercedes-Benz. The future of air bags looked great. When Locke decided to sell off Aerospace, he removed the air-bag work from the Aerospace Group, with the result that no one was interested in buying a solid rocket and ordnance products company that, in just a few years, was losing its largest source of revenue. Locke was disturbed at the performance of MTI stock over the past few years; it was quite apparent to everyone that the bad publicity from *Challenger* had depressed the stock. More important, the public perception was that Morton Thiokol was basically an aerospace company. Stocks from American aerospace companies were selling at a much lower price-to-earnings multiple than commercial companies at that time. It became quite apparent that the rash of bad publicity from *Challenger* made people totally forget about Morton Thiokol's salt and specialty chemicals business, which were starting to show some recovery from a general recession in the chemical industry. Unsuccessful in selling the Aerospace Group, Locke was convinced that he could spin off this portion of the company, to be called Thiokol Corporation, and that the remaining part of the corporation, to be called Morton International, Inc., would demand a higher price-to-earnings multiple in the marketplace, effectively offsetting losing half the sales of the existing company.

When Locke decided this was the proper course of action, he indicated that a search for a new CEO for the newly formed Thiokol Corporation was under way; he himself planned to remain as the CEO of Morton International. As it turned out, Locke was no more successful in finding a new CEO than he was in selling the Aerospace Group, so he finally appointed Garrison as CEO of the newly formed Thiokol Corporation. What Locke did not reveal at the time was that he was keeping the air-bag business that had been developed by the Aerospace Group and that he was transferring most of the liability of the old Morton Thiokol, Inc. to Thiokol Corporation. He was keeping that portion of the old MTI pension fund that was above the minimum requirements for the company that he would lead—that is, Morton International. At the time of the spin-off, this pension fund was significantly overfunded.

This left the newly formed Thiokol Corporation with a lot of debt, very little cash, a marginally funded pension fund, loss of its largest contributor to future revenues (air bags for automobiles), and possible loss of its major aerospace program over the next seven years, with the RSRM being replaced by the ASRM for the Space Shuttle.

In the spring of 1989, Garrison briefed the senior management of the surviving Aerospace Group. I was at that briefing when he tried to paint a very rosy picture of the spin-off, which was to become official on July 1, and make it appear as if it were a good deal for those of us who were still in Aerospace. Following a viewgraph presentation, he opened the floor to questions. Immediately he got inundated: "Why aren't we keeping the very promising air-bag business that we developed as part of Thiokol Corporation? Why are we absorbing nearly all of the debt of the old MTI Corporation?"

In particular, Dick Davis, Vice President and Assistant General Manager of Space Operations, asked some very pointed questions. Garrison snapped back at Davis, as if Davis wasn't being a team player. It became quite apparent to all of us that this was a done deal, and Garrison really didn't want anyone questioning his remarks or the basis for the decisions that Charlie Locke had already made. Locke had set all the rules, and Garrison bought into them to be the CEO of the new company. It was not surprising that Locke in turn was selected by Garrison to be one of the members of the newly created board of directors for Thiokol Corporation.

Apparently, several key financial executives in the rocket divisions of the Aerospace Group discussed the situation after Garrison's briefing and decided to do something about the poor cash situation that Locke was leaving for the new Thiokol. They decided to prepay some of our vendors and other expenses to reduce the amount of cash position, which was subject to loss to the new Morton International. They also chose to delay billings to the government for these goods and services so that this money would flow into the company after the spin-off on July 1; they apparently did this without Garrison's or Locke's knowledge. This strategy significantly improved the cash position of the company, the new Thiokol Corporation.

For me, it was rather ironic that the last official day of Morton Thiokol, Inc. occurred on June 30, 1989, because it was the exact day of my thirtieth anniversary with the company. Some two weeks earlier I had received a mantel clock as my thirty-year service award—nice clock, the only problem was its brass plate, which misspelled my name and had "Thiokol Corp." on the nameplate. I returned the clock to our Human Resources organization to correct the inscription. A couple of weeks later, I received a new clock with the correct information. For me, the clock came to symbolize the chapter in my life on the Shuttle program—one that was traumatic and troubled but also one that I forever treasure.

I had been the lone survivor of the *Challenger* accident. All of the other key individuals at the time of *Challenger* at Morton Thiokol and at NASA, with the exception of Arnie Aldrich at JSC, had long ago disappeared from the Space Shuttle scene as a result of being fired, retiring, or being reassigned to other programs or organizations within MTI or NASA.

No longer was I involved in the day-to-day trials and tribulations of the Space Shuttle program, but in my new position within Thiokol Space Operations, I was still required to participate as an active member of the Thiokol Flight Readiness Review Board for the Space Shuttle RSRM—though I had requested to be removed from this board.

I had survived a great national catastrophe and felt good about my personal contribution to restoring manned space flight in the United States. But it came at a high personal cost—as it did to many, many others: the very long hours, over 3,000 hours of noncompensated overtime, very little vacation, the pressures and stresses of the very difficult job of returning the Space Shuttle once again to safe flight, all the time away from my beloved family, not being home to watch my little girl Meghan go from a three-year-old to almost a seven-year-old. I frankly could not remember what she had done for the past three and a half years. My oldest daughter, Lisa, had graduated from college two years ago, and it seems like just yesterday that my wife and I had dropped her off as a freshman all the way across the country in a strange place called Boston College. My son, Greg, would graduate the next year from Westminster College in Salt Lake City, and my other daughter, Lora, would be starting her junior year at Loyola Marymount University in Los Angeles. Where had all the time gone? My wife, Linda, never complained, because she knew how important it was to me to "fix" the Space Shuttle, but at what price? I finally realized that I had lost over three good years of my life with my family that I could never replace. Yes, I had survived all of this, but not without a great sacrifice and great apprehension about my future at the new Thiokol Corporation. My work on the return-to-flight program not only turned my hair completely gray, it hadn't even earned a "thank-you" from my own corporate management.

One of the more enjoyable experiences I had after these most difficult years occurred in the summer of 1989. Thiokol, in conjunction with the Business Department of Utah State University had developed a weeklong management-teamwork program that was held in the mountainous terrain of Logan Canyon. The program was called the "Outdoor Management Challenge," and it was just that. The program developed great teamwork and trust with your coworkers and also pushed individuals out of their comfort zone into a more risk-taking area that stretched their capabilities. I found myself walking across rope bridges suspended several hundred feet across deep gorges, walking on planks and ropes attached to limbs of large pines some forty to fifty feet in the air, and climbing sheer rock faces on the side of a mountain. I certainly was out of my comfort zone, but I received great support from my work team members.

I enjoyed the team project challenges that were also part of the program, such as scaling walls, crossing ponds, and building a raft for a dozen people from a couple of old oil barrels, two inner tubes, some rope, and a few boards. After a set time limit, the rafts were transported to a small, cold mountain lake called Tony's Grove for a race across the lake with a competitive team of raft builders. The good news was that our raft won the race and crossed the lake in one piece without getting anyone wet—it was so strong it could have survived an Atlantic hurricane. The bad news was, unbeknownst to us, each team had to swap rafts for the race. The other team's raft basically disintegrated with all of our team aboard about fifty yards from shore. We all managed to make it to shore, but boy was it cold as the sun was starting to go behind the mountains. It was one of the most rewarding and enjoyable experiences that I had at Thiokol.

The corporate management of the newly formed Thiokol Corporation soon underwent a number of changes. Garrison replaced my former boss, John Thirkill, as Vice President and General Manager of Space Operations with Bob Lindstrom, ex-NASA Marshall employee. Lindstrom had been Director of the Space Shuttle Project at MSFC prior to his retirement in the fall of 1985, *before* the *Challenger* accident. MTI had hired him as a consultant immediately after *Challenger* and later employed him on a team that investigated the fatal Peacekeeper fire. Lindstrom was a good manager, but he never really took off his NASA hat. It was alleged that he had convinced Garrison that Thirkill was too confrontational with the people at Marshall and that John should be replaced. Garrison apparently accepted his advice and replaced Thirkill with Lindstrom. Though Garrison tried to make it look like Thirkill chose to retire, John was effectively fired from his job—and Thirkill so stated to the local press. I was very disappointed about this change because I had a lot of respect for Thirkill's integrity and abilities as an engineer. I had worked for John in one capacity or another for most of my thirty years with the company and considered him a true friend. I knew that if this could happen to Thirkill with all the time and energy he had put into this company, it could also happen to me.

My pride in the fantastic success of our Space Shuttle has never waned. In the 1990s, it was great to see our "redesigned" solid rocket motor finally indeed become the "reusable" solid rocket motor we wanted it to be—one that was 100 percent successful for 23 years prior to the retirement of the Space Shuttle fleet in 2011.

In 2005, I was happy to see Dr. Michael Griffin, the NASA Administrator, select—with the unwavering support of the astronaut corps—a five-segment version of our cost-effective and reliable RSRM as the booster for the new Ares I crew launch vehicle with a new crew exploration spacecraft named *Orion*, which after 2010 would replace the Shuttle for flying to the International Space Station by using a single modified Apollo J-2X engine for its upper stage. Griffin's NASA also moved ahead with plans to use the crew launch vehicle in conjunction with a new heavy-lift cargo launch vehicle (named Ares V), one that employs two five-and-a-half-segment versions of our RSRM as its boosters, for returning astronauts to the Moon and their eventual voyaging on to Mars.

The 110 successful flights of the Space Shuttle using the RSRM never erased the memory of the *Challenger* accident, for me or anyone else, but they were a fitting tribute for carrying on the work of the *Challenger* crew and allowing our space program to continue after this great national disaster. I feel that that precious crew sacrificed their lives to make it safer for their children and grandchildren to follow in their footsteps in humankind's quest to explore the heavens, and I may have in some small way helped them to achieve that goal.

40

Requiem

Eventually, our redesigned solid rocket motors earned the privilege of being named "reusable." They did so by performing flawlessly in 110 flights since the return-to-flight of the Shuttle with *Discovery* in September 1988.

The outstanding condition of our motors also finally resulted in cancellation of the ASRM project in 1993, in spite of many years of political support from NASA management and Congress for what really amounted to a pork-barrel project. J. R. Thompson, the NASA Marshall Director during the redesign program, became NASA's Associate Administrator, and Dr. James Fletcher, who had returned for the second round as NASA Administrator, was replaced by Admiral Richard Truly. Thompson had been the strongest NASA supporter for the ill-conceived ASRM project, dictated by Congress for NASA Marshall's backyard in northern Mississippi, with the main rocket plant in Iuka, Mississippi, the home district of the powerful chairman of the House Appropriations Committee, Congressman Jamie Whitten.

In the four years ARSM had been funded, the program costs had escalated from $1 billion to over $3 billion, and the anticipated first launch of the ASRM had slipped more than four years. That meant the launch date slipped more than a year for every year the project was funded. On top of all this, there were some very serious technical concerns with the ASRM materials, design, and manufacturing processes. Finally, it became totally apparent that the ASRM would be far less safe and less reliable than the RSRM; at least it had proved it would cost a lot more. What a waste of the taxpayers' money!

A member of my carpool, Bob Lyles, our Vice President of Business Development, and I had been actively trying to undermine the ASRM program since its inception. We told Bob Lindstrom that it wasn't in the best interests of the employees or shareholders of the company to allow the ASRM to replace the RSRM at some future date. Besides being a waste of the taxpayers' money, it was also a safety risk to the astronauts.

Lindstrom agreed with us but related that if Garrison found out we were trying to kill the ASRM program, we'd be in a heap of trouble. Lyles and I essentially went underground, meeting with people from the Office of Management and Budgets at designated McDonald's restaurants in Washington, D.C., revealing to them the real costs and problems with the ASRM that they were not hearing from Lockheed, Aerojet, or NASA. I also talked to some members of the NRC and of NASA's Aerospace Safety Advisory Panel, and had behind-the-scenes support from many astronauts. I think our clandestine campaign was finally successful and helped get the ASRM program canceled.

Many of the problems involved with ASRM were uncovered by Joseph Sutter, the chair of NASA's Aerospace Safety Advisory Panel and a former member of the Rogers Commission. Sutter recommended terminating the project early, but high-level NASA managers under political pressure told the Aerospace Safety Advisory Panel that it wasn't their charter to recommend whether NASA conduct the program, they were only to recommend how to do it more safely. In support of Sutter's position for canceling the ASRM, I passed along a poem to him through his ASAP Vice-Chair Norm Parmet. I had received a similar poem and modified it for the ASRM program to read:

In the beginning was the Plan, and then came the Assumptions, and
    the Assumptions
were without form, and the Plan was completely without substance.
    And the darkness was upon the face of the engineers.
And they spoke amongst themselves, saying:
"It is a crock of shit, and it stinketh."
And the engineers went unto their lab directors and sayeth:
"It is a pail of dung and none may abide the odor thereof."
And the lab directors went unto their center director and sayeth unto
    him:
"It is as a container of excrement and it is very strong, such that none
    may abide by it."
And the center director went unto the local politicians and sayeth:
"It is a vessel of fertilizer, and none may abide its strength."
And the local politicians amongst themselves, saying one to another:
"It contains that which aids plant growth, and it is very strong."

And the local politicians went unto the deputy administrator to sayeth
  unto him:

"It promotes growth and is very powerful."

And the deputy administrator went unto the Congress and sayeth unto
  it:

"This new Plan will actively promote growth and efficiency of the
Space Program, and these areas in particular."

And the Congress looked upon the Plan and knew that it was good, and
  the Plan became Policy.

This is how shit happens

—ANONYMOUS

I couldn't give this to Joe Sutter directly, or I might have been fired, because Thiokol senior management was supporting the ASRM program at
this time because we were a team member with a subcontract from Aerojet
for the ASRM nozzle assembly.

The ASRM fiasco not only raised questions about the charter and mission of the ASAP, but it also tarnished the image of four of the most important people at NASA—Dick Truly, J. R. Thompson, John Thomas, and
Royce Mitchell—who had been responsible for the very successful RSRM
program. Truly, himself a Mississippian, was under great pressure from
President Bush's administration to reduce NASA's budget while trying to
please the Democratic Congress, which was pushing the ASRM program.
While Thompson was NASA's chief ASRM defender to Congress and the
media, John Thomas was the chief technical proponent for Lockheed, the
contractor. Thomas had left NASA and was managing the Shuttle assembly
operations for Lockheed Space Operations at KSC under contract to NASA.
Shortly after the Lockheed-Aerojet team won the ASRM contract, Lockheed
promoted Thomas to vice president and sent him to Iuka, Mississippi, to
manage the facility construction and the development activities associated
with the ill-conceived ASRM. Royce Mitchell, NASA program manager for
the RSRM, served in a similar capacity for the ASRM.

Things did not go so well for Roger Boisjoly either, the primary spokesman for Thiokol engineering against launching *Challenger*. As already
mentioned, Roger left the company shortly after the release of the Rogers
Commission report and filed an unsuccessful lawsuit against the company
for over $2 billion. He told me he was blackballed in the industry by MTI
management and, indeed, was unsuccessful in finding an engineering job

for over a year before he decided to become an expert technical witness for a group of lawyers in California. Roger moved from Utah to Arizona and had a very difficult time emotionally and financially for several years. He was mentally hurt by the *Challenger* accident and had a tough time making a living. He successfully passed his professional engineer's examination and eventually became a relatively successful expert witness before retiring.

Arnie Thompson, the other MTI engineer who tried his best to halt the launch of the *Challenger*, worked in engineering on the redesign of the solid rocket motor. Arnie retired shortly after the Shuttle returned to safe flight.

So what about me? In my new position as Vice President of Advanced Programs and Technology in Space Operations, I was deeply involved in the preparation of a $200-million modernization plan for manufacturing the RSRM and identifying further enhancements to the design of the RSRM to improve its performance, reliability, and cost. My job also involved identifying and developing new solid rocket motor technology for new space booster systems.

One new SRB technology study involved a new air force/NASA launch vehicle program called the Advanced Launch System. USAF management for the program in Los Angeles decided there was an environmental problem associated with the launch of solid rocket boosters and had eliminated consideration of SRBs for any new-generation launch vehicle unless the solid propellant chemistry was totally changed to be more environmentally benign. I asked the folks at the Advanced Launch System Project Office where the data was indicating that this was really an environmental problem; they didn't have any. I went to one of the best propellant chemists in our Development Laboratories, Dr. Bob Bennett, and asked him to find out if there was any significant environmental damage to the atmosphere from the use of solid rocket motors in space launch vehicles. I found out there was no such data, but, rather, the Russians were leaking information to the international press that it was a serious problem with the Western world's space launch vehicles, because they all used solid rocket strap-on boosters to a liquid rocket core vehicle whereas all Russian launchers used only liquid rockets as boosters.

The publicity wasn't good, and I brought the issue to Bob Lindstrom, who agreed we needed to spend some time researching the problem and identifying new propellant chemistries if the environmental data indicated a problem with conventional solid rocket motor propellants.

After checking with some of his sources at NASA and reviewing the situation with Ed Garrison, Lindstrom called me in to his office. "Al, I'm giving you a new assignment. I'm relieving you of your current position and putting you on my own staff as Vice President and a member of the Senior Staff so you can spend full time for the next year or two working this environmental issue."

Reluctantly, I accepted the assignment. It was tremendously important to do this, I knew, but I told Lindstrom that I was concerned about what position I would have after this temporary assignment was over. "Not to worry, Al. This is a very important assignment for the entire corporation. I'll find a place for you when it's all over."

I went to work to find out as much as I could about the alleged problem of environmental damage to the atmosphere from the use of solid rockets. It quickly became apparent from the data there was no significant environmental problem associated with use of solid rockets and that there were a number of avenues to pursue changes in the solid propellant chemistry—albeit at great expense—if it became necessary to do so.

To settle the matter, I arranged through the AIAA a workshop comprised of different experts from the atmospheric science community, from the propulsion community, from environmental regulating agencies such as the Environmental Protection Agency, from international atmospheric researchers, from NASA and DOD, and even from several environmental groups that had been advocating discontinuing the use of solid rockets. When the workshop was over, all participants, including the environmental activists, agreed that the problem was rather insignificant but that further work should be done to verify the atmospheric computer models that indicated there was no significant environmental problem with either solid or liquid rockets—that, in fact, there was no real significant difference between the two.

The environmental issue was obviously an emotional issue, not a technical problem. It became an emotional issue because solid propellant chemistry contained chlorine, and at the time there was an international ban on use of chlorinated fluorocarbons, chemicals that were being used as refrigerants, as cleaning agents, as fire-suppression agents, and in other manufacturing activities. These chemicals were considered to be compounds that depleted our planet's stratospheric ozone. I spent the next few years working on this issue while still supporting the launch of the Space Shuttle through participation in the flight readiness review boards at Thiokol. The Aerospace Corporation, with funding from the air force and Thiokol, measured ozone depletion in rocket exhaust plumes at stratospheric altitudes with highly instrumented aircraft, which confirmed that the problem was insignificant. This program, called Rocket Influence on Stratospheric Ozone (RISO), was under the direction of the Aerospace Corporation's Dr. Marty Ross. In the company of Dr. Ross and Dr. Bob Bennett of Thiokol, I spent considerable time presenting the RISO data to various technical organizations and government agencies in the United States and abroad. The environmental issue had quickly become an international concern, and I presented the data on the limited environmental impact of rocket propulsion at various conferences in Norway, France, the Netherlands, Brazil, Japan, and Australia.

Everything about my temporary assignment was going well, but then two years later something happened that proved my concerns about taking the task were well founded. Bob Lindstrom retired and turned over the reins of Space Operations to one of his old retired cronies from NASA Marshall, Dr. Joe Lombardo, who decided that my position as a member of his staff was not important enough to warrant a Vice President's position or, for that matter, the executive bonus I had been receiving. A year later, Dr. Lombardo, in agreement with our new corporate Vice President of Human Resources, Jim McNulty, decided I should be demoted from a Vice President to a Director. In the new job, I was supposed to work for the fellow who had taken over my old job when Lindstrom put me on special assignment.

It was a real blow and a real insult to me at this point in my career, especially after all I had done for the company, and I told Dr. Lombardo just that. I reflected on my earlier concern, that once I had finished helping to get the Shuttle flying again and we were back in production, the big bosses would have no use for me, and I would be "encouraged" to quit by making it uncomfortable for me to stay. I briefly considered calling Congressman Markey from Massachusetts to tell him I was finally being demoted from my current position and that maybe he should consider reinstating that House bill that he had drafted to ban Thiokol from any NASA contracts in the event that I was ever demoted. However, I did not do that. A few days later, Lombardo called me in: "McDonald, Bob Lund has asked for you to be transferred out of Space Operations and into Science and Engineering. You'll work with Lund as his Deputy. Both Dee Jay Hammon [Lund's Chief Project Engineer] and Dr. John Bennett [Lund's Chief Scientist] are retiring in the next few months, and Bob needs a strong technical person with a lot of experience in solid rocket motor design and manufacturing to replace them."

I accepted the offer to transfer, but it certainly wasn't what I had planned to do in the final days of my career with the company. I had no animosity toward Lund; in fact, I had always felt sorry for him for being put in such a bad position relative to the launch decision for *Challenger*. However, I found somewhat uncomfortable the idea of working for the guy who, when he had been asked to change his engineering hat to his management hat, had also changed his mind and supported the launch of *Challenger* that I had worked so hard to prevent.

The last few years of my career at Thiokol were spent in advanced technology as the Deputy to the Vice President of Science and Engineering and Technical Director for Thiokol. I was also the Chairman of the Technical Review Board (TRB). The TRB was established to review the technical issues associated with every program in the plant. It consisted of technical experts in various disciplines in science and engineering that would either approve or recommend changes in various design approaches or test programs to minimize technical risk. My position involved development of all of the solid rocket motor, pyrotechnic, and commercial technology for Thiokol operations in Utah. It involved interfacing with senior managers from the army, navy, air force, and NASA relative to research and technology development. I thoroughly enjoyed the work, especially the people who worked for me. Many of them were young lions who did not have a great deal of experience but who brought with them a great deal of energy and a desire to succeed.

I also became very concerned about our future capability to maintain and develop a new strategic missile system if required. The Strategic Arms Limitation Treaty II (SALT II) with Russia had allowed Russia to develop a new system but specifically prevented us from doing the same; as a result, we were regraining forty-year-old Minuteman motors to maintain our strategic deterrence. I spent considerable effort and time visiting many offices in the administration, NASA, DOD, the Army War College, the National Security Council, the National Reconnaissance Office, the Central Intelligence Agency, and Congress trying to get them to use large solid rocket motors for space launch applications to maintain this capability in the United States for our future national defense.

Another one of my responsibilities during this time period was to remain as a member of the Space Shuttle RSRM Flight Readiness Review Board at Thiokol. I had been an active member of this board since the return-to-flight of the Shuttle in September 1988, and by the time I retired from Thiokol in 2001, I had been involved in a total of eighty successful Shuttle launches. I was particularly grateful to still be serving on the board when John Glenn returned to space after thirty-six years as part of STS-95 in late October 1998. As Group Leader on the Minuteman ICBM flight test program at the Cape, I had witnessed Glenn firsthand become the first American to orbit the Earth in February 1962. I had sent a postcard to my parents in Montana the day after his Mercury *Friendship 7* launch describing what a fantastic experience it was to witness this historic event. I was unable actually to attend Senator Glenn's Shuttle flight, so I had one of the folks in our field office at KSC ask Glenn to sign the backside of the original postcard that I had sent to my parents in 1962. Not only did he honor my request, the senator also sent me an autographed first-day-of-issue envelope honoring his return to space in the Shuttle *Discovery*.

In the early to mid-1990s, Thiokol Corporation acquired two other companies, Huck International and Howmet Corporation; a few years later, the company changed its name from Thiokol Corporation to Cordant Technologies. Shortly after the acquisition of Howmet in 1995, Ed Garrison announced his pending retirement as chairman of the board. I immediately called Garrison and asked him if I could retire now at the age of fifty-eight and receive my retirement supplement just as he had provided to Bob Ebeling. Ed told me that the arrangement provided to Ebeling was a special situation to help Bob in his early retirement and his recovery from his mental and physical problems. Contrary to what Ebeling and Thiokol's attorney from Chicago had told me, Garrison stated he had no intention to provide similar benefits to others. In 2000, Cordant Technologies was sold to Alcoa. All the key executives Garrison had hired off the street less than ten years earlier with no propulsion business experience, and who made no contribution to the success of the company, walked away with millions of dollars worth of benefits for selling the company.

One of the highlights of my professional career came in 2001 when I was invited to speak in Moscow at a United Nations–sponsored International Space Forum honoring the fortieth anniversary of the first manned spaceflight by Yuri Gagarin. I was asked to speak about the *Challenger* accident:

what caused the failure, how did we verify the failure mode, how did we fix the problem, and how did we verify we'd really fixed the problem. The title of my presentation was "The Space Shuttle Reusable Solid Rocket Motor: 20 Years of Flight/15 Years after *Challenger*." My wife and I really enjoyed our visit to Russia and the opportunity to visit my friend Dr. Igor Karol in St. Petersburg. Dr. Karol was a Russian atmospheric scientist who had supported our conclusions on the environmental impact of rockets.

I retired three months later, shortly after the Thirty-seventh AIAA Joint Propulsion Conference held in Salt Lake City. An added incentive to retire came in the acquisition of Thiokol by our major competitor, Alliant Tech-Systems, and the creation of ATK Thiokol Propulsion. This was the third change in company ownership in the past twelve years and the second time in just one year; it was clearly time for me to leave.

The AIAA Joint Propulsion Conference was followed by the Fiftieth Joint Army/Navy/NASA/Air Force (JANNAF) Propulsion Conference, also held in Salt Lake City. I served as technical chairman for the AIAA conference and as the chairman of a government panel on "The Future of the U.S. Solid Rocket Industry" for JANNAF. Thiokol's president at the time was former astronaut Robert Crippen, a past president of the AIAA who had also served as executive chairman for the Thirty-seventh AIAA Joint Propulsion Conference. Crippen was also retiring as a result of the recent acquisition of the company, and he arranged a fantastic retirement party for me sandwiched between these two propulsion conferences. Knowing that many of my friends and colleagues from around the country and other parts of the world would be there, "Cripp" arranged a "roast" at a lavish affair in the Salt Lake Marriott Hotel. Several hundred people attended, including my good friend Moshe Gill from Israel and Alain Davenas from France, making it a fantastic ending to my forty-two-year career at Thiokol.

One of my most treasured gifts that night was a scale model of the Space Shuttle autographed by Crippen and John Young, the first two astronauts to fly the Shuttle some twenty years earlier, on April 12, 1981—exactly twenty years to the day after Gagarin's historic flight. I also received a nice letter from Utah Congressman James V. Hansen, along with the flag he had flown for me over the Capitol in Washington, D.C., and several mementos from the air force, navy, and army.

However, my fondest memories of the evening were to hear my wife and four children all speak at the end of the festivities. Linda never speaks in front of an audience, but she did this time and was as funny as she was beautiful. My children made me very proud, and a few teardrops formed in my eyes.

The loss of the STS-107 *Columbia* and its crew (Commander Rick Husband, Pilot William McCool, Payload Commander Michael Anderson, Mission Specialist David Brown, Mission Specialist Kalpana Chawla, Mission Specialist Laurel Clark, and Payload Specialist Ilan Ramon) on February 1, 2003, during reentry brought back many terrible memories. I followed the press briefings and read the final report of the *Columbia* Accident Investigation Board, quickly realizing that this accident, too, did not have to happen. Many of the lessons that should have been learned from *Challenger* some seventeen years earlier had clearly been forgotten.

A few days after the accident, I received a phone call from a reporter with the *New York Times* asking if I had been following the NASA briefings on the *Columbia* accident. I told him I did not agree with many of the statements being made. NASA had indicated that it did not think the problem was attributable to foam impact on the orbiter, because its analysis had indicated that was OK. NASA's stated approach to determining the cause of the accident was to try to ignore the obvious under the guise that it wanted to be totally open-minded and not eliminate any possible cause. I told the reporter that I believed just the opposite: "I would say you have to prove first that it's not the tiles, and then move on to other theories. There's a great temptation in these cases to search for some other, stranger explanation. In the first days after *Challenger*, I sat through meetings where people listed thirty things it could be, and 90 percent of them were ludicrous. I wanted to find reasons that it wasn't my booster rocket that caused the problem, but I could not. I think it's far more productive to first take what is obvious and prove that wasn't the cause by finding some data that would conflict with that conclusion."

I also told the reporter: "The *Columbia* disaster may turn out to be rooted in a different kind of negligence—a failure not just to properly model, but also to do physical testing to validate models." I made this suggestion after reading that several experts had supported NASA in its modeling accuracy. I also stated: "The chunk of insulation that flew off the Shuttle's external tank and hit the Shuttle's left wing might have caused enough damage to doom it. NASA really cannot have analyzed that threat realistically, because, although it modeled the possibility, it did not do physical laboratory tests to prove that foam at these speeds could not do serious damage." On a legal pad, I ran the same back-of-the-envelope-type kinetic energy calculation for the bipod foam ramp that had hit the left wing of *Columbia* that I had conducted for the piece of cork alleged to have come off the RSRM that almost doomed *Discovery* in the first launch after *Challenger*. The cork impact on *Discovery* could have been equivalent to a 30.06 rifle bullet from 100 yards; the impact of the foam on the left wing of *Columbia* may have been *more than six times as severe*—and no one in NASA management seemed to be concerned.

The losses of both *Challenger* and *Columbia* were a result of self-imposed schedule pressure by NASA to meet its planned Shuttle launch manifest, which represented the lion's share of the agency's budget. Both accidents were preventable and caused by well-known problems that were never fixed. NASA's "can-do" attitude resulted in many remarkable accomplishments in the past, but in these two cases, it resulted in launches that never should have occurred. In spite of these mistakes, it was NASA's changing from a "can-do" philosophy to a "can't-do" philosophy that sealed the fate of the *Columbia* crew. This was clearly the biggest mistake ever made by the agency in its fifty-year history. Several NASA engineers were very concerned that the bipod foam strike witnessed on the day of launch, in spite of the analysis that said it was OK, could have severely damaged the left wing thermal protection system; they recommended that the space agency request that images from DOD's spy satellites be taken to assess the damage. NASA officials concluded that there was nothing they could do about it if severe damage was revealed, and therefore denied the request for such photos. The *Columbia* Accident Investigation Board, under the leadership of Admiral Harold W. Gehman, concluded the satellite images would have revealed the severe damage to the leading edge of *Columbia*'s left wing. Furthermore, the board concluded that there was a reasonable chance that NASA could have rescued the crew of *Columbia* by extending its stay in orbit by discontinuing its payload experiments and sending up the next planned Shuttle flight, which was nearly ready for launch at the time.

At the end of July 2005, the Shuttle returned to flight with the successful launch of STS-114 *Discovery*. Much like *Challenger*, it took two and a half years to recover from the *Columbia* accident, even though the amount of redesign, testing, and analysis activities were not nearly as extensive as what was accomplished over the same period of time after *Challenger*.

It was also very disappointing to see another large piece of ramp-foam come off the external tank during the ascent of STS-114 *Discovery*. NASA was very fortunate that this piece of foam did not impact the orbiter, or we'd probably had another lost orbiter. The incident grounded the Shuttle again for another year. It was a different piece of foam than the one that had hit *Columbia*, so NASA did not repair or redesign the area because it concluded that no problems had been experienced with this particular piece of foam in the past. This explanation was ludicrous, because NASA had very limited data to make such a judgment: the camera coverage prior to *Columbia* was very poor, and none of the ET hardware was ever recovered. To not have fixed this piece of ramp foam prior to return-to-flight made about as much sense as Morton Thiokol not fixing the forward and center field-joints of the SRM because they had never failed in the past—only the aft field-joint had failed.

I sincerely hope that we will have learned the powerful lesson that the *Challenger* and *Columbia* accidents should have taught us all. That lesson is that problems in technically complex systems must be addressed with total honesty by highly competent scientists and engineers of high integrity. There is no place for bureaucratic solutions to technical problems, and any attempt to do so should be resisted by ethical engineers and scientists.

The *Challenger* crew was unique and should be a lasting reminder that technical errors can be very unforgiving. The crew represented a genuine cross section of American society—so much so that it almost felt like the crew was selected by a much higher authority. The crew consisted of five men and two women; Caucasians, an Asian, and an African American; Christians (including a Catholic and a Protestant), a Jew, and a Buddhist; military personnel, scientists, engineers from industry and civil service, and, of course, a schoolteacher. Each one of these astronauts deserves to be considered a hero. I dedicate this book to the memory of the *Challenger* crew and their families and the powerful legacy they have left behind. I am proud to be a charter member of the *Challenger* Center founded by the families of the *Challenger* crew.

"We will never forget them nor the last time we saw them that morning as they prepared for their journey and waved good-bye and slipped the surly bonds of earth to touch the face of God."

President Ronald Reagan
January 28, 1986

GOD SPEED *CHALLENGER*!

They shall never be forgotten: the crew of Space Shuttle *Challenger*: *front row*, Pilot Michael J. Smith, Commander Francis R. Scobee, and Mission Specialist Ronald E. McNair; *back row*, Mission Specialist Ellison S. Onizuka, schoolteacher Sharon Christa McAuliffe, Payload Specialist Gregory B. Jarvis, and Mission Specialist Judith A. Resnik. (Courtesy of NASA.)

STS-26R: *Discovery*
September 29, 1988
18:37:00 GMT

Commander Rick Hauck and Pilot Dick Covey, two of the three astronauts that I had known personally before the *Challenger* accident, signed this photograph for me after the first Shuttle launch with the redesigned solid rocket motors in September 1988—STS-26R *Discovery*. (Courtesy of NASA/Morton Thiokol.)

STS-27R: *Atlantis*
December 2, 1988
14:38:00 GMT

Commander Robert "Hoot" Gibson was a member of the redesign team for two and a half years and a close friend of mine; his orbiter suffered the worst debris impact damage prior to the loss of *Columbia* in 2003. This photograph featuring the second launch of the redesigned solid rocket motors in December 1988 on STS-27R *Atlantis* was actually signed by "Hoot" in May 1987, three months before the first test of the RSRM and over a year and a half before his flight. (Courtesy of NASA/Morton Thiokol.)

A team of seven Morton Thiokol managers and engineers received awards from NASA for returning the Shuttle to safe flight: *left to right*, Allan McDonald, Dick Davis, Joe Pelham, John Thirkill, Charlie Bown, Stan Graves, and Scott Stein. (Courtesy of Morton Thiokol.)

QM-8
January 1989
Cold conditioning with max. lift-off and max. Q loads applied

In January 1989, the final qualification test (QM-8) of the redesigned solid rocket motor (at a propellant mean bulk temperature of 39° F and a skin temperature as low as 29° F) was conducted in Thiokol's snow-covered test bay in Utah with Shuttle prelaunch, launch, and flight loads imposed during the static test. (Courtesy of NASA/Morton Thiokol.)

MOON, MARS, AND BEYOND

NASA's next-generation launch vehicle family, Ares, will be boosted into space by more powerful five- and five-and-a-half-segment RSRMs. Hopefully, Ares will be returning astronauts to the Moon and eventually transporting them to Mars. (Courtesy of NASA.)

The advanced solid rocket motor (ASRM) that was to replace the RSRM was nothing more than a pork-barrel government project supported by powerful politicians who were only looking out for the economic health of their own voting districts. (Cartoon by Cal Grondahl, 1993, *Standard Examiner*, reprinted with permission.)

Through all the stress and sadness of the Shuttle era, my family always stayed close, as they did at my retirement roast in July 2001: *upper left*, oldest daughter, Lisa; *upper right*, son, Greg; *center*, wife, Linda; *lower left*, youngest daughter, Meghan; and *lower right*, middle daughter, Lora. (Courtesy of ATK Thiokol Propulsion.)

Acknowledgments

I noted in the preface that writing a book is not rocket science—it is harder than that. What I really found out was that writing a book is not so difficult at all—getting it to a point where someone is willing to publish it is the hard part. I wrote the entire manuscript in three months from the twenty-year-old notes that I retrieved from my attic. It took more than three years to get the book published, and this would not have been possible without some help and advice of several people. I owe all of these people a great debt of gratitude.

I initially contacted Susan Edsall, who had just written a successful book titled *Into the Blue* (St. Martin's Press, 2004). The book is a very inspiring story of how she and her sister helped their father in Montana recover from a severe stroke, enabling him to return to flying old pre–World War II airplanes that he restored. I read about the book in an alumni newsletter from my alma mater, Montana State University. Susan Edsall suggested that I not pursue publication of my manuscript by sending it to all the publishers and editors I could find on the Internet, as she originally had done. She suggested, instead, that I contact one of the literary agents in New York City. In the New York City Yellow Pages, I found nearly 150 literary agents listed. I didn't know whether I should make a dartboard with all these names or send a request to all of them.

I then contacted a friend of mine, Pete Evanoff, who lived in the small rural community of Friendship, Maryland, near the nation's capitol. Pete had once told me that he had seen Tom Clancy talking to his pharmacist in a small drugstore in Friendship; the pharmacist was a friend and neighbor of Clancy's. Pete asked the pharmacist if Tom Clancy would mind if I sent him an e-mail about my manuscript soliciting his advice on how to go about finding a literary agent for the book. The pharmacist, Leo Mallard, sent me an e-mail with Clancy's e-mail address and suggested I contact him; Leo had talked to Clancy, and the author said he would be glad to help another writer. Clancy suggested to me that I contact his original literary agent, Rob

Gottlieb, of the Trident Media Group. After talking with Rob, I sent his agency a book summary, table of contents, and a few sample chapters. After a few weeks, they told me that the material I had sent possessed one basic flaw: it was written above an eighth grade reading level, which was the level most representative of the mass media market. I was told that the material in my book, though well written, was probably more suited to a quasi-academic clientele than the more general-type readership with whom they normally dealt.

I then contacted several other literary agents and either got a similar response or no response at all. I began to seriously consider self-publishing, as the Internet was full of these opportunities. In fact, the list of self-publishing agencies on the Internet was more overwhelming than the list of literary agents in the NYC telephone directory.

The day after Thanksgiving 2005, I was in New York City with my wife, Linda; youngest daughter, Meghan, from Boston; and my oldest daughter, Lisa, and her family from Barrington, Rhode Island. We decided to experience the excitement of Christmas shopping in New York City on the busiest shopping day of the year. While my family members were shopping for gifts, I went into a local bookstore and purchased *First Man: The Life of Neil A. Armstrong* (Simon & Schuster, 2005). The book, which had just been released and had already become a *New York Times* bestseller, was the authorized biography of professor, engineer, test pilot, and astronaut Neil Armstrong. Its author was Dr. James R. Hansen, a professor of history at Auburn University in Alabama and a former NASA historian. I read the book over the holidays and thoroughly enjoyed it.

Neil Armstrong had been the Vice-Chairman of the Presidential Commission on the *Challenger* accident and later became a member of the board of directors of Thiokol. Since I had appeared before the Presidential Commission on five different occasions, I felt like I really knew him. I sent Neil an e-mail message, telling him of the problems I had encountered and asking him for his help and advice. He told me that James Hansen, his authorized biographer, was first-rate—one of the preeminent aerospace historians in the world. Neil suggested that I contact Hansen and provided me with the professor's telephone number and address.

I contacted James Hansen shortly thereafter, and he was very interested in my story. I was making an AIAA Distinguished Lecture titled "*Challenger* Remembered in Light of *Columbia*" at the University of Tennessee Space Institute in Tullahoma in early February 2006, and was spending one night with my good friends Carver and Martha Kennedy near Atlanta, Georgia, so I took the opportunity to visit Hansen at his home in Alabama while I was in the area. I was hoping that he could help me in some small way or at least point me in the right direction. I was absolutely ecstatic when Jim offered to help me rework and improve the manuscript and find a publisher for the book. We shook hands and agreed to be partners in producing the book that you have just read.

Without Jim's help, this book may never have been published, and if it were published, it would not have been as well written, balanced, or nearly as interesting reading as the final product turned out to be. I cannot thank Jim enough for making all this possible and for his outstanding contributions to the book, as well as his untiring dedication and effort to get it published.

I would like to thank General Don Kutyna for his critical review of a chapter of my manuscript that resulted in some needed corrections and changes and for his support of me when he was a member of the Presidential Commission during the hearings on the *Challenger* accident.

Heartfelt thanks also to my two sons-in-law, Ted Fischer and John Gilmartin, for updating and maintaining my computer system while I was trying to write this book, and especially John, who graciously accepted all of my late-night telephone calls and fixed all of my software problems with my e-mail and for training me on the proper use of Microsoft Word. Without their help, I wouldn't have made it through this process.

I hope that you have enjoyed the book and have learned some of the powerful lessons the tragic accident of *Challenger* should have taught us all. *Challenger* was an accident that was totally avoidable and should never have happened. The two most important lessons in this story are, one, that there are times when one should take "no" for an answer, and, two, that a person should always stand up for the truth and be prepared to defend it no matter what the consequences may be. As I had learned the hard way, revealing the truth and disagreeing with your colleagues do not make you popular, but, if you ever find yourself in a position like the one I was in, perhaps you can take solace, as I have, in a statement made by John Adams, the second U.S. president and a signer of the Declaration of Independence. Adams had just completed a book on his political views and the lessons he had learned in the long, hard struggle for independence. In his biography, *John Adams* (Touchstone, 2001), author David McCullough relates that Adams had written to James Warren at the start of 1787, the year before the first presidential election, to say that he had just completed a book that was almost certain to make him unpopular. Adams wrote, "Popularity was never my mistress, nor was I ever, or shall I ever be, a popular man. But one thing I know, a man must be sensible of the errors of the people, and upon his guard against them, and must run the risk of their displeasure sometimes, or he will never do them any good in the long run."

The only regret that I have is that I didn't publish my book many years ago, because many of my cherished family members and friends have now passed away. I lost my brother-in-law, Jim Vollmer, in 2001; my oldest brother, Arthur, in 2003; and my only sister, Juanita, in 2005. In addition to my family members, the list of those who never lived long enough to read this book includes two members of the Presidential Commission that I considered as my friends—Chairman William Rogers and Nobel Laureate Dr. Richard Feynman; my newfound friend Wiley Bunn, Director of Reliability and Quality Assurance Office at the NASA Marshall Space Flight Center; several members of the National Research Council on the Redesign of the Solid Rocket Boosters for the Space Shuttle, including Lawrence Adams, Robert Anderson, Robert Watt, and Dean Hanink; Joe Pelham, my mentor and friend and Chief Scientist for Thiokol; my esteemed colleague Dr. Robert Bennett, who was killed in a horrific bicycle accident; Dr. Karl Klager, one of the pioneering German "rocket scientists" at Aerojet Solid Propulsion Company; Charles A. Sinclair, Vice President and General Manager of Pratt and Whitney Chemical Systems Division; Dr. Clark W. Hawk, Director of the University of Alabama-Huntsville Propulsion Research Center, and Dr. Harold W. Ritchey, ex-CEO of Thiokol and visionary founder of Thiokol's rocket fortunes.

Big Sky Values

A Biography of Allan J. McDonald

JAMES R. HANSEN

It is curious that physical courage should be so common in the world, and moral courage so rare.

—*Mark Twain*

A smooth sea never made a skilled mariner.

—*Old English Proverb*

What are the sources of a person's ethical behavior? Why does one human being stand so resolutely to defend fundamental principles while another moves to easy compromises? Why do some cling faithfully, desperately, to the truth while others lie, cheat, and cover up? What enables the ethical weave of our human fabric to turn and twist so wildly? Is it our different biologies, our variant family values and upbringing, our educations, our spiritual nurturings?

Before the *Challenger* accident in 1986, Allan J. McDonald, like most engineers, never gave much thought to such riddles of human personality. After the disaster, he pondered them more and more frequently. What he came to fathom about the subject conformed well to what the ancient Greek philosopher Aristotle taught over 2,300 years ago: Human beings are what we repeatedly *do*. Moral virtue is not so much a willful act as it is a habit. Once ingrained in our childhood and adolescence, it is the architecture that drives who we are and how we must live.

Al McDonald learned his ethical values from his father, John William McDonald Sr. (1897–1975). The offspring of Scotch-Irish immigrants, Al's dad never failed to pay any bill the very day he received it. He never owed a dime to anyone. He always saved enough money to pay cash for his cars, but seldom had enough money to make a down payment on a home. The family in which Al McDonald grew up mostly lived in rental houses. That's why they resided in twelve different houses and only owned one of them by the time Al graduated from Montana's Billings Senior High School in 1955. Finishing the sixth grade, he had attended five different schools.

Al's father was a very proud man and never let anyone know that his own schooling had ended at grade four. Ashamed, his dad never told his family. Al only learned it from his mother when he was in high school.

Al's dad was good at math and eventually became the deputy county assessor for Gallatin County in Bozeman. He was truly a self-made man. Both he and his older brother Allan, for whom Al was named, were orphans at a very young age: John McDonald was nine, Uncle Allan eleven. Al's paternal grandmother had been killed in a horse-and-buggy accident in 1906; the buggy turned over and broke her neck. A few years earlier, her husband, a conductor, had been killed in a train accident. Totally alone, the McDonald boys were taken in by two different families. John had to quit school and go to work on the McCrae farm for his room and board. When the McRaes moved from Kalispell to Klamath Falls, Oregon, John went with them. Allan went with his adopted family, the Galbreths, to Canada.

John totally lost contact with his brother for forty-five years; neither had any idea where the other lived or for that matter whether the other was still alive. One day in the summer of 1951, when little Al was about to enter the ninth grade in Billings, his dad, John, went out to meet the mailman walking up the driveway. Handing John a few letters, the mailman said, "You know, Mac, I just ran into a guy during my vacation in Canada that looked and talked just like you, and his name was also McDonald." The mailman said it happened in a pub in a little logging town in Ontario, the name of which he couldn't remember. Later that day, the mailman telephoned to say that his wife remembered that it was Dryden, Ontario.

John wrote a letter to the Dryden Chamber of Commerce and asked if an Allan McDonald lived there. It wasn't long before the reply came, listing the address and telephone number. It was, in fact, his brother.

Later that summer, John took his family to Regina, Saskatchewan, to meet his brother; Allan's wife, Gertrude; and their daughter, Nancy. It was a tearful reunion. John and Allan almost looked like twins. It was truly a one-in-a-million chance that brought them together.

John's wife, Eva Marie Gingras (1899–1976), was one of the sweetest and kindest women in the world. She wasn't five feet tall in her high-heel shoes. A French Canadian born in Iron Mountain, Michigan, her family had moved with a host of other French Canadian iron miners to work in the higher paying copper mines in Butte right after the turn of the century. Her dad was a tailor and followed the miners wherever they went. Eva Marie graduated from Butte Central Catholic High School and became a teacher. She was a good piano player who played the background music in the "pit" for the silent movies. She never had a driver's license or learned to drive a car her entire life. Neither did her first child, Juanita, young Al's older sister. Eva Marie's father had died long before Al was born, and Al's only grandmother had died when he was just one year old, so, in effect, he never had grandparents.

Al did have three siblings: two brothers and a sister. Juanita (1919–2005) was eighteen years older than Al. His oldest brother, Arthur (1922–2003), was fifteen years older. Both were born in Butte. His other brother, John, and he were both born during the Great Depression in Cody, Wyoming, a year and a half apart. John was unplanned. Eva Marie was thirty-six at the time and decided that John needed a playmate, so eighteen months later she had Al. He was the spoiled "baby" of the family. Like John, he was born at home. When his dad first saw Al, he said, "Hi, Pete." The nickname stuck with the boy for his entire life, and he is still known as Pete to all members of his immediate family.

Al never knew his oldest brother, Arthur, very well because Arthur quit college and enlisted in the U.S. Marine Corps right after Pearl Harbor, when Al was only four years old. Art became a Marine pilot in the Pacific theater flying Hellcats and Corsairs. He was recalled to active duty when the Korean War broke out in 1950, so "Pete" seldom saw Art while he was growing up. Pete always carried a picture of him and his airplane in his pocketbook. As a Marine captain and pilot, Art was awarded the Distinguished Flying Cross and several Air Medals in World War II and Korea, and earned a Purple Heart for getting shrapnel through his mouth and chin during a bombing raid in Korea.

When Al was five years old, he pulled his little red wagon up and down both sides of the street collecting grease for the war effort. He would take it down to the local butcher shop and receive tokens for his family to buy meat; he usually got a nickel for an ice-cream cone. He picked up some of the grease from a local hamburger joint called The Burger Inn; with his speech still a little undeveloped, he called it "The Burglar Inn." His dad would look at the dirty black grease that he brought home for his mother to render and would accuse him of getting it at the gas station next door to the hamburger joint. Even today Al can remember walking down Bozeman's Main Street on VJ Day in August 1945. His dad carried an American flag, and Al and his brother John strutted right behind him.

One of the most pleasurable times of young Al McDonald's life occurred when his family moved from Bozeman to Kalispell in the fall of 1945. Bozeman lay in the southwestern part of the state, just ninety miles north of Yellowstone National Park, while Kalispell was in the northwestern part of the state, just thirty miles south of Glacier National Park and ten miles from beautiful Flathead Lake, the largest body of fresh water between the Mississippi River and the Pacific Ocean. Al's parents were unable to rent a home in Kalispell, so one of his dad's coworkers at the Nash Finch Company, a wholesale grocery distributor, allowed them to stay in his summer cabin on Flathead's western shore.

Al and John walked three miles each way, along the shore of the lake, to a one-room schoolhouse in Lakeside. Al was eight years old and learned to swim by diving off the dock, the water well over his head, to catch frogs. John and Al took their poles to school and always fished on the way home. Back home, they'd climb into a two-man kayak and paddle around the inlet. Eva Marie was always worried about her boys drowning, since they never wore life vests. She was also concerned about the quality of their education in the one-room schoolhouse, with all eight grades in one small room and a single teacher for all classes. She worried that her boys would be well behind their class when the family moved into Kalispell in December. As it turned out, both John and Al proved to be well ahead of everyone else. The teacher they had in the one-room schoolhouse along the shore of Flathead Lake was the most remarkable teacher they ever had.

Young Al and his brother John were inseparable. They were best friends. John protected his little brother and pulled him out of many fights. Al never backed down from an injustice and, since he was the smallest kid in class, all the bullies wanted to take him on. He can't remember ever winning a fight and often came home with a bloody nose. John would find out who gave it to him and usually returned the favor.

Al's most traumatic boyhood experience was losing his best friend, Carl McEvony. Carl was a gifted athlete, probably the best junior high basketball player and baseball pitcher in Billings. He had only one major flaw; he loved to fly-fish but couldn't swim. One day while fishing with his father, Carl slipped on some slick rocks and fell in the river. His hip boots filled with water so quickly that he drowned. John and Al and several others from their baseball team were pallbearers at his funeral.

A few years later, Al nearly suffered the same fate. Fly-fishing with his brother John, he waded out to the middle of the Madison River above a deep hole formed by a fork in the river. The current was swift. Al didn't weigh very much, and as he approached the fork, the current started carrying him downstream. The small rocks on the bottom moved under his feet, and he couldn't hold his ground as the water started to flow over the top of his hip boots. In a panic, Al yelled for help, but when he saw John starting toward him, he yelled at him to go back; Al felt he was lost and that John would quickly suffer the same fate. Being dragged down into the deep hole, Al's boots nearly filled with water and weighed a ton. John pushed forward through the swift current and held out his fishing rod. As Al grabbed tight, John started backing up and eventually pulled him out.

For fifty years, Al never again wore fishing boots. He'd wade into a river with his shorts and tennis shoes, thinking that he could swim out of anything in that condition.

Of all the McDonald children, Al seemed to contract more diseases than all the rest combined. By the time he got out of grade school, he had made it through red measles, German measles, chicken pox, mumps, scarlet fever, pneumonia, ear infections, tonsillitis, and numerous cases of flu. However, he became very ill during the early spring of his senior year of high school when he came down with rheumatic fever. He woke up one morning and could not move his right leg. Even the weight of the sheet on his leg was very painful. His mother called the family doctor, who told her to take Al to the hospital right away. There, the emergency room doctor said that Al either had polio, spinal meningitis, or rheumatic fever—it would take a few days of tests before they could really know. After a few days in the hospital, they concluded he had rheumatic fever and started treating him with a new wonder drug called cortisone.

Graduating from Billings Senior High School in June 1955, Al was fortunate to receive an honors scholarship to attend any of the state colleges in Montana; he received the scholarship for graduating seventh in a class of 368 students. He chose to attend Eastern Montana College of Education in Billings (now called Montana State University at Billings) so he could live at home. His scholarship only covered basic tuition—no fees, books, room, or board—and he had to maintain a B average to keep it.

Eastern Montana College was a four-year teachers college but provided two-year diplomas in most academic areas. Brother John also went there, to study prelaw, but he transferred to the University of Montana in Missoula, where he received a B.S. in business before ultimately earning his law degree. For a while, John ran his own law practice in Livingston. Eventually, he returned to the University of Montana as a law professor.

At Eastern Montana, Al also started in prelaw but, figuring that two lawyers in the family were too many, transferred into pre-engineering, which required his moving on to Montana State University in Bozeman, where he graduated with honors with a bachelor's degree in chemical engineering in 1959.

John and Al worked every summer in Billings while they were going to college. Al worked for Lambs Distributing Company, packaging merchandise for shipping; for Murphy Oil Corporation, an oil exploration company, as a draftsman; and (two summers) for Russell-Miller Milling Company, as a chemist in the laboratory of a flour mill. During those last two years, John worked as a sales clerk at Sears, and for one summer he and his younger brother rented a basement apartment.

They didn't have a car, so they walked to work—Al, down the railroad tracks to the flour mill about a mile away. One day leaving the mill, he saw a tremendous black cloud billowing up out of the west. One of the women from the mill came running down the tracks yelling at him to hurry back inside. "That isn't a tornado out there, is it?" Al asked. "Yes, it is," she screamed, "and everybody has been ordered to go to the office and vacate the flour mill because they're concerned a power line might fall on the mill. The grain and flour dust might ignite. There could be an explosion."

Not five minutes later, the twister came tearing right by the mill. Al watched a huge oak tree get ripped out of the ground and disappear as the cyclone passed right in front of the office. The storm tore a path fifty feet wide, destroying everything in its path. The mill lost power. Power lines snapped, sparks showering everywhere. Luckily, there was no fire.

This was the first and only time Al ever saw a tornado, and it was too close for comfort. If it had arrived only a few minutes later, he would have been halfway home in the middle of nowhere, walking alone down the tracks.

During his senior year, to obtain a little spending money, Al worked a night shift part-time at a char plant. The plant was on campus and operated by the Department of Chemical Engineering, under Dr. Lloyd Berg, the department head. It was a processing plant for extracting various chemicals from Montana coal. Al's job was to monitor and record the various instruments and keep shoveling coal into the reactor. Al's parents had moved from Billings to Bozeman prior to the start of his senior year, so he lived at home. His mom always left a pile of old rags at the front door for Al to wipe off with before entering the house. From head to toe, he would be covered with black coal dust and soot.

A frightful thing happened during the spring of his senior year at Montana State. Having joined Sigma Chi Fraternity (Al was fortunate to have obtained a loan from the fraternity, or he would not have been able to finish his schooling that year), he drove his date and another couple to a party up at a small ski area called Bear Canyon. At the party, they consumed a little beer, which, granted, was risky for the drive home. Steering down a series of narrow switchbacks on a steep mountain road, the electrical system of his dad's car shorted out, and he lost the headlights. Immediately Al hit the brakes as hard as he could. The car started skidding on the slick road and literally was heading over the edge of a cliff when the rear axle got hung up on a large rock, stopping the car. Looking out the front window into a dark chasm below them, the young couples could see the car pointing down over the cliff as it tilted over slightly on its right side. Ever so carefully, the fellow in the backseat along with his date opened the left rear door and exited the car, stepping on a rock. Al and his date crawled over into the backseat and climbed out of the car the same way. Some of their buddies that came by later gave them a ride back into town.

It wasn't until the next day that Al told his dad and brother-in-law, Jim Vollmer (1921–2001), what had happened. Jim drove Al up to the ski area in his Jeep truck. Jim worked in his family's meat-packing business, and his truck had a winch for hoisting the dead cows that they would frequently pick up from ranchers for disposal. He parked the truck behind two large trees and hooked the winch cable around the rear axle of the Plymouth. As soon as the cable tightened, the car fell farther over the cliff. It hung suspended by the cable over the 500-foot-deep canyon.

Al stood there in a state of shock, recognizing how close he had come to killing himself and his friends. His mother told him that his guardian angel had saved her "baby." Al was really worried what his dad would do because the car was his pride and joy, and there was some significant damage. When the serviceman told him that the whole electrical system had shorted out, his father knew Al had told him the truth and did not get angry with him.

Before graduating in June 1959, Al received several job offers. Two of his interviews took place in New York City, with Standard Oil of New Jersey and with Esso Research and Engineering. Climbing into a taxi at what was then called Idlewild Airport, Al told the driver to take him to the Statler-Hilton, where Standard Oil had made his reservation. The taxi driver asked where Al had come from. When Al told him Billings, Montana, the driver said he used to work in the copper mines in Butte before he got laid off. He asked Al why he was in New York City. Al told him he was interviewing for a job there. "Are you nuts?" he asked. Passing several large construction sites, the driver said, "These buildings are to be brand new insane asylums, and you have to be crazy to leave Montana and come to work here."

Registering at the hotel, Al noticed a bearded guy along with a large entourage all dressed in army fatigues walking through the lobby. It was Fidel Castro, who had just come to power in Cuba; he was scheduled to address the United Nations. Castro later got evicted from the hotel because some of his colleagues were cleaning chickens in their hotel room. They left the hotel and moved over to a hotel in Harlem.

The next morning, Al tried to take a train from Grand Central Station to Linden, New Jersey, to Standard Oil's Bayway Refinery, but he made the mistake of giving up his seat to a older woman, and, along with several others who were also standing, he got thrown off the train because it was overloaded. He finally made it to his interview, but he was late. The next day he interviewed with Standard Oil in Rockefeller Center and was then taken out to an elegant dinner before going to the Broadway play *My Fair Lady* starring Rex Harrison and Julie Andrews. The play had been sold out for months, but Standard Oil had a standing reservation for their VIPs and guests. In truth, it was the only pleasant experience Al had in New York.

Al received job offers from several other companies, but he decided to accept the offer from Thiokol for four important reasons. First, it was located in the Rocky Mountains and was close to home, less than 400 miles from Bozeman. Second, most of the people he met at Thiokol weren't much older than him. Third, it was a new and exciting business without any experts—no one knew much about rockets in the spring of 1959, just a year and a half after the Soviet Union had launched *Sputnik I*. Finally, Thiokol had just received a contract to develop solid rockets for the air force's Minuteman intercontinental ballistic missile—the highest priority program in the Department of Defense. It was an exciting opportunity in clearly what was to be a pioneering effort.

Al followed in his father's footsteps by paying off his entire student loan the very first year he worked. At first he received $500 per month from Thiokol and almost felt ashamed to be making so much money. It was more than his father had made in a month in his entire lifetime.

Al didn't stop his education when he went to work at Thiokol. In 1965, six years after beginning there, he enrolled in a master's program at the University of Utah. By taking classes eight straight quarters and having to drive ninety miles from work to some of the classes taught on the University of Utah campus in Salt Lake City, it took Al only the regular two years to graduate with a master's in Engineering Administration from the Department of Mechanical Engineering. He was presented his diploma by Dr. James Fletcher, University of Utah's president, who left a short time later to become the Administrator of NASA. Thiokol was awarded the original Space Shuttle solid rocket motor contract in 1974 while Fletcher was the Administrator. At the request of President Reagan, he returned to head up NASA again after the *Challenger* accident in 1986. When Al's son, Greg, graduated from Westminster College in Salt Lake City in 1990 (with a bachelor's in Airline Operations and Pilots Option), Dr. Fletcher was the commencement speaker. Al had the opportunity to introduce his son that day to Dr. Fletcher. Greg now flies a lead plane for the U.S. Forest Service, fighting wildfires.

Al has always been a strong believer in the importance of a good education, which was probably reinforced by his father never having the opportunity to go beyond the fourth grade. He convinced his wife, Linda, to go back to school and get a college education; he told her it was far more valuable than any life insurance policy that he could afford. Linda enrolled in Weber State University in Ogden and graduated cum laude with a bachelor's in business education in 1980. Al is proud that all of his children and their spouses have earned college degrees, even though none of them decided to become a "rocket scientist" like him. Al and Linda currently have seven beautiful grandchildren and plan on doing everything they can to make sure all of them—and those yet to come—get a college education.

Some people may doubt that a person can be ethical and do the right thing for his family and fellow humans without being devoutly religious or at least adhering to some system of religious belief, but Al McDonald has never looked at it that way.

Al didn't really think much about religion until he went to work at Thiokol in Brigham City, a town that was nearly 100 percent Mormon before Thiokol built its rocket plant there in 1957. At first, Al was still a bachelor. Most of the secretaries were local women, and the first question they always asked him was, "Are you a member of *the* Church?"

It wasn't long before his roommate, Jerry Dyer, and he were visited by two female Mormon missionaries. The men allowed them to come in because one of them was rather attractive. The other one looked like a matronly bodyguard. After hearing several sessions of their sales pitch, Al decided that he had heard enough. The Mormon doctrine was too foreign for Al to accept, but these sessions with the missionaries did reveal to him that he knew very little about his own religion. He decided to take some classes at St. Joseph Catholic Church in Ogden. When he completed the classes, he made his first communion and was confirmed in St. Henry's Catholic Church in Brigham City. Strange as it may seem, Al had the Mormon missionaries to thank for making him a better Catholic.

The circumstances that brought Al and his wife, Linda, together were quite interesting. Al had the opportunity to visit the Shrine of Guadalupe in Mexico City in November 1961. He stopped over in Mexico City on his return from Cape Canaveral, where he had been involved in a launch of a Minuteman ICBM with his roommate, Jerry Dyer. He said not to ask him how they ended up in Mexico City in returning from Orlando to Salt Lake City, because that was a long and complicated story. While visiting the shrine, Al kneeled down and said a prayer in the church asking the Virgin Mary to help him find a good Catholic girl for a wife; this seemed almost impossible in this predominately Mormon community of Brigham City, Utah, where he lived.

Just a year later, Al was attending mass at St. Henry's Church with an attractive young woman, an ex-stewardess from Braniff Airlines by the name of Jan Jones, who worked at Thiokol. They decided to go downstairs to the social center after mass and have a cup of coffee and a doughnut. As they were sitting there, Al noticed another very attractive young woman sitting at a nearby table. He asked Jan if she knew her, and Jan replied, "Sure, that's Linda Zuchetto. She's from Detroit, and she works over in Plant 78." Al asked Jan if she would introduce him to Linda, and she did. The rest was history, because he married that beautiful Italian woman about six months later in that very same church.

Al has never considered himself a very religious person, even though he went through the motions of being a good Catholic. Everyone knows there can be a big difference between being religious and being spiritual. Al Mc-

Donald's Catholic roots run deep, but his spirituality has led him more toward a transformed understanding of himself and his faith, which is deeply different from the professional journey he began early in his engineering career.

Like many others, Al started at a point of his own unexamined ignorance. His journey of self-realization turned into a deeper understanding not only of what to know and how to know it but also what to believe and how to put forward and defend those beliefs. More and more for Al, the two became inextricable—the search for knowledge and the statement of belief. What a person is told to believe versus what one truly knows is definitely not always the same. Sometimes a person must conjure up the courage to stand up for what he knows to be true despite what he has been told to believe.

An essential element of Al McDonald's philosophy of life is summed up in the following statement he made to me during my work on his memoir:

> Our personal journey, if the journey genuinely involves our experiencing of life in a thoughtful and accepting way, teaches us more than we could ever hope to learn from any one teacher, priest, textbook, or doctrine. Spiritually, as well as intellectually, our journey should teach us that we must stand up for what we believe to be true no matter the consequence of what might befall us. Truly "knowing" must be grounded in solid ideas and information, but standing up for what we believe comes from an inner voice, a voice that lets us know without reservation, if we are accustomed to hearing the voice, that what we are attesting to is true and that any variance with that truth is unacceptable.

As a young boy under the gorgeous Big Sky of Montana, "Pete" McDonald had already come to think this way, I am sure. Over the years, it became the essence of who he was, reinforced mainly not by formal thoughts about religion or philosophy, but by the values and the love centered in his feelings for his family, and theirs for him. It was largely due to the solidarity and wondrous strength provided by his family—by his parents, his siblings, his wife, and his own children—that Al was able to act so courageously during the terrible trauma of *Challenger* and ride out in such a constructive way the very tough times and situations he faced.

Al's moral courage and tenacity combined with his remarkable engineering knowledge allowed him to weather the *Challenger* storm and to eventually return the Shuttle to safe flight thirty-two months after the accident. Al could have easily walked away from the *Challenger* accident by saying, "I told you so!" but he focused his energies on fixing the problem to make it safer for future astronauts to explore the heavens rather than lamenting on what could have been.

His is a story that should not be forgotten.

Abbreviations

ABC	American Broadcasting Company
AFB	Air Force Base
AFPRO	Air Force Plant Representative Office
AIAA	American Institute of Aeronautics and Astronautics
ASRM	advanced solid rocket motor
CELV	complementary expendable launch vehicle
CDR	critical design review
CEI	contractor-end-item
CSD	Chemical Systems Division
DCR	design certification review
DDT&E	design, development, test, and evaluation
DM	development motor
DOD	Department of Defense
ECP	engineering change proposal
ELV	expendable launch vehicle
ET	external tank
ETM	Engineering Test Motor
FRF	flight readiness firing
FSS	fixed service structure
FWC	filament-wound case
GD	General Dynamics
GOCO	government-owned-company-operated
GOX	gaseous oxygen
HOSC	Huntsville Operations and Support Center
ICBM	intercontinental ballistic missile
IG	Inspector General
IR	infrared
JES	Joint Environment Simulator
JSC	Johnson Space Center
KSC	Kennedy Space Center

LAX	Los Angeles International Airport
LCC	launch commit criteria
LOX	liquid oxygen
Max Q	maximum dynamic pressure
MIT	Massachusetts Institute of Technology
MMT	Mission Management Team
MSFC	Marshall Space Flight Center
MTI	Morton Thiokol Inc.
NASA	National Aeronautics and Space Administration
NBC	National Broadcasting Company
NRC	National Research Council
OBR	outer-boot-ring
OMS	orbital maneuvering system
OPT	operational pressure transducer
PAS	Problem Assessment System
PMBT	propellant mean bulk temperature
psia	pounds-per-square-inch absolute
PV	Production Verification
QM	qualification motor
RID	review item discrepancy
RFP	request for proposal
RSRM	"redesigned" solid rocket motor and "reusable"
S&E	Science and Engineering
SMRB	Senior Materials Review Board
SRB	solid rocket booster
SRM	solid rocket motor
SSME	Space Shuttle main engine
STA	Structural Test Article
STS	Space Transportation System
TPTA	Transient Pressure Test Article
TRB	Technical Review Board
USAF	United States Air Force
USBI	United Space Boosters Inc.
VAB	Vehicle Assembly Building

Bibliographical Essay

JAMES R. HANSEN

Prior to reading McDonald's memoir, I was not terribly dissatisfied with the books and articles covering the *Challenger* accident. As a professional historian who had written about aerospace history and followed contemporary aerospace events for more than twenty-five years, I had read all of the books on *Challenger* and had reviewed a few of them for academic journals and trade magazines. I thought the coverage was insightful and largely complete. Naturally, the earliest books were exposés by investigative journalists who opportunistically produced them for trade presses as quickly as they could. From a scholar's viewpoint, these books turned out to be significant enough statements of public understanding of what lay behind the subject not long after the time of the accident but, as to be expected, were largely deficient in any lasting value as "dispassionate, critical, and technically and programmatically informed analysis."[1] Still, I eagerly read the books, gleaning what I could from them about the causes of the *Challenger* accident and pondering what our society could learn from such a tragic mistake so that it would not be repeated.

Racing to be first into the bookstores in the spring of 1987 were *Challenger—A Major Malfunction: A True Story of Politics, Greed, and the Wrong Stuff* (Doubleday) by Malcolm McConnell, and *Prescription for Disaster: From the Glory of Apollo to the Betrayal of the Shuttle* (Crown) by Joseph J. Trento, with reporting and research by Susan B. Trento. The book by McConnell, a young journalist who had been present at Cape Canaveral for the launch of *Challenger*, stressed the immediate causes of the accident by highlighting the political pressures to launch. *A Major Malfunction* argued that NASA leaders caused the disaster by pressing operations officials at the Cape to launch on January 28 so that President Ronald Reagan could mention Christa McAuliffe, the highly publicized "Teacher in Space" who had a special lesson from orbit planned for millions of schoolchildren, in that evening's scheduled State of the Union address.

If McConnell had gone no farther, his analysis might have been compelling, but he went on to sensationalize the disaster by building a conspiracy theory involving a Mormon connection between NASA Administrator Dr. James Fletcher and NASA's chief congressional supporter, Senator Frank Moss (D-Utah), both Mormons, who had helped select the Utah company, Morton Thiokol, for the Shuttle's solid rocket motor contract, even though Thiokol, according to McConnell, had "an inferior design." The conspiracy theory focused on Fletcher's having been the president of the University of Utah when President Nixon selected him in 1971 to head NASA and on Fletcher's wife being a Mormon from Brigham City, where Thiokol was located. McConnell blamed Nixon for appointing Fletcher and for not stopping the award of the SRM contract to Thiokol. He also blamed President Reagan for applying schedule pressure to the Shuttle program to support his "Star Wars" missile defense program, a charge that seemed very possible to many people at the time but which since has not been borne out by any solid historical evidence.[2]

The prime competitor to McConnell's book, *Prescription for Disaster*, was written by reporter Joseph J. Trento and his wife, Susan B. Trento, and was their first book. Earlier, Joseph Trento had worked for a newspaper in Wilmington, Delaware, for CNN's Special Assignment Unit, and as a researcher for investigative columnist Jack Anderson. In their *Challenger* book, the Trentos laid the blame squarely on NASA's decision to proceed with the Shuttle program in the first place. Once NASA made the fatefully erroneous decision, the agency compounded its mistake by using solid rocket motors for boosters rather than liquid rockets (against Wernher von Braun's recommendation), and then made everything worse by selecting Morton Thiokol to produce the SRMs. Their book argued—persuasively enough, if not with much nuance—that the decision to build the Shuttle had been a political decision by the Nixon administration and that the Shuttle program had never been funded sufficiently by Congress or by any presidential administration. Underlying all of the fundamental mistakes, the Trentos argued, was NASA's loss of its "giants of the 1960s"—the experienced engineers who had successfully managed the *Apollo* lunar landing program. In their place sat far less skilled and technically adept government bureaucrats who, unlike their predecessors, were willing to play the political game. As a result, they sold the Space Shuttle to the country as an inexpensive program, thereby sowing the seeds for disaster. *Prescription for Disaster* painted a picture in which the *Challenger* disaster was caused not by defective O-rings but by a political system that produced the Shuttle program in the first place. When I first read the book, I found *Prescription for Disaster* compelling and regularly referred to its argument in the space history classes I taught at Auburn University.[3]

Dominating the discussion surrounding *Challenger* in the years after the accident were provocative journal articles written by leading space program analysts who were also examining the causes of the accident in order to make broader points about the character of the entire U.S. space program. The main combatants in this debate were Alex Roland, a former NASA historian who taught military history and the history of technology at Duke University, and John M. Logsdon, a political scientist at George Washington University whose 1970 book *The Decision to Go to the Moon: Project Apollo and the National Interest* (University of Chicago Press) effectively created the field of space policy studies in the United States.

Their clashing of swords began in the May 1987 issue of the British journal *Space Policy*. Fresh off his seemingly clairvoyant November 1985 *Discover* magazine article, "The Shuttle: Triumph or Turkey?" Roland contributed "Priorities in Space for the USA," a thorough criticism of the Space Shuttle program for being highly oversold and impractical.[4] NASA's promise that the operation of the Shuttle would be cost-effective had been not just an illusory dream but "delusionary," as there was no way, Roland explained, that such a complicated recoverable launch vehicle could turn out to be more economical than an expendable rocket. The agency may indeed have cared deeply about astronaut safety, but those concerns were seriously compromised by NASA's thinking about low Shuttle costs. From the start, Roland posited, the STS should have been regarded as an experimental vehicle rather than as a refined cargo carrier that could provide routine access to space. Shuttle launches provided space "spectaculars" but in truth were useful only for a limited number of specialized missions. Making no less than nine specific references to the *Challenger* accident, Roland assailed the entire Shuttle program as misguided, uninspired, and wasteful. The sooner it was scratched from operation and replaced by a new generation of reliable unmanned boosters, the better. Roland felt much the same way about the Space Station; it could not be justified on a cost-effective basis either, and should also be canceled. In succeeding years, the Duke history professor expanded his arguments to the point where he considered the American preoccupation with human spaceflight to be one of the most serious and large-scale societal mistakes in U.S. history, equating the eras of Mercury, Gemini, Apollo, and Space Shuttle together as "Barnstorming in Space: The Rise and Fall of the Romantic Era of Spaceflight, 1957–1986."[5]

The editor of *Space Policy* sought a response from John Logsdon for the same May 1987 issue. In "Priorities and Policy Analysis: A Response to Alex Roland," Logsdon assailed Roland's analysis for being based on "shaky historical evidence" and representing "a view that is not widely shared among either those leading the development of space policy in the USA and other major space-faring countries, or the general public in the USA"—that view being that human spaceflight was not a socially valuable endeavor and should not be considered to be so by the public unless it could prove itself to be more practical. Logsdon countered by citing poll data following *Challenger* that showed an *increase* in public support for human spaceflight in America, not a decrease, which one should expect following Roland's thesis. Logsdon agreed that NASA had invented a number of dubious justifications for the Shuttle's existence, but rather than blaming NASA for inappropriately and foolishly cultivating America's romance with human spaceflight and thus setting the stage for the Shuttle disaster, Logsdon blamed the White House, Congress, and space program administrators for not bringing the resources, energies, and focus required to make NASA's Shuttle program a success, as had been the case for the Moon landing program of the 1960s.[6]

The quarrel didn't end there—not even in that particular issue of *Space Policy*, wherein Roland was given the chance to offer a rejoinder, which he did, calling Logsdon's article "a bemusing and gratuitous piece of editorializing." In Roland's view, Logsdon was attacking him for heresy, for his not keeping faith with the orthodox American sentiment that the dream of humans in space has been and always should be a legitimate driver of what happens in the American space program. "Such a standard would be strange at any time," Roland wrote, "it is almost laughable in the wake of the collapse of the Shuttle program."[7] Several times over the following years, Roland and Logsdon would spar over the proper ideological interpretation of NASA and its history and the place of the *Challenger* accident in it.[8]

Into the mid-1990s, the loss of *Challenger* was usually ascribed to NASA's decision to accept a safety risk to meet a too ambitious launch schedule. But in 1996, *The Challenger Launch Decision: Risky Technology, Culture, and Deviance at NASA* (University of Chicago Press) was published. In this provocative book, Dr. Diane Vaughan, then a professor of sociology at Boston

College (since 2005 at Columbia University), argued that the disaster's roots were to be found in the very nature of modern institutional life. All organizations develop cultural beliefs that shape their actions and outcomes—not all of them for the good. In Vaughan's view, NASA's institutional development throughout the 1970s and 1980s—the period when the STS program came to life—reflected an organization whose dominant underlying perception was that it was competing for scarce resources. This special sense of challenge fostered an organizational structure that accepted taking risks and cutting corners as *norms*. It was not political pressure, conspiracy, incompetence, or defective engineering that led to the *Challenger* accident; rather, it was a long series of small and seemingly harmless modifications to technical and procedural standards that took place within NASA over the course of several years. No specific rules had been broken—not at NASA or at Morton Thiokol. Instead, it was the accumulated weight of mundane organizational norms and group dynamics that ultimately made both organizational entities vulnerable to making a very bad decision.

Vaughan's interpretation was compelling: some level of "deviant" behavior becomes "normalized" in all organizations. Normalization of deviance builds error into all human systems. Only an institution that stays very in tune with itself, with what it is, with what it should be, and with its historical development has any chance of avoiding a "culture of complacency" that can lead to terrible decisions and catastrophic mistakes. The conclusion to be drawn from Vaughan's book for the fate of all institutions everywhere was rather pessimistic and disturbing.

The cumulative force of Vaughan's analysis was impressive, providing, as it did, a rare view into the working-level realities of NASA as a federal organization. No book published up to that time had contained more thorough research of all the hearings and interview testimony taken in conjunction with the *Challenger* accident—and certainly none offered more arresting insights into the basis for NASA's "group think" and decision making.

Yet Vaughan's book troubled me from the start. While her conclusion that a "normalization of deviance" had occurred within both NASA and Morton Thiokol—certainly concerning the O-ring problems that were known prior to *Challenger*—was likely on the mark, I found numerous technical errors in her book. Moreover, she had conducted forty-nine personal post-*Challenger* interviews, but thirty-nine of them had been with personnel from NASA Marshall Space Flight Center in Alabama, with the other ten all being with Roger Boisjoly, the ex-Thiokol employee who had sued his former employer for over $2 billion. Her book reflected a bias favoring both parties. The book bothered me even more when I learned that it had been accepted by both NASA and ATK Thiokol as the "Bible" on the causes of the *Challenger* accident, a convenient new party line for the two organizations, as Vaughan's *Challenger Launch Decision* concluded that no one in either place should be held responsible for "deviant" actions. Though never the type of person to search for scapegoats, I believed that someone associated with *Challenger* must have done something wrong at some point, somewhere, even if it was after the fact, in the form of a cover-up.

When I first got my hands on Allan McDonald's memoir, I realized my uneasiness with Vaughan's book was justified. Not that her book didn't deserve the high praise it had received (including the Rachel Carson Prize, Robert K. Merton Award, Honorable Mention for Distinguished Contribution to Scholarship of the American Sociological Association, and nominations for the National Book Award and Pulitzer Prize). But it concerned me that her book had itself become a norm for NASA. Following her persuasive analysis of the causes of the *Challenger* accident, Vaughan was asked to testify before the *Columbia* Accident Investigation Board in 2003; she even became a member of the board's research staff, where she worked to help analyze and write the section of the report identifying the social causes of the *Columbia* accident.[9]

Also in 1996, in a publication timed to coincide with the tenth anniversary of the *Challenger* accident, *No Downlink: A Dramatic Narrative about the Challenger Accident and Our Time* (Farrar, Straus, & Giroux) appeared. The work of Danish writer Claus Jensen, who claimed no advanced technical training but a deep and abiding personal interest in space exploration, *No Downlink*—originally published in Denmark as *Challenger, et teknisk uheld* (Challenger: A Technical Accident[10])—offered an effective survey of what had already been written about *Challenger* but very little that was fresh or original. The first two-thirds of the book was not even about *Challenger*, but rather was a summary of early rocketry from Dr. Robert Goddard's pioneering work of the 1920s, through the German V-2 missile of World War II, to the Mercury, Gemini, and Apollo programs that preceded the Shuttle. Only the last third of the book actually focused on the *Challenger* accident, with most of its information based upon the Rogers Commission investigation report, the congressional hearings, plus what had appeared in the popular press and previous books written on the subject between 1986 and 1996. Nothing in the book was based on interviews that Jensen himself conducted.

Jensen's central thesis was that when modern society combined advanced technology with large bureaucratic organizations and political decision making, the outcome can easily turn tragic. His book emphasized why the American space program from the start took shape in relation to political decisions (for example, on foreign policy and federal budget) that were routinely being made at higher levels of government with little or no regard to the details of space science and technology. Jensen focused on the impact of NASA public relations and of enthusiastic media coverage on the NASA mentality, writing that NASA in its heyday of the 1960s had "that famous 'can do' spirit and felt it could do anything." Sadly, NASA folks "began to believe their own public relations department—it was a picture created for external use, but it bounced back on them. It was dangerous to believe in this myth." After reading Jensen's book, I determined to set off one of my graduate students on a study of the history of NASA's public relations office.[11]

Another principal feature of Jensen's book was its contribution to the legend of Dr. Richard P. Feynman, the Nobel laureate theoretical physicist from Caltech who had become a figure of great public acclaim while serving on the Rogers Commission. From the moment on national television when Feynman dropped a rubber O-ring into a glass of ice water, thereby demonstrating the hardening effects of cold temperature on the pliability of O-rings, the legend of the eccentric, feisty scientist mushroomed: Feynman was the Rogers Commission's Don Quixote, its one great savior, its giant killer, the nation's best hope at getting at the truth behind the *Challenger* accident. An adoring media reported that Feynman, motivated as he was by an uneasy conscience due to his participation in helping to build the first atomic bomb during World War II, was singularly determined to ensure there would be no institutional cover-up of the negligence that led to the *Challenger* tragedy. Reinforcing this heroic iconography was the fact that Feynman became sufficiently dissatisfied with the Rogers Commission's final findings that he insisted on issuing his own "minority report."

As remarkable a scientist and human being as Feynman was (I had read his wonderful autobiography, *Surely You're Joking, Mr. Feynman! Adventures of a Curious Character* [W. W. Norton, 1985] even before his appointment to the *Challenger* Commission), the truth was that Feynman was *not* the only member of the Presidential Commission actively seeking to uncover the real problems with the Shuttle or with NASA in general. Feynman may well have gotten the idea for his cold O-ring experiment during his visit with one of his colleagues on the Commission, General Don Kutyna. Kutyna was showing Feynman his old Opel GT and mentioned to Feynman that the seals in the carburetor leak when they get cold. Furthermore, the Vice-Chairman of the Commission, Neil Armstrong, the first man on the Moon, was the individual who most effectively, but nearly invisibly, ran the operational side of the *Challenger* investigation. And contrary to stories that the Commission tried hard to suppress the publication of Feynman's "Minority Report" because it was allegedly "anti-NASA," both Armstrong and Chairman William Rogers were "okay" with the colorful physicist expressing his unique take on the subject and attaching it to the Commission's final report as an appendix.[12]

In 1988, Feynman's *What Do You Care What Other People Think?* (W. W. Norton) was published posthumously (he had just died in February 1988). The latter half of the book dealt almost exclusively with his involvement on the Rogers Commission. Along with a gripping firsthand account of his famous bit of theater with the O-ring dipped in ice water, the memoir detailed Feynman's battles during the *Challenger* investigation between science and logic, on the one side, and politics and expediency on the other. Purposefully acting the part of the innocent in the convoluted land of the unnecessarily complex, the self-consciously nutty professor revealed to his readers exactly how he used a quick idea or sharp question to cut through all the machinations and messy thinking. His was the first insider account of the *Challenger* investigation published by one of its leading participants and, because he had become virtually a household name by the late 1980s, the memoir deeply influenced both the public and scholarly take on the factors behind the demise of *Challenger*. In his book, Feynman credits McDonald's testimony for enabling the Presidential Commission to find the real cause of the *Challenger* accident:

> Then something happened that was completely unexpected. An engineer from the Thiokol Company, a Mr. MacDonald [*sic*], wanted to tell us something. He had come to our meeting on his own, uninvited. On the night before the launch, they told NASA the Shuttle shouldn't fly if the temperature was below 53°—the previous lowest temperature—and on that morning it was 29°. Mr. MacDonald said NASA was

"appalled" by that statement. Thiokol (management) reversed itself, but MacDonald refused to go along, saying, "If something goes wrong with this flight, I wouldn't want to stand up in front of a Board-of-Inquiry and say that I went ahead and told them to go ahead and fly this thing outside what it was qualified to." That was so astonishing that Mr. Rogers had to ask, "Did I understand you correctly, that you said . . . ," and he repeated the story. The whole Commission was shocked, because this was the first time any of us had heard this story; not only was there a failure in the seals, but there may have been a failure in management, too. Mr. Rogers decided that we should look carefully into Mr. MacDonald's story and get more details before we made it public.

Indeed, Richard Feynman was a highly laudable and brilliant man—some have called him a "contemporary Leonardo"—who made important contributions to what was learned about the causes of the *Challenger* disaster. But he was definitely not the drama's solitary hero—or even its main protagonist.

Also in 1988, veteran space journalist Richard S. Lewis, a former science writer with the *Chicago Sun-Times*, published *Challenger: The Final Voyage* (Columbia University Press). It was a short and heavily illustrated survey based largely on the *Report of the Presidential Commission on the Space Shuttle Challenger Accident* (Washington, D.C.: Presidential Commission on the Space Shuttle Challenger Accident, 1986), on transcripts of the Commission hearings, and on articles appearing in *Aviation Week and Space Technology*, but its clarity, technical accuracy, and evenhandedness distinguished the book favorably from the opportunistic and rapidly composed works of 1987 by McConnell and the Trentos. Given that Lewis had also once been managing editor of the *Bulletin of the Atomic Scientists*, it was no surprise that the book also featured Feynman in the pivotal investigative role.

Throughout the 1990s, very few books specifically about *Challenger* actually appeared. NASA, which from its inception in 1958 had administered a very active and highly respected history program, published a number of books during the decade, but none of them dealt primarily with *Challenger*; few broached the subject very directly at all. Undoubtedly, the lack of attention had something to do with the controversial nature of the topic within and around the space agency; after all, in the wake of *Challenger*, some critics called for the end of the Space Shuttle program, with a few calling NASA "an astronaut killer" and going so far as to demand the total abolition of the agency. But under the direction of its strong-minded and academically oriented chief historian, Dr. Roger D. Launius, those in the NASA History program fully understood the significance of exploring the causes of the accident and worked to lay a strong foundation for scholars to eventually produce penetrating, reliable, comprehensive, and critical histories of *Challenger* and the entire Shuttle program.

With that goal in mind, Launius (with the assistance of Aaron K. Gillette) began by compiling *Toward a History of the Space Shuttle: An Annotated Bibliography*, which NASA published as its inaugural "Monograph in Aerospace History" in 1992. Chapter 7 of the ninety-seven-page publication focused on the "*Challenger* Accident and Aftermath" and cited seventy-one different bibliographical items. Nearly a third of the entries entailed U.S. government reports or white papers (from NASA, the General Accounting Office, House of Representatives, U.S. Senate, and Congressional Budget Office) that came out in the immediate aftermath of the accident. Fifty-seven of the seventy-one citations involved books or articles appearing by 1988, with four of the referenced works dating prior to *Challenger*. Fifty-two of the items referred to journal, magazine, and newspaper articles. The bibliography included substantial annotations of the McConnell, Trento, and Lewis books, as well as the Feynman autobiography. It also briefly mentioned five other books:

> *Challengers: The Life Stories of the Seven Brave Astronauts of Shuttle Mission 51-L* (New York: Pocket Books, 1986), "a relatively standard journalistic account" written by the staff of the *Washington Post*;
> *Heroes of the Challenger* (London: Archway Paperbacks, 1986), a compact "media approach" to covering the lives of the dead astronauts by Daniel Cohen and Susan Cohen;
> *Keeping the Dream Alive: Putting NASA and America Back in Space* (San Diego: Earth Operations, 1987), "a slim volume . . . chiefly interesting [for] its discussion of the difficulties NASA has experienced in meeting the challenge of using and exploring space," by Michael C. Simon, a General Dynamics engineer who later became President of San Diego's International Space Enterprises in San Diego;
> *Shuttle Challenger* (London: Salamander Books, 1987), basically a "picture book" with "assorted tidbits" about the ill-fated Space Shuttle, its technology, and crew, by space writer David Shayler;[13] and

Space Shuttle: The Quest Continues (London: Ian Allen, 1989), a popular
account for "buffs" of the reassessment of the Shuttle program following *Challenger*, containing no references to where the information
came from, by British writer George Forres.

Interestingly, three of the books were published in England, almost as if
the publishing industry in the United States wanted to stay away from the
unfortunate subject of the *Challenger* disaster.[14]

NASA chief historian Launius also sponsored the preparation of an ambitious, multivolume documentary history of the U.S. civilian space program.
Commissioned in the early 1990s for a team of researchers at George Washington University headed by John Logsdon, the hefty inaugural volume of
this important series, entitled *Exploring the Unknown: Selected Documents
in the History of the U.S. Civil Space Program*, Vol. 1, *Organizing for Exploration* (NASA SP-4407), appeared in 1995. It was followed by Volume
2, *External Relationships*, in 1996; Volume 3, *Using Space*, in 1998; Volume
4, *Accessing Space*, in 1999; Volume 5, *Exploring the Cosmos*, in 2001; and
Volume 6, *Space and Earth Science*, in 2004. Inside the six large volumes
rested a massive collection of letters, reports, memos, and other primary
historical records not just belonging to NASA but also from presidential
libraries, corporate and university archives, and a variety of other sources,
accompanied by introductory essays written by Logsdon and other leading aerospace scholars. *Accessing Space* dealt directly with Shuttle history
(notably chapter 2, "Developing the Space Shuttle," and chapter 3, "Com-
mercializing Space") and included over seventy important Shuttle-related
documents from the era 1966 to 1995. In that volume rested no less than 137
references to *Challenger*, with the full text of such documents as Executive
Order 12546 by which President Reagan had set up the Presidential Commission on the Space Shuttle *Challenger* Accident on February 3, 1986; a
March 1986 memorandum from Associate Administrator for Space Flight
Richard H. Truly, on a "Strategy for Safely Returning the Space Shuttle to
Flight Status"; as well as a memo from John W. Young, Chief of the Astro-
naut Office in Houston, on the *Challenger* accident, with Young's attach-
ment "Examples of Uncertain Operational and Engineering Conditions, or
Events Which We 'Routinely' Accept Now in the Space Shuttle Program."

In the early 1990s, the NASA History Office also supported the publication of *Inside NASA: High Technology and Organizational Change in the U.S.
Space Program* by American University political science professor Howard

E. McCurdy. Published in 1993 in the *New Series in NASA History* by the Johns Hopkins University Press, the book investigated the relationship between the declining performance of the American space program in the 1970s and 1980s and the fundamental changes in NASA's organizational culture that had occurred during that time. As one reviewer of McCurdy's book noted, *Inside NASA* showed "how NASA became the Post Office," that is, how it had moved away from its strong technical orientation of the 1960s, which had accomplished the Moon landing and when engineering decisions mostly overrode political considerations, to become increasingly political and bureaucratic, as did the U.S. government as a whole in that time period.[15] This erosion of the organizational culture caused by bureaucratization, in conjunction with an increase in subcontracting to industrial partners like Morton Thiokol, maker of the Shuttle's solid rocket motor—which worked generally, in McCurdy's view, to compromise innovation and reliability at NASA—ultimately destroyed *Challenger*.[16]

Only at the very end of the 1990s did a book sponsored by the NASA History program come out that dealt explicitly with the Space Shuttle. This was T. A. Heppenheimer's *The Space Shuttle Decision: NASA's Search for a Reusable Space Vehicle* (1999), later republished by the Smithsonian Institution Press in two volumes.[17] In his acknowledgments to the book, Heppenheimer, an accomplished freelance writer with an impressive engineering background, recounted his book's genesis. Visiting the NASA History Office in 1996 to conduct research on an article for *Air & Space*, a Smithsonian magazine, Heppenheimer met Roger Launius for the first time. When he asked the head of NASA History about the lack of Space Shuttle coverage, Launius replied with "a tale of woe." Launius had tried to interest aerospace historians in the project, "but to no avail."[18]

Working under contract to NASA History, Heppenheimer produced a very coherent and readable narrative having extremely detailed coverage of the Shuttle's formative period (1965 to 1981), but which did not go far enough ahead chronologically to examine the immediate backdrop to the *Challenger* accident. Nor was his book highly interpretive or scholarly, with many of its assertions and quotations left wholly undocumented. In his review of the book, Alex Roland wrote that Heppenheimer did not quite exonerate NASA for its Shuttle decisions—in fact, the author was often "refreshingly critical of all the parties involved." But, in the end, the book withheld "the critical scrutiny" of the decision to go ahead with the Space Shuttle that Roland strongly thought was warranted.[19]

For the later edition of his book, Heppenheimer did add a new final chapter entitled "The Working Shuttle." This coda examined the *Challenger* accident and its aftermath, albeit briefly. The author's conclusion was that *Challenger* had been destroyed "due to the pressures of tight schedules, which led NASA to cut corners." Only the hard way did people learn that the Shuttle could not be all things to all people. The naive hope that the Shuttle would provide frequent and low-cost space operations was dead. "Instead, the nation would treat the Shuttle as a rare and valuable commodity, to be flown only on special occasions."[20] Heppenheimer's abbreviated treatment offered nothing new or original about the *Challenger* accident or the follow-up investigations.

On the other hand, in 1999, the same year that the original version of Heppenheimer's book came out, another NASA History publication offered a strongly revisionist interpretation of the demise of *Challenger*. This book was *Power to Explore: A History of Marshall Space Flight Center, 1960–1990* (NASA SP-4313), by University of Alabama–Huntsville (UAH) history professors Andrew J. Dunar and Stephen P. Waring.

Prior to the appearance of this impressive book, most interpretations of the *Challenger* disaster had been highly critical of the role played by Alabama's NASA Marshall Space Flight Center. In his 1998 autobiographical survey of American space history, *Liftoff: The Story of America's Adventure in Space* (NASA/Grove Press), even the fair-minded Michael Collins, command module pilot for the historic *Apollo 11* mission, had come down hard on Marshall. "Huntsville has become a castle whose occupants have pulled up the drawbridge, leaving the moat full of serpents," Collins wrote. MSFC was "the only part of NASA that did not cooperate fully" with the Rogers Commission investigating the *Challenger* accident. "In an organization, healthy pride can come very close to destructive arrogance. The attitude that 'we'll handle our own problems, thank you' can be taken too far."[21] Collins, as others had already done, suggested that a "cult of arrogance" had pervaded Huntsville and was the major factor in the organizational and communications failure leading to the *Challenger* incident.

In *Power to Explore*, Dunar and Waring questioned the accuracy of this view and saw much of the post-*Challenger* "Huntsville-bashing" as scapegoating. Discrepancies between the findings of the Rogers Commission and documentary sources in the Shuttle program led the UAH historians to question what had become the standard belief that Marshall grossly mishandled the *Challenger* launch. The Rogers Commission reported four major findings, the authors reminded the reader. First, the cause of the accident was frozen rubber O-rings in the SRB joints, which allowed a leak of burning fuel. Second, engineers working at Marshall and at Morton Thiokol in Utah, the SRB contractor, knew that the joint design was dangerous, especially in cold temperatures. Third, Marshall project managers had known for some time that the joints were hazardous but failed to communicate that understanding to chief Shuttle officials at Johnson Space Center in Houston and to NASA headquarters during preflight reviews in Washington. Fourth, MSFC officials botched the last-minute teleconference with Thiokol, held the evening prior to launch. They pressured Thiokol's top managers to overrule their engineers and recommend launch even in the cold weather expected the next morning at Cape Canaveral.

The revisionist analysis by Dunar and Waring faulted the Rogers Commission findings and offered possible alternative interpretations. Regarding the first finding, the authors acknowledged that O-ring failure had something to do with *Challenger*'s destruction, but they suggested a more complicated view, which was that excessive compression on the O-rings could have been the principal cause. (The NASA Accident Analysis Team following *Challenger*, in fact, reported this condition, as the McDonald memoir details.) If so, the source of the problem might have been faulty assembly procedure at Kennedy Space Center involving recycled SRB segments, which an assembly crew forced together and left out of round, forming the fateful joint. Dunar and Waring admitted there was no way to know for sure which technical scenario explained what caused the accident, but they made their readers wonder if all the attention should really have been on O-ring design, cold weather, and Marshall.

As for the second finding of the Rogers Commission, Dunar and Waring claimed that the O-rings became highly significant only in hindsight. NASA and Thiokol engineers believed that the joints, if properly assembled, were safe. Other problems, especially with the Shuttle's main engines, consumed most of Marshall's attention. Limited reliable engineering data existed at the time of the accident to support a strong correlation between O-ring anomalies and low temperatures. As for the Rogers Commission's third finding, the UAH historians denied that Marshall project engineers knew the joints to be dangerous and withheld that judgment from their superiors. The historical record indicated rather that the responsible MSFC officials believed that the joints were safe and communicated that consensus to chief Shuttle officials. The Shuttle catastrophe shocked them as much as anyone.

Dunar and Waring agreed with the Rogers Commission's conclusion that the prelaunch teleconference was mishandled. But it was unfair, they argued, to condemn the involved MSFC officials for *believing* that Thiokol, when they were "off-line" from the teleconference, had properly validated the launch's continuation. The Rogers Commission charged that Marshall officials knew that the validation had been coerced, because they had done the coercing themselves, by putting pressure on Thiokol's upper management. Dunar and Waring asked readers to ask what would have happened if MSFC officials *had* informed Houston and Washington of the teleconference's initial "no-launch" recommendation. Might the top NASA officials have overruled the "no-launch" recommendation themselves? Given the positive publicity and hubbub surrounding "Teacher in Space" Christa McAuliffe and the fact that President Reagan's State of the Union address was scheduled for the night of the launch, might NASA have gone for launch anyway? In the scenario of a "no-launch" recommendation communicated from Marshall (or Houston) to Washington, what role might the White House have played in the matter? Many analysts at the time speculated that the White House had already been exerting pressure on the launch schedule. McDonald's memoir addresses this scenario and rightfully points out that if this would have happened the Mission Management Team at least would have been sensitized about the concerns of operating the O-rings in a temperature as low as 29 ° and would have cancelled the launch when they actually measured 7°–9° in the joint area of the right-hand SRB during the Ice Team's inspection.

In the end, the UAH historians suggested in two lengthy chapters on *Challenger* that the disaster had at least as much to do with NASA's organizational patterns, technological decisions, and politics dating back to the early 1970s as it did with immediate circumstances related to O-ring design, cold temperature, and problems at Thiokol or arrogance in Huntsville. They concluded their discussion of *Challenger* with a description of the psychological hurt suffered by many at MSFC. Individuals already deeply sorry for letting down their country felt even more pain as the Rogers Commission, the rest of NASA, and the media used the Alabama center as a scapegoat, which deflected blame away from others who might also have deserved some of the responsibility, including Morton Thiokol management and political leaders in Washington.

A few reviewers thought that Dunar and Waring, local historians under contract to NASA to write the MSFC history, were simply apologists for Marshall. But those who knew the whole story of what the two professors had gone through with Marshall management while writing their book could not regard them as anything but outstanding scholars whose professional integrity and doggedness had enabled them to complete an honest and extremely significant scholarly book, one amazingly free of direct influence from NASA Marshall, even though some in MSFC management had, in an effort to protect the center's image from further historical scrutiny, subjected the hardworking UAH professors to something close to Dante's ninth circle of hell. For their successful defense of their work against censorship, the Alabama Library Association in 2003 awarded its Intellectual Freedom Award to the authors. Earlier, in 2001, the American Institute of Aeronautics and Astronautics presented Dunar and Waring with its prestigious AIAA History Manuscript Award.

In the July 2002 issue of the *Alabama Review*, I published a detailed account of *Power to Explore*'s tortured path to publication.[22] "Without question," I concluded, "the book represents one of the most outstanding products in forty years of publishing by the NASA history program." Today, I still feel that way about the book, though I now look at their *Challenger* analysis much more critically, due to what I have learned from the McDonald memoir. While the authors did everything they could consciously to be objective, their analysis does reflect a pro-Marshall bias. Several of their points about the *Challenger* accident in relation to Marshall are highly rationalized.

Most significant from the point of view of the McDonald memoir, *Power to Explore* contains several errors and misinterpretations of technical data—some of which may derive from oral history interviews conducted by the authors with individuals who had worked in key Shuttle positions at MSFC during *Challenger*. One of the most glaring misinterpretations of this sort arose as part of the book's sympathetic explanation as to why, in the face of some worrisome data related to O-rings, Marshall management still supported the launch decision. Contrary to what Dunar and Waring stated in their book about "no one" having performed a statistical analysis correlating past O-ring performance with either temperature or leak-check pressure, and that "lacking such analysis" Marshall managers "rightly questioned" Thiokol's demonstration of the statistical correlation between temperature and erosion of an O-ring, McDonald makes it clear that Thiokol engineers had never thought that *erosion* was the critically important correlation between temperature and O-ring distress; rather, it was *blowby*. Temperature had absolutely no influence on any of the O-ring's material properties that impacted the magnitude of erosion, but it had major impact on several factors that contributed to the severity of O-ring blowby, which could, and did, prove more catastrophic. On page 359 of *Power to Explore*, Dunar and Waring reproduced an engineering chart entitled "Plots of Incidence of O-Ring Distress as a Function of Temperature," which seems to have been provided by Bob Marshall, the individual who had served as manager of MSFC's Shuttle Project Office immediately after the *Challenger* era. McDonald's memoir makes clear that this chart is "bogus," a red herring used by Marshall officials to defend MSFC by muddying the water and demonstrating Thiokol's inability to understand the influence of temperature statistically. This same chart with the same erroneous conclusion also appears in Diane Vaughan's book, *The Challenger Launch Decision*.

Although their book was in most respects thoroughly researched, Dunar and Waring relied far too heavily on sources internal to MSFC. Without question, for accuracy in their *Challenger*-related chapters, they should have interviewed a number of outsiders, including Thiokol's Allan McDonald and Roger Boisjoly, both of whom were star witnesses during the Presidential Commission hearings in 1986. In dozens of emotional talks given around the country following *Challenger*, Boisjoly had been charging that MSFC officials played "fast and loose" with the astronauts' lives, "absolutely abdicating their professional responsibility" in pressuring Thiokol to reverse its original recommendation not to launch. In Boisjoly's view, stopping the launch of *Challenger* was a "no-brainer," requiring "only common sense."[23] *Power to Explore* reported that technical data on the effects of cold weather on O-rings, both from ground tests and previous Shuttle flights, had been "inconclusive," even though Boisjoly, the man most responsible for analyzing the data at the time, persistently claimed that the data was absolutely definite and had provoked the "most unified company position" ever put forward to NASA to stop a launch.[24] NASA officials, in Boisjoly's view, were so determined to launch *Challenger* that its top Shuttle experts forced Thiokol to prove beyond any doubt that it was not safe to do so—when in most flight readiness reviews officials had to prove just the opposite. Dunar and Waring treated Boisjoly's role in *Challenger* accurately, but they clearly didn't accept his personal judgment about the degree of Marshall's culpability, largely because the authors had swallowed the MSFC party line that "Thiokol's data were inconclusive" and that MSFC project managers had genuinely believed the SRB joints were not hazardous.[25] If Dunar and Waring had interviewed Boisjoly or given McDonald a chance to present his version of the events to them directly, maybe they would have come to some very different conclusions, as I did following my initial reading of the Mc-Donald memoir.

None of the conversations that Allan McDonald had with MSFC's Lawrence Mulloy or Stanley Reinartz during or after the infamous prelaunch caucus was mentioned in *Power to Explore*; it appears as if the authors intentionally ignored it. In fact, McDonald's name appeared only a couple times in their chapter on *Challenger*. One of those references contained the subsequently controversial comment made by McDonald during the teleconference about the secondary O-ring being in a better position to seal than the primary, as if McDonald was in fact saying, as MSFC officials in the aftermath of *Challenger* would have it, that the secondary seal provided redundancy. Readers of *Power to Explore* inferred from the author's reference to that statement that McDonald indeed supported MSFC's prolaunch decision. The authors also reported that McDonald made his comment *before* the caucus was requested, when the truth of the matter was that he made this statement *after* the caucus had been requested by Thiokol's Joe Kilminster, as noted in both Larry Mulloy's and George Hardy's sworn testimony. It was NASA's George Hardy who *first* made this observation about the secondary O-ring being in a better position to seal during the teleconference. McDonald just reiterated what Hardy had said earlier, because no one had commented on it at the time when Hardy had mentioned it, and he thought it was an important consideration.

One of the unique features of the McDonald memoir is that it is the first and only book ever written on the *Challenger* accident that fundamentally agrees with the findings of the Presidential Commission that investigated the accident. Dunar and Waring's history of NASA Marshall definitely doesn't support the Rogers Commission findings, declaring, "In summary, the conclusions of the Presidential Commission were a mix of fact and fallacy. The Commission engaged in scapegoating that put unfair blame on a few individuals . . . , did not listen sympathetically . . . , made conclusions too early in the investigation, put too much emphasis on cold weather as the technical cause of the accident, paid too little attention to assembly factors, and then made unfair accusations against Marshall managers."[26] McDonald's version of *Challenger* history stands adamantly in opposition to this defensive Marshall interpretation and squarely on the side of the Rogers Commission.

Anyone reading McDonald's memoir also must question whether Dunar and Waring should really have given the lion's share of credit for the successful redesign of the Shuttle's solid rocket motors to NASA Marshall. "Marshall members of the SRM redesign team," their book states, "deserve the greatest credit for the successful return to flight. Not only did Marshall personnel determine the technical cause of the accident and analyze the weaknesses in the motor joints, but the Center also conceived the solution." If the authors had not relied so heavily just on MSFC's documentation on the SRM and on personal interviews with selected MSFC employees, they might have discovered more information about the critical role played by Allan McDonald and his engineering staff at Thiokol in the diagnosis of the causes of the *Challenger* accident and in the successful redesign and recertification, which in some cases were in direct conflict with MSFC's conclusions and recommendations.

Surprisingly, not until 2002 did a *Challenger*-specific memoir by someone who had been directly involved with either the accident itself or the investigation appear. *Challenger's Shadow* was written by John Macidull, a member of the staff assigned to the Presidential Commission, with the help of Lester E. Blattner, an author for various magazines on the U.S. space program who had published books on various aspects of aerospace history. The book, which focused primarily on the process of the accident investigation, concluded that NASA tried its best to cover-up both the cause of the accident and the ill-fated decision to launch against the objections of Thiokol.

Although *Challenger's Shadow* includes numerous technical errors, its conclusions relative to the mishandling of the decision to launch by both NASA and Morton Thiokol management basically aligns with what McDonald relates in his memoir. It also acknowledges McDonald's singular contribution to the accident investigation: "There was Allan McDonald, Morton Thiokol's Solid Rocket Motor Director, who took on both managers and adamantly presented his engineering and personal concerns, only to be slapped down like some inconsequential fob." "In my eyes," wrote Macidull, "McDonald was a hero in the *Challenger* drama. He thought he would be fired because of his opposition was too vehement with Thiokol's largest customer. Still he did try to stop the launch, even though Thiokol senior management subsequently sold him out."[27]

Published by a small vanity press in Florida, Macidull's book received very little attention. The same was true for *Some Trust in Chariots: The Space Shuttle Challenger Experience*, a 2006 memoir by James A. ("Gene") Thomas (Xulon Press).[28] Thomas was the NASA Launch Director at Kennedy Space Center at the time of the *Challenger* launch, which makes his account immediate and firsthand. In actuality, however, Thomas possessed no knowledge about the cold O-ring concerns leading up to the decision to launch, had no prior knowledge of the O-ring problems, and was an extremely minor player in the subsequent Presidential Commission's hearings relative to the accident. *Some Trust in Chariots* primarily told the story of Thomas's personal religious experience as a result of being "involved" as the person who finally pushed the button to launch *Challenger*. What little Thomas discusses about the accident itself or its causes are somewhat defensive but primarily quite vanilla—neither new nor controversial. Instead of "trusting in chariots," Thomas concluded, the only sure trust for a person is that which he or she places in Jesus Christ. Unfortunately, that sort of faith did nothing to save the lives of the Christian believers onboard *Challenger*.

A memoir that did cause some stir when it was published in 2006 was *Challenger Revealed: An Insider's Account of How the Reagan Administration Caused the Greatest Tragedy of the Space Age*, by Richard Cook (Thunder's Mouth Press). Shortly after the *Challenger* disaster, Cook, a budget analyst at NASA headquarters, leaked a July 1985 internal NASA memo to the *New York Times* in which he considered a request from Morton Thiokol to add a capture feature to the field joint to help the O-ring sealing problem. Cook had no technical training, but based on conversations he had at the time with technical people at NASA headquarters, he believed the O-ring sealing problem could be a significant problem from a budgetary perspective. When his memo was published in the *Times*, the result was the first public hearing conducted by the Presidential Commission, on February 11, 1986.

In reality, Cook's memo overstated the financial magnitude of the problem. "The potential impact of the problem depends on the *as yet undiscovered cause*" of the O-ring problem, his memo stated. "A worst case scenario . . . would lead to the suspension of Shuttle flights, redesign of the SRB, and scrapping of existing stockpiled hardware."

What the McDonald memoir makes clear is that the Thiokol engineers already *knew the cause* of the O-ring erosion and blowby problem at the time; that was why they had recommended proceeding with incorporating the capture feature that was already being used on the FWC-SRM. Contrary to Cook's warning, the problem would not have resulted in the scrapping of stockpiled hardware because each casting segment was made from two case sections, one of which would have remained unchanged.

Cook's memo illustrated a problem that would become common in many interpretations of Shuttle history: without a detailed understanding of the engineering involved, it was easy for a person to arrive at misleading or incorrect conclusions.

The most provocative aspect of Cook's *Insider's Account* was the support the book gave to the claim made by Senator Ernest ("Fritz") Hollings (D-S.C.) in the aftermath of *Challenger* that President Reagan had insisted on launching the Shuttle on January 28, 1986, because his State of the Union address that evening was going to acknowledge the historic moment of schoolteacher Christa McAuliffe going into orbit. In his memoir, Cook claimed that White House Chief of Staff Donald Regan had been the person who recommended the establishment of a Presidential Commission, and that Regan suggested that Reagan's personal friend, William Rogers, be the chairman. Moreover, all of the other Commission members were allegedly handpicked by another personal friend of the president's, Dr. William Graham, then the acting NASA Administrator. This was done to protect the Reagan administration and NASA from the penetrating charges that would inevitably come out of any authentically independent investigation into the causes of the accident. Cook claimed that the administration was in the process of militarizing the Space Shuttle program in support of Reagan's Strategic Defense Initiative, or "Star Wars" program, and that this was the real reason why Reagan had replaced former NASA Administrator James Beggs with Bill Graham, a nuclear weapons expert. Cook even claimed that Allan McDonald provided a key piece of testimony in this regard: that NASA was reluctant to increase its launch commit criteria to 53° because the possibility of temperatures colder than that were higher for air force launches from Vandenberg AFB in California than for NASA at the Cape. Before Cook started writing his book, he tried to get U.S. Attorney General Edwin Meese to establish an independent prosecutor to investigate the role of President Reagan and members of his administration in "forcing" the launch of *Challenger*.

Unfortunately for this particular conspiracy theory, neither Cook, Senator Hollings, the Attorney General, nor anyone else (not just the Presidential Commission) was ever able to find any solid evidence that the Reagan administration applied any direct pressure on NASA to launch.

One of the more fanciful conjectures in Cook's memoir concerned the curiosity shown by President Reagan and especially his wife, Nancy, in astrological forecasts and the connection of it to the *Challenger* launch decision. Stories about the possible ramifications of the Reagans' affinity for astrology first captured public attention in 1988 following the publication of Donald Regan's autobiography, *For the Record: From Wall Street to Washington* (Harcourt Brace Jovanovich). Regan, who had left the White House a year earlier, disclosed that the First Lady had planned presidential travel, press conferences, and even the president's cancer surgery in 1985 based on information she received from stargazer Joan Quigley, who had become the Reagans' principal astrologer in 1976 after Nancy Reagan dropped Jean Dixon because of Dixon's prediction that Reagan would not gain the presidency, at least not that year. More revelations about the Reagans and astrology later came out in *Nancy Reagan: The Unauthorized Biography* (Simon & Schuster, 1991) by the gossipy author Kitty Kelley.

Having followed the news about the Reagans' attraction to astrology with interest, Richard Cook conducted some of his own research into the subject for his book. He contacted Ed Helin, an astrology instructor at the Carroll Righter Institute in Los Angeles, who, along with Carroll Righter, the institute's founder and well-known "Astrologer to the Stars," had been doing star charts for the Reagans since the 1950s. With Righter himself deathly ill with cancer (he would die in 1988), President Reagan purportedly began calling Helin once or twice a month from Camp David, asking the astrologer to "determine the best timing for invading Grenada, for bombing Libya, for launching *Challenger*—things like that."[29] Cook tracked down Helin and learned that President Reagan had called Helin a week or so before *Challenger* and that Helin had told him that, astrologically speaking, "January 28 was not a good day to launch." (As a matter of fact, as noted in McDonald's memoir, it was not known a week or so earlier that the *Challenger* would be launched on January 28, because at that time it was scheduled for January 26 and was later scrubbed again late in the countdown on January 27.) Reagan personally ordered the launch to go ahead on the 28th, anyway, because of growing pressure from the media about the Shuttle's frequent launch delays. "The television networks were down there in Florida for the Teacher-in-Space," Cook quoted Helin as telling him, "and they told [Reagan] that they were spending a million dollars a day keeping their crews and equipment down there. . . . They were calling him and pressuring him to launch."

As historian Douglas Brinkley makes clear in the editorial commentary accompanying *The Reagan Diaries* (Harper Collins, 2007), many people, especially those who disliked Reagan or his politics, still believe the "nonsense" about the influence of astrology on the president. Just as with the theory suggesting that the decision to launch was controlled by the White House because of its ambition to militarize space, solid evidence simply does not exist to show that Reagan's use of astrologers, whatever it really amounted to, in any way affected the decision to launch *Challenger*. In fact, there are entries in Reagan's diaries that clearly negate that alleged association.

At any rate, Cook's memoir greatly overstates Cook's personal knowledge about and involvement in the *Challenger* accident and investigation. The book contains several errors and many unsubstantiated rumors and allegations that previously had appeared in the press and were summarily dismissed. The best that can be said about Cook's *Insider's Account* is that he does provide some interesting observations concerning what happened to him after he leaked the memos to the *New York Times* and appeared on several radio and television programs.

A book bordering on a *Challenger* memoir is *Apollo, Challenger, Columbia: The Decline of the Space Program* (Roxbury, 2005; Oxford University Press, 2007[30]), by Phillip K. Tompkins. Rather than memoir, however, Tompkins's study is more like a sequel to Diane Vaughan's *Challenger Launch Decision*, but without Vaughan's personal interviews with key players. For Tompkins, a professor emeritus of communication at University of Colorado at Boulder, the *Challenger* accident is but one of five "data points" within NASA history from which to conduct a study of pathological communication within an organization—the other four data points being: (1) from 1967, when Tompkins served as a summer faculty consultant at NASA Marshall assigned to MSFC Director Dr. Wernher von Braun to look into issues related to the *Apollo* fire of January 27, 1967; (2) from 1968, when Tompkins served in the same capacity to help reorganize the MSFC organization; (3) from 1987, when Tompkins did research into NASA's highly successful Aviation Safety Reporting System; and (4) from 2003, when Tompkins interpreted the communication failures leading up to the catastrophic failure of Space Shuttle *Columbia*. From his "longitudinal" case study of process changes in "communication-as-organization," Tompkins identified ten "communication transgressions," the most serious of which, he argued, was *ignorantia affectata*, or a cultivated ignorance of the problems growing inside an organization. In contrast to communication pathologies inherent to failing organizations like NASA, Tompkins offered sketches of select healthy organizations that had successfully applied ethical values in the workplace and thereby communicated "in concert." The main lesson to be learned from NASA's disasters, in Tompkins's view, was that even in a large and complex organization, individuals must take responsibility for their own actions.

Unfortunately, the book does not live up to the expectations built into its title. Readers expecting a deep examination of the accident board findings and conclusions from the three major space program failures (*Apollo 204* capsule fire, *Challenger* disaster, and *Columbia* accident) will be disappointed. Even though Tompkins was consulting for NASA within a few months after the *Apollo* fire investigation report was released, in his book he never discusses the findings of the NASA failure board. Given some strong evidence of NASA management cover-up activities in the aftermath of the *Apollo* fire (one of the reasons why both the White House and Congress were so adamant about having a more independent failure board investigate both *Challenger* and *Columbia*), one should question the correctness of Tompkins's very favorable portrayal of "open" communication at NASA Marshall under Von Braun's direction. Certainly, one of Von Braun's key assistants at the time, Dr. William R. Lucas, the Director of Propulsion and Head of the Vehicle Engineering Laboratory at MSFC, would not be known for his open communication after he became Marshall's Center Director in 1974. Allan McDonald's memoir denies Marshall's openness and charges that Lucas was largely responsible for developing the closed and arrogant MSFC management culture that helped lead to the *Challenger* accident.

As for *Challenger* specifically, Tompkins's book refers to it only in relationship to the NASA communication problems that were identified by the Rogers Commission. The same is true for what Tompkins wrote about *Columbia*. His coverage here relies heavily on the *Columbia* Accident Investigation Board Report and is basically a summary of the conclusions contained in that report. Tompkins had no personal knowledge about either the *Challenger* or the *Columbia* accident to draw on, and his book also contained several technical errors. Still, his resulting analysis does support McDonald's conclusion that NASA management made serious mistakes leading to the *Challenger* accident and that some of the responsible parties escaped without the blame they deserved. Tompkins's book also acknowledges that McDonald was one of the very few people who, immediately after the *Columbia* accident, disagreed with NASA's original conclusion that foam from the external tank did not cause the problem leading to the *Columbia* accident.[31]

Regrettably, with the exception of McDonald's memoir, none of the key people most directly involved in the *Challenger* accident or its aftermath have chosen to write their memoirs. Unfortunately, a great number of key individuals have died without offering much firsthand testimony. Hopefully, soon, in the years to come, a number of them will at least agree to comprehensive interviews about their role in *Challenger* history.

Notes

1. Roger D. Launius's review of Malcolm McConnell, *Challenger—A Major Malfunction: A True Story of Politics, Greed, and the Wrong Stuff* (New York: Doubleday, 1987), http://www.amazon.com/Challenger-Major-Malfunction-Story-Politics/dp/0385238770/ref=sr_1_2/002-6430155-5697655?ie=UTF8&s=books&qid=11852020 80&sr=1-2, accessed July 21, 2007.

2. Now the author of roughly twenty books, McConnell's most recent publications are as coauthor of *American Soldier* (HarperCollins, 2006), the autobiography of U.S. General Tommy Franks, and *My Year in Iraq: The Struggle to Build a Future of Hope* (Simon & Schuster, 2006), a memoir by American ambassador to Iraq L. Paul Brenner III.

3. Joseph Trento went on to publish (with William Corson and wife, Susan Trento), *Widows: Four American Spies, the Wives They Left Behind, and the KGB'S Crippling of American Intelligence* (New York: Crown, 1989); in it, Trento argued that the CIA essentially lost its war with the Soviet KGB. This book was followed by *Renegade CIA: Inside the Covert Intelligence Operations of George Bush* (New York: Putnam Publishing Group, 1993); *The Secret History of the CIA, 1946–1989* (Roseville, Calif.: Forum, 2001); and *Prelude to Terror: The Rise of the Bush Dynasty, the Rogue CIA, and the Compromising of American Intelligence* (New York: Basic Books, 2005). His newest book (with David Armstrong) is *America and the Islamic Bomb: The Deadly Compromise* (South Royalton, Vt.: Stearforth Press, 2007).

4. Alex Roland, "Priorities in Space for the USA," *Space Policy* 3 (1987):104–11.

5. Roland's article on "Barnstorming in Space" appeared in Radford Byerly Jr., ed., *Space Policy Reconsidered* (Bolder, Colo.: Westview Press, 1989), 33–52. Following the *Columbia* accident in 2003, Roland made many of the same arguments in his statement before the Subcommittee on Science, Technology, and Space of the Senate Committee on Commerce, Science, and Transportation (April 2003), http://history. nasa.gov/columbia/Troxell/Columbia%20Web%20Site/Documents/Congress/Senate/FEBRUA~1/roland_statement.html. See also Alex Roland, "The Shuttle's Uncertain Future," *Final Frontier*, April 1988, 24–27. For a critique of Roland's analysis of space history, see Dwayne A. Day, "Professor Grinch," *Space Review*, December 24, 2004, http://www.thespacereview.com/article/291/1. Both Internet sites accessed July 21, 2007.

6. Previous to his rejoinder, Logsdon had already laid out his argument in "The Space Shuttle Program: A Policy Failure," *Science* 232 (1986): 1099–105. Logsdon also wrote the opening editorial in the May 1987 issue of *Space Policy*, "Reconstituting the U.S. Space Programme," 86–88. In 1987, George Washington University in Washington, D.C., established its Space Policy Institute, with Logsdon as its head.

7. Alex Roland, "Rejoinder," *Space Policy* 3 (1987): 114.

8. See, for example, the letter from Roland (and Edward Tenner) responding to the article that Logsdon had written for *Issues in Science and Technology*, "A Sustainable Rationale for Human Spaceflight" (Winter 2004), following President George W. Bush's announcement of a new human spaceflight initiative: Edward Tenner and Alex Roland, "Human Spaceflight Forum, with Two Comments on John M. Logsdon's Article," *Issues in Science and Technology* (Spring 2004), http://pagebang.com/cgi/nph-proxy.cgi/111011A/http/www.issues.org/20.3/forum.html.

9. Prior to her *Challenger* book, Diane Vaughan had published *Controlling Unlawful Organizational Behavior: Social Structure and Corporate Misconduct* (Chicago: University of Chicago Press, 1983) and *Uncoupling: Turning Points in Intimate Relationships* (New York: Oxford University Press, 1986). She is currently at work on a book entitled "Theorizing: Analogy, Cases, and Comparative Social Organization," whose goal is cross-case analysis of similar events, activities, or phenomena across different organizational forms in order to elaborate general theory or concepts.

10. The term "uheld" used in Danish literally means "glitch." The English translation of Jensen's book was done by Barbara Haveland.

11. The first part of this larger study has been undertaken in the Auburn University History Department by Kristen Starr, whose 2008 doctoral dissertation is entitled "NASA's Hidden Power: Public Affairs and the Cold War, 1946–1967."

12. See James R. Hansen, *First Man: The Life of Neil A. Armstrong* (New York: Simon & Schuster, 2005), 616.

13. David Shayler has been a prolific contributor of books about space exploration, both for adults and children. For the well-known series of books on space exploration published by London-based Springer-Praxis, Shayler has authored or coauthored no less than ten works, including *Disasters and Accidents in Manned Spaceflight* (2000), *Skylab: America's Space Station* (2001), *NASA's Scientist-Astronauts* (with Colin Burgess) (2001), and *Apollo* (2002).

14. All quoted material is from chapter 7 of *Toward a History of the Space Shuttle: An Annotated Bibliography* (Washington, D.C.: NASA History Office, 1992), also published as *NASA Technical Memorandum 108647*. The monograph is available at the NASA Web site: http://www.hq.nasa.gov/office/pao/History/Shuttlebib/cover.html. Internet site accessed on July 21, 2007.

15. Noortje Marres, "Did NASA Become the Post Office Gone to Space," *European Association for the Study of Science and Technology* (March 1999), http://www.easst.net/review/march 1999/marres, accessed July 21, 2007.

16. Prior to *Inside NASA*, McCurdy had published *The Space Station Decision: Incremental Politics and Technological Change* (Baltimore: Johns Hopkins University Press, New Series in NASA History, 1990). He went on to write *Space and the American Imagination* (Washington, D.C.: Smithsonian Institution Press, 1997) and to edit (with Roger D. Launius), *Spaceflight and the Myth of Presidential Leadership* (Urbana: University of Illinois Press, 1997). In each of these books there are insights relevant to the *Challenger* accident.

17. T. A. Heppenheimer, *History of the Space Shuttle*, Vol. 1, *The Space Shuttle Decision, 1965–1972*, and Vol. 2, *Development of the Space Shuttle, 1972-1981* (Washington, D.C.: Smithsonian Institution Press, 2002).

18. Heppenheimer, *Development of the Space Shuttle*, v. Actually, as Roger Launius well knew, a very good, straightforward history of the Space Shuttle had been published in 1992. This was Dennis R. Jenkins, *Space Shuttle: The History of Developing the National Space Transportation System: The Beginning through STS-75* (Mareceline, Mo.: Walsworth, 1992; 2nd ed., Minneapolis, Minn.: Motorbooks International, 1996). Jenkins's book was a narrowly technical history, however, that paid little to no attention to the broader political, economic, and managerial aspects of the Space Shuttle's development. Written by an engineer who had worked in the Shuttle program for fifteen years, the book was incredibly detailed with numerous technical drawings for illustration, and simply was not written to draw connections to larger issues. But for what it was, the book stands as the definitive technical history of the STS.

19. Alex Roland, review of T. A. Heppenheimer, *The Space Shuttle Decision: NASA's Search for a Reusable Space Vehicle* (Washington, D.C.: NASA, 1999), in *Technology and Culture* 42, no. 2 (2001): 386–88.

20. Heppenheimer, *Development of the Space Shuttle*, 413.

21. Michael Collins, *Liftoff: The Story of America's Adventure in Space* (New York: NASA/Grove Press, 1988), 235.

22. James R. Hansen, "Review Essay: A Battle over the Historian's Power to Explore," *Alabama Review: A Quarterly Journal of Alabama History* 55 (July 2002): 192–99.

23. The Boisjoly quotations are from a talk he gave at the Auburn University School of Business on January 25, 2001.

24. Andrew J. Dunar and Stephen P. Waring, *Power to Explore: A History of Marshall Space Flight Center, 1960–1990* (Washington, D.C.: NASA History Office, 1999), 373.

25. Ibid., 199.

26. Ibid., 406.

27. John Macidull and Lester E. Blattner, *Challenger's Shadow* (Coral Ridge, Fla.: Llumina Press, 2002), 77–78.

28. James A. Thomas, *Some Trust in Chariots: The Space Shuttle Challenger Experience* (Longwood, Fla.: Xulon Press, 2006). Xulon describes itself as a "Christian self-publisher" that provides "affordable print-on-demand publishing."

29. On the Reagans and astrology with references to *Challenger*, see Richard C. Cook, *Challenger Revealed: An Insider's Account of How the Reagan Administration Caused the Greatest Tragedy of the Space Age* (New York: Thunder's Mouth Press, 2006), 474–76; and Kitty Kelley, *Nancy Reagan: The Unauthorized Biography* (New York: Simon & Schuster, 1991), 570n.

30. The books formerly published by Roxbury Publishing were acquired by Oxford University Press, effective February 2007.

31. Phillip K. Tompkins, *Apollo, Challenger, Columbia: The Decline of the Space Program* (Cary, N.C.: Roxbury, 2005), 39–40.

Further Reading

Columbia Accident Investigation Board Report, Volume I. Washington, D.C.: Government Printing Office, August 2003.

Hearing before the Subcommittee on Science, Technology, and Space of the Committee on Commerce, Science, and Transportation of the United States Senate, 100th Congress, First Session on Oversight of the National Aeronautics and Space Administration's Space Shuttle. January 20, 1987.

Hearings before the Committee on Science and Technology House of Representatives, Volume I. Report No. 137. 99th Congress, 2nd sess., June 10, 11, 12, 17, 18, and 25, 1986.

Hearings before the Committee on Science and Technology House of Representatives, Volume II. Report No. 139. 99th Congress, 2nd sess., July 15, 16, 23, and 24, 1986.

Hearings of the Presidential Commission on the Space Shuttle Challenger Accident, February 6, 1986 to February 25, 1986, Volume IV. June 6, 1986.

Hearings of the Presidential Commission on the Space Shuttle Challenger Accident, February 26, 1986 to May 2, 1986, Volume V. June 6, 1986.

Investigation of the Challenger Accident, Report of the Committee on Science and Technology House of Representatives, House Report 99-1016. 99th Congress, 2nd sess., October 29, 1986.

McDonald, Allan J. "*Challenger* Remembered in Light of *Columbia.*" Presented at the AIAA/ASME/SAE/ASEE 41st Joint Propulsion Conference, Tucson, Ariz., July 10–13, 2005.

Report of the Presidential Commission on the Space Shuttle Challenger Accident, Volume I, Appendices A, B, C, and D. June 6, 1986.

Report of the Presidential Commission on the Space Shuttle Challenger Accident, Volume II, Appendices E, F, G, H, I, J, K, L, and M. June 6, 1986.

Report of the Presidential Commission on the Space Shuttle Challenger Accident, Volume III, Appendices N and O. June 6, 1986.

Index

Page numbers in *italics* refer to illustrations.

Allan J. McDonald retired as Vice President and Technical Director for Advanced Technology Programs at ATK Thiokol Propulsion in July 2001. He was the Director of the Space Shuttle Solid Rocket Motor Project at the time of the *Challenger* accident in 1986 and later the Vice President of Engineering for Space Operations during the redesign and requalification of the solid rocket motors in the return-to-flight program for the U.S. Space Shuttle program.

James R. Hansen is Professor of History and Director of the Honors College at Auburn University. He is the author of a wide range of aeronautical and space history books, including *First Man: The Life of Neil A. Armstrong* (Simon & Schuster, 2005), an award-winning biography of Neil Armstrong.